PROCEEDINGS OF THE FORTIETH COLLOQUIUM ON THE LAW OF OUTER SPACE

INTERNATIONAL INSTITUTE OF SPACE LAW
OF THE
INTERNATIONAL ASTRONAUTICAL FEDERATION

OCTOBER 6-10, 1997
TURIN, ITALY

Published and Distributed by
AMERICAN INSTITUTE OF AERONAUTICS AND ASTRONAUTICS
Alexander Bell Drive, Suite 500
Reston, VA 20191-4344

ISBN 1-56347-278-3

Printed in the U.S.A.

INTRODUCTION

by

N. JASENTULIYANA
PRESIDENT

During the course of 1997, we witnessed several important anniversaries in the area of international space law and policy. The first was the thirtieth anniversary of the Treaty on Principles Governing the Activities of States in the Exploration and Use of Outer Space, Including the Moon and other Celestial Bodies, the cornerstone of international space law. We also witnessed an historic occasion for the IISL. The 40th Colloquium on the Law of Outer Space was convened in Turin, Italy in October which was devoted to the celebration of the thirtieth anniversary of the Outer Space Treaty. On this occasion, a commemorative dinner was organized which had many prominent space lawyers in attendance, including some who had taken part in the drafting of the Outer Space Treaty thirty years before.

The fortieth anniversary session of the Committee on the Peaceful Uses of Outer Space was convened in June. This landmark achievement demonstrates the commitment of countries from around the world to pursuing the goal of peaceful activities in outer space. The other was

When I first took office, one of my main goals was to forge a closer relationship between the Institute and the Committee. One of the ways we have established stronger ties with the Committee has been through the organization of symposia during the annual sessions of the Legal Subcommittee. This series of symposia aim to discuss contemporary space law issues as well as highlight potential areas the Committee, through its Legal Subcommittee, might wish to consider in the future. Thus the IISL is able to play a role in bringing developing space law matters to the attention of the Committee.

During last year's session of the Legal Subcommittee of COPUOS, the International Institute of Space Law, in cooperation with the European Centre for Space Law, organized a symposium entitled "Celebrating the 30th Anniversary of the Outer Space Treaty" which was attended by numerous space law experts. At the Symposium, Dr. Yuri Kolossov spoke on the background and history, Dr. Ram Jakhu spoke on space applications, and Mr. Frans von der Dunk spoke about the future. Prof. V. Kopal was the commentator. Dr. Fasan, as in the past, was the Coordinator of the Symposium. Based on the success of the Symposium, the Legal Subcommittee invited IISL to organize a similar symposium on a space law subject at next year's session as well.

To further celebrate the 30th anniversary of the Outer Space Treaty, I am pleased to note that during the past year, the Institute was able to acquire the copyrights to the first three *Colloquia on the Law of Outer Space*. Since acquisition of these copyrights, the Institute has, in cooperation with the UN Office for Outer Space Affairs, republished the Proceedings of the first three IISL Colloquia (1958-1960). The sets of three books may be ordered from the IISL Secretary (the price per set of US$30 covers the costs of printing).

These publications represent an historical record of the early work of the international space law community as they struggled with defining a legal regime for outer space. At that time, the possibilities offered by space technology were the stuff of dreams and space lawyers worked extremely hard to set-up a regulatory system that could withstand the test of time. The first three *Colloquia* detail some of the considerations and concerns of these early pioneers and clearly reflect the visionary work accomplished by these individuals.

The IISL has also made unique contributions to other substantive discussions of the Committee in many ways. The IISL has submitted legal and policy statements to COPUOS and its subsidiary bodies through the International Astronautical Federation, the parent body of the Institute. The statements aim to provide up-to-date information on the latest developments in space law and policy as well as charting the course of future space law-making opportunities. These statements have been well received by delegations attending COPUOS sessions. The Institute also continues to make an annual contribution to the review publication "Highlights in Space". This publication seen as one of the most important products of the Committee, gives the Institute the opportunity to describe the important space law issues that have arisen during the course of the year.

A second goal of my administration was the expansion of available time for substantive discussions of important issues during the IISL Colloquia and to foster stronger interaction among participants in these annual events. Not only has this been accomplished but I am encouraged by the results of these measures. Participants are provided with ample opportunity to present their papers during the Colloquium. Since separate sessions for our meetings have also been allocated, participants have been extremely involved in the proceedings and have thoroughly examined the issues under discussion. This has led to a comprehensive understanding of the issues we face and a greater understanding of the necessary steps that deserve to be taken in order to push the issue to a strong conclusion.

I am also satisfied to note that the Institute now makes greater use of the Internet and e-mail in taking care of its daily organizational tasks. It is imperative that we continue our efforts to place online, publications such as the IISL Newsletter, the Space Law Bibliography and the History of the Institute. This will enable Institute Members to remain in contact with their colleagues around the world and serve as an effective tool in our common goal of promoting the work of the Institute.

Among the other items the Institute will consider during the course of the next few years within its existing mandate, I would like to highlight the following:

(i) The continuing efforts of the Institute in the area of membership. The Institute needs to attract more individuals and institutions, and in particular young lawyers who are taking greater interest in the work of the Institute;

(ii) The expansion of the Manfred Lachs Space Law Moot Court Competition to cover all continents. In this respect, at this year's competition, teams from North America, Europe and for the first time, Asia participated. Efforts to further expand this competition to a truly world level will continue with the eventual inclusion of teams from law schools in Africa, Central and Latin America and the Caribbean.

(iii) The forging more cooperative links with academic space law and policy institutions. The IISL has cooperated closely with ECSL in many areas including the organization of the annual symposia during Legal Subcommittee sessions. With this type of cooperation as an example, the Institute should aim to establish ties with similar institutions and academic departments in Europe, the Americas, Africa and the Asia-Pacific region. As one of the leading space law Non-Governmental organization in the world, it is one of its main responsibilities to bring together educational institutions which are interested in space law;

(iv) Closer links with space scientists and engineers who can bring their knowledge and experience to bear in discussions with space lawyers. Through joint cooperation between technicians and policy-makers, the Institute can engage in more comprehensive discussions that may ultimately lead to forward-thinking and technically sound policy directives; and

(v) Establishing solid ties with national and regional space agencies and with private space industry as it continues to grow by leaps and bounds with each passing year.

The continuing pursuit of promoting the development of space law and policy as the prime directive of the Institute is of paramount interest. As a group and through our parent body, we provide the world with information on the benefits of space science and technology. As President of the Institute, I look forward to the up-coming events that will be organized by the Institute.

In particular, with the coming of the Third United Nations Global Conference on the Exploration and Peaceful Use of Outer Space, otherwise known as UNISPACE III, we must focus our efforts in order to have the opportunity to spread the message of the Institute to that Conference as it examines issues that are in need of negotiated rules and regulations. I look forward with great interest to the active participation of the Institute in that meeting. I am sure that the results of our work during that Conference will form a part of many of its recommendations and decisions that will provide a blueprint for the activities of mankind in outer space in the next millennium.

I. Background

The International Institute of Space Law (IISL) was founded by the International Astronautical Federation (IAF) in 1960. The IISL re-placed the Permanent Committee on Space Law which the IAF had been created in 1958 under the chairman-ship of Andrew G. Haley. Since 1958, the IISL has held annual colloquia on space law in many nations. The IISL presently has some 250 elected members who are distinguished for their contributions to space law development. While a component of the IAF, the IISL is authorized to function autonomously in accordance with its Statutes.

The purposes and objectives of the Institute include the cooperation with appropriate international organizations and national institutions in the field of space law, the carrying out of tasks for fostering the social science aspects of astronautics, space travel and exploration, the organization of meetings and colloquia on juridical and sociological aspects of the space sciences, and the publication of proceedings, reports and other publications.

The IISL recognizes that activities in the new dimension are both Earth and space oriented, and that in both domains laws must be formulated for future developments. It is necessary to provide for the coordination and regulation of space activities by law in order to promote international cooperation and avoid conflicting claims and competition which would jeopardize beneficial uses for all mankind. Considering the multidisciplinary nature of problems and situations arising from the exploration and uses of outer space, the Institute plans programs which afford opportunities to analyze national and international legal matters in the total context of developing space science, technology and the social sciences. The IISL may play an important role in rallying distinguished specialists who interpret existing principles and rules of the law of outer space and prepare proposals for states to consider in concluding treaties or accepting other instruments of cooperation or otherwise contribute to the development of the law of outer space.

The financial resources of the Institute are derived from membership dues, research grants, remuneration received from the sale of its publications, and donations received from individuals, associations, institutions, foundations, and other organizations with interests in the development of law governing space activities.

The headquarters of the International Institute of Space Law are located at 3-5 rue Mario Nikis, 75015 Paris, France.

II. IISL and the United Nations

IISL members have been appointed by the International Astronautical Federation as Official Observers to annual sessions of the United Nations (UN) Committee on the Peaceful Uses of Outer Space and its Legal Subcommittee. The 1997 session of COPUOS' Legal Subcommittee and the Plenary session of COPUOS were both held in Vienna. Dr. E. Fasan, Prof. Back Impallomeni and Mr. Vivod were appointed as official IAF observers for the COPUOS Legal Subcommittee meeting, whereas Dr. Steven Doyle and Prof. Böckstiegel were appointed as official IAF observer for the Plenary COPUOS session. Only one predesignated person at each session serves as official spokesman for the IAF.

III. 40th IISL Colloquium

The 40th Colloquium on the Law of Outer Space was held in Turin, Italy, during the 48th Congress of the International Astronautical Federation (6-10 October 1997). The Colloquium was attended by some 75 participants who attended the four sessions whose results are published in this volume:

- Session 1: Background and History of Outer Space Treaty; Chairman: A.A. Cocca; Rapporteur, J.F. Renaud.
- Session 2: Concepts of Space Law and the Outer Space Treaty; Chairmen: E. Galloway and G. Catalano-Sgrosso; Rapporteur, M.L. Longo.
- Session 3: Applications & Implementation of the Outer Space Treaty; Chairmen: S. Doyle and G. Lafferranderie; Rapporteur, F. von der Dunk.
- Session 4: The Future Applications of the Outer Space Treaty; Chairman: K.-H. Böckstiegel; Rapporteur, O. Ribbelink.
- A Scientific-Legal round table was organized in cooperation with the IAA on SETI and Society.

A Dinner Celebrating the 30th Anniversary of the 1967 Outer Space Treaty, graciously offered by SAGAT Turin Airport, was held on 7 October. Over 90 persons attended, including officials of the IAF and IAA and many prominent space lawyers some of whom had actually taken part in the drafting of the Treaty thirty years earlier. Prof. Bin Cheng, who was awarded an IISL Award for his writings on air and space law, gave a dinner speech.

The 6th Manfred Lachs Space Law Moot Court Competition was held at the University of Turin on 9 October 1997. The 1997 problem (Openskey vs. Antipapadia), written by H. Tuinder, M. Ferrazzani and F. von der Dunk, dealt with Very High Resolution (VHR) remote sensing systems. The University of Paris XI and the University of North Carolina (USA) competed before a bench composed of Judge a. Koroma, Judge F. Rezek and Judge V. Vereshchetin of the International Court of Justice. The team of the University of Paris won the competition. The University of Turin hosted a reception after the Finals. A report on the competition, the case and the winning briefs are published in this volume. The competition will be held for the 7th time in Melbourne.

IV. Awards

A lifetime achievement award was given to Hon. E. Finch, Jr.. A Book Award for all his publications on aerospace law was offered to Prof. Bin Cheng.

VI. 41st IISL Colloquium

The following sessions will be held during the 41st IISL Colloquium, which will take place in Melbourne, Australia, from 28 Sept.-2 Oct. 1998.

Session 1: Institutional approaches to

managing space resources, Chairmen: T. Masson-Zwaan (The Netherlands) & Representative of Monash University (tbd).
Session 2: Legal and policy aspects of Confidence Building Measures using space technology, Chairmen T. Kosuge (Japan) & S. Doyle (USA).
Session 3: Legal aspects of navigation satellites and Global Positioning Systems, Chairmen P. Sterns (USA) & He Qizhi (China).
Session 4: Other Legal Matters, Including The 30th Anniversary of the Rescue Agreement Of 1968, Chairmen M. Komar Kantaatmadja (Indonesia) & F. Lyall (UK)

The 7th Manfred Lachs Space Law Moot Court Competition will be held. The case deals with the commercial exploitation of the moon (Lunar Rover Project).

VII. <u>42nd IISL Colloquium</u>

The 42nd Colloquium will be held in Amsterdam, The Netherlands, in 1999. The session topics for this colloquium will de decided in the Spring of 1998.

VIII. <u>Elections</u>

Board of Directors

The result of the Board elections was:

<u>Re-election as Board Member</u>: Dr. V. Kopal (Czech Republic), Dr. N.M. Matte (Canada), Mrs T. Masson-Zwaan (The Netherlands), Prof. Dr. K.-H. Böckstiegel (Germany), Dr. S.E. Doyle (USA), Dr. He Qizhi (China), Prof. T. Kosuge (Japan) and Dr. A. Terekhov (UN/Russia)

<u>Election as Board Member</u>: Dr J. Monserrat, Filho (Brazil)

<u>Re-election as Vice-Presidents</u>: Dr. V. Kopal (Czech Republic) and Dr. N.M. Matte (Canada)

<u>Re-election as Secretary</u>: Mrs T. Masson-Zwaan (The Netherlands).

The results of new IISL members elected in April and October 1997 were:

<u>Institutional Members</u>:

CSP Japan
YS Bld. 2nd Floor
2-2-6 Kaigan, Minato-ku
Tokyo 105, JAPAN
(Mr Suzuki, Mr Yoshida)

Law Offices of Sterns and Tennen
849 North Third Avenue
Phoenix, Arizona 85003-1439, USA
(Dr. Sterns, Dr. Tennen)

<u>Individual Members</u>:

Dr R.I.R. Abeyratne
ICAO, Suite 1134
299 University Street
Montreal H3C5H7, CANADA

Dr. M. Bahrami
Mechanical Engineering Dept.
Amir Kabir University of Technology
424 Hafez Avenue
Tehran 15, IRAN

Mr. Michel Bourbonniére
Associate Professor
4445 Rue Saint Louis-Roy
St Léonard, Quebec H1R 3V9, CANADA

Prof. Chia-jui Cheng
Dean, Faculty of Law
Soochow University
125, Roosevelt Road, 6F-2, section 3
Taipei, TAIWAN ROC

Mr. Peter A. Iiskola
Kanaalweg 96
2584 CN Den Haag, THE NETHERLANDS

Mr. Kresno Putro
National Aeronautic and Space Institute (LAPAN), Jalan Cisudane 25 Cikini
Jakarta Pusat, INDONESIA

Mr. Andrew L. Quiat
Attorney at Law
8100 W Crestline Ave, Ste A-15341
Littleton, Colorado 80123-1200, USA

Mrs Sriwulan Rios, SH
National Aeronautic and Space Institute (LAPAN), Jalan Cisadane 25 Cikini
Jakarta Pusat, INDONESIA

Mr. David W. Sagar
Senior Attorney, INMARSAT
99 City Road
London EC1Y 1AX, UNITED KINGDOM

Dr. K-U. Schrogl
DARA (German Space Agency)
P.O. Box 30 03 64
D-53183 Bonn, GERMANY

H.E. Judge Chr. Weeramantry
International Court of Justice
Peace Palace, Carnegieplein 2
2517 KJ Den Haag
THE NETHERLANDS

TABLE OF CONTENTS

CONCEPTS OF SPACE LAW AND THE OUTER SPACE TREATY

APPLICATIONS & IMPLEMENTATION OF THE OUTER SPACE TREATY

ADDENDA

International Institute of Space Law
October 6-10, 1997
Turin, Italy

Mr. Jasentuliyana welcoming the 90 guests to the IISL dinner celebrating the 30th anniversary of The Outer Space Treaty, sponsored by SAGAT, Turin Airport.

The recipients of the reprints of the 1st three IISL proceedings, all present at the very first colloquium in 1958. Left to right: Dr. V. Kopal, Dr. E Fasan, Dr. E. Galloway, Prof. I Diederiks-Verschoor, Prof. S. Gorove.

Mr. Jasentuliyana expressing thanks to dignitaries of the University of Turin for their hospitality.

THE 1967 SPACE TREATY: THIRTY YEARS ON*

Bin CHENG**

Mr President, Members of the Board of Directors of the Institute, Distinguished Representatives of SAGAT Turin Airport, Honoured Guests, Friends, Colleagues, Ladies and Gentlemen:

I. Introduction

Thank you, Mr President, for your very kind and generous words about me. I don't know how deserved they are, but perhaps that make them all the more appreciated!

It is an immense honour for me to be asked by the International Institute of Space Law to attend this special dinner, co-hosted by SAGAT, to mark the 30th anniversary of the 1967 Outer Space Treaty,[1] and to say a few words on the theme of the Institute's 40th Colloquium on the Law of Outer Space which is dedicated to the celebration of this happy and worthy anniversary. I wish to thank the Institute most sincerely for this overwhelmingly gracious and wholly unanticipated invitation.

Ever since receiving this invitation, I have been trying hard to work out the job description of what I have been asked to do. Now, especially since, on the one hand, we have already heard both this morning and this afternoon, on this the first day of the Colloquium, a good number of extremely authoritative, valuable and learned papers on so many aspects of the Treaty, and since, on the other hand, we are all looking forward to enjoying the culinary delights of Turin that are awaiting us, it occurs to me that perhaps my piece may not be unlike the sort of minuscule dish that some restaurants like to offer you with the chef's compliments after you have chosen your food and wine. It is just something to keep you occupied while the food is being prepared, without holding up the dinner once it is ready.

That reminds me of the advice that used to be given. A speaker should follow the concept of a miniskirt. Now, now! I did not say miniskirt *tout court*. I said the concept of! The advice was that it should be short enough to be interesting, but long enough to be decent. However, that was said years ago when miniskirts first came on to the scene. Fashion has since moved on. If we have a look at the London fashion show last week, and the one that is going on this week a stone's throw away in Milan, minimalism is strictly *de rigueur*.

This being the case, I thought that, in order to meet the current requirement of decency, it would be sufficient if I were just to cover a single strategic point and then sit down. However, my wife, who has kindly agreed to

---

* Keynote address at the special dinner held by the International Institute of Space Law and SAGAT Turin Airport on 7 Oct. 1997 to celebrate the 30th anniversary of the 1967 Outer Space Treaty on the occasion of the Institute's 40th Colloquium on the Law of Outer Space held in Turin, Italy, 6-10 Oct. 1997.

** Licencié-en-droit, Ph.D., LL.D., Hon. LL.D., FRAeS; Emeritus Professor of Air and Space Law, University of London; Visiting Professor of Law, University of Detroit Mercy.

come with me to Turin and who is with us this evening, quite rightly admonished me that this is not some catwalk celebration, but a gathering of the most eminent, distinguished and respected practitioners and scholars in the field of space law in the world, and that I should respectfully cover at least a couple more points.

Even so, in view of the fact that we already had, and shall continue to have, in the course of this Colloquium, a large number of extremely learned papers on the subject, I hope you will allow me this evening to simply share with you a few random thoughts on the 1967 Space Treaty, 30 years on, without going into details. Such details and any supporting arguments, if I may follow Steve Doyle's excellent example this morning and slip in a commercial, will probably all be found in my *Studies in International Space Law* which Oxford University Press is bringing out next month under the Clarendon Press imprint.

To start with, while we are here to celebrate the 30th anniversary of the 1967 Treaty, I think we all wish to take this opportunity of congratulating the Institute on its 40th Colloquium, which happily coincides with the 40th anniversary of man's first entry into space. In a masterpiece of planning, the Institute has scheduled this celebratory dinner, not only on the first day of the 40th Colloquium, but also, with perfect even-handedness, three days after the launching of Sputnik I, and three days before the coming into force of the Space Treaty.

II. The Space Treaty: 30 Years on

The consensus which emerged clearly from the various speeches at today's first two sessions of the Colloquium, with which I entirely agree, is that the 1967 Space Treaty is a truly remarkable instrument. It has successfully provided an indispensable legal framework for the exploration and use of outer space from practically the beginning of the space age. It was a great political and legal achievement.

On 10 October 1967, when the Treaty actually came into force, having been ratified by all the powers that mattered and more, everyone was able to utter a sigh of relief, and to rejoice that the superpowers were finally able, ten years after Sputnik I, to agree, first, that the agreed principles would take the form of a legally binding treaty instead of just a General Assembly resolution, secondly, that at least celestial bodies would be used exclusively for peaceful purposes, thirdly that no nuclear weapons or any weapons of mass destruction would be stationed anywhere in outer space, fourthly that there would be no race for colonies in outer space, and fifthly that all contracting States would assume direct State responsibility for national activities in space, protect the environment, pay for any damage caused, be helpful to one another, and try to do everything for the good of all. The Treaty met, if not entirely, at least in appreciable measure, some of the deepest concerns and keenest aspirations of the world at the time.

What one has to remember, however, is that that was 1967. In fact, apart from Article IV, much of the Space Treaty had been agreed upon in 1963 in the form of the Declaration of Legal Principles Governing the Activities of States in the Exploration and Use of Outer Space[2]. By then, not a single satellite had yet been launched into the geostationary orbit, and even the Interim INTELSAT had not been established. Both came only a year later. Even by the time the Space Treaty was opened for signature in 1967, only France had joined the then space club of two with the successful launching of Astérix in 1965. The Chinese did not do so until 1970, and the United Kingdom only in 1971. Landsat I, the first remote sensing satellite, was launched only in 1972. Often it is not easy to remember how far we have come in the thirty years since 1967.

It is, therefore, hardly surprising if today the

1967 Treaty needs a thorough review in the light of all the changes in circumstances. This is not the time or place to go into details. The most important changes may perhaps be simply enumerated:

(i) phenomenal *advance in space technology*, as exemplified, for instance, by the current US Martian exploration with the spacecraft Pathfinder with its Sojourner rover vehicle, and Alpha the international project of a permanently manned International Space Station;
(ii) rapid development of the *commercial exploitation* of space and space-related activities, such as in the field of remote sensing, not to mention telecommunication and direct television broadcast by satellites; and
(iii) increasing participation of *private enterprise* in all aspects of space activities, including, for example, actual launching of space objects.

As Ambassador Jankowitsch was saying this morning, we have now entered the third phase in man's exploration and use of outer space. However, notwithstanding all these changes, what needs to be done is not a root and branch operation radically to transform and still less to replace the 1967 Treaty. Rather it is a case of judicious adjustments. These may be grouped mainly into three categories:

(i) *authoritative* and more precise or systematic differentiation, classification, clarification or *definition* of various terms and concepts;
(ii) *closer co-ordination* of the provisions, as well as these terms and concepts, of not just the 1967 Treaty itself, but in all the UN treaties and declarations on space;
(iii) specific amendments and *supplementary provisions* to take into account changes in circumstances since the signing of the Treaty.

Among the many issues that may be raised in reviewing the Treaty for updating, I shall limit myself this evening to merely three areas:

(i) Terminology;
(ii) Main areas of concern;
(iii) Conditions governing the successful making of international treaties and rules.

In order to save time, I hope that I may simply enumerate some of the points without further elaboration.

III. Terminology

1.Filling in *Lacunae*, e.g., "Outer Void Space"

Under the heading of terminology, one can mention first of all the task of filling in the many gaps that have revealed themselves over the years in the vocabulary of space law. For example, owing to a lead given by the 1967 Treaty, there is at present no convenient expression to describe the space in between all the celestial bodies. Thus, while the 1963 UN Declaration of Legal Principles Governing the Activities of States in the Exploration and Use of Outer Space[3] speaks of "outer space *and* celestial bodies", making a distinction between outer space and celestial bodies, the 1967 Treaty and after it all the other UN treaties and declarations always use the expression "outer space, *including* the moon and other celestial bodies", which means that the term "outer space" *includes* "the moon and other celestial bodies." As a result, whenever we refer to "outer space", we will be understood to refer also to all the celestial bodies in it, excluding perhaps the earth. There is no longer a simple expression to designate the space in between the celestial bodies. This is why I have been calling this space the "outer void space"[4]. I hope that this name will find general acceptance.

2. Clarifying and Defining Various Technical Terms

There is a long list of terms used in the 1967 Treaty and other UN treaties and declarations on outer space that are crying out for clarification and definition. Example include “astronauts”, “appropriate State”, “debris”, “national activities”, “space objects”, and a host of others. They have already received much attention in the literature of space. law.[5] There is no need to elaborate the point here, even if we may be referring to one or two of them later.

3. Two Perennials: “Outer Space” and “Peaceful Purposes”

Then there are those two perennial controversies, the definition and delimitation of outer space, and the proper interpretation of the meaning of “peaceful”.[6] I think it is high time that they should be resolved, and these terms and concepts authoritatively defined. The ca'canny and obscurantism involved in delaying a definition of outer space, and the deliberate distortion of the word “peaceful” to mean not “non-military” but “non-aggressive” are the work of politicians and diplomats done, one suspects, at the behest of the military, who, at least in the latter case, may well have based their conclusions on some misreading of the law.[7] Speaking of the antics of some politicians and diplomats, one is reminded of what Sir Henry Wotton wrote in 1604 when he was on his way from England to Venice to take up his post as King James I's ambassador there:

> “A diplomat is an honest man, sent to lie abroad for the good of his country”.

It seems to me that the time has come when we space lawyers have to make a determined effort to convince the powers that be that (a) clarity, precision and accuracy in the use of these and other terms are of paramount importance in the future development of space law and of space activities as a whole, and (b) shielding behind equivocation and the distorted meaning of words is no longer a healthy option.

IV. Four Areas of Concern

If we look back at some of the concerns at the beginning of the space age, the thoughts uppermost in people's minds towards space can probably be divided into four categories:
(i) the arms race and the military use of outer space;
(ii) possible scramble for colonies or resources;
(iii) worries over responsibility and control, as well as over potential harm or damage; and
(iv) international co-operation and mutual assistance.

By and large, the concerns remain much the same today, although the perspectives may have changed over the years.

1. The Arms Race and the Military Use of Outer Space

First, the military use of outer space. For those in the 'fifties and the 'sixties who had only just witnessed the awesome role of air power and air supremacy in the relatively recent conflict, the first and foremost concern was from the military and strategic angle. To use an apt American expression, outer space brought with it a whole new ball game. For the protagonists, it was a question of how to contend and to contain. For third parties, it was how to prevent and to avoid a space war in which they might be embroiled, or of which they might become the victims. Hence, there was this tremendous popular clamour that outer space should be used exclusively for peaceful purposes. To this call the space powers paid lip-service, but with a great deal of mental reservation, inasmuch as their space efforts were then, perhaps even more intensely than now, directed primarily towards military ends.

Insofar as the demilitarisation of outer space is concerned, President L. B. Johnson hailed the 1967 Treaty as “the most important arms control development since the limited test ban treaty of 1963"[8]. 1963 was of course also the year when the General Assembly adopted the Declaration of Legal Principles,[9] the precursor of the 1967 Treaty. 1963 was moreover the year when the

General Assembly in resolution 1884 (XVIII) welcomed the statement of the two superpowers that they would not station nuclear weapons or other weapons of mass destruction in outer space, and called upon other States to follow suit. It will not escape notice that the only really substantive and important provision in the Treaty that is not in the Declaration of Legal Principles is Article IV. Yet Article IV(1) corresponds basically to resolution 1884. What this means is that what the two superpowers were unable to agree in 1963, and managed to do so only three years later in 1966 was Article IV(2), which restricted celestial bodies for use exclusively for peaceful purposes. This was then the breakthrough referred to by President Johnson.

However, much confusion surrounds the subject of the military use of outer space. Thus, only too often one hears and finds the assertion that under the Space Treaty, the whole of outer space, including celestial bodies, has been reserved for exploration and use exclusively for peaceful purposes. If we examine Article IV and the rest of the Treaty carefully, we will find that this is not true. Only the moon and other celestial bodies have been restricted by Article IV(2) to use "exclusively for peaceful purposes", but not outer void space. Under the Treaty, apart from the ban to station there nuclear weapons and other weapons of mass destruction, and, as they are reminded by Article III, subject to the ordinary rules of international law, including the Charter of the United Nations, contracting States are perfectly entitled to use outer void space for whatsoever military purpose they wish. They can put up there reconnaissance satellite, anti-satellite satellites, early-warning satellites, geodetic satellites, and any other weapon as long as it is not nuclear or capable of mass destruction. There is nothing in the Space Treaty as such which would, for instance, preclude projects like the United States' "Strategic Defense Initiative" (SDI), unless it turned nuclear or were to cause mass destruction.

What has befuddled the discussions on the military use of outer space during all these years has been the distorted use of the term peaceful to mean non-aggressive instead of its usual meaning of non-military. It is largely this double-talk which has allowed the false impression to be propagated that the whole of outer space, including both the celestial bodies and the outer void space, has been reserved by the Space Treaty for solely peaceful, i.e., to the uninitiated, non-military, purposes. This has in turn caused those concerned with using outer void space for military purposes, who have either not read Article IV of the Space Treaty, or have not read it properly, stubbornly to defend this abuse of the language.

If peaceful means non-aggressive, this would make the first sentence of Article IV(2), which provides that the moon and other celestial bodies can only be used "exclusively for peaceful purposes", completely meaningless, because, except for the specific prohibitions in the second sentence, the legal status of the moon and other celestial bodies would then be exactly the same as the outer void space, which, as we have just seen, can be used, and is being extensively used, for all kinds of activities for military purposes except of course for aggressive purposes. Under general international law, and especially Article 2(4) of the United Nations Charter, there is no place in the whole universe that States may lawfully use for aggressive purposes. Thus, to say that peaceful means non-aggressive is to deprive not only the first sentence of Article VI(2) of the Space Treaty, but also the word peaceful itself of all meaning.

The sooner the United States and all the others who embrace this spurious interpretation of the word peaceful can be persuaded to abandon it, and the sooner the proper interpretation of peaceful can be embodied in an authoritative definition in the Treaty the better. It would greatly contribute to clear thinking and to saving this useful term from being reduced to utter meaninglessness.[10]

As to the substantive issue of arms control in outer space, what with the latest plan to test ground-based anti-satellite laser weapons,[11] the UN General Assembly in its resolution 51/123 last December quite rightly expressed its concern "about the possibility of an arms race in outer space".[12] Now that the "Evil Empire" has collapsed, and the Cold War has finally been banished with the signing this May of the Founding Act of Mutual Relations between NATO and the Russian Federation,[13] there is obvious room for agreement on drastic arms control not only in outer void space, but also in general; for it is well recognised from the beginning, arms control in space cannot be divorced from disarmament on earth.[14] With outer space now providing hitherto unavailable means of verification, it is earnestly to be hoped that outer space can act as the catalyst that will greatly facilitate and accelerate the task of reaching agreement.

There is, however, obviously some demarcation problem between COPUOS and the Conference on Disarmament.[15] At the beginning of COPOUS, it was the Soviet Union that insisted on excluding all disarmament issues from COPUOS. Nevertheless, as we have seen, between the 1963 Declaration of Legal Principles and the 1966 negotiations, the Soviet Union was finally persuaded, no doubt by the United States, to include Article IV in the Space Treaty. One can only hope that the two bodies can co-operate to prevent an arms race in outer space, and, if at all possible, to reduce and regulate the use of outer void space for military purposes. A proper definition of the term peaceful would obviously be a step in the right direction.

<u>2. Scramble for Colonies or Resources: Occupation v. Appropriation</u>

Secondly, at the beginning of the space age, there was a strong demand especially from the non-space powers that outer space and celestial bodies should not be subject to national appropriation. Some were motivated by anti-colonialism, others wished to discourage a colonial war, and yet others did not want to see the spoils of outer space fall irretrievably and exclusively into the hands of the space powers.

The principle of non-appropriation of outer space and of celestial bodies quickly met the agreement of the superpowers, who at the time were far from sure which of them was to land on the moon first. The principle was adopted unanimously in General Assembly resolution 1721 A (XVI) in 1961. Only the insistence of the United States prevented it from taking the form of a treaty. Eventually it was the fear that the Soviet Union might be the first to make a landing on the moon that prompted the United States to change its attitude towards resolutions versus treaties in space-law making, and in May 1996 actually to call for the conclusion of a treaty to prevent any nation from claiming sovereignty over the moon or any other celestial bodies. It was this that led to the conclusion of the 1967 Treaty.[16] Article II on non-appropriation is thus one of the Space Treaty's chief *raisons d'être*, and what brought it into *de facto* existence in the astonishingly short time of a little over seven months, the text having been adopted by the General Assembly on 19 December 1966.

The complete freedom of outer space and of celestial bodies for exploration and use by all thus established in Article II is supplemented by Article I which inter alia specifies "free access to all areas of celestial bodies".

Insofar as the principle of non-appropriation is concerned, while Article II speaks of no national appropriation by means of use or occupation, and Article I of free access to all areas of celestial bodies, the line in fact, if not in law, between occupation and appropriation is often difficult to draw. This applies to orbits in outer void space as well as to celestial bodies. The problems arising from the continuous occupation of prime slots in the geostationary orbit, and the phenomenon of paper satellites fall into this category. It is to be hoped that the current efforts in COPUOS and in ITU to resolve them may reach early fruition. In due course, the same problem will apply to occupation of portions of celestial bodies when exploitation becomes possible. The Moon Treaty has not really resolved it. Further consideration is required.

3. Worries Over Responsibility and Control, as well as Over Potential Harm or Damage

Thirdly, especially at the beginning, excitement over space was tempered with grave apprehension of the unknown. While space activities were greeted with wonderment, and astronauts were treated almost as superhuman, there were also serious qualms whether some space activities might irreparably damage the space environment, or grievously contaminate, or even destroy the earth, and wipe out all life on it. Moreover, on a more down-to-earth level, having been brought up to believe that what goes up must come down, people were uncomfortable with the thought that tons and tons of metal objects were to whirl round and round over their head, and were worried over the damage which such objects might caused them or their property when they were to fall down. The feeling was that everything connected with space needed to be strictly controlled by States, which should also be made responsible for any adverse consequences.

When it comes to control of and responsibility for space activities, these are covered basically by the revolutionary principle in Article VI of the Space Treaty which makes the contracting States directly responsible internationally for national space activities, by whomsoever carried on. At the same time, Article VII makes all the contracting States responsible for the launching of a space object directly liable for any damage which the space object may cause to third parties. Both these principles already appeared in the 1963 Declaration, as well as the rule in Article VIII which places space objects and their personnel under the jurisdiction of the State of registry. Article IX of the Treaty now adds a specific duty on contracting States to avoid harmful contamination of either outer space or the earth. Moreover, Article XIII makes it clear that the Treaty provisions apply to contracting States whether they carry on space activities individually or jointly with other States.

Since 1967, the rise in non-governmental space activities has been beyond belief. Whilst the need for governmental control as envisaged in Article VI of the Space Treaty will always remain necessary, there is need to define which State is responsible for whose and which space activities. At present, both the term national activities, and the term appropriate State in Article VI give rise to a great deal of uncertainty.[17]

Moreover, the extent of the concept of international responsibility is far from clear.[18] We know that contracting States have to assure that national activities conform to provisions of the Treaty, and that they must subject non-governmental national activities to authorisation and continuing supervision. But does their responsibility for non-governmental national activities extend to beyond compliance with the Treaty and through Article III of Treaty with rules international law, including all treaty obligations, to compliance with rules of municipal law, both civil and criminal, including even contractual obligations? And, notwithstanding the fact that Article VI speaks of the "appropriate State" in the singular, does international responsibility fall in fact on all the States which may qualify as launching States?

Switching from Article VI to Article VIII of the

Treaty, one finds that Article VIII on jurisdiction is by no means free from ambiguity. Thus, one may well ask whether, according to Article VIII, the jurisdiction of the State of registry of a space vehicle extends to persons who do not form part of that vehicle's personnel? For instance, what is the legal position of an astronaut of State A nationality, part of the crew of a spacecraft registered in State B, assaulting an astronaut of State C nationality, part of the crew of a spacecraft registered in State D, after the two spacecraft docked in outer space and the former astronaut was visiting the latter spacecraft? Or someone visiting a moon station operated by another State? And how about a space tourist?

Moreover, the interpretation which has been given to Article II of the Registration Convention further weakens the link between the State of registry of a space vehicle and the vehicle itself, as well as all those on board. At present, flags of convenience can be easily established, and it may sometimes be difficult to ascertain which State exercises jurisdiction over which space object and over which persons on board.[19]

Furthermore, the effect of Article VIII is further eroded by some of the other subsequent UN treaties on outer space, which often resort to different connecting factors. Launching, including all the different aspects of it, nationality of the astronaut, and ownership of the space object, and possibly even employment can all come into play.[20] There is probably much to be said for reverting to the traditional concept of nationality, while at the same time tightening the rules on registration.[21]

Consideration needs also to be given to the legal status and registration of installations and manned or unmanned stations on celestial bodies, as well as regulation of and liability for activities carried on in them. There should be better co-ordination between control and responsibility, and generally some redefining and perhaps adjustment of the extent of international responsibility especially for non-governmental activities, in view of the almost phenomenal increase in private commercial space activities.

On the other hand, because of the rapid proliferation of space activities, the suggestion of setting up machinery and procedures for the elaboration of standards and recommended practices along the lines of ICAO (International Civil Aviation Organisation), WHO (World Health Organisation) and IMO (International Maritime Organization) to regulate and co-ordinate especially the technical aspects of international space activities is to be welcomed.[22] What is needed is probably a high-powered and compact unit, where the different interests are duly represented.. To such a body, subjects such as the use of nuclear-powered sources, space debris, and collision now being considered by COPUOS can perhaps with advantage be entrusted.

If a quasi-legislative opting-out procedure is adopted, States which do not opt-out of specific regulations, should be made responsible for their compliance, implementation and enforcement. In order that such measures are effectively implemented and enforced in the case of non-governmental space activities, States may need to be reminded, encouraged and perhaps even bound by treaty to extend their domestic laws to persons and objects under their jurisdiction in outer space, just as aviation found it necessary to adopt the 1963 Tokyo Convention on Offences and Certain Other Acts Committed on Board Aircraft.[23]

4. International Co-operation and Mutual Assistance

Finally, in mankind's exciting new venture into the hitherto unknown, while there was obviously rivalry between, and later among, the space powers, there was also probably on the part of all, a recognised need for mutual assistance and co-operation. Of course, what everyone had in mind may not have been the same. The space powers would like to see the maximum of co-operation in their space activities from everyone, affording them tracking and other facilities, including, if need be, the right of passage for their space objects, and assistance to their astronauts or space objects in distress. Near-space powers would like a helping hand to get into space, whilst they and non-space powers would all like to share in the benefits of space.

In the Space Treaty, Article IX clearly stipulates that contracting parties should be guided by the principle of international co-operation and mutual assistance, in all their activities in space. Article V provides for assistance to astronauts in distress, Article IX for consultation in order to avoid harmful interference with one another's peaceful space activities, Article X for opportunities to observe one another's space flights, and Article XI for at least in theory the spontaneous total disclosure by the contracting parties of the nature and even results of all their peaceful space activities. Finally to cap it all is the famous Article I which enjoins that all space activities shall be carried out "for the benefit and in the interests of all countries", and that contracting States, in scientific investigations in outer space "shall facilitate and encourage international co-operation".

From this point of view and speaking of the extension of domestic law into space, it is of interest to note that, although, for example, Article V(2) of the Treaty specifies an obligation on astronauts of contracting States to render all possible assistance to astronauts of other contracting States, the Treaty does not specifically lay a duty on any particular contracting State through its domestic law to ensure that such assistance is always given. What space law can do would be, for example, to take a leaf from international air law where Annex 12 to the Chicago Convention (Search and Rescue) lays down specific rules and procedures, which contracting States that do not opt out have to implement, setting out what pilots-in-command of aircraft of their registry will have to do when they observe either another aircraft or a surface craft in distress, or intercept a distress call from another aircraft.

In this connection, apart from possibly adopting the quasi-legislative procedures of ICAO, WHO, and IMO, space law might think of further borrowings from related fields of international law and activities, particularly in the field of international co-operation and mutual assistance, on which the 1967 Treaty has laid much stress. A great deal can be learnt, for example, from international air law, and international nuclear law.[24]

At the same time, it seems to me that what one needs to guard against is being unrealistic. The first years of the Space Treaty coincided with a period when the economic consciousness and expectations of the developing countries were steadily rising. There was a degree of militancy which manifested itself in successive UN General Assembly resolutions on natural resources, culminating in the proclamation of the New International Economic Order and the Charter of Economic Rights and Duties of States.[25] What it amounted to was an attempt to translate the kind of social conscience and pressure which may exist in homogeneous domestic societies and which cause legislation to be made providing for assistance and benefits to be given to the disadvantaged sections of society, to an essentially heterogenous horizontal society like the international political and legal system. In international society, members are sovereign, and are motivated almost exclusively by self-interest. They are not subject to any higher authority which can compel them to be munificent towards their poorer and weaker brethren without discrimination. The United Nations General Assembly certainly has no such power. They need to be persuaded that it is in their interest to do so, and at the moment this can probably be done more easily in individual and concrete cases than in a blanket fashion.

In Article I of the Space Treaty on the subject of international co-operation, the space powers paid lip service to the surging expectations of the time of the developing countries. Some countries and commentators have ever since tried very hard to give Article 1 an excessively literal interpretation involving a legally binding obligation. Such efforts can hardly be said to have succeeded, although there have been, as we all know, a good number of bilateral and plurilateral co-operative and even collaborative arrangements based on mutual interests and mutual consent. Thirty years on, and after ten years of discussion, a more realistic attitude seems to have manifested itself in the 1996 Declaration on International Co-operation in the Exploration and Use of Outer Space for the Benefit and in the Interest of All States.[26] The key provision is probably to be found in the first part of paragraph 2:

> "States are free to determine all aspects of their participation in international cooperation in the exploration and use of outer space on an equitable and mutually acceptable basis. Contractual terms in such cooperative ventures should be fair and reasonable . . ."

In other words, international co-operation is to be voluntary, and it is to be on "fair and reasonable" terms to be negotiated and agreed upon between the parties on either a multilateral or bilateral basis. This said, one should on the other hand equally remind States with space capabilities that it must be in their long term interests that space technology and space benefits are shared to the widest extent possible. Even the most mercantilist State must realise that in the long run one must be better off living in a world not populated by hungry mouths, but by well-to-do clients. From this point of view, both the '67 Treaty and the United Nations can doubtless play useful roles in promoting international co-operation in order to achieve that objective.

V. Conditions Governing the Successful Making of International Agreements and Rules

This brings me to my last point that I hope to make. The study of space law has taught me that in a horizontal legal system like international law, in order successfully to forge an international agreement or to build up a rule of international law, three conditions have to be met.

(i) perceived need on the part of the States concerned;

(ii) due representation of the dominant section of international society having special concern in the subject-matter; and

(iii) a propitious political climate.[27]

First, the States concerned must feel a need for the agreement or rule. They must feel that it is in their interests to do so. All that one can hope is that all the persons concerned will understand and pursue their countries' long-term and broader interests, and will not seek short-term success or personal glory to the detriment of those interests, and that the dominant section of international society will not abuse their dominant position. Insofar as we lawyers are concerned, while naturally those who represent clients, whether governmental or private, have to protect their clients' interests to the best of their ability, I think we all also owe a duty to our profession to ensure that whatever is done is, to use the phrase of the International Co-operation Declaration, "fair and reasonable". If I recall correctly, it was a famous son of this beautiful country, and of this illustrious city we are in, Count Camillo Benso di Cavour, who once said, What scoundrels would we not be if we do for ourselves what we do for our country? We are, however, no longer in the turbulent days of fighting for national unity. I do not think it is in a State's long term interests in whatever negotiations to drive for what in domestic law might be branded as unfair contract terms. In particular, representatives of powerful States would do well to remember that international negotiations are not like a domestic adversarial process where there is a judge to curb forensic excesses. This said, one needs also to remember, on the other hand, the ancient proverb as put by Dr Samuel Johnson, "One man may lead a horse to the water, but twenty cannot make him drink". This saying is particularly apt when it comes to attempts to change the behaviour of States, big or small, through treaties, when they see no national interest in accepting the obligations the treaties involve, however desirable or even imperative the objectives of the treaties in question may be from the international point of view. The number of poorly ratified treaties unfortunately testify to this sorry tale.

As for due representation of the dominant section, experience has shown that, in any international treaty-making or rule-making, due weight has inevitably to be given to the views of those whose co-operation is indispensable to the working of the treaty or rule, including those which are, in the words of the International Court of Justice in the *North Sea Continental Shelf Cases*, "specially affected".[28] It is only a truism to say that in any society, the law always represents the will of the dominant section.[29]

Finally, as regards a propitious political climate, experience has also shown that even where a given rule or treaty is felt by all the States specially affected as reasonable or even desirable, it is unlike to come to fruition unless the international political situation is propitious for it to be born. Each of the five treaties relating to outer space drawn up by the United Nations proves this.[30] At this moment, as mentioned before, the Cold War has been officially buried. Despite some local turmoils, the overall political barometer reads "Fair-set".

On this auspicious 30th anniversary of the Space Treaty, with Mir circling over us, and the International Space Station being on its way, both outstanding exemplars of international co-operation in space, redoubled efforts ought to be made by all, both in and outside the United Nations, to prepare space law, including all the existing UN treaties and declarations, for the New Millennium. May we wish this enterprise Fair Wind and Godspeed.

Notes

1. Treaty on Principles Governing the Activities of States in the Exploration and Use of Outer Space, including the Moon and Other Celestial Bodies. London, Moscow, and Washington, 27 Jan. 1967. Adopted by UN General Assembly resolution 2222 (XXI) on 19 Dec. 1966, opened for signature on 27 Jan. 1967, entered into force on 10 Oct. 1967. 610 UNTS p. 205.

2. Resolution 1962 (XVIII).

3. Resolution 1962 (XVIII).

4. See B. Cheng, *Studies in International Space Law* (Oxford: Clarendon Press, 1997), e.g., p. 327.

5. See, e.g., the many papers on the subject in successive IISL Colloquia on the Law of Outer Space in Cumulative Index of the Proceedings 1958-94 (UN Office of Outer Space Affairs, *Space Law: A Bibliography*, UN Doc. A/AC.105/636. New York: UN. 1996), under "Definitions". See also Cheng, *op. cit.* in note 4 above, Ch. 18 (pp. 492-509), Ch. 23, s. II (pp. 598-602), Ch. 24, s. III.C (pp. 634-6).

6. Regarding the former, see *op. cit.* in note 4 above, Ch. 14 (pp. 425-56), and *Space Law: A Bibliography*, cited in note 5 above, under "Definition/Delimitation of Airspace and Outer Space". Regarding the latter, see Cheng, *op. cit.*

in note 4 above, Ch. 19 (pp. 513-22), and *Space Law: A Bibliography*, under "Definition".

7. See further Cheng, *loc. cit.* in previous note.

8. 55 *Dept. of State Bull.* (1966), p. 952; statement released on 8 Dec. 1966, when agreement was actually reached between the major space powers. See *op. cit.* in note 4 above, p. 215.

9. Resolution 1962 (XVIII).

10. See further *op. cit.* in note 4 above, Chs. 19-20 (pp. 513-38); see also B. Cheng, "Military Use of Outer Space: Article IV of the Space Treaty Revisited", paper presented to the International Conference on Air and Space Policy, Law and Industry for the 21st Century: The Utilization of the World's Air Space and Free Outer Space in the 21st Century, Seoul, 23-25 June 1997, organised by the Korean Association of Air and Space Law. The proceedings are being published.

11. See "US ready to unleash Star Wars laser", *The Times* (London), 3 Oct. 1997, p. 12, cols. 1-8. MIRACL (Mid-Infra-Red Advance Chemical Laser) will be tested, firing two pulses at a military satellite 260 miles above the earth. Cf. also 8(38) *Space News* (6-12 Oct. 1997), headline on p. 1: "Pressure from Senate Majority Leader Trent Lott (R-Miss.) might result in a quintupling of the budget for the U.S. Air Force's Space Based Laser program. See page 4.".

12. International Co-operation in the Peaceful Uses of Outer Space, resolution 51/123, 13 Dec. 1996, Preamble, para. 5.

13. Paris, 27 May 1997, 36 *International Legal Materials (ILM)* (1997), p. 1006.

14. Cf. US Senator Gore before the UN First Committee, 3 Dec. 1962: "The question of military activities in space cannot be divorced from the question of military activities on earth", cited in *op. cit.* in note 4 above, p. 515.

15. Cf. UN, *Report of the Committee on the Peaceful Uses of Outer Space,* GAOR, 52nd Sess., Supplement No. 20 (A/52/20), 1997, para. 25.

16. See *op. cit.* in note 4 above, Ch. 9, ss. II and III.A (pp. 215-20).

17. See *op. cit.* in note 4 above, Chs. 23 and 24 (pp. 598-640).

18. See *op. cit.* in note 4 above, Ch. 24 (pp. 621-40).

19. See further *op. cit.* in note 4 above, Ch. 24, ss. H and I (pp. 626-30).

20. See B. Cheng, "Space Objects and Their Various Connecting Factors", in Gabriel Lafferranderie and Daphné Crowther (eds.), *Outlook on Space Law Over the Next 30 Years* (The Hague: Kluwer, 1997), pp. 203-15.

21. Cf. *op. cit.* in note 4 above, Ch. 17 (pp. 475-91).

22. Cf. N. Jasentuliyana, "Celebrating Fifty Years of the Chicago Convention: Twenty-Five Years after the Moon Landing: Lessons for Space Law", 19-2 *Annals of Air and Space Law* (1994), p. 429; B. Cheng, *The Law of International Air Transport* (London: Stevens, 1962), Ch. 3, s. 2 (pp. 63-7).

23. 704 UNTS, p. 219; ICAO Doc. 8364. See further B. Cheng, "Aviation Criminal Jurisdiction and Terrorism: The Hague Extradition/Prosecution Formula and Attacks at Airports", in B. Cheng and E. D. Brown (eds,), *Contemporary Problems of International Law: Essays in honour of Georg Schwarzenberger on his eightieth birthday*. (London: Stevens. 1988), pp. 25-52.

24. Cf. *op. cit.* in note 4 above, Ch. 3: From Air Law to Space Law, pp. 31-51; Ch. 4: International Co-operation and Control: From Atoms to Space, pp. 52-69. See also Cheng, *op. cit.* in note 22 above, Ch. 4, s. D: International Co-operation and Facilitation, pp. 145-70.

25. UN General Assembly resolution 3281 (XXIX), 12 Dec. 1974, Charter of Economic Rights and Duties of States, 14 *ILM* (1975), p. 251. Before that, UN General Assembly resolution 1803 (XVII), 14 Dec. 1962, UN *Yearbook* (1962), p. 503; resolution 2185 (XXI), 25 Nov. 1966, 6 *ILM* (1967), p. 147; resolution 3016 (XXVII), 18 Dec. 1972, 12 *ILM* (1973), p. 226; and resolution 3171 (XXVIII), 17 Dec. 1973, 13 *ILM* (1974), p. 238, all on Permanent Sovereignty over Natural Resources; resolutions 3201 (S-VI) and 3202 (S-VI), 1 May 1974, Declaration and Programme of Action on the Establishment of a New International Economic Order, 13 *ILM* (1974), p. 715.

26. Declaration on International Cooperation in the Exploration and Use of Outer Space for the Benefit and in the Interest of all States, Taking into Particular Account the Needs of Developing Countries, General Assembly resolution 51/122, UN GAOR, 51 Sess. (1996). See J. S. Thaker, "The Development of the Outer Space Benefits Declaration", XXII-I *Annals of Air and Space Law* (1997) 537-58.

27. See further *op. cit.* in note 4 above, Epilogue, s. IV (pp. 687-94).

28. *ICJ Rep. 1969*, p. 3, at p. 43.

29. Cf. B. Cheng, "How Should We Study International Law?", 13 *Chinese Yearbook of International Law and Affairs* (1994-5), pp. 214-18, s. II: What is Law? On p. 215, eight lines from base, "Savigny" was a slip of the pen for "R. v. Ihering".

30. See *op. cit.* in note 4 above, Epilogue, s. IV.B (pp. 690-1).

Session 1:
BACKGROUND AND HISTORY OF THE OUTER SPACE TREATY

Chairmen: A.A. Cocca, Argentina
N. Hosenball, USA

Rapporteur: J. F. Renaud, France

CONCEPTS OF SPACE LAW BEFORE SPUTNIK[1]

by Stephen E. Doyle, Consultant
3431 Bridget Brae Road
Shingle Springs CA 95682

ABSTRACT

It is sometimes written by commentators on space law that, because there was no activity in space prior to the launch of the first man made orbital object (Sputnik-1), there was no need for space law and, consequently, no attention was given to the subject before 1957. A careful reading of the available legal literature before October 1957 not only gives clear evidence of considerable thought devoted to space law problems, but also shows that many concepts embodied in the early international instruments purporting to create space law reflected the thought of numerous precursor commentators. Many concepts contained in early declarations of principles and treaties adopted by the United Nations General Assembly after 1958 had been developed before Sputnik by visionary lawyers and legal pundits.

EARLY SEGREGATION BY LANGUAGE

The legal and scholarly commentary on space law in the first half of the 20th century was kept segregated in many cases by curtains of language. Almost none of the commentary produced prior to 1950 contains references to other space law commentary that had been previously published in other languages. A notable exception is in the work of Vladimir Mandl (1932). Mandl's work was both comprehensive in the scope of its consideration of space law and well researched in the legal literature of Western Europe available in languages known to Mandl.

When the several national astronautical and rocket societies assembled in Paris for the first international astronautical congress, held there in 1950, the isolation of national commentaries behind the linguistic curtains separating them was irreversibly ended. Emergence of the International Astronautical Federation, with annual international meetings in London (1951), Stuttgart (1952), Zurich (1953), Innsbruck (1954), and thereafter, provided a necessary international forum for an expanding exchange of views and information on all aspects of astronautics, including the perceived early problems of space law. In addition, the broader international distribution of legal and policy journals originating in Europe and the Americas provided documentary vehicles for international information flow.

One can readily discover the relationship between the formative space law work at the United Nations in the 1960s and 1970s and the works of the pioneering commentators, if one examines the early UN declarations of principles and the later treaties to determine how much of their content can be found in the prior literature. Although the specific wording of particular passages may or may not be identical, the basic concepts contained in many of the early declarations were long known in the literature, to at least those who were conversant with sources published in their original languages. As the 1950s proceeded, and greater international exchange of ideas occurred, many of the early sources became better known, although they were relatively rarely translated or produced in alternative publications in different languages.

A bibliography containing all of the works referenced herein follows the text of this paper. Where multiple works of a single author are referred to, the indicated date of reference differentiates the respective sources.

Neither time nor space permits the duplication and incorporation of texts drawn from the many UN declarations and resolutions that emerged in the decade from the launching of Sputnik-1, in October 1957, and the coming into force of the *Treaty on Principles Governing the Activities of States in the Exploration and Use of Outer*

Space, Including the Moon and Other Celestial Bodies, in October 1967. Herein, is a selected survey of some of the ideas, concepts and proposals that were in the legal and the general literature, and thus were available to interested law makers, before that historic day, October 4, 1957, when Sputnik-1 was launched.

ON THE QUESTIONS OF THE NEED FOR AND DEFINITION OF SPACE LAW

Emil Laude (1910), V. A. Zarzar (1926), Vladimir Mandl (1932), Arthur C. Clarke (1946), Alex Meyer (1952), and others, had written that new law is required to govern new juridical relationships, because of the unique aspects of spaceflight and the areas, technologies, speeds and altitudes involved. Laude (1910) and Mandl (1932) declared that a new set of legal terms with clear definitions would be required to deal with the issues of spaceflight because it is a unique pursuit. The new body of law will be called space law (or the law of outer space). In the Russian SSR, Zarzar wrote (1926) that for interplanetary travel there will be an interplanetary transport law. Mandl also observed (1932) that space law shall apply equally to all flight devices, manned or unmanned.

Mandl (1932), Meyer (1952) and C. Wilfred Jenks (1956) had declared that technology will advance continually until many capabilities will be demonstrated using spacecraft. As significant new capabilities are added, voices will be raised calling for special regulation based on genuine space law, that is on legislation which would do justice to all the peculiarities of spaceflight which come to light at that future stage of development. Oscar Schachter (1951), Cyril Horsford (1955) and Jenks (1956) declared that the unique features of spaceflight will have to be addressed by a body of space law, developed for that purpose with reference to modern international law and customary law based on accepted legal principles.

ON INTERNATIONAL COOPERATION IN THE USE OF OUTER SPACE

John Cobb Cooper (1951), Schachter (1951), Meyer (1952), Welf Heinrich, Prince of Hanover (1953) and Joseph Kroell (1953) had all urged that international cooperation is the only means to achieve and guarantee peaceful use of space and travel therein. They urged that immediate steps be taken toward achieving such cooperation. Jenks (1956) wrote that it would be entirely fitting that control of space activities should be a world responsibility and every effort should be made to apply such a solution to the problem from the earliest stages of development. Jenks suggested that if legislative authority over human activity beyond the atmosphere of the Earth were to be regarded as vested in the United Nations General Assembly (UNGA), it would be possible for the UNGA to evolve progressively the necessary rules on the subject. In Jenks 1956 view, at some future time it may be necessary to have an international code to regulate space.

Horsford (1956), Cooper (1956) and H. Alberta Colclaser (1956) wrote that the political consequences of any substantial conquest of space are so far-reaching that an international body would seem to be essential, so great would be the need. It certainly should not be impossible to one create through international cooperation. David Lasser (1931) had written that an International Interplanetary Commission, comprising delegates from nations interested in spaceflight, could bring together all the talent, knowledge and resources necessary to assemble an interplanetary spacecraft and handle all the details necessary for design and production of the spacecraft. Aldo Armando Cocca (1954) had similarly urged that international cooperation be the basic approach to spaceflight so that the interests of all men would be collectively represented. As Cocca wrote: "... if the studies, plans, tests, and knowledge of today are under a universal public dominion, the vehicle that emerges from these studies should obtain the same juridical status. In which case the conquest of interplanetary space will be a conquest by humanity."

Musto (1956) wrote that the attempt of man to explore space, even if attempted by individuals, is an attempt to explore by the collectivity of mankind and will benefit the collectivity. Some basic principles are derived from this observation, wrote Musto: (1) the absolute freedom of overflight, (2) freedom of flight after landing, (3) financial support by all states, and (4) the obligation of all states to assist, protect and favor the flyer [astronaut/cosmonaut] no matter what state he represents. These principles give rise to the need for an International Center for Interplanetary Flight, financed through the contributions of all states. As for discoveries and accomplishments, there shall be immediate devolution (*i.e.* release to governments), exclusively for peaceful and scientific purposes, of all principles and discoveries which shall be acquisitions by all states, except for the moral paternity of those who discover the new laws. [Thus, recognition of individual achievement would be preserved. -ed]

ON DETERMINING THE UPPER LIMIT OF NATIONAL SOVEREIGNTY

The *International Convention relating to Aerial Navigation,* adopted in Paris in October 1919, provided in Article 1 that "The High Contracting Parties recognize that every Power has complete and exclusive sovereignty over the airspace above its territory." This tenet was reflected in the *Convention on International Civil Aviation,* adopted in Chicago in 1944, Article 1 of which reads: "The contracting States recognize that every State has complete and exclusive sovereignty over the airspace above its territory." With this fundamental and universally accepted air law in place, the emergence of spaceflight raised the issue of whether or not states had a right to control flight over the airspace above their national territories. Commentators held widely varying views on this subject, but there was no lack of proposals to solve the problem.

Zarzar (1926), Mandl (1932), Clarke (1946), Cooper (1948, 1951) Schachter (1951), Heinrich (1953), Kroell (1953) and Jenks (1956) had predicted that there will be an issue as to altitude at which the international regime of spaceflight begins and replaces the concept of sovereignty over superjacent airspace. Andrew G. Haley (1955) noted that it would be repugnant to the concept of space travel to have an object traveling in space treated as if it were passing constantly through the legal jurisdictional influence of subjacent states; therefore it is necessary to have a new philosophy of law to apply to space travel. This was not an original observation. Merignhac (1914), Zarzar (1926), Mandl (1932), Clarke (1946), Cooper (1951), Schachter (1951), Heinrich (1953), Kroell (1953) and USN Rear Admiral Chester Ward (1957) had observed that state sovereignty must have some kind of limit; sovereignty cannot reach into infinity. The consensus among these pundits was that sovereignty ends where the airspace ends.

Cooper (1948, 1951) and Kroell (1953) proposed that national sovereignty extend outward in space to the limit of the Earth's gravitational influence. In a similar vein, Catellani (1912), Eugene Korovin (1934) and Haley (1955) believed that state sovereignty radiates without limit into the heavens as far as is necessary to assure or protect the state's sovereignty over the subjacent territory. Or, as Korovin (1934) and Haley (1955) stated, it is sufficient in international law to recognize in each state, the territorial or aerial frontier of which has been violated, the undeniable right to take all necessary measures for its protection, including the seizure of the crew of an airship (in the case of a landing) to reprisals of any kind. As Catellani (1912), Zarzar (1926), Korovin (1934) and Haley (1955) contended, governments have the right to protect their citizens from threats to their safety and to the safety of their territories from above, regardless of the altitude of a superjacent flight.

There was no disagreement with the observations of Mandl (1932), Meyer (1952), Danier (1952), Heinrich (1953) and Colclaser (1956) that while transiting the atmosphere where aeronautical vehicles operate, a spacecraft, although not itself subject to aeronautical law, must operate in cognizance of and in compliance with the applicable aeronautical law. The formidable group of Mandl (1932), Schachter (1951), Meyer (1952), Heinrich (1953) and Jenks (1956) asserted with conviction that where the airspace ends, there at the same time ends that zone of sovereignty above the territory which states have mutually recognized by treaty and custom. Mandl (1932) and Meyer (1952) explained that airspace is not an independent state territory. Airspace is only qualified as an object of sovereignty as an appurtenance to the subjacent earth's surface, and when it appears desirable as a showplace in which to demonstrate sovereignty. The *de facto* resolution of this continued academic debate about upper limits on national sovereignty was clearly seen and reported by Haley in 1955, when he wrote that states might have objected to passage of planned spaceflights above their national territories; but in 1955, when the United States and the USSR announced their intentions to orbit scientific satellites around the Earth, no nation complained or objected to the overflight of their territories. Haley noted, "The scientists benefited mankind as a whole in a field where lawyers might well have failed." Taoka (1956) noted that until recently, aircraft had not flown in the upper atmosphere, but with development of modern rocketry, questions have arisen concerning satellites passing at high altitudes in their orbits. Such overflight can be considered as not violating sovereign rights of subjacent states.

Nations chose to ignore the suggestion of Aaronson (1955) to the effect that it would be highly desirable, before the implementation of proposed earth satellite projects, to convene an intergovernmental conference of all the nations in order to define the limits of airspace. [For all intents and purposes, the matter was already practically resolved by the customary practices, *i.e.,* the manifest tolerance, of states. -ed]

ON REGULATING THE USE OF OUTER SPACE FOR MILITARY PURPOSES

Vladimir Mandl (1932) wrote that military considerations will present the most pertinent and most important motives for government support of spaceflight. Mandl (1932) also wrote, followed by US Army AF Gen. Henry (Hap) Arnold (1945) and Ralph A. Smith (1949), that international warmaking regulations will be expanded to cover spacecraft and these will be integrated into normal warmaking resources without anyone seriously thinking that the employment of spacecraft in warfare is to be excluded. Mandl (1932) also believed that the rules of land, sea or air warfare will be binding on spacecraft according to their use to support ground forces, or sea or air formations. In fact, Mandl (1932), Gen. Arnold (1945), and Cyril Horsford (1955) had written that outer space, as in the case of any sea area not subject to state sovereignty, would offer itself as a theater of war to anyone who wished to use it.

Opposing these early views of the inevitability of military use of outer space were Oscar Schachter (1951) and Alex Meyer (1952). Schachter wrote that it may be expected that there will be a demand that the use of outer space for military purposes will be outlawed. Meyer (1952) urged that the exploration of outer space be restricted for peaceful purposes. Kroell (1953) urged that interplanetary travel and the use of fixed or moving space stations should not be permitted if for purposes of political domination or of war; even studies or preparations of warlike ends should be prohibited. Wernher von Braun (1952) wrote that cameras located on a space station could be focused on the Earth below; then there would be no "iron curtain" because the entire Earth could come under the observation of the space station.

Myres McDougal (1956) and R. Adm. Ward (1956) were very pragmatic in their assessments that if a vehicle in flight is believed to pose a threat to a particular nation, and if that nation has the capacity to shoot it down, that nation is going to shoot it down. Thus, there was clearly a substantial division of opinion between pundits on the one hand, who believed military use of space was inevitable and acceptable, and those on the other hand, who opposed military use and firmly recommended against it, seeking the prohibition of such use of space. It remained for that issue to be brought before an international institution with the authority to address and resolve it.

Hester (1955) wrote that overall international security would be served if the UN would draft a plan offering a basis for exploratory discussions among immediately interested states. Hester thought the UN draft should be based on a number of realities: (1) the conduct of human spaceflight is inevitable, (2) spaceflight offers a means for one nation to threaten all nations, (3) controls should avoid hazards to people and property, and (4) a unified general astronautical policy should apply to all nations. But Hester was realistic enough to observe that a number of thorny problems would have to be resolved by the UN in the process of defining a regulatory regime, including: (1) the international composition of a commission to coordinate space R&D, (2) the creation of methods to pool information and exchange experts, (3) allowing international developments while limiting private ventures, (4) establishing controls and coordination of flight operations, (5) finding means for reducing operational hazards, (6) prohibiting weapons and war-heads in spaceflight, (7) establishing penalties for defaulting actions which may:
- misapply astronautics for military uses,
- involve refusal to make required disclosures, or
- disregard agreed safeguards and procedures.

Jenks (1956) declared that there would clearly be a great advantage in vesting the necessary authority in an international body, if this should be politically practicable, but the difficulties of doing so may well be formidable, particularly in view of the relationship between the exploration and exploitation of space and the question of defense. Schachter (1956) believed that international regulation of activities in outer space is required for: (1) protection of national security, (2) the maintenance of peace, (3) enabling scientific investigations, and (4) use of resources of outer space. Schachter thought and wrote that scientific research and exploitation of outer space would best be served by a regime of free use such as applies to the high seas.

ON THE REGULATION OF SPACE TELECOMMUNICATION

Laude (1910), Zarzar (1926), Sterling (1954), and Cooper (1956) noted that a separate body of law is required to deal with the ownership and use of the radiomagnetic spectrum. Sterling (1954) wrote that the use of radio frequencies for transmitting to and receiving information from rockets raises technical and regulatory problems involving national and international considerations. The selection of frequencies for use in space will be affected by propagation characteristics of the frequencies involved. Although frequencies desirable for astronautical communication are not now [1954] internationally well allocated or regulated, global coordination for such uses will be important to successful use of radio. Space communication will require clear channels and agreements not to employ jamming techniques; all these requirements create the need for unprecedented cooperation and planning among nations on an international basis. Haley (1955) noted that in moving toward spaceflight, science [meaning the scientific community] and government must advance more rapidly in the field of telecommunication than in any other field. Haley also noted that substantial machinery was already in place to meet the regulatory needs in the form of the International Telecommunication Union.

ON LIABILITY FOR DAMAGE CAUSED BY SPACEFLIGHT ACTIVITY

Mandl (1932) wrote that liability for damage to third parties should be absolute (no fault liability) because of the novelty of the subject, restricted familiarity with the nature of the technology on the part of the public and the courts, and the likely unexpected nature of any damage to be suffered by third parties. Mandl believed liability may be reduced if there is a contributing fault or negligence of the injured party [contributory negligence]. Mandl explained that liability for damage to the crew, a cargo owner, or other party in a contractual relationship with the operator of a spaceship should be dealt with in a manner consistent with the applicable contract. Accident insurance may cover damage to the crew or vehicle. In the absence of any special arrangements, noted Mandl, liability of the promoter to parties involved would exist only in cases of premeditation or gross negligence, because the involved parties can be said to have assumed the risks involved. Possible passenger or even cargo transportation contracts will exhibit the character of liability-free speculative contracts rather than real production contracts, accordingly the various contract points must be clarified with regard to disputes.

Mandl (1932) was of the opinion that in the case of a landed spacecraft causing damage on the ground, normally a lien should apply to the spacecraft for the damages suffered, for payment prior to the removal of the spacecraft. Provision probably will be made by the legislature that a person entitled to damages caused by an emergency landing will not be able to prohibit the removal of the spacecraft, but there will be some general salvage and rescue obligation which will permit a claim for salvage or rescue compensation as well as for damage. Liability for damage to third persons will have to be met through compulsory liability insurance, posting of bonds, or by some such means. As if endorsing all that Mandl had written, but without reference to it, Quincy Wright (1956) said that issues of liability for damage by spaceflight activity will arise early in the era of spaceflight and these issues should be resolved early. However, cautioned Wright, no state should be allowed to establish laws that cannot be enforced.

ON THE AUTHORITY, LOCATION, AND ELIGIBILITY TO CONDUCT SPACEFLIGHT

Mandl (1932) obviously thought a great deal about this area of concern. He wrote that an entity licensed to conduct aerial flight as its only statutory purpose may not extend its field of activity to spaceflight on its own authority. Even in the absence of statutory direction, a spacefarer is responsible to exercise due care in dealing with an explosive and potentially dangerous vehicles, and owes a duty of prudence and reasonable care to the community in which spaceflight operations are undertaken. For example, prior to departure of a flight, the equipment (flightcraft and related ground equipment) and crew should be examined to ensure that they pass a reasonable test of capability to perform the operations contemplated. It would not be prudent to undertake a spaceflight with an untrained crew and an untried or untested vehicle.

Mandl (1932) wrote that the capability and appropriate precautions to handle explosives, high levels of electric power, flammables and installations of the flight facility should be suitable to the type of propulsion used in each case. The launch area, including the ground infrastructure, must be set up in full consideration for the overall safety. Prior to launch, the spacefarer must either provide aeronautical interests with notice to airmen of a launch operation so that aviators can avoid the launch area, or be prepared to comply with all regulations applicable to aeronautical flight while transiting the atmosphere. When a prohibited air zone is established, it must be respected by the spacefarer also, just as he must comply with the prohibitions against using cameras and using radio transmitting equipment. As a rule, national or federal authority will be the appropriate authority to regulate and license spaceflight activity.

A senior administrative board, or a ministry or agency will have to be established to handle all the regulatory requirements such as all the questions of space law, the effect on the environment of the launch area, the manifold licenses, oversight activities, crew health and readiness and spacecraft qualification and readiness, the various administrative jurisdictions and the like. The flight support organization on the ground must also demonstrate required levels of performance.

Mandl (1932) declared that the proper use of public or donated monies to support spaceflight operations must be assured through public financial disclosure as a responsibility of the project manager. Corrupt procedures, the exhaustion of public confidence, and thereby the degradation of the whole concept of spaceflight will be prevented by requiring official scrutiny and state control. Mandl wrote that the legislature will perhaps consider it to be its duty to make provisions in the law for the case in which a years-long spaceflight is undertaken. The spacecraft commander will have command authority and even penal power over his crew. International law will provide that during passage through non-sovereign zones occurrences onboard spacecraft are to be adjudicated according to the laws of that state to which the spacecraft belongs. The principle of "missing in space" will have to be introduced for the case wherein a spacecraft is lost and its fate unknown.

Mandl (1932), Schachter (1951), Heinrich (1953) and Horsford (1955) believed that nationality of a spacecraft probably shall be determined by the nationality of the owner or of a majority of partners.

Mandl (1932) noted that international agreements established for the internationally agreed operation of aircraft would not apply to spacecraft because of their entirely different nature. If a launch occurs on the high seas, the clearance procedures will become simplified, without, however, becoming unnecessary. Similar to the case of the launch site, the prospective landing area will have to be checked out, for the spacecraft and for all parts which are released during the flight, booster rockets and the like, to the extent that these objects do not reach the Earth's surface in a harmless form, completely disintegrated.

Jenks (1956) declared that at some future time it may be necessary to have an international code to regulate space navigation, space radio communication, spacecraft flight worthiness, and flight crew qualifications, space rescue, the carriage of passengers and goods, and law applicable to legal transactions in space. Jenks believed there would eventually be a need to agree on law to regulate:

- jurisdictional issues and questions of applicable law,
- personal legal status of individuals in space,
- rights of property beyond the Earth,

- contract law relating to contracts made in space,
- tort law applicable to tortious activity in space, and
- appropriate and applicable criminal law.

Jenks (1956) also wrote that if legislative authority over human activity beyond the atmosphere of the Earth were to be regarded as vested in the United Nations General Assembly, it would be possible to evolve progressively the necessary rules on the subject through the Assembly. Jenks believed one could conceive of the United Nations governing extra-mundane settlements, directly or by some special agency created for the purpose. The ideal arrangement would appear to be that sovereignty over unoccupied territory on the Moon or on other planets or satellites should be regarded as vested exclusively in the United Nations.

ON THE RESCUE AND RETURN OF SPACE OBJECTS AND OF ASTRONAUTS

Mandl (1932) wrote that in the case of a landed spacecraft causing damage on the ground, normally a lien should apply to the spacecraft for the damages suffered, for payment prior to the removal of the spacecraft. Provision probably will be made by the legislature that a person entitled to damages caused by an emergency landing will not be able to prohibit removal of the spacecraft, but there will be some general salvage or rescue obligation which will permit a claim for salvage or rescue compensation as well as for damages.

Carlo Musto (1956) believed that every state has an obligation to protect and assist space crews, even to immediate release of returned crews regardless of whether or not damage has been suffered by the state landed in, provided: (1) politico-customs controls have been complied with and (2) just compensation has been paid to any injured parties suffering property or personal injury. Damages incurred should be verified by a special state commission established in every state for this purpose, including representation from a proposed International Center for Interplanetary Flight.

ON THE LEGAL STATUS OF OUTER SPACE AND CELESTIAL BODIES

Zarzar (1926), Mandl (1932), Clarke (1946), Schachter (1951), Meyer (1952), and Heinrich (1953) wrote that in an area beyond the airspace appurtenant to the Earth, there is a regime in which spacecraft traffic is completely free from terrestrial jurisdictions. But Cooper (1951 and 1956) and Kroell (1953) believed that only in the area beyond the reach of Earth's gravity would there be a regime in which spaceflight would be completely free from terrestrial jurisdictions.

Mandl (1932) wrote that erection of some kinds of stations in space, or artificial moons, will require appropriate international administration and regulation of such resources, including access to space stations. Meyer (1952), Heinrich (1953), and, to some extent, Kroell (1953) wrote that space stations beyond the atmosphere could not be classified as res communes, or common property; their legal nature would be determined by their modes of construction. They believed that space stations would come under the dominion of the country which had constructed them or subject to whose order or protection they had been constructed. Meyer (1952) and Heinrich (1953) believed that there would be no obligation to open space stations to public access, but once opened there should be no discrimination of any kind in access. Kroell (1953) thought that space stations may be freely established for the use of space travel but they should be internationalized and be open to all. Horsford (1955) believed that a space station is a concept with enormous potentiality, unfortunately for evil as well as for good, and it is to hoped that mankind can achieve the proper regulation of the whole question of the law to regulate spaceflight.

Cooper (1951), Schachter (1952) and Jenks (1956) explicitly believed that outer space should be free for use by nations for peaceful and scientific purposes. Schachter (1951) and Danier (1952) wrote that outer space, like the high sea, is not and should not be under the sovereignty of any nation state. However, this does not mean that outer space would be in a state of lawlessness and anarchy; it would be governed by rules of international law to provide a degree of legal order while preserving the principle of its freedom. R. A. Smith (1949) wrote that the Moon is not the property of any state; if it is untenanted, it is the common heritage of mankind. Lionel Laming (1949) and Schachter (1951) declared that all the solar system, not only the Earth, deserves to be considered as the heritage of mankind. Oscar Schachter (1951) was even more explicit, saying that the Moon and other celestial bodies could be considered as *res nullius*, free from occupation, like the then undiscovered parts of the Antarctic continent. Schachter (1951) and Meyer (1952) proposed that outer space, the Moon, and other celestial bodies should remain free for use by all; the special rules that may be required could be established under the aegis of the United Nations. Heinrich (1953) had concluded that the entire area beyond the atmosphere would have to be considered free territory both on technical grounds founded on the law of nature and for reasons of legal construction and policy.

SOME CONCLUDING OBSERVATIONS

There are many additional areas we could survey here to show the variety and extent of proposed concepts of space law that were recorded in the literature well in advance of the successful first orbiting of a spacecraft. The content of this paper comprises selected extracts from the concluding chapter of a book now in preparation, entitled *Concepts of Space Law Before Sputnik*, which includes the complete texts of many of the earliest writings on space law, collected, with some translated into English, with historical introductions and annotations, where appropriate. The book is expected to be published in 1998 and should be available at the next Colloquium on the Law of Outer Space.

The conclusion one can draw from this survey is that there was no failure on the part of the legal community to see or to openly discuss the legal issues given rise to by the emergence of spaceflight. Perspicacious jurists in many countries had their eyes and minds open to the potentials of the future. They addressed the foreseeable problems admirably, proposing solutions to anticipated problems well before the problems actually existed. When the United Nations Committee on the Peaceful Uses of Outer Space turned its attention to the many issues addressed in the 1960s and 1970s, there was a wealth of prior work on and thinking about the problems then faced. Unfortunately there were still curtains of language and a lack of general access to the legal literature's of the several states. Those were the conditions in the infancy of space travel. Today, the access problem is substantially alleviated and forums such as this meeting provide valuable means for the effective international interchange of ideas. We may feel we have come a long way from those humble beginnings, but in my opinion we still have a great deal to learn by studying the substantially unexploited works of the pre-Sputnik pioneers of space law.

-O-

Bibliography of Sources

Aaronson, M., "Earth Satellites and the Law," 220 *Law Times* 115-116, August 26, 1955.

Arnold, Gen. H. H., US Army AF, *Third Report of the Commanding General of the Army Air Force to the Secretary of War*,* 12 November 1945, US Army Air Force, Wash., DC, 72 pp.

Braun, W. von, see von Braun.

Catellani, E., *Air Law*, Paris, 1912.

Clarke, A. C., "The Challenge of the Spaceship,"* in *Journal of the British Interplanetary Society*, 1946-47, pp. 66-78, discussion following thereafter.

Cocca, A. A., *"Al Encuentro de un Derecho Nuevo: El Interplanetario; Naturaleza Juridica del Espacio Interplanetario"* [To Meet a New Law: Interplanetary Law; the Juridical Nature of Interplanetary Space],** 4 *Revista de Jurisprudencia Argentina, Ano* XVII, No. 5820 Buenos Aires, October 15, 1954. This is the same paper as *"Die Rechtliche Natur des Weltraums"* [The Juridical Nature of Space], presented in Innsbruck, Austria, August 1954, and in *the Proceedings of the Vth International Astronautical Congress*, 283-299, Springer-Verlag, Vienna, 1955; see also Cocca, A. A., *"Metodo para el Estudio de los Problems Juridicos que Plantes la Conquista del Espacio Interplanetario"* **[A Method for the Study of Juridical Problems Posed by the Conquest of Interplanetary Space], in Spanish, in *the Proceedings of the VIIth International Astronautical Congress*, 155-160, *Associazione Italiana Razzi*, Rome, 1956.

Colclaser, H. A., Remarks on Cooper, J. C., "Legal Problems of Upper Space," address to the 50th Annual Meeting of the American Society for Int'l Law, April 1956, in *Proceedings of the American Society of International Law*, 85-93, Wash., DC, 1956. See discussion following the paper at pp. 99-102 of the Proceedings, reproduced in US Congress, *Space Law: A Symposium*, prepared at the request of Sen. Lyndon Johnson, Chairman, Senate Special Committee on Space and Astronautics, 85th Cong., 2nd Sess., Committee Print, GPO, Wash., DC, Dec. 31, 1958, pp. 122-149. These *Proceedings* contain a discussion of Cooper, J. C., "Legal Problems of Upper Space," by a distinguished panel of lawyers, academicians and other experts conversant with air law and emerging space law issues.

Cooper, J. C., "International Air Law,"* an address at the US Naval War College, Newport, R. I., Dec. 20, 1948; see also "High Altitude Flight and National Sovereignty," an address to the *Escuela Libre de Derecho*, Mexico City, Jan. 5, 1951, published in 4 *Int'l Law Quarterly* 411 (1951); then in 13 *Int'l Air Transport Association Bulletin* 46 (June 1951); then as *"Les Vols a haute altitude et la Souverainete nationale,"* in 5 *Rev. Francaise du Droit Aerian* 123 (1951); and as *"Der Flug in Grosse Hohen und die Nationale Staatsgewalt,"* in 1 *Zeitschrift fur Luftrecht* 237 (1952). An early, if not the first, example of multilingual presentations of a single work on space law to break through the curtains of language and to stimulate more international thinking about the issues emerging. An excellent anthology of the original works of John Cobb Cooper on air and space law is contained in Vlasic, I. (ed), *Explorations in Aerospace Law*, McGill University Press, Montreal, 1968, 480 pp. See also Cooper, J. C., "Legal Problems of Upper Space," address to the 50th Annual Meeting of the American Society for Int'l Law, April, 1956, in *Proceedings of the American Society of International Law*, 85-93, Wash., DC, 1956.

Danier, E., *"Les Voyages Interplanetaires et le Droit"* (Interplanetary Travel and the Law), in 15 *Revue Generale de l'Air* 422-425, Paris, 1952.

Haley, A. G., "Basic Concepts of Space Law," a paper presented to the 25th Annual Meeting of the American Rocket Society, November 1955, Chicago, Illinois; published in a revised version under the same title in *Jet Propulsion*, vol. 26, no. 11, Nov. 1956, pp. 951-957 and 968. *Jet Propulsion* dropped an introductory abstract, references to a future book, an appendix, and selected portions of the historical elements of the paper, but maintained intact all its essential arguments. See also his remarks after the Sept. 19, 1956, morning Technical Session of the VIIth International Astronautical Congress, in the *Proceedings of the VIIth International Astronautical Congress*, 907-908, *Associazione Italiana Razzi*, Rome, 1956.

Heinrch, W., Prince of Hanover, 1953 doctoral dissertation, *Air Law and Space*, translated and reproduced in *Saint Louis University Law Journal*, Spring 1958, pp. 11-69; reproduced in US Congress, *Space Law: A Symposium*, prepared at the request of Sen. Lyndon Johnson, Chairman, Senate Special Committee on Space and Astronautics, 85th Cong., 2nd Sess., Committee Print, GPO, Wash., DC, Dec. 31, 1958 at pp. 18-76. See also Heinrich, W., *"Die Rechtsprobleme des Weltraumes"* [Legal Problems of Outer Space], in *Weltraumfahrt*, vol. 4, no. 4, October 1953, containing a summary of the content of his doctoral dissertation submitted to the Georg-August University at Goettingen, Germany earlier that year.

Hester, W. B., "Some Political Implications of Spaceflight,"* *Journal of the British Interplanetary Society* 314-319, Nov.-Dec. 1955.

Horsford, C. E. S., "The Law of Space," in *Journal of the British Interplanetary Society*, May-June 1955, pp. 144-150; reproduced in US Congress, *Space Law: A Symposium*, prepared at the request of Sen. Lyndon Johnson, Chairman, Senate Special Committee on Space and Astronautics, 85th Cong., 2nd Sess., Committee Print, GPO, Wash., DC, Dec. 31, 1958 at pp. 77-83; also in US Congress, *Legal Problems of Space Exploration: A Symposium*, prepared for the Senate Committee on Aeronautical and Space Sciences, 87th Cong., 1st Sess., Doc. No. 26, GPO, Wash., DC, March 1961 at pp. 20-26.

Jenks, C. W., "International Law and Activities in Space," *Int'l and Compar. Law Quarterly*, January 1956, pp. 99-114; reprinted in US Congress, *Legal Problems of Space Exploration: A*

Symposium, prepared for the Senate Committee on Aeronautical and Space Sciences, 87th Cong., 1st Sess., Doc. No. 26, GPO, Wash., DC, March 1961 at pp. 33-45.

Korovin, Y. A., *"La Conquete de la Stratosphere et la Droit International,"** 41 *Revue Generale de Droit International Public* 675-686, Paris, Dec. 1934.

Kroell, J., *"Einem Weltraumrecht Entgegen"* [Towards a Space Law], 1 *Zeitschrift fur Luftrecht*, 246-255, 1952; and Kroell, J., *"Elements createurs d'un Droit Astronautique"* [Creative Elements of a Space Law], *16 Revue Generale de l'Air* 222-245, Paris, 1953.

Laming, L., *L'Astronautique*,* Presses Universitaires de France, Paris, 1950. [The publication bears the date 1950, although it is shown within to have been printed in 1949.]

Lasser, D., *The Conquest of Space*, Hurst & Blackett, Ltd., London, 1931, 288 pp.

Laude, E., "Questions Pratiques,"** *Revue Juridique Internationale de Locomotion Arienne* 16-18, Paris, 1910.

Mandl, V., *Das Weltraum-Recht: Ein Problem der Raumfahrt*** (Space Law: A Problem of Spaceflight), J. Bensheimer, Mannheim, Berlin, Leipzig, 1932, 48 pp.

McDougal, M., Remarks on Cooper, J. C., "Legal Problems of Upper Space," address to the 50th Annual Meeting of the American Society for Int'l Law, April 1956, in *Proceedings of the American Society of International Law*, 85-93, Wash., DC, 1956. See discussion following the paper at pp. 108-109 of the Proceedings.

Merignhac, A., "The Private and Public Aerial Domain and the Laws of Aviation in Times of Peace and of War," in *Revue Generale de Droit International Public*, vol. 21, no. 2, 1914, at p. 205.

Meyer, A., "Legal Problems of Flight into the Outer Space,"** an address delivered before the Third International Astronautical Congress, Stuttgart, Germany, 5 September 1952; later published in the *Annual Report of the British Interplanetary Society*, 1952 pp. 353-54; also in *Zeitschrift fur Luftrecht*, in English at p. 31; in German at p. 43; and reproduced in US Congress, *Legal Problems of Space Exploration: A Symposium*, prepared for the Senate Committee on Aeronautical and Space Sciences, 87th Cong., 1st Sess., Doc. No. 26, GPO, Wash., DC, March 1961 at pp. 8-19. See also his remarks on Cooper, J. C., "Legal Problems of Upper Space," address to the 50th Annual Meeting of the American Society for Int'l Law, April 1956, in *Proceedings of the American Society of International Law*, 85-93, Wash., DC, 1956. See discussion following the paper at pp. 96-98 of the Proceedings.

Musto, C., Remarks* after the Sept. 19, 1956, morning Technical Session of the VIIth International Astronautical Congress, in the *Proceedings of the VIIth International Astronautical Congress*, 903-905, *Associazione Italiana Razzi*, Rome, 1956.

Schachter, O., "Legal Aspects of Space Travel," a paper presented at the Hayden Planetarium Symposium on Space Travel, October 12, 1951, published in *the Journal of the British Interplanetary Society* ,* vol. 11, no. 1, January 1952, pp. 3-16; subsequently enlarged and included as a separate chapter in Ryan, C. (ed), *Across the Space Frontier*, New York, The Viking Press, 1952, pp. 118-131. See also his remarks on Cooper, J. C., "Legal Problems of Upper Space," address to the 50th Annual Meeting of the American Society for Int'l Law, April 1956, in *Proceedings of the American Society of International Law*, 85-93, Wash., DC, 1956. See discussion following the paper at pp. 105-106 of the Proceedings.

Smith, R. A., A comment entitled "Man and His Mark,"** in "Correspondence," vol. 8, no. 3, *Journal of the British Interplanetary Society* 131-132, May 1949.

Sterling, G. E., "Utilization of Radio Frequencies in Connection with Rockets,"** *Jet Propulsion*, vol. 24, no. 5, Sept.-Oct. 1954, pp. 322-23.

Taoka, R., *Kuiki no ryoyuken** (On the Right to Air Space), *Kuho*, No. 2, 1-30, Tokyo, Oct. 1956.

von Braun, W., "The Space Station," *Colliers* magazine, March 22, 1952; also see "Prelude to Space Travel," included as a separate chapter in Ryan, C. (ed), *Across the Space Frontier*, New York, The Viking Press, 1952, pp. 12-70.

Ward, C., "Projecting the Law of the Sea into the Law of Space,"** in *The JAG Journal* (Navy) 3-8, March 1957; reproduced in US Congress, *Legal Problems of Space Exploration: A Symposium*, prepared for the Senate Committee on Aeronautical and Space Sciences, 87th Cong., 1st Sess., Doc. No. 26, GPO, Wash., DC, March 1961 at pp. 120-128.

Wright, Q., Remarks on Cooper, J. C., "Legal Problems of Upper Space," address to the 50th Annual Meeting of the American Society for Int'l Law, April, 1956, in *Proceedings of the American Society of International Law*, 85-93, Wash., DC, 1956. See discussion following the paper at pp. 107-108 of the Proceedings.

Zarzar, V. A., "Mezhdunarodnoye Publichnoye Vozdushnoye Pravo" (Public International Air Law),* in *Voprosy Vozdushnogo Prava, Sportnik Trudov Sektsii Vozdushnogo Prava Soyuza Aviyakhim (Soyuz Obshchestv Druzhey Aviatsionnoy i Khimicheskoy Oborony i Promyshlennosti)* [Problems of Air Law, a Symposium of Works by the Air Law Sections of the USSR and RSFSR Unions of Societies for Assisting Defense and Aviation and Chemical Construction], vol. 1, pp. 90-103, SSSR i Aviakhim RSFSR, Moscow, 1927. The conference was held in 1926, the proceedings appeared in 1927. Translation in NASA TM-76913, NASA, Wash. DC, dated June 1982.

[All entries in this bibliography designated by a single asterisk (*) are included in part, and those with a double asterisk (**)are included in total, in English or in English translation, in the forthcoming publication by Doyle, S. E. *Space Law Concepts Before Sputnik*, Univelt, San Diego (1998).]

THE USSR AND THE 1967 TREATY ON OUTER SPACE

Yuri Kolosov, Ph.D., Professor of International Law, Moscow State Institute of International Relations (University) of the Foreign Ministry of Russia.

ABSTRACT

The beginning of codification and progressive development of international outer space law was laid down by the United Nations General Assembly Resolution 1721 (XVI) of 20 December, 1961, in which the first principles were formulated. The USSR actively contributed to this achievement.

The USSR was the first Sate to submit to the Legal Sub-Committee of the Outer Space Committee on 6 June, 1962, a draft declaration on the basic principles governing the activities of States pertaining to the exploration and use of outer space. Most of the provisions of the Soviet draft were included into the Declaration of Legal Principles Governing the Activities of States in the Exploration and Use of Outer Space, of 13 December, 1963.

On May 30, 1966, the USSR proposed to conclude an internationally binding treaty incorporating the principles of the Declaration of 1963.

In June 1966 the USSR and the USA submitted to the UN draft treaties on outer space activities which fact testified the race in political and legal aspects of outer space activities. Collaboration of the USSR and the USA made the Outer Space Treaty a reality. To-day the leadership in the development of outer space law has no importance.

The new agenda item of the Legal Sub-Committee ("Over-view of the Status of the Five International Legal Instruments on Outer Space") might bring to the necessity of evaluating the interpretations of the Outer Space Treaty of 1967 by scholars. It was impossible in 1966 to agree on prohibition of military use of outer space. Some countries would like to amend the Treaty to that end.

The ever increasing role of NGO's may have its impact on an over-view of the Outer Space Treaty.

Collaboration between the USA and Russian Federation might play the same role in future development of outer space law as in the days of adoption of the Outer Space Treaty.

* * * * * *

Outer space law in the USSR takes its origin form academic teachings. Professor of international law Evgheni Korovin published his article "The Conquest of the Stratosphere and International Law" as early as 1934[1] .

Since then Soviet lawyers published many hundreds of scientific works in the field of outer space law[2] .

The launching of the first man-made satellite by the USSR on October 4, 1957, not only gave the new impetus to legal research but also made the use of outer space an acute political problem. The issue at stale was the legal regime of outer space.

On 18 November 1958 the Soviet Union submitted in the First Committee of the UN General Assembly a draft resolution on setting up an international agency for co-operation

---

Copyright

in research in outer space for peaceful purposes[3]. In the course of discussions the UN General Assembly drafted and adopted Resolution 1348 (XIII) of 13 December 1958 entitled "Question of the peaceful use of outer space". It established an ad hoc Committee on the Peaceful Uses of Outer Space which got the mandate to report to the General Assembly at its next session on the nature of legal problems which may arise in the carrying out of programmes to explore outer space.

The composition of the ad hod Committee was unacceptable to the USSR and its representatives did not participate in its work.

The UN General Assembly Resolution 1472 (XIV) of 12 December 1959 adapted the composition of the Committee to the Soviet interests (7 socialist States out of 24 became its members) and since then Soviet representatives have been actively involved in its deliberations and contributed to the success in developing international outer space law.

The XVI-th session of the General Assembly agreed on a number of principles guiding the exploration of outer space: international law, including the Charter of the United Nations, applies to outer space and celestial bodies; outer space and celestial bodies are free for exploration and use by all States in conformity with international law and are not subject to national appropriation. Besides, the same resolution called upon States launching objects into orbit or beyond to furnish information promptly to the Committee on the Peaceful Uses of Outer Space, through the Secretary - General, for the registration of launchings[4]. Those were the first bricks laid in the foundation of the future Outer Space Treaty.

In his message to the President of the USA dated 20 March 1962 the Chairman of the Council of Ministers of the USSR welcomed the agreement on the first principles of outer space law and suggested that an agreement on refraining from potentially harmful experiments in outer space be reached[5].

On 6 June 1962 the USSR submitted to the Legal Sub-Committee of the Outer Space Committee a draft declaration on the basic principles governing the activities of States pertaining to the exploration and use of outer space[6]. Later the United Arab Republic, the United Kingdom and the USA submitted their own draft declarations.

The respective Declaration was adopted by the UN General Assembly on 13 December 1963[7] after the second attempt of the USSR to formulate a compromise text[8].

The comparison of the texts of the draft and of the adopted Declaration testifies that most of the provisions of the Soviet draft were incorporated into the Declaration. However, three Soviet proposals were not included into the Declaration. One dealt with prohibition of use of outer space for propaganda of war, national or racial hatred and enmity among peoples. Another one suggested that only States were allowed to undertake all activities in outer space. The third was intended to prohibit the use of satellites for military reconnaisance[9].

To a large extent the text of the Declaration was used in the process of drafting the Outer Space Treaty as a legally binding instrument.

In Resolution 1963 (XVIII) the General Assembly recommended that consideration be given to incorporating in international agreement form, in the future as appropriate, legal principles governing the activities of States in the exploration and use of outer space. This recommendation was repeated in the UN General Assembly resolution 2130 (XX) of 21 December 1965.

On May 30, 1966, the USSR proposed to include in the agenda of the XXI-st session of General Assembly an item "On concluding an international agreement regarding legal principles of activities of States in exploration and use of the Moon and other celestial bodies"[10]. There was no draft treaty attached to the letter of the Foreign Minister of the USSR to the Secretary-General of the United Nations, but four principles were formulated.

The draft treaty governing the exploration of the Moon and other celestial bodies was submitted to the Outer Space Committee by the United States[11]. Its scope was limited to the Moon and other celestial bodies. But one day earlier the USSR submitted a draft which was more relevant to respond the recommendations of the UN General Assembly, i.e. Draft Treaty on principles Governing the Activities of States in the

Exploration and Use of Outer Space, the Moon and Other Celestial Bodies[12].

Thus the race in outer space was accompanied with the race in political and legal aspects of outer space activities. Already in 1976 the Committee on Aeronautical and Space Sciences of the United States Senate made the conclusion that "current Soviet policy is built around international cooperation in space activities. This has led to treaties governing the use of outer space... At the same time, Soviet jurists believe U.S. representatives are trying to dominate the writing of rules and of treaties to fit U.S. interests"[13].

The Outer Space Committee had the choice of two draft proposals. The Soviet proposal was closer to the recommendations of the UN General Assembly.

In July-September 1966 the Working Group of the Legal Sub-Committee of the Outer Space Committee drafted several articles of the future Outer Space Treaty. On October 4, 1966 the USSR, in the course of the XXI-st session of the General Assembly submitted a revised draft of the treaty which incorporated the already agreed by the Working Group articles. This effort enabled the UN General Assembly to adopt unanimously the text of the treaty entitled "Treaty Governing the Exploration and Use of Outer Space, including the Moon and Other Celestial Bodies" on 19 December 1966[14].

Although the Soviet and the USA representatives to the United Nations expressed during the period of drafting the Outer Space Treaty controversial views on the leading role of the two States in the development of outer space law[15], to-day after 30 years since its entry into force it has no importance. Very many UN members actively contributed to the successful adoption of the Treaty.

Since the Legal Sub-Committee of the Outer Space Committee decided to include into its agenda a new item, i.e. "Over-view of the Status of the Five International Legal Instruments on Outer Space"[16], the background of the Outer Space Treaty may be important not only historically and academically. To-day it might be relevant to go back to the evaluation of the Treaty made three decades ago.

The USSR Minister for Foreign Affairs in his statement upon signature of the Outer Space Treaty on 27 January 1967 expressed firm belief that the Outer Space Treaty would be a step on the way of further development of co-operation and mutual understanding among States and peoples and would contribute to solving most important international problems which the humanity was facing on the Earth[17]. This hope has become the reality of life.

Unfortunately it was not possible to agree on prohibition of military use of outer space by analogy with the Moon and other celestial bodies (Art. IV of the Treaty).

Therefore the Second United Nations Conference on the Exploration and Peaceful Uses of Outer Space (UNISPACE 82) mentions in its report the prevention of an arms race and hostilities in outer space as an essential condition for the promotion and continuation of international co-operation in the exploration and use of outer space for peaceful purposes[18].

Some delegations at the Conference expressed the urgent need to supplement the 1967 Treaty on Outer Space and to elaborate norms aimed at banning the arms race from outer space[19]. Thus we might be facing a new splash of discussions on this vital issue.

In Resolution 1721 (XVI) the General Assembly requested the Outer Space Committee in co-operation with the Secretary-General to maintain close contact with governmental and non-governmental organizations concerned with outer space matters. During the last decade the role of NGO's has been rapidly increasing in the field of human rights. By analogy, the role of NGO's, including the International Institute of Space Law, the International Astronautical Federation and the International Academy of Astronautics, may be increasing in the course of the over-view of the five outer space treaties.

In conclusion, the UN success resulting in the adoption of the first multilateral Treaty on Outer Space in history of mankind became possible due to the successful adoption in 1963 of the Declaration of Legal Principles

Governing the Activities of States in the Exploration and Use of Outer Space". As Nicolas Matte put it, the Declaration was "the result of direct cooperation between the United States and the Soviet Union"[20]. And it is noteworthy that the Declaration contained "nine points, most of them reappearing in the Treaty on the Peaceful Uses of Outer Space"[21]. This last remark means that the Treaty became possible due to direct cooperation between the two major space powers.

Collaboration between the United States and the Russian Federation might play not lesser role in future codification and progressive development of outer space law. Positions of the two States will be even closer in connection with the increasing interest of Russia for outer space commercial activities[22].

REFERENCES

1. Eugene Korovine. La conquete de la stratosphere et le droit international. "Revue generale de Droit International Public", 1934, № 6, Paris, Editions A.Pedone, p.p. 675-686.

2. See: Международное право. Библиография 1917-1972 гг. Москва, 1976, стр. 250-264. (International Law. Bibliography 1917-1972. Moscow, 1976, p.p. 250-264 - in Russian).

3. General Assembly. Official records. XIII-th session. Annexes. Doc. A/C. 1/L. 219/Rev.1.

4. UN General Assembly Resolution 1721 (XVI) of 20 December 1961, General Assembly. Official Records. XVI-th session. Add. 17 (A/5100).

5. "Pravda", 22 March 1962 (in Russian).

6. UN Doc. A/AC. 105/C. 2/L. 1.

7. Declaration of Legal Principles Governing the Activities of States in the Exploration and Use of Outer Space. UN General Assembly Resolution 1962 (XVIII) of 13 Dec. 1963. General Assembly. Official Records. XVIII-th Session. Add. № 15 (A/5515).

8. UN Doc. A/AC. 105/C. 2/L. 6 of 16 April 1963.

9. None of these ideas have ever been reflected in the UN instruments relating to outer space activities.

10. "Pravda", 1 June 1966 (in Russian).

11. UN Doc. A/AC/ 105/32 of 17 June 1966.

12. UN Doc. A/6352 of 16 June 1966.

13. Soviet Space Programs, 1971-75. Goals and Purposes, Organization, Resource Allocations, Attitudes Toward International Cooperation and Space Law. Staff Report prepared for the use of the Committee on Aeronautical and Space Sciences, United States Senate. Vol. II, U.S. Government Printing Office, Wash., 1976, p. 6.

14. General Assembly Res. 2222 (XXI). General Assembly. Official Records. XXI-st Session. Add. № 16 (A/6316).

15. See: Manual on Space Law. Vol. III, 1981, Oceana Publications, Inc./Dobbs Ferry, New York. Compiled and edited by N. Jasentuliyana and Roy S.K.Lee, p.p. 37-102.

16. See: Report of the Legal Sub-Committee on its 36-th session. UN Doc. A/AC. 105/674, April 14, 1997.

17. "Izvestia", 27 January 1967 (in Russian).

18. UN Doc. A/CONF. 101/10, para. 14.

19. Ibid. para. 521.

20. Nicolas Mateesco Matte. Aerospace Law. Canada, 1969, p. 106.

21. Ibidem.

22. See: K.Gorove, E.Kamenetskaya, in "Beyond Confrontation, International Law for the Post-Cold War Period", ed. by L.F.Damrosch,G.M.Danilenko, R.Mullerson, Westview Press. Inc., 1995, p. 230. Outer space private commercial activities have been legalized in Russia by the "Law of the Russian Federation on Outer Space Activities" of 20 August 1993 ("Rossijskaya Gazeta", 6 October, 1993) as amended by the "Federal Law of the Russian Federation on Amendments to the Law of the Russian Federation on Outer Space Activities" of 29 November 1996 ("Rossijskaya Gazeta", 10 December 1996).

The UNITED STATES AND THE 1967 TREATY ON OUTER SPACE

By
Dr. Eilene Galloway*

Abstract

Faced in 1957 with rocketry that could be used for war and peace, we have managed for 40 years to preempt outer space for peaceful exploration and uses. The United States contributed to the concepts in the 1967 Treaty which was formulated during previous years with the purpose of guiding States in the conduct of their space activities. The NASA Act of 1958 includes some concepts designed to maintain peace and avoid war, others were developed in the United Nations where the U.S. played a leading role in organization of the Committee on the Peaceful Uses of Outer Space. Engineers, scientists and legal experts combined to lay the foundation for regulations essential for maintaining the space environment for efficient operation of satellites. As we celebrate 30 years of the 1967 Treaty and consider the future, we need to identify and safeguard our successful policies, organizations and programs and continue to build harmoniously on our regime of international space law.

Introduction

On October 10, 1997 it will be 30 years since the 1967 Treaty on Outer Space[1] went into force, and we can also celebrate 40 years since the space age began on October 4, 1957. For four decades we have been successful in achieving the main goal: preserving outer space for peaceful space exploration and uses and preventing the new environment from becoming an arena for orbiting weapons and international conflicts. We need to analyze the reasons for this success in order to protect the future by maintaining the regime of international space law that has been created.

*Honorary Director, IISL; Trustee Emeritus, International Academy of Astronautics; Fellow, AIAA; American Astronautical Society. Member NASA Advisory Committee on International Space Station.

The 1967 Treaty on Outer Space should not be analyzed merely from the date it went into force because the Treaty was actually 10 years in the making, and during that period nations complied with guiding principles that were incorporated in the Treaty. The influences that shaped legal provisions during the decade from 1957 to 1967 should be identified so we can better understand the foundation of the 30-year history and know what to carry forward into the future.

Nations were motivated by a number of forces directing their decisions and actions: there was fear of orbiting weapons of mass destruction; the same rocket technology promised amazing benefits to all mankind: space science and technology are inextricably international and generate patterns of cooperation among nations; space activities are risky and must be regulated; there were international organizations staffed with expert personnel to cope with new and related problems; the diversity of space applications increased opportunities for new services and expanded existing activities, notably in communications, meteorology and navigation.

The decade 1957-1967 was a period of amazing scientific and technological space development with political repercussions demanding decisions on national and international policies, organizations and programs. The United States played a leading role in the development of basic concepts both in national space laws and participation in United Nations negotiations on drafting the text of the Treaty's provisions and furthering their coverage in additional international space agreements.

Growth of Concepts for the 1967 Treaty

Among the significant movements that influenced the selection of concepts for the Treaty were the International Geophysical Year, the Antarctic Treaty, and Arms Control deliberations in the United Nations.

International Geophysical Year (IGY), July 1, 1957 - December 31, 1958.[2]

On July 29, 1955, the United States announced that "the President has approved plans by this country for going ahead with the launching of small, unmanned, earth-circling satellites as part of the United States participation in the International Geophysical Year... This program will, for the first time in history, enable scientists throughout the world to take sustained observations in the regions beyond the earth's atmosphere"[3] The U.S.S.R. Committee for the IGY stated on September 11, 1956 that "The U.S.S.R. intends to launch a satellite by means of which measurements of atmosphere pressure and temperature, as well as observations of cosmic rays, micro-meteorites, the geomagnetic field and solar radiation will be conducted."[4]

The IGY covered a period when peak sunspot activity offered opportunities for scientists and engineers to make worldwide interdisciplinary studies of the Earth's total environment. The international scientific community had organized two previous synoptic studies, known as Polar Years, in 1882-1883 and 1932-1933. On April 5, 1950, the American scientist Lloyd V. Berkner, suggested during an informal meeting of scientists, that advances in aviation and communications technology had developed new methods for extending knowledge of the Earth, oceans, atmosphere and outer space in connection with rocket and satellite research. The proposal was welcomed by the scientific community which was organized by scientific disciplines supported by 67 national governments, and coordinated by the International Council of Scientific Unions (ICSU).[5]

Dr. Joseph Kaplan, Chairman of the U.S. National Committee for the IGY, stated on September 11, 1956 that while the rocket-satellite program was only one part of the IGY effort, "the earth satellite is unique ... revolutionary, an historic event without parallel in terms of man's relationship to his cosmic environment ... men are taking this historic step jointly under the aegis of the IGY, that unprecedented, cooperative exploration of our physical environment."[6]

The IGY experience developed basic concepts that are reflected in principles and methods developed for their implementation: (1) the natural worldwide scope of data obtained from outer space; (2) the matching international views and practices of scientists and engineers in working on interdisciplinary global influences; (3) the models developed for organization and management of scientific, local, regional, and national aspects coordinated by an international organization; (4) the ability of scientists and engineers to obtain government support for funding and technical assistance; (5) their concern for preserving the environment and preventing contamination; (6) the necessity for exchange of scientific and technical information in the interests of mankind; and (7) the effectiveness of the scientific community in developing peaceful space benefits, working with political decisionmakers, and continually augmenting their ranks by educating young scientists and engineers from all countries.

The Antarctic Treaty, 1959[7]

When President Eisenhower addressed the United Nations General Assembly on September 22, 1960, on the use of the space environment, he pointed out that "The nations of the world have recently united in declaring the continent of Antarctica "off limits" to military preparations. We could extend this principle to an even more important sphere. National vested interests have not yet been developed in space or in celestial bodies. Barriers to agreement are lower than they will ever be again. The opportunity may be fleeting. Before many years have passed, the point of no return may have passed ... We must not lose the chance we still have to control the future of outer space."[8]

The Antarctic Treaty made significant advances in international relations in melding scientific, political and legal elements involved in devising a regulatory regime. Strong agreement on mutually beneficial objectives motivated States to identify situations likely to cause international conflicts and defuse them in advance. Similarly, the requirements for success were formulated, and the process of negotiation gave nations the experience necessary for conducting fruitful international cooperation. Basic concepts in the Antarctic Treaty were carried over to the 1967 Treaty on Outer Space which also has a pattern of prohibitions and permissions to ensure peace and prevent war.

Main provisions of the Antarctic Treaty which became models for the 1967 Treaty included (1) the interest of all mankind that Antarctica be used "exclusively for peaceful purposes"; (2) setting aside conflicting claims of sovereignty so positive work could proceed; (3) freedom of scientific investigation with exchange of information and results;
(4) application of the United Nations Charter;
(5) consultation between parties to settle disputes;
(6) prohibition of military bases and fortifications, maneuvers, and testing of weapons; (7) permission for military personnel to engage in scientific research or other peaceful purposes; (8) prohibition of nuclear explosions; and (9) personnel made subject to the jurisdiction of which they are nationals.

The timing of this treaty in 1959 gave a psychological thrust for the next seven years to those who shaped the consensus for orderly guidance of the exploration and use of outer space, the Moon and other celestial bodies.

United Nations and Arms Control for Outer Space

Even before scientists and engineers hailed rockets and satellites as new tools for advancing scientific research, the United Nations focused on preventing their potential for use as weapons of mass destruction. The United States approached this problem as part of general disarmament and on January 12, 1957 sent a memorandum to the First Committee of the UN General Assembly that "... if this advance into the unknown was to be a blessing rather than a curse, the efforts of all nations in this field need to be brought within the purview of a reliable armaments control system."[9] Again on January 12, 1958, President Eisenhower called for action dedicating outer space to peaceful uses for mankind and denying the new technology for purposes of war.[10]

On September 22, 1960, President Eisenhower addressed the United Nations General Assembly and identified as the basis for international space cooperation some basic concepts that found their way into the 1967 Treaty:[11] He proposed to nations that:

1. We agree that celestial bodies are not subject to national appropriation by any claims of sovereignty.
2. We agree that the nations of the world shall not engage in warlike activities on these bodies.
3. We agree, subject to appropriate verification, that no nation will put into orbit or station in outer space weapons of mass destruction. All launchings of spacecraft should be verified in advance by the United Nations.
4. We press forward with a program of international cooperation for constructive peaceful uses of outer space under the United Nations. Better weather forecasting, improved worldwide communications, and more effective exploration not only for outer space but of our own earth – these are but a few of the benefits of such cooperation. Agreement on these proposals would enable future generations to find peaceful and scientific progress, not another fearful dimension to the arms race, as they explore the universe.

The United States and the United Kingdom played leading roles in United Nations' negotiations culminating in the 1963 Treaty Banning Nuclear Weapons Tests in the Atmosphere, in Outer Space and Under Water.[12] This treaty, now ratified by 112 nations, provided a concept that was later elaborated in Article IV of the 1967 Treaty prohibiting nuclear and other kinds of orbiting weapons of mass destruction.

The United States pursued policies in the United Nations that strengthened peaceful uses and implemented U.S. national space laws.

United States Laws and Programs

Nine years before the 1967 Treaty came before the United States Senate to consider advice and consent to ratification, the United States passed the National Aeronautics and Space Act of 1958[13] whose declaration of policy and purpose begins with, "The Congress hereby declares that it is the policy of the United States that activities in space should be devoted to peaceful purposes for the benefit of all mankind (Sec. 102(a))."

Section (7) calls for "Cooperation by the United States with other nations and groups of nations in work done pursuant to this Act and in the peaceful application of the results thereof."

Section 205 provides that, "The Administration, under the foreign policy guidance of the President, may engage in a program of international cooperation in work done pursuant to this Act, and in the peaceful application of the results thereof, pursuant to agreements made by the President with the advice and consent of the Senate."

When President Eisenhower signed the NASA Act, he stated that Section 205 authorized treaties but did not preclude less formal arrangements for cooperation in appropriate cases.[14] This made it possible for NASA to launch a diversified space program of international cooperation.[15]

International agreements were made for tracking stations. Additional cooperative activities were undertaken in the following categories: manned flight, scientific satellites, optical satellites, transportable telemetry and command, data acquisition, and deep space. NASA published annual reports listing countries and details about the investigations being pursued. The 1967 report lists 84 countries, and this type of activity began immediately upon the establishment of NASA. In addition there were personnel exchanges of scientists and engineers from many nations. Such actions emphasized the concept that space exploration and uses involved all countries and individuals in any country.

The orbiting by the U.S.S.R. of the 184-pound Sputnik on October 4, 1957 was perceived as creating a defense problem, an impression that was reinforced when a second Sputnik weighing 1,120 pounds was launched on November 3, 1957. Such rocketry was evidence of the capability of launching intercontinental ballistic missiles. The IGY Vanguard satellite, planned by the United States, weighed only 3.25 pounds. The first Congressional hearings were held by the U.S. Senate Committee on Armed Services' Preparedness Investigating Subcommittee of which Senator Lyndon B. Johnson was Chairman. This "Inquiry into Satellite and Missile Programs" began on November 25, 1957 and continued into November and January 1958.[16] It became apparent from this investigation that special committees would be required to assist in planning the civilian organization of the government to carry on a comprehensive space program which could develop beneficial uses. On February 6, 1958 the Senate established the Special Committee on Space and Astronautics under the leadership of Senator Lyndon B. Johnson and on March 5, 1958 the House of Representatives created the Select Committee on Astronautics and Space Exploration with Hon. John W. McCormack as Chairman.

On May 13, 1958, Congressman McCormack introduced a House Concurrent Resolution calling for the Peaceful Exploration of Outer Space and expressing the sense of the Congress ---[17]

> That the United States should seek through the United Nations or such means as may be most appropriate an international agreement providing for joint exploration of outer space and establishing a method by which disputes arising in the future in relation to outer space will be solved by legal, peaceful methods, rather than by resort to violence.

The Concurrent Resolution was passed by the House on June 2, 1958 and by the Senate on July 23, 1958.

In the Final Report of the Senate Committee on Space and Astronautics,[18] Senator Lyndon B. Johnson emphasized the concept of the common interest of mankind in outer space and included in the recommendations that –

> The Congress should be kept informed of progress being made in studies undertaken by the United Nations Ad hoc Committee on the Peaceful Uses of Outer Space. Particular attention should be paid to preserving and extending the patterns of cooperation which were formed during the International Geophysical Year. The special committee commends the National Aeronautics and Space Administration for establishing an Office of International Programs and appointing a Director of International Cooperation.

The Special committee concurred with the policy expressed in the President's message to the Congress of April 2, 1958 that 'it is of great importance to have the fullest cooperation of the scientific community at home and abroad in moving forward in the fields of space science and technology.'

After the NASA Act was passed, the Congress created new permanent committees with jurisdiction over space activities: the Senate Committee on Aeronautical and Space Sciences on July 24, 1958 with Senator Lyndon B. Johnson as Chairman; and the House of Representatives established the Committee on Science and Aeronautics on July 21, 1958 with Hon. John W. McCormack, Chairman.

The first significant development of space technology was in global communications. NASA developed communications' satellites and the U.S. moved quickly to establish the ownership, operation and regulation of a commercial communications satellite system. The Communications Satellite Act of 1962 continued the policy and purpose of international cooperation of using space technology for the benefit of all countries:[19]

(a) The Congress hereby declares that it is the policy of the United States to establish, in conjunction and in cooperation with other countries, as expeditiously as practicable a commercial communications satellite system, as part of an improved global communications network, which will be responsive to public needs and national objectives, which will serve the communication needs of the United States and other countries, and which will contribute to world peace and understanding.

(b) The new and expanded telecommunications services are to be made available as promptly as possible and are to be extended to provide global coverage at the earliest practicable date. In effectuating this program, care and attention will be directed toward providing such services to economically less developed countries and areas as well as those more highly developed, toward efficient and economic use of the electromagnetic frequency spectrum, and quality of services and charges for such services.

This was the step that developed into the International Telecommunications Satellite Organization (INTELSAT) and anticipated concepts included in the 1967 Treaty. It led the way toward the functional development of space activities rather than toward the path of a world space agency.

United States and United Nations Space Organization

U.S. policy for international space cooperation as stated in the NASA Act of 1958, was quickly implemented in the United Nations. On September 2, 1958 President Eisenhower requested the UN General Assembly to consider a U.S. draft resolution proposing the creation of an Ad hoc Committee on the Peaceful Uses of Outer Space. The committee was to report on United Nations activities, resources and organizational practices which could facilitate international space cooperation, including a study of the legal problems involved.[20]

Then occurred one of the most dramatic episodes in American history. President Eisenhower wished to indicate the unity of the U.S. Government, Executive and Legislative, on the values of international space cooperation, and at a time when the President was a Republican and the leader of the U.S. Senate a Democrat, Lyndon B. Johnson. The President sent a plane to Texas and Senator Johnson flew to the United Nations to lend his support for establishing the Ad hoc Committee on the Peaceful Uses of Outer Space. On November 17, 1958, Senator Johnson addressed the United Nations and urged the adoption of the resolution initiated by the United States:[21]

> We of the United States have recognized and do recognize, as most all men, that the penetration into outer space is the concern of all mankind ... If nations proceed unilaterally, then their penetration into space becomes only extensions of their national policies on earth. What their policies on earth inspire – whether trust or fear – so their accomplishments in outer space will inspire also ... Today outer space is free. It is unscarred by conflict. No nation holds a concession there. It must remain that way.

Nineteen other nations joined in sponsoring the resolution which was adopted by the General Assembly on December 13, 1958. Many of the basic concepts in this resolution became principles in the 1967 Treaty. There was some delay in moving from an ad hoc to a permanent committee, but after agreement was reached on making committee decisions by consensus, the permanent Committee on the Peaceful Uses of Outer Space (COPUOS) was established on December 12, 1959 by UN Resolution 1472 (XIV). The result was that all the elements that were necessary to work on drafting the basic space treaty on principles were in place by the end of 1959: the policy for international cooperation for peaceful exploration and uses; the organization of members and supporting United Nations staff; and the method of reaching decisions by consensus. The organization of COPUOS into two subcommittees: the Scientific and Technical to ensure that principles are grounded in the factual requirements of space technology, and the Legal Subcommittee to shape realistic guidelines, created a situation which brought successful results. During the next several years, issues were debated and decisions cast into UN resolutions that contained the legal concepts which were to be included in the 1967 Treaty.

United Nations Resolutions Prior to 1967 Treaty. On December 2, 1963, at a meeting of UN Committee 1 (Political and Security), U.S. Ambassador Adlai Stevenson stated the United States position on the draft text of the Declaration of Legal Principles Governing Activities of States in the Exploration and Use of Outer Space:[22]

> We believe these legal principles reflect international law as it is accepted by the Members of the United Nations. The United States, for its part, intends to respect these principles. We hope that the conduct which the resolution commends to nations in the exploration of outer space will become the practice of all nations.

Consensus was finally reached and on December 13, 1963 the UN General Assembly passed, by unanimous vote in plenary session the Declaration of Legal Principles Governing Activities of states in the Exploration and Use of Outer Space.[23] The document refers to "the common interest of all mankind" and "the betterment of mankind and for the benefit of States irrespective of their degree of economic or scientific development."

The Declaration identified guiding principles which were to become part of the 1967 Treaty. Space exploration and use were to be free for all States on the basis of equality and in accordance with international law; outer space and celestial bodies not subject to national appropriation by claim of sovereignty, use, occupation or any other means. International law, the United Nations Charter and international cooperation were specified, and States become internationally responsible for national space activities. They were to be guided by mutual assistance, cooperation, and consult regarding potentially harmful interference, retain jurisdiction and control over space objects carried on their registries, and be internationally liable for damage. Astronauts were to become "envoys of mankind."

On the same day, December 13, 1963, the UN General Assembly passed the Resolution on International Cooperation in the Peaceful Uses of Outer Space.[24] This was a specific list of tasks that needed to be performed in order to implement the principles that were being formulated. The list reveals the careful planning and attention to detail required by the COPUOS and the UN Secretariat. COPUOS was to work out an international agreement on space exploration and use, and draft agreements on liability for damages and assistance and return of astronauts and space vehicles. Working papers were to be prepared on the resources of the United Nations and its specialized agencies, particularly the International Telecommunication Union (ITU) and the World Meteorological Organization (WMO). Summaries were wanted for national and international cooperative space activities.

The General Assembly called particular attention to the fact that the Secretary General, in accordance with General Assembly Resolution 1721 (XVI)[25] had established a public registry of objects launched into orbit and beyond. The United States and Soviet Union were already registering their launchings because space scientists and engineers required this information to keep track of their projects.

Four days after passing the resolutions on Legal Principles and International Space Cooperation, the United Nations General Assembly took action to prevent the spread of the arms race in outer space. On the question of General and Complete Disarmament, Resolution 1884 (XVIII) welcomed the expression of the United States and the U.S.S.R. of "their intention not to station in outer space any objects carrying nuclear weapons or other kinds of weapons of mass destruction." The Resolution called upon all States[26]

(2) to refrain from placing in orbit around the earth any objects carrying nuclear weapons or other kinds of weapons of mass destruction, installing such weapons on celestial bodies, or stationing such weapons in outer space in any other manner;

(b) to refrain from causing, encouraging or in any way participating in the conduct of the foregoing activities.

This was another principle that was incorporated in the 1967 Treaty.

<u>United States Ratification of the 1967 Treaty on Outer Space</u>

On May 7, 1966, President Johnson proposed that international negotiations begin on the preparation of a treaty providing rules and procedures for the peaceful exploration of outer space and celestial bodies. Considerable progress had been made by the 1963 UN resolutions on Legal Definitions and International Space Cooperation adopted unanimously by the General Assembly "as a result of United States Initiatives" as explained by Amabassador Goldberg,[27] but the President wanted action in the form of a binding legal instrument. The U.S. Ambassador informed the Chairman of COPUOS, Ambassador Kurt Waldheim, of the President's proposal. The Legal Subcommittee, under the distinguished Ambassador Manfred Lachs, met in Geneva from July 12, to August 4 and again in New York, finally reaching consensus in early December. The UN General Assembly passed the 1967 Treaty by acclamation on December 19, 1966.

On January 27, 1967, the Treaty was opened for signature in London, Moscow, and Washington, and at the White House ceremony 60 nations signed the Treaty. On February 7, 1967, President Johnson sent the Treaty to the Senate, whose advice and consent to ratification is required by the U.S. Constitution; he stated that[28]

> ...The Treaty is an outstanding example of how the law and political arrangements can keep pace with science and technology. We hope by this treaty to establish an open and rational regime for outer space which will permit the greatest benefit to be derived from man's exploits there ... And we should act now to provide against the extension of the nuclear arms race to this new environment.

The Treaty was referred to the Senate Foreign Relations Committee whose Chairman, Senator J.W. Fulbright, opened hearings on March 7, 1967.[29]

The Treaty did not come before the Committee as a newly-minted document as the Senators, officials in the Department of State and NASA had been participating in the drafting and consensus process of the United Nations, and particularly the Committee on the Peaceful Uses of Outer Space, ever since 1957, and even before that in the Arms control negotiations. Some of the legislators had served on the U.S. delegations to the COPUOS, including members of the House of Representatives. The Government was united and bipartisan on the objective of establishing a peaceful legal international regime for outer space and celestial bodies. The witnesses at the hearings were the Secretary of State, Dean Rusk, the U.S. Ambassador to the United Nations, Arthur Goldberg, members of the Joint Chiefs of Staff and other officials, all of whom endorsed the Treaty.

During the hearings, however, questions arose that required clarifying answers, particularly with regard to Articles I, IV, and VII. The Committee analyzed all the provisions but required the specific meaning of general provisions.

Senator Albert Gore asked: "To just what are we binding ourselves by this Article I, paragraph 1?" Does it obligate the United States to make the use of outer space by the Communications Satellite

Corporation available to all nations? Article I, paragraph 1 provides:

> The exploration and use of outer space, including the moon and other celestial bodies, shall be carried out for the benefit and in the interests of all countries, irrespective of their degree of economic or scientific development, and shall be the province of all mankind.

Ambassador Goldberg undertook to dispel Senator Gore's concern. His analysis was that the "provision is based on a declaration by the Congress in Section 102(a) of the National Aeronautics and Space Act of 1958. That section states:

> The Congress hereby declares that it is the policy of the United States that activities in space should be devoted to peaceful purposes for the benefit of all mankind. (76 Stat. 419).

He pointed out that the wording was a policy statement in the 1962 United Nations Declaration on Legal Principles Governing Activities of States in the exploration and Use of Outer Space After the U.S. introduced this provision, two additions were made to that declaration: (1) a reference to include the moon and other celestial bodies, and (2) the phrase proposed by Brazil – "irrespective of their degree of economic or scientific development." Ambassador Goldberg said the Article does not create legal obligations with respect to the terms of international cooperation on any existing or future space projects." The fact that the Communications Satellite Act leading to the creation of the International Telecommunications Satellite Organization (INTELSAT) was established was an example of ability to make specific agreements. Nevertheless, the concern of the Committee was such that it was decided to clarify the understanding in the Committee report (as hereinafter cited).

Senator Gore's question did not include the meaning of "the province of all mankind" in Article I, a question posed by Senator Frank J. Lausche. He wondered why the word "province" was chosen. Ambassador Goldberg replied that "province" was derived from the Declaration of Legal Principles in UN Resolution 1962 (XVIII), December 13, 1963, and that the U.S. insisted on "province" because the negotiations dealt with many languages and it was considered "there is no difference in conception between "benefit" and "province." Senator Clifford P. Case said" "it is a different word, too. 'Province' means anybody can get into it [outer space] rather than receiving benefits that come from it [outer space]."

Actually, the word "province" is not included in the 1963 Legal Principles but they do provide that "The exploration and use of outer space shall be carried on for the benefit and in the interest of all mankind". An explanation of considering "province" and "benefit" to be the same concept could be that benefits to all mankind inure from using province and are in contrast to using outer space for destructive purposes. There was no mention of the common heritage of mankind in the hearings.

The second paragraph of Article I includes the moon and other celestial bodies, provides for the rule of freedom in space on the basis of equality and in accordance with international law", a provision taken from the 1963 Declaration of Legal Principles for which the U.S. had exercised leadership in its formulation and adoption.

The third paragraph of Article I provides that –

There shall be freedom of scientific investigation in outer space, including the moon and other celestial bodies, and States shall facilitate and encourage international cooperation in such investigation.

This provision came from Article II of the 1959 Antarctic Treaty, and it was evident that the Committee was very favorable to the provisions made for the Antarctic. The Ambassador stated that "the United States has been leading the way in practicing international cooperation and felt it was appropriate for all parties to "facilitate and encourage cooperation."

While the Committee was concerned about Article I in applying the general principle to specific cases, the members did not want to go so far as to insist on amending the Treaty, and it was decided to clarify their understanding in the report to the Senate so there would be no misunderstanding about the U.S. position.

The Committee report on Article I states:[30]

> Article I of the treaty provides that "the exploration and use of outer space * * * shall be carried out for the benefit and in the interests of all countries, irrespective of their degree of economic or scientific development, and shall be the province of all mankind." The committee raised the question whether the language of this general principle might imply a fixed treaty obligation on the part of the United States to share the benefits and results of its space activities, particularly in the communications satellite field.
>
> After a full discussion of this point with administration witnesses, the committee was assured that no such specific treaty obligations would result. Nevertheless, the committee wishes to make its position clear on its understanding of the obligations the United States will accept under article I, paragraph 1 of the treaty. It is the understanding of the Committee on Foreign Relations that nothing in article I, paragraph 1 of the treaty diminishes or alters the rights of the United States to determine how it shares the benefits and results of its space activities.

It is interesting to note that the same interpretation is embodied in the UN General Assembly Resolution adopted on December 13, 1996 on the Declaration on International Cooperation in the Exploration and Use of Outer Space for the Benefit and in the Interests of all States, Taking into Particular Account the Needs of Developing Countries.[31]

> States are free to determine all aspects of their participation in international cooperation in the exploration and use of outer space on an equitable and mutually acceptable basis. Contractual terms in such cooperative ventures should be fair and reasonable and they should be in full compliance with the legitimate rights and interests of the parties concerned, as, for example, with intellectual property rights.

The Committee also had serious questions about Article VII on liability for damage. Ambassador Goldberg explained that a launching State is liable for physical damage caused by its space activities, such as loss of life, personal injury and destruction or damage to property. He stated that Article VII covers only damage caused by an impact of a space vehicle or object and does not cover electronic damage which would come under Article IX on interference. States Parties are required to consider the interests of other States, and if they are concerned about interference of any kind to consult about any situation that might cause potential damage. Ground-based space activities are not covered by Article VII.

The Committee was informed that work was already underway on drafting a specific convention on liability for damage, as well as one on assistance and return of astronauts and space objects. More definitive provisions on liability could be expected in the new draft convention. The Committee report therefore concluded:[32]

> Questions were also raised about the nature and extent of international liability assumed by a signatory state under article VII. That article provides in part that each state party to the treaty "from whose territory or facility an object is launched, is internationally liable for damage to another state party to the Treaty or it its natural or juridical persons by such object or its component parts. * * *."
>
> Administration witnesses assured the committee that a treaty on international liability in outer space is in the process of negotiation and will, if successfully negotiated, in due course be submitted to the Senate. Nevertheless, the committee concluded that a preliminary clarification of the liability provision was necessary. The question was raised during the testimony whether the liability provision of article VII includes nonphysical damage, such as electronic jamming and interference. In the course of the testimony, administration witnesses informed the committee that electronic interference by one spacecraft with another is not covered by the treaty. The committee wishes to record its understanding that article VII pertains

only to physical, nonelectronic damage that space activities may cause to the citizens or property of a signatory state. Article VII establishes the general proposition of international liability for damage caused by space vehicles. A separate convention devoted wholly to liability is needed to establish detailed rules. Such a convention is now under negotiation in the United Nations Outer Space Legal Subcommittee. The convention would provide procedures for determining liability and a mechanism for effectively resolving any differences. It would have to fix a limitation on liability, if there is to be one, and provide for allocation of liability among participants in a joint space venture causing damage.

The other clarification concerned a point in Article IV of the Treaty.[33]

> Another prime concern to the committee was the implications for American security of the first sentence of article IV: "States Parties to the Treaty undertake not to place in orbit around the Earth any objects carrying nuclear weapons or any other kinds of weapons of mass destruction, install such weapons on celestial bodies, or station such weapons in outer space in any other manner.' Because the inspection privileges, as defined in article XII, do not apply to objects in orbit, the possibility existed that the United States, for the first time, was committing itself to an arms control measure that was not safeguarded from violation by either the right of physical inspection or an effective national detection system.

During the public hearing on March 7 Secretary Rusk, in referring to the detection problem, said: 'We have no doubt we can monitor effectively a weapons system placed in outer space.'

The Senate gave its advice and consent and the United States ratified the Treaty on October 10, 1967.

<u>Expanding the 1967 Treaty</u>
During the negotiations on the 1967 Treaty, it became apparent that some of the general guiding principles dealt with problem areas where more specific directives were required, and information was available for determining solutions. In fact, the Legal Subcommittee began working on these subjects while completing the 1967 Treaty and the United States actively participated in shaping the results.

This was the situation with the first three international space agreements that represent extensions of articles in the basic "Magna Carta" 1967 Treaty: Agreement on the Rescue of Astronauts, the Return of Astronauts and the Return of Objects Launched into Outer Space. Adopted by the UN General Assembly on December 18, 1967 (Resolution 2345 (XXII)) and entered into force on December 3, 1968.

Convention on International Liability for Damage Caused by Space Objects Adopted by the UN General Assembly on November 29, 1971 (Resolution 2777 (XXVI)) and entered into force on September 1, 1972.

Convention on Registration of Objects Launched into Outer Space. Adopted by the UN General Assembly on November 12, 1974 (Resolution 3235 (XXIX) and entered into force on September 15, 1976.

The reasons why consensus was achieved on these three treaties should be examined in order to determine the factors that make for success in achieving international agreements.

First, the problems were real and imminent and it was apparent that solutions could be formulated. Astronauts and space objects were already being launched into outer space where they would encounter known dangers and their status needed to be defined. The need of engineers and scientists to know what objects were orbiting caused a registry to be established by the United Nations even before the registration convention was completed. Potential damage to lives and property raised questions of liability.

Second, early legal problems were all connected with conditions essential for successful operation of space vehicles, and COPUOS was organized to combine these factors by the work of its Scientific and

Technical Subcommittee followed by its Legal Subcommittee.

Third, those who were formulating the draft texts could include the relevant general principles of the 1967 Treaty in each of the specific three cases and be assured that a framework of harmonious space law would develop.

Fourth, the subject matter of these three treaties did not include political, economic, philosophical ideas and differences which can erect barriers to consensus, especially when they veer away from the facts of efficient operation of space technology.

This author attended many sessions of COPUOS and its Legal Subcommittee, and in observing the process of decisionmaking by consensus, came to the following conclusion:[34]

It is evident that consensus is a highly desirable way of achieving international accord because (1) the process of seeking agreement continues with patience and is not cut off suddenly by a vote which may defeat what might have come to fruition had more time been taken with the give and take process of consensus; (2) the situation may be such that a majority vote could not result in the adoption of a course of action, particularly if implementation of the decision in terms of funding, personnel, and technological expertise, depended upon nations which had voted against the measure; and (3) group solidarity in decisionmaking ensures maximum compliance in establishing and maintaining an activity of general benefit. There is also a positive psychological effect when members of a group feel together with sympathy for differing viewpoints, motivated by a desire to bring about harmony in their collective judgment. If a member has not objected, a proposal can be adopted but this unspoken consent should not be interpreted as negativism; there is a positive willingness to settle the issue in question.

Principles Adopted by the UN General Assembly

Although the space principles adopted by the UN General Assembly do not have the legal status of treaties, they provide guidance for activities that need governance, and have the potential of becoming treaties at a later time. For example, the work of the U.S. Delegation on the1962 Declaration of Legal Principles benefited the formulation of articles included in the 1967 Treaty.

After COPUOS was organized, the United Nations passed three more declarations on space principles, all of which cover subjects included in the U.S. space program:

Principles Governing the Use by States of Artificial Earth Satellites for International Direct Television Broadcasting. Adopted December 10, 1982 Res. 27/92.

Principles Relating to Remote Sensing of the Earth from Space. Adopted on December 3, 1986, Res. 41/65.

Principles Relevant to the Use of Nuclear Space Power Sources in Outer Space. Adopted December 14, 1992. Res. 47/68.

Principles covering Nuclear Power Sources remain on the agenda of the Legal Subcommittee, thus providing the place and time to monitor developments that could have the potential of becoming a treaty. In any event, this UN procedure affords opportunity for keeping abreast of this vital force.

The Moon Agreement: Definition of the Problem.[35]

The Agreement Governing the Activities of States on the Moon and Other Celestial Bodies was adopted by the UN General Assembly on December 9, 1979 and opened for signature on December 18, 1979. Eighteen years have passed and only nine nations have ratified this treaty: Australia, Austria, Chile, Mexico, Morocco, The Netherlands, Pakistan, Philippines and Uruguay. Five nations signed but have not ratified: France, Guatemala, India, Peru and Romania. This record is in sharp contrast to that of the four previous UN-formulated space treaties which encountered no such delay. The fact that none of the space powers has ratified the Moon Agreement is evidence of a problem that requires identification so that realistic solutions can be proposed. We have been successful since 1957 in establishing space law as a recognized branch of international law, and in this necessarily continuing process, it is not prudent to let the Moon Agreement drift without trying to understand the reasons for general nonacceptance.

An historical perspective should contribute to an analysis of the nature of this problem. During the nine years when the Moon Agreement was being formulated, 1970-1979, we can distinguish four different climates of opinion based on different assumptions of the nature of the problem with consequent variations in proposals for solutions. The United Stated had landed a man on the Moon on July 21, 1969 and the Soviet Union had obtained samples from the Moon which led to the perception that exploitation of the Moon's natural resources was beginning although this was not the case. There was, however, a strong motive for continuing to ensure that the Moon would be used only for peaceful purposes. This was the psychology during the first period, 1970-1971.

The second period began in 1972-1973 when developing countries were motivated by concern that the natural resources of the Moon and other celestial bodies would be exploited to their disadvantage. However, by this time lunar exploration by the United States and the Soviet Union had practically ceased, so the problem was no longer primarily that of providing space law in tandem with space science and technology, which had been the case with the three previous space treaties spun from articles in the 1967 Treaty; instead, the motive was to establish the concept of the common heritage of mankind which would eventually require an international regime with jurisdiction over the exploitation of natural resources.

A third period emerged by 1978 when delegates to the Legal Subcommittee became impatient with the length of time this treaty's process was consuming and wanted either to settle something or lose priority for this agenda item. The Legal Subcommittee, however, could not reach a consensus at their 1979 session but hoped that the full COPUOS would be able to bring about consensus. This result was achieved on July 3, 1979 when major compromises were reached. The Soviet Union, which had objected to the common heritage of mankind concept since the beginning of negotiations, agreed to accept Brazil's proposal that the CHM concept apply only to the Moon Agreement. The developing countries gave up their demand for a moratorium on exploitation of the Moon's natural resources prior to establishment of an international regime.

U.S. Ambassador Richard W. Petree addressed the UN General Assembly Special Political Committee on November 1, 1979 on the COPUOS report of the Moon Agreement. He explained the U.S. understanding that first, references to the moon are intended also to be references to other celestial bodies within our solar system other than the earth; secondly, that references to the moon's natural resources are intended to comprehend those natural resources to be found on these celestial bodies; and thirdly, that the trajectories and orbits referred to in Article I, paragraph 2, do not include trajectories and orbits of space objects between the earth and earth orbit or in earth orbit only.

The U.S. endorsed the Committee's understanding that Articled VII does not prohibit exploitation of natural resources on celestial bodies but is intended to ensure that there will be minimum disruption and adverse changes to the environment.

The Ambassador pointed out that –

> The common heritage concept, which was initially suggested by Argentina, but formally proposed by the United States in 1972, is set forth in Article XI, paragraph 1, which makes clear that its meaning for purposes of the Moon Treaty is to be found within the Moon Treaty itself.

A fourth climate of opinion developed in the United States when the Moon Agreement became available for ratification. Although the United States is only one among other space powers that has not ratified the Moon Agreement, nevertheless, some background may be helpful in achieving an up-to-date factual definition of the problem, and lead to realistic proposals for solutions.

Certain provisions in the Moon Agreement aroused strong pro and con reactions, largely because of the ambiguities and uncertainties of the common heritage of mankind concept, which gave rise to a variety of interpretations and definitions, and the proposal for an international regime to exercise control over national space activities. To assess the situation, official actions were taken by the Executive Branch and the Congress. The Department of State surveyed aerospace and extractive industry companies and trade organizations. An Inter-agency review group was charged with analyzing the Treaty and estimating its effects. An indepth research effort resulted in a 3-volume study initiated by the Senate Committee on Commerce, Science and Transportation: Parts 1 and 2 are an historical study and analysis by Eilene Galloway. Part 3 is by the Office of Technology Assessment and covers current and foreseeable technologies related to exploitation of nonterrestrial resources, relevant plans to use the deep seabed, a summary of explicit and implied constraints on U.S. activities, and alternative legislative options for action. Part 4 by the Congressional Research Service is an interdisciplinary study of technological, foreign policy and legal issues.

On July 29 and 31, 1980 the Senate Subcommittee on Science, Technology and Space of the Committee on Commerce, Science and Transportation held hearings on the Moon Treaty to obtain testimony from government officials and other experts to determine how the treaty would affect the future use of space. Ronald F. Stowe testified as chairman of the Aerospace Law Committee of the Section of International Law, American Bar Association, in favor of ratifying the treaty but with specified interpretations of the common heritage of mankind, and that the United States "reserves to itself the right to decide how and to what extent it will share the benefits it derives from lunar resources..."

These serious, objective analytical approaches in the United States did not result in a clear go-ahead signal for ratification. It would be necessary to add the objections of other spacefaring nations in order to compile a complete record of the issues that need to be resolved. When France signed the Agreement on January 20, 1980, it was with a clarification of Article 3, but ratification has not followed.

This author's personal interpretation of the situation is as follows. The first three treaties spun from articles in the 1967 Treaty were based on then current problems with factual information from scientists and engineers which could be closely aligned with legal solutions. The issue of exploitation of the natural resources of the Moon arose from historical fears of colonialism experiences and was not based on factual reports identifying the Moon's natural resources, the technology involved for exploitation, and the funding which would require billions of dollars. Based on assumptions, which were inadequate, and in some instances factually incorrect, the solutions proposed were more general principles: the common heritage of mankind and an international regime, concepts that were being interpreted in various ways. This was different from the pattern followed before when we advanced from a general guiding principle to specifics for an existing project. General principles are an advantage during negotiations when differences must be reconciled to produce a consensus. But when there is a specific problem, especially one involving the operation of space technology, it is necessary to match proposed regulations with the function to be performed. In formulating space law for the future, it is essential to base solutions on scientific and technological facts, and indeed the COPUOS is organized for this approach by its two subcommittees linking technology and law. In the case of the Moon Agreement, as time went on there were conflicting definitions for common heritage of mankind and especially of the international regime, while there was no movement toward exploitation of lunar natural resources or plans for funding either by governments or private industry. When the UN General Assembly reviewed the Moon Treaty in 1984, ten years following its entry into force after ratification by five nations, there was no evidence of feasibility for exploitation and no action was taken.

The probability that the Moon Treaty might need revision, however, was foreseen in Article 18 which provides that a review conference take account of any relevant technological developments. Commercial developments of potential relevance have now emerged. Emphasis of the Moon Treaty on natural resources eliminates consideration of other resources; for example, its provisions do not apply to solar power, and its application to asteroids might be questioned. Ideas about the international regime range all the way from less formal arrangements similar to those for the Antarctic to an institution with a system of unspecified controls over national activities.

It is time to organize a workshop to study all related aspects of any proposed international regime and formulate an agreed definition of terms in harmony with the operational effectiveness of space technology as well as commercial aspects. The exploitation of natural resources is only one function among many global space applications. It is necessary to form a consensus among nations on common objectives and how they are to be achieved. We must keep in mind the implications of the Moon Treaty's provisions that "apply to other celestial bodies within the solar system, other than the earth, except in so far as specific legal norms enter into force with respect to them." On the basis of established space law, with which nations are already in compliance, we already have what amounts to an international regime organized by functions and managed by institutions, e.g. communications, meteorology, navigation and remote sensing.

Conclusions

The most remarkable achievement in the first 40 years of the space age is that nations cooperated to preserve outer space for peaceful uses and succeeded in denying space wars and other forms of international conflict. The United Nations became the forum for the convergence of space policies, organizations and methods for achieving decisions on beneficial objectives. During the first 10 years, 1957-1967, the United States actively contributed concepts that are included in the 1967 Treaty, both in the conduct of foreign policy and in national space legislation providing for international cooperation. At the time the Treaty entered into force, the United States had a variety of cooperative arrangements with 84 nations. Many nations were involved with the adoption of this "Magna Carta" treaty which helps to account for universal compliance by the international community with its guiding principles.

Certain articles in the Treaty have been expanded into new treaties to provide solutions for specific problems. Each new treaty repeats main provisions of the 1967 Treaty so that a consistent body of space law has been established. The result is that space law is now a recognized branch of international law.

As we evaluate the situation with an eye for the future, it is evident that the United Nations' organizations and procedures for outer space matters are effective: the Committee on the Peaceful Uses of Outer Space with its practice of analyzing problems by the Scientific and Technical Subcommittee, followed by the Legal Subcommittee, the Office for Outer Space Affairs as a continuously functioning secretariat; and the United Nations' specialized agencies, particularly the International Telecommunication Union and the World Meteorological Organization. The practice of making decisions by consensus has proved remarkably effective. Resolutions have also assisted in providing guidance for space activities. Criticism is voiced when actions on proposals are slow, but some subjects take more time than others to shape into a form that can attract general acceptance.

Consideration should be given to providing a better understanding of some general terms: mankind, benefits, access to space, resources, province, and the common heritage of mankind. Each term can be interpreted in several ways often depending on differing assumptions on political or economic philosophy. We have mankind in general that benefits from world peace, the absence of war, and a clean environment; and portions of mankind benefit in unique circumstances such as disaster relief or protection from the spread of contagious diseases. Some persons think of benefits only in terms of making a profit, and unless they are making money conclude that they have no benefits. They overlook the fact that the major benefit from space activities is information to solve problems on the Earth and contribute to an understanding of the Universe. Nations are willing to accept "province of all mankind" where opportunities are available to participate in a variety of space projects, but there is not general acceptance of the "common heritage of mankind" if that means property rights or a specific institution to control national space programs. Then there are those who conclude that unless they get the specific benefit they want, then they get no benefit, evidently taking for granted such blessings as global space communications and a healthy environment.

It has been 25 years since the common heritage of mankind concept was introduced as an agenda item for the Committee on the Peaceful Uses of Outer Space, and in 18 years only nine nations have ratified the Moon Agreement, and during this time the original definition of the problem, for which legal remedies were formulated, became outdated. New scientific, technological, political and commercial trends have developed so we now need a current assessment of the problem. This is particularly important became the Moon Agreement specifies that its provisions "shall also apply to other celestial bodies within the solar system, other than the earth, except in so far as specific legal norms enter into force with respect to any of these celestial bodies." (Article I (1)). This provision is not a sufficient guideline for formulating future space law for all celestial bodies and, indeed, creates uncertainty, unpreparedness, and inaction.

An objective study of current facts and options for the future should be undertaken. This could be by a Workshop with broad representation of scientists, engineers, industry and government personnel. Or the task could be assigned to a Standing Committee of the International Institute of Space Law combined with the International Academy of Astronautics. In either case, the terms of reference should include the requirement for factual information identifying natural resources, the technology required to exploit them, the cost, estimates of whether private industry or governments (or a combination of both) would undertake this venture.

References

[1] Treaty on Principles Governing the Activities of States in the Exploration and Uses of Outer Space, including the Moon and Other Celestial Bodies. 18 UST 2410; TIAS 6347; 610 UNTS 205. The United Nations adopted the treaty on December 19, 1966 by UN Resolution 222 (XXI); opened for signature January 27, 1967, and entered into force on October 10, 1967.

[2] Jones, Sir Harold Spencer, The Inception and Development of the International Geophysical Year. Annals of the International Geophysical Year, Vol. I. London, New York, Pergamon Press, 1959, pp. 383-413. See also Walter Sullivan, Assault on the Unknown: the International Geophysical Year. New York, McGraw Hill, 1961, pp. 8-19.

[3] Plans for Launching of Earth-Circling Satellites. White House Press Statement by James C. Hagerty. Department of State Bulletin, August 8, 1955, p. 218.

[4] Statement by the U.S.S.R. Committee for the IGY. CSAGI Conference on Rocket and Satellite Observations. Report by Dr. Joseph Kaplan, Chairman of U.S. National Committee for the IGY. September 11, 1956, p. 661.

[5] Yearbook of the International Council of Scientific Unions, 1961. London, Spottiswoode, Ballantyne & Co., 1961. 110 p.

[6] Ibid., p. 300-313.

[7] Antarctic Treaty 402 UNTS 71; 12 UST 794; TIAS 4780.

[8] Documents of Disarmament, Department of State Publication 7009, 1960, Vol. II, p. 938-940.

[9] Ibid., p. 733.

[10] Ibid., p. 938-940

[11] Department of State Bulletin, October 10, 1960. P. 554-555.

[12] 14 UST 1313; TIAS 5433; 480 UNTS 43. See also International Cooperation and Organization for Outer Space which includes a resume of UN General Assembly Actions and Related U.S. Negotiations. U.S. Senate document No. 56, 89th Congress, First Session. August 12, 1965, p. 163-352.

[13] National Aeronautics and Space Act of 1958. Public Law 85-568, 85th Congress, H.R. 12575, July 29, 1958.

[14] White House Press Release, July 29, 1958.

[15] United States International Space Programs. Texts of Executive Agreements, Memoranda of Understanding, and other International Arrangement, 1959-1965. U.S. Senate Document No. 44. 89th Congress, First Session. July 30, 1965. Indexed. 575 p.

[16] Inquiry into Satellite and Missile Programs. Hearings of the Preparedness Investigating Subcommittee of the Senate Armed Services Committee. 85th Congress, First and Second Sessions, November and December 1957, January 1958, Parts, 1, 2, 3. 2476 p.

[17] House concurrent Resolution 332, 85th Congress, Second Session. June 2, 1958.

[18] Final Report of the Senate Special Committee on Space and Astronautics, Pursuant to S. Res. 256, 86th Congress, First Session. Senate Report No. 100. March 11, 1959 76 p. This report contains a history of passage of the NASA Act, Congressional organization, and related legislation.

[19] U.S. Communications Satellite Act of 1962. Public Law No. 624. 87th Congress, Second Session. 76 Stat. 419. Approved August 31, 1962.

[20] Department of State Bulletin, October 6, 1958, p. 529.

[21] Address on the Peaceful Uses of Outer Space by Senator Lyndon B. Johnson, UN Committee No. 1 (Political and Security). November 17, 1958. Final Report of the Senate Special Committee on Space and Astronautics, Senate Reort No. 100. 86th Congress, First Session, p. 58-62.

[22] United Nations Document A/C.1/PV. 1342. December 1963, p. 12.

[23] United Nations Resolution 1962 (XVIII), December 13, 1963.

[24] United Nations Resolution 1963 (XVIII), December 13, 1963.

[25] United Nations Resolution 1721 (XVI), December 20, 1963.

[26] Question of General and Complete Disarmament. UN Resolution 1884 (XVIII), December 17, 1963.

[27] Statement by Ambassador Arthur J. Goldberg before the Senate Committee on Foreign Relations on the Outer Space Treaty. Hearings on Executive D, 90th Congress, First Session. March 7, 1967, p. 149.

[28] Treaty on Outer Space. Hearings before the Senate Committee on Foreign Relations. 90th Congress, First Session of Executive D. March 7, 13, and April 12, 1967, 162 p.
[29] Ibid.
[30] Treaty on Outer Space. U.S. Senate Executive Report No. 8. 90th Congress, First Session. Report. April 18, 1967. 6 p.
[31] United Nations General Assembly. Resolution Adopted on Declaration on International Cooperation in the Exploration and Use of Outer Space for the Benefit and in the Interest of all States, Taking into Particular Account the Needs of Developing Countries. A/RES/51/122, February 4, 1997. 3 p.
[32] Op. Cit., No. 30.
[33] Ibid.
[34] Galloway, Eilene. Consensus Decisionmaking by the United Nations Committee on the Peaceful Uses of Outer Space. Journal of Space Law. Vol. 7, Spring 1979, No. 1. P. 5.
[35] The Moon Treaty. Hearings before the Subcommittee on Science, Technology and Space of the U.S. Senate Committee on Commerce, Science, and Transportation. 96th Congress, Second Session. Serial No. 96-115. July 29 and 31, 1980. 267 p. ; Agreement Governing the Activities of States on the Moon and Other Celestial Bodies. U.S. Senate Committee on Commerce, Science, and Transportation. 96th Congress. Second session. 3 committee prints. Part 1 and 2 by Eilene Galloway, May 1980, p. 1-264. Part 3 by Office of Technology Assessment, August 1980, p. 265-379. Part 4 by the Congressional Research Service, November 1980, p. 381-516.

IISL-97-IISL.1.04

NEGOTIATING ISSUES IN FORMING THE 1967 TREATY ON OUTER SPACE*

By
Paul G. Dembling, Esquire+

Before
40th International Colloquium on the Law of Outer Space
IAF Congress, Turin, Italy
October 7, 1997

ABSTRACT

This paper considers the expression of views, international treaties, and other events which are pertinent to the establishment of principles governing exploration and use of outer space and celestial bodies and which led to the final agreement by the 28 members of the Legal Subcommittee of the United nations General Assembly's Committee on the Peaceful Uses of Outer Space to the Outer Space Treaty.

INTRODUCTION

An announcement was made 8 December 1966, that agreement had been achieved among the 28 members of the United Nations Committee on the Peaceful Uses of Outer Space (COPUS) on the text of a treaty establishing principles governing the activities of states in the exploration and use of outer space, the moon, and other celestial bodies. Approval of the Treaty was recommended unanimously by the Political Committee of the General Assembly on 17 December 1966. Two days later, the Treaty was endorsed by a unanimous vote of the General Assembly and on 27 January 1967 it was open for signature.

A remarkable endeavor of great significance to international law and politics had reached fruition. Nations often in conflict with one another and adhering to widely divergent political philosophies had agreed on the first Treaty of general applicability governing activity in outer space.[1]

The principles set forth in the Treaty had been advanced previously in the form of General Assembly resolutions, analogous international agreements, domestic legislation, statements by government officials, articles by scholars in the field and other expressions of vies. Final agreement on the Treaty, however, was primarily the product of the labors of the 28 members of the Legal Subcommittee of COPUS.

+Member, U.S. Delegation to Legal Subcommittee, 1964-1969.

[1] The Treaty is officially entitled "Treaty on Principles Governing the Activities of States in the Exploration and Use of Outer Space, Including the Moon and Other Celestial Bodies," and is annexed to a resolution of the General Assembly. U.N. Doc. A/C.1/L.396 (1966). 18 UST 2410,TIAS 6347, 610 UNTS 205. As of June 3, 1997, the U. S. Dept. of State Office of Treaty Affairs lists 92 countries parties to the Treaty.

PRINCIPLES APPLICABLE TO CELESTIAL BODIES

Although the scope of Treaty as eventually agreed upon includes both outer space and celestial bodies, an important aspect of the deliberations leading to agreement on the Treaty is the extent to which the nations and individuals were concerned, for the first time, with the formulation of realistic principles which might govern activity on celestial bodies in addition to, but as distinct from, outer space.[2] This consideration of celestial bodies was based upon a body of thought and action that preceded the Fifth Session (1966) of the Legal Subcommittee. Even prior to 1960, a considerable amount of commentary existed on the question of whether it was possible for a terrestrial nation-state to acquire sovereignty over all or part of a natural celestial body, and what would be required under existing law to make such a claim legally valid. Analogies were drawn to the manner in which nations had previously sought to exert legal claims to sovereignty over portions of the earth's surface, *e.g.*, through discovery, occupation, annexation and contiguity. Considerable discussion arose over the legal effect of the reported striking of the moon by an early Soviet satellite carrying the Soviet flag. However, the Soviet Union did not seek to exert any claim of sovereignty based upon this occurrence.

Although writers regarded the legal principles derived from exploration of the earth's surface as potentially applicable to exploration of celestial bodies, they did not consider such applicability to be desirable. The suggestion was made that both public and private groups formulate standards and procedures that will guarantee access by all to these resources on equitable terms and prevent interference by one State with the scientific programs of another.

[2] See Dembling and Arons, *The United Nations Celestial Bodies Convention*, 32 J. Air L. Com. 535 (1966) for a discussion of principles applicable to celestial bodies, analogies to other treaties, prior activity in the United Nations, and events giving rise to the Fifth Session of the Legal Subcommittee of the United Nations Ad Hoc Committee on the Peaceful Uses of Outer Space.

As early as 1959, the American Bar Association passed a resolution declaring "that in the common interest of mankind . . . celestial bodies should not be subject to exclusive appropriation." The United Nations Ad Hoc Committee on the Peaceful Uses of Outers Space took the position that "serious problems could arise if States claimed, on one ground or another, exclusive rights over all or part of a celestial body," and suggested that "some form of international administration over celestial bodies might be adopted."[3] In an address before the General Assembly in September 1960, President Eisenhower proposed that:

"1. We agree that celestial bodies are not subject to national appropriation by any claims of sovereignty.

"2. We agree that the nations of the world shall not engage in warlike activities on these bodies.

"3. We agree, subject to verification, that no nation will put into orbit or station in outer space weapons of mass destruction. All launchings of spacecraft shall be verified by United Nations."[4]

The Ad Hoc Committee, however, concluded that "since 'early exploration of celestial bodies . . . was not likely in the near future,' no priority treatment was necessary." With the formation of COPUS in 1960, attention was directed primarily to problems associated with the launching of spacecraft, their revolving in earth orbit, and their return to earth.

If one includes principles applicable to the exploration of celestial bodies under those pertaining to the exploration of outer space generally, the practice developed during the International Geophysical Year and further developed by subsequent space flights would support the view that, as a principle of customary international law, anything outside the earth's atmosphere, except an item launched from earth, is not subject to claim of national sovereignty.

[3] U. N. Doc. A 4141/25 (1959).

[4] Address by President Dwight D. Eisenhower to the U.N. General Assembly, 10 Sept. 1960, 43 DEPT STATE BULL. 554(1960)

ANALOGIES TO OTHER TREATIES

The draft conventions tabled by the United States and the Soviet Union at the Fifth Session of the Legal Subcommittee, contain provisions quite obviously based upon analogous provisions in the Antarctic Treaty,[5] Article 1 provides that Antarctica shall be used only for peaceful purposes. Article II provides for freedom of scientific investigation in Antarctica and cooperation in that regard. Article III provides for exchange of scientific information and personnel. Article IV, paragraph 2, prohibits nations from making additional claims of sovereignty, although it does not require renunciation of existing claims.

Another treaty which afforded some precedent to agreement on the use of outer space and celestial bodies for peaceful purposes is the Nuclear Test Ban Treaty. The Treaty bans any nuclear weapon test explosion or any other nuclear explosion in the atmosphere, in outer space, and under water, or in any other environment.

Whether one regards the moon and other celestial bodies as included in "outer space," or "in any other environment," as referred to in Article 1 of the Nuclear Treaty, nuclear explosions are effectively prohibited from being carried out on celestial bodies. Thus, the negotiation and drafting of principles providing for the peaceful exploration and use of outer space and celestial bodies proceeded from the standpoint that an activity of immense military significance had already been banned.

PRIOR U. N. ACTIONS

Although the Fifth Session of the Legal Subcommittee provided an opportunity for intensive examination, in the United Nations, of principles governing the exploration and use of outer space and celestial bodies, it was not the first time that the U. N. had ever considered this matter. At the first meeting of the COPUS (1961), the nations represented agreed on a draft resolution, originally proposed by the United States, which, as adopted by the General Assembly on 20 December 1961 as G. A. Res. 1721 (XVI), commended to States for their guidance in the exploration and use of outer space the following principles:

(a) International law, including the Charter of the United Nations, applies to outer space and celestial bodies;

(b) Outer space and celestial bodies are free for exploration and use by all States in conformity with international law and are not subject to national appropriation.

Proposed elaborations of, and additions to, the principles stated in that Resolution 1721 were further discussed in 1962 and 1963. This discussion of "basic principles," together with discussions of draft conventions and resolutions covering assistance to, and return of, astronauts and space vehicles, and of liability for damages caused by space vehicles, led to the unanimous adoption by the General Assembly, on 13 December 1963, of Resolution 1962 (XVIII) entitled *Declaration of Legal Principles Governing the Activities of States in the Exploration and Use of Outer Space.*

Although the *Declaration*, like other General Assembly resolutions, does not having the contractually binding characteristics of a treaty, the *Declaration* does reflect a certain international understanding of the principles which ought to govern the exploration and use of outer space and celestial bodies and, therefore, provided evidence of the customary international law in that regard. Thus, over two and one-half years prior to the 1966 (Fifth) Session, a general consensus had been obtained among the nations involved in space exploration that outer space and celestial bodies should be governed by the principles of international law and free for peaceful exploration and use without being subject to claims of national sovereignty.

During its previous four sessions, the Legal Subcommittee had been primarily concerned with subjects of assistance to and return of astronauts and space objects and liability for damages caused by

[5] The Antarctic Treaty signed at Washington on 1 Dec. 1959, by the seven Antarctic sector States (Argentina, Australia, Chile, France, New Zealand, Norway and the United Kingdom) and Belgium, Japan, Union of South Africa, the Soviet Union and the United States.

space vehicles. By October 1965, agreement had been virtually achieved on a draft convention covering the former subject, and considerable progress had been made on the latter. However, the activities of the Legal Subcommittee were not limited to these two subjects. Under the mandate governing its activities during the Fifth Session, the Subcommittee was not only "urged" by the General Assembly to prepare draft international agreements on "assistance and return" and "liability" but also to consider incorporating in international agreement form the legal principles governing the activities of States in the exploration and use of outer space. The consideration and use of outer space and celestial bodies come within this last part of its mandate.

EVENTS GIVING RISE TO FIFTH SESSION

That a sense of urgency had developed concerning the need for an international agreement on the exploration of the moon and other celestial bodies was made clear in a statement by President Lyndon B. Johnson on 7 May 1966. He emphasized the need to "take action now . . . to insure that explorations of the moon and other celestial bodies will be for peaceful purposes only" and "to be sure that our astronauts and those of other nations can freely conduct scientific investigations of the moon."[6] The President suggested a treaty containing the following elements:

1. The moon and other celestial bodies should be free for exploration and use by all countries. No country should be permitted to advance a claim of sovereignty.
2. There should be freedom of scientific investigation, and all countries should cooperate in scientific activities relating to celestial bodies.
3. Studies should be made to avoid harmful contamination.
4. Astronauts of one country should give any necessary help to astronauts of another country.
5. No country should be permitted to station weapons of mass destruction on a celestial body. Weapons tests and military maneuvers should be forbidden.

[6] For full text, see 54 Dep't State Bull. 900 (1966).

Two days after the President made his statement, the United States Ambassador to the United Nations addressed a letter to the Chairman of COPUS requesting an early convening of the Legal Subcommittee to consider the treaty proposed by President Johnson. On 30 May 1966, the Soviet Ambassador transmitted to the Secretary-General of the United Nations a letter from the Minister for Foreign Affairs of the U. S. S. R., requesting the inclusion of an item on the agenda for the 21st Session of the General Assembly entitled "Conclusion of an International Agreement on Legal Principles Governing the Activities of States in the Exploration and Conquest of the Moon and Other Celestial Bodies." In his letter, the Minister suggested that such an international agreement be based on four principles, which appeared to be quite similar to the principles stated by President Johnson.

On 16 June 1966, both the U. S. Ambassador and the Acting Permanent Representative of the U. S. S. R., forwarded the draft treaties proposed by their countries. In diplomatic discussions that followed, agreement was reached that 12 July would be the date on which formal consideration would commence and that the meeting would be held at Geneva, the date being the preference of the United States, and the place being the preference of the Soviet Union.

THE FIFTH SESSION OF THE LEGAL SUBCOMMITTEE

Scope of the Treaty

During the first few days of the Fifth Session, the various delegations discussed the urgent need for the Treaty, whether its scope should be limited to activities on celestial bodies or should include outer space as well, and whether its provisions should state general principles or should provide specific rules for the conduct of activity in outer space and on celestial bodies.[7] There was a belief that a treaty regulating

[7] All 28 members of the Legal Subcommittee were present, representing Albania, Argentina, Australia, Austria, Belgium, Brazil, Bulgaria, Canada, Chad, Czechoslovakia, France, Hungary, India, Iran, Italy, Japan, Lebanon, Mexico, Mongolia, Morocco, Poland, Rumania, Sierra Leone,

the conduct of States on celestial bodies should be agreed upon as soon as possible. It was apparent that the delegations regarded the prospect of manned lunar landings by both the United States and the Soviet Union as necessitating regulation before such landings. Also, there was particular desire to prohibit the use of celestial bodies, if not outer space as well, for military purposes. In this regard, there was also general agreement that a critical need existed to include a provision banning nuclear weapons and other weapons of mass destruction from outer space.

The belief that agreement must be reached as soon as possible affected the matter of whether the agreement should be limited to a statement of general principles or whether it should establish more specific regulation of space activity. As noted above, previous sessions of the Subcommittee had devoted considerable attention to the detailed draft treaties on assistance to and return of astronauts and space vehicles and liability for damages caused by space vehicles. Various delegations expressed a desire that the Subcommittee continue its work on these drafts during the Fifth Session, and were not satisfied with the inclusion of general provisions on those subjects as items in a treaty as broad as those suggested by the United States and the Soviet drafts. However, the Subcommittee was interested in obtaining "maximum results in a minimum time" and believed it should limit itself strictly to settling essential and urgent issues.

Most of the delegations felt that the principles set forth in the United States and Soviet drafts were a starting point and would be applied in practice later - in particular in the field of liability and the return of astronauts. It was therefore essential to define and codify the largest number of points of agreement. As stated by the head of the Soviet delegation to the Fifth Session, and later agreed to by the members of the Subcommittee, the inclusion in the Treaty of two broadly phrased articles on assistance and return and liability respectively was not intended to prejudice the efforts already being made in the Subcommittee to conclude a special agreement on those matters.[8]

A further matter to which considerable ;discussion was devoted during the general debate was whether the Treaty should establish rules governing activity on celestial bodies or should include all of outer space as well. The most obvious difference between the Soviet and United States drafts was that the Soviet draft would have applied to celestial bodies *and* outer space while the United States draft would have applied only to celestial bodies. As expected, the delegate from the Soviet Union and the representatives from Communist bloc countries of Eastern Europe advocated the Soviet version. In addition, however, several delegations from non-aligned and pro-Western nations supported the Soviet position on this matter. Cogent arguments were advance to the effect that the implementation of several of the proposed treaty articles would be extremely difficult, if not impossible, should the scope of the Treaty be limited to activities on celestial bodies to the exclusion of outer space.

In view of the various statements made concerning the scope of the treaty, the United States delegation recognized that a consensus had been reached on the broad proposition that "the Treaty should not be limited to celestial bodies alone but should include outer space along the lines of the U. S. S. R. draft" and agreed to work towards the conclusion of such a treaty.[9] In return, the Soviet delegate stated that his delegation was prepared "to consider the possibility of including, in the draft treaty to be prepared by the Subcommittee, provisions which did not appear in the Soviet text, including certain points from the United States draft."[10] The Soviet delegate was referring particularly to the provisions in the United States draft that provided for reporting of scientific information and free access to all areas of celestial bodies. As a comparison of the Soviet and United States drafts readily indicates, there were not many substantive points of difference between the Soviet

---

Sweden, Sweden, United Arab Republic, U. S. S. R., United Kingdom and the United States.

[8] Sum. Rep. 57 at 13.

[9] Sum. Rep. 63 at 2.

[10] Sum. Rep. 62 at 11.

and United States positions on the matters sought to be covered.

Thus even before the Subcommittee began its article by article analysis of the respective drafts, a reasonable amount of agreement existed between the two major space powers, and among all the members of the Subcommittee, on the general cope and purpose of the Treaty. The remainder of the discussions during the Session concerned specific matters to be covered in the Treaty.

The Preamble and Articles I, II and III of the Treaty state broad principles which, from the outset of discussion, were generally acceptable to the Subcommittee members and provoked little disagreement as to the wording. The text of these provisions was taken almost entirely from the Soviet draft though the same general principles appeared in the United States draft. The first three articles are, in large part, a codification of the first four paragraphs of the *Declaration of Legal Principles*, and are analogous to the principles set forth in the Antarctic Treaty. Thus, quick agreement on these provisions was not surprising. The second paragraph of Article I, provides that "there shall be free access to all areas of celestial bodies." This provision, however, must be read in the light of Article XII, which provides that "All stations, installations, equipment, and space vehicles shall be open to representatives of other States Parties to the Treaty on a basis of reciprocity." So Article I is therefore to be read as granting free access at all times to all areas of outer space and celestial bodies, except as provided in Article XII.

The text of Article II, which prohibits national appropriation of outer space and celestial bodies, provoked little debate. The banning of national appropriation reinforces the free access language. if an individual nation cannot claim sovereignty to any particular area, it cannot deny access to that area.

Article III, by making international law, including the U. N. Charter applicable to outer space and celestial bodies, further reinforces Article I. Article III is important if viewed in the light of the consensus reached earlier that this Treaty is intended to establish basic principles applicable to conduct in outer space and on celestial bodies.

Article IV of the Treaty provides that no weapons of mass destruction shall be placed in orbit or on celestial bodies or stationed in outer space in any other manner; celestial bodies shall be used exclusively for peaceful purposes. The Article reflects principles previously agreed upon in the Nuclear Test Ban Treaty and in U. N. Resolution 1884 (XVIII) and is quite similar to Antarctic Treaty language.

There was debate on the use of the terms "installations" and "military equipment." The Soviet delegate insisted on use of the word "installations" since in Russian translation "bases" and "fortifications" connote facilities for military purposes while "installations" might apply to a facility used for peaceful purposes but constructed or inhabited by military personnel. Finally, the United States accepted the use of the term "installations" while the Soviets agreed to the inclusion of a provision which would not ban the use of military equipment on celestial bodies. Thus, the placement of a weapon or other items of military equipment would appear to be prohibited unless it can be shown that the item of military equipment will be devoted solely to peaceful exploration or use of the celestial body. Agreement on Article IV was not reached until after the close of the second (New York) portion of the session when the few outstanding differences were resolved so announcement of treaty agreement could be made.

Article V of the Treaty contains two distinct though related principles. The first two paragraphs provide for the assistance and return of astronauts, which had been discussed in considerable detail during previous sessions of the Legal Subcommittee and were already accepted by its members. The third paragraph, derived from a United States proposal that the U. N. Secretary-General shall be promptly notified of any information relating to the physical safety of astronauts. Article XI of the Treaty, a provision for reporting of activities in outer space and on celestial bodies, originated with the United States draft. This draft took the position that parties to the Treaty should be under a mandatory obligation to notify the U. N. Secretary-General. Originally, the Soviets argued against the mandatory obligation, favoring a voluntary basis for such reporting. The United States and its supporters were seeking to embody in treaty form a requirement that there be full dissemination of scientific and technical information for peaceful purposes. A compromise working paper was proposed by the United Arab Republic which retained the Soviet proposal for voluntary reporting but also provided that "all information shall be promptly submitted, preferably in advance or at the carrying out of these activities or immediately after." While one might read this as a mandatory reporting provision, a fair reading of the first two paragraphs would be that the parties to the Treaty agree to report voluntarily but it must do so promptly if it chooses to report. Agreement was reached when it was further agreed that the Secretary-General would disseminate the information immediately and effectively.

Article VI of the Treaty assures that the parties cannot escape their international obligations under the Treaty by virtue of the fact that activity in outer space or on celestial bodies is conducted through the medium of non-governmental entities or international organizations. While the general concept was agreed to without too much debate, the more difficult question was whether or not international organizations should be made responsible for compliance with the Treaty in the conduct of their outer space activities. The compromise reached, as reflected in Articles VI and XIII, appears to require States which are parties to the Treaty, when they conduct activities through an international organization, to use their best efforts to secure compliance by that organization with the obligations set forth in the Treaty.

Although recognizing that the Legal Subcommittee was in the process of drafting a detailed treaty on liability, there was no objection raised to the mere inclusion of an article stating the general principle in the Outer Space Treaty. The subject of international liability for damage caused by space vehicles is indeed one involving a multitude of problems. Since Article VII is essentially a repetition of Paragraph 8 of the *Declaration of Legal Principles*, these problems were hardly touched during the Session. The meaning of the word "internationally" as used to modify "liable," however, did cause some concern. Many stated that the article would be acceptable only if "internationally" meant "absolutely." Other delegations noted that the concept of "absolute liability" was still being refined in discussions of the detailed drafts on liability and would have to await the outcome.

Article VIII consists of three sentences, two of which state general rules concerning control and ownership of personnel and objects while in outer space and on celestial bodies. The third sentence imposes an obligation upon the parties to return found objects to the party to the Treaty on whose registry they are carried. The State of registry is required to furnish identifying data if so requested. With some minor drafting changes, agreement was reached on the Article.

The parties to the Treaty must conduct their activities to avoid harmful contamination of outer space or celestial bodies and adverse changes in the environment of Earth according to Article IX. The Article also imposes a mandatory obligation upon a party planning a potentially harmful experiment to consult with other parties. It also provides each party with the right to request consultations concerning a potentially harmful activity or experiment planned by another State in outer space or on a celestial body. While agreement was reached on reporting outer space activities to the Secretary-General, it was understood that there was a distinction between the mandatory consultations in advance of the event, under Article IX, and what is regarded as voluntary reporting after the event under Article IX.

Article X of the Treaty pertains principally to the establishment and use of tracking facilities by parties to the Treaty on the territory of other parties. Protracted disagreement among the delegations proved to be the major stumbling block to agreement on the Treaty as a whole. The genesis of the provision was in the second sentence of Article I of the Soviet draft which provided that "The parties to the Treaty undertake to accord equal conditions to States engaged in the exploration of outer space." Essentially, this would have included a most-favored nation clause with respect to the availability of tracking facilities. The United States delegation and its supporters strongly opposed this proposal because it appeared to be for the benefit of the space powers alone. It "would give a space power the right to require of a non-space power equivalent facilities in regard to the tracking of space objects if the non-space power had previously granted facilities of that kind to another State. Thus, the State would be bound to accord tracking facilities without reference to any bilateral negotiations.... Moreover, the proposal put a premium on non-cooperation... Only if State A had extended facilities to a third party was it obliged to make the same facilities available to State B. Besides, a country having tracking facilities and using them exclusively for its own space programs would have no obligation at all towards other countries.... Finally, the installation of tracking facilities in the territory of a host country raised many technical and political questions which could only be dealt with bilaterally."[11] Notwithstanding the strenuous objections of the United States and its supporters, the Soviets tabled a revised working paper which reiterated its earlier position, but stated that any expenses incurred by a party to the Treaty in rendering assistance to another party for the purpose of observing the flight of space objects would be reimbursed by the party receiving the assistance.

The New York (Second) portion of the Fifth Session adjourned without an accommodation on the use of tracking facilities. Agreement was finally reached on the text of Article X after extensive bilateral negotiations by the United States, the Soviet Union, and other States, particularly those which had already granted tracking facilities to the United States.

Parties to the Treaty which afford tracking facilities to other parties are only obligated to "consider on a basis of equality any requests by other States Parties to the Treaty to be afforded an opportunity to observe the flight of space objects launched by those States." This language recognizes that there must be agreement between the parties concerned for the establishment of a tracking facility.

Article XII of the Treaty is another provision which reflects a compromise of United States and Soviet positions. The United States draft initially provided that "All areas of celestial bodies, including all stations, installations, equipment and space vehicles on celestial bodies, shall be open at all times to representatives of other States conducting activities on celestial bodies"[12] The Soviet draft provided that "there shall be free access to all regions of celestial bodies "but this related more to the broad principle of freedom of scientific investigation on celestial bodies which was eventually covered by Article I of the Treaty. The Soviet delegation accepted the United States draft subject to deleting the words "all areas of celestial bodies, including" and the words "at all times" and the addition of the phrase: "on the basis of reciprocity under the conditions that the time of the visit is to be agreed between the parties concerned." The United States was seeking a treaty provision providing for an unlimited right of access. The Soviets, while accepting the principle of open access, was seeking to impose conditions upon the ability of individual nations to exercise that right. The Soviet position that the right of access to stations, etc. should not be so absolute as to endanger the lives of astronauts or to interfere with normal operations caught favor with other delegations. However, the Soviet suggestion that access should be on the basis of "reciprocity" provoked considerable discussion whereupon the United States delegate acceded to including "reciprocity" language in the treaty provision covering access to stations, etc. subject to certain interpretative caveats.[13] The issue was resolved when the United States introduced a revised version of Article 6, omitting the phrase "at all times" and adding the language that was eventually adopted as the second sentence of Article XII of the Treaty.

The first twelve articles of the Outer Space Treaty prescribe, more or less, the general rules governing the conduct of the parties to the Treaty. Article XIII seeks to establish the applicability of the substantive principles to actions by the parties whether taken singly, jointly, or within the framework of international organizations. It does not provide a mechanism whereby international organizations can become, for all practical purposes, parties to the

---

[11] Sum. Rep. 73 at 4-5.

[12] Article 6 of the United States draft was based on Article VII, para. 3, of the Antarctic Treaty.

[13] This acceptance was based upon the understanding that the right of access by one state is not conditioned upon whether a second state wishes to exercise its right of access.

Treaty. It does, however, provide that the provisions of the Treaty shall apply to space activities carried out by parties to the Treaty within the framework of international intergovernmental organizations.

Miscellaneous Matters

The Treaty does not include a provision for the resolution of disputes arising between parties to the Treaty over matters covered therein. Both the United States and Soviet drafts contained articles on settlement of disputes. Earlier sessions of the Legal Subcommittee, however, had revealed an inability for the two delegations to compromise their differences on this matter. In the interest of expediting agreement on the Treaty as a whole, neither pressed for inclusion of a specific provision on this topic.

The only remaining provision which caused controversy was the one that provides that the Treaty shall be open to all States for signature. This was the position advocated by the Soviet Union. The United States proposed that the Treaty "be open for signature by States Members of the United Nations or of any of the specialized agencies or parties to the Statute of the International Court of Justice, and by any other State invited by the General Assembly of the United Nations to become a party." This latter formulation would have excluded certain non-United Nations members from being permitted to become parties to the Treaty, notably Communist China and East Germany. Because of the very broad geographical coverage on the Treaty, the United States agreed to the Soviet formulation but subject to the understanding that accession to the Treaty by a regime or entity not recognized by the United States does not, without more, amount to recognition of that regime or entity by the United States.

Little need to be said about the remaining provisions of the Treaty. There was some debate over what agency would constitute the depositary authority. Other portions of the Treaty concern the mechanics and legal effect of ratification and deposit of the appropriate instruments; the method for amending the Treaty, how a party may withdraw from the Treaty; and that Chinese, English, French, Russian, and Spanish texts of the Treaty are equally authentic.

CONCLUSION

The Treaty was approved by acclamation on 19 December 1966 by the United Nations General Assembly. The Treaty was open for signature in Washington, London, and Moscow on 27 January 1967. Sixty nations signed the Treaty on that date. Since the signing and ratification of the Treaty, the activities of human beings in outer space and on celestial bodies have been subjected to a regime of law. The Treaty reflects a broad international consensus that outer space and celestial bodies are to be free for exploration and use for the benefit of all mankind; that the principles of international law are applicable thereto; that celestial bodies are to be devoted exclusively to peaceful purposes, and weapons of mass destruction are to be banned from outer space; that assistance is to be rendered to astronauts; that States are to be held internationally responsible for their activities in outer space, and held liable for damages caused thereby; that ownership of objects is not changed by their presence in outer space and on celestial bodies; that harmful contamination of the environment of earth, outer space, and celestial bodies shall be avoided; that information gathered from activities in outer space and on celestial bodies is to be broadly disseminated; and that stations, installations, etc., on celestial bodies are to be open for inspection.

In establishing these principles in treaty form, the parties are now contractually obligated to carry out their activities in outer space and on celestial bodies in accordance with accepted norms and goals validated in a legal form.

From Cold War to Détente in Outer Space: The Role of the United Nations in Outer Space Law Development

Peter Jankowitsch
Past Chair (1972 - 1991), COPOUS

In order to understand and appreciate the major role and importance of the United Nations in Outer Space Law Development, it is essential to put it into the international context of a difficult period of Post-World-War II history, a period characterised by super-power rivalries and the chilly atmosphere of the Cold War.

Confrontation in the Cold War became increasingly dangerous as it successively left the European theatre in which it had started and rapidly developed into a global phenomenon, Its major players were constantly in search of new areas and fields were advantage over the adversary could be gained and military technology was certainly one of the most tvpical areas in which this contest took place.

The rapid development of nuclear arms was a clear sign that in this confrontation no avenue would be left unexplored. While thus land, air and sea had already been subjected to military uses the question remained to what extent the arms race would also move into new media: and indeed early ballistic weapons developed by Nazi Germany towards the end of the Second World War - forerunners of today's ballistic missiles - had already begun to infringe upon humanity's last frontier.

When, finally, in October 1957 a first man-made object was launched into Outer Space, it became clear that a new arena of competition between the two super powers of the day had been opened. The question remained, however, to what extent this competition would be limited to the civilian field or whether it would also become a military one.

It is not easy to speculate, even today, on the intentions and motives of these two major players in regard to Outer Space. If, in the end, there was a clear turn towards more peaceful uses of Outer Space, we can assume that next to political considerations there must also have been economic ones, essentially the cost, even more prohibitive in those early days, of moving (and maintaining) large military structures in Outer Space.

As early as 1963 therefore, a few years before the conclusion of the Outer Space Treaty, a general understanding was reached between the USA and the USSR to ban the deployment of nuclear weapons or other weapons of mass destruction in Outer Space. Originally in the form of a bilateral agreement, it was later endorsed by the General Assembly of the United Nations.

The way thus was open for entering into a wider agreement on the principles that should henceforth govern the activities of states in the exploration and uses of Outer Space. The history of the birth of the Outer Space Treaty, leading to its signing, in January 1967 in London, Moscow and Washington, has been told many times and therefore is not in need of a new version.

Much has also been said and written about its legal significance and there is general agreement that this is and remains the cornerstone of an entirely new branch of international public law. Of an innovative nature in many respects it is setting tight limits to the exercise of state sovereignty in Outer Space and creates a new ethic and spirit in relations between states rarely to be found in the traditional pages of international law which is much stronger marked by "realpolitik" as pages devoted to Outer Space.

Unlike the continents newly discovered by Europeans from the 16th to 19th centuries, "Outer Space", including the moon and other celestial bodies, is not subject to national appropriation. And unlike the high seas, which since Salamis and Actium have been the arena of decisive military engagements, the exploration and use of space is to be "for peaceful purposes".

It is innovative also in the sense that, to this day, it has attempted, albeit not always successfully, to move ahead of technological developments and to try create a secure legal environment for future scientific or economic activities. This characteristic is perhaps best exemplified by the visionary dispositions of such follow-up treaties to the Magna Charta of Outer Space as the 1979 Moon Treaty.

By designating in its Article II the Moon itself, as well as its natural resources a "common heritage of mankind" (echoing, incidentally a similar description for natural resources in the deep sea-bed) a step was certainly made towards a future, more broadly designed regime for such resources. The scope for such a regime would be even wider, as the provisions of the Moon Treaty are also applicable "to other celestial bodies within the solar system, other than the earth". Not surprisingly this Treaty, although adopted finally by the General Assembly of the UN has to this day, found only few states willing to ratify it and thus endorse the principles it contains.

The United Nations & Space Law Development

In developing the broad principles on which space rests the United Nations had to contend, from the outset, with opposing philosophies which its member states brought to this new subject matter. Thus, the United States and the Soviet Union which for many years governed all major space activities, were primarily motivated by national security concerns and were aiming to allow some military uses of Outer Space some of which set in from the very beginning of the "Space Age". Satellites soon became indispensable for military communications, reconnaissance or military weather forecasting and it is estimated that up to 75 % of all satellites launched have some military applications.

Thus, even in the Outer Space Treaty, its rules are guarded in their restraints on national military activities. Article IV, the key provision, states that "The moon and other celestial bodies shall be used exclusively for peaceful purposes". As for outer space generally, the only provision restricting military activities forbids the placing "in orbit around the Earth of any objects carrying nuclear weapons or any other kinds of weapons of mass destruction or station(ing) such weapons in outer space in any other manner". The "peaceful purposes" rubric applied to the moon and other celestial bodies is never defined in the Treaty, but presumably comprehends more than the simple prohibition applied to outer space generally.

The reason for the different treatment of "celestial bodies" and "outer space" generally was to accommodate nuclear ballistic missiles, which were just entering the arsenals of the US and the Soviet Union as the treaty was being negotiated. A major portion of the trajectory of such missiles is in outer space, but they do not go into orbit. The language of Article IV was carefully chosen to ensure that the general principle of "peaceful uses" would not interfere with the testing of these weapons.

The treaty also remains silent on the use of military satellites for reconnaissance, surveillance, early warning, and communications.

In any case, it is clear from this history that reconnaissance and other "passive" military satellites are not prohibited by the Outer Space Treaty. This conclusion has since been confirmed by the provisions of the ABM treaty and other arms-control agreements in which the United States and the Soviet Union endorse the use of "national technical means of verification" to assure compliance, and agree not to interfere with them.

Although only a few provisions of the Outer Space Treaty deal specifically with military activities, and those that do leave much ground uncovered, the affirmation of the basic principles of peaceful purposes and international co-operation in exploration and use nevertheless remained important for the construction and application of more specific agreements governing outer space activities.

On the other hand, the space for military activities left open by the 1967 Treaty created numerous controversies over the years as efforts were made to complete its provisions so as to avoid what appeared, especially in the hotter years of the Cold War, a growing militarization of Outer Space.

These efforts were motivated by efforts of the early space powers, the US and the USSR to use space not only for purposes of information and communication but also to develop "conventional" space weapons: the first of those weapons were anti-satellite weapons whose development started in the late 1950's and which were brought to some perfection in the 1980's. An even more menacing perspective was opened by the idea, proposed by President Reagan in 1983 to build a space based system of Ballistic Missile Defence using all kinds of new and sophisticated technology and weaponry.

Had this idea been realised it would have eliminated one of the pillars of the arms control system of the Cold War era that also had its relevance for space law, namely the bilateral, Soviet-American AMB-treaty of 1972 that was motivated by a judgement that security is enhanced and the stability of the strategic balance strengthened if both sides in the Cold War forswear defensive systems. This plan would have undermined the widely accepted doctrine of nuclear deterrence, given rise to an enormously expensive escalation of the arms race and introduced weapons into a realm which had been largely peaceful, or at least non-violent.

It is not difficult to understand, therefore, that in the work of the UN Outer Space Committee militarization of Outer Space was one of the most contentious issues and the only one that threatened to seriously disrupt its work in the mid 80's. This issue also raised questions about the purpose of the Committee and the United Nations.

The United States, with some support from other Western countries hoped to keep this question out of the Committee and confine it to the 40-nation Conference on Disarmament, where it was less likely to attract less attention in the context of a variety of other arms control questions. A majority of countries, while agreeing that the Conference on Disarmament was the appropriate body for negotiating formal agreements on the question, insisted that the militarization of Outer Space was a political issue of general concern and should therefore be discussed in a number of relevant bodies.

This was also an example of the different attitudes of the Third World and the West towards international organisations such as the United Nations. The West considered the United Nations to be a mechanism for reaching agreement on issues where agreement was possible and could serve a useful purpose. For the developing countries, the United Nations were a unique forum in which they could let their views be known to the world and exert the pressure of their numbers, even on questions where clearly there would be no practical effect.

The fact that military and security concerns of the two initial major space powers had a strong influence on the work of the UN Outer Space Committee, not least in its legal work also limited its membership. After the People's Republic of China had been restored to UN membership in 1971 it first refused to occupy its seat in the Committee as it felt that it was too largely dominated by Soviet-American concerns. Albania, that in this period was a close ally of China, followed its example. It was only some years later and in view of the increasing importance that developing countries devoted to the work of the Committee that China finally participated in its deliberation.

While it were thus the security concerns of the major space powers that put severe limitations on the development of space law, the "new majority" of the UN that became dominant in the early 60s brought a different concern to the deliberations of COPOUS: developing nations saw a need to use this new technology for the benefit of their economic and social development. There was, in particular, a fear that space benefits would remain limited to a small number of advanced countries. This view was clearly reflected in a memorandum that U Thant, as Secretary General of the United Nations submitted to the 1968 Vienna Conference on the Exploration and Peaceful Uses of Outer Space. Participants in this first global UN Space Conference were warned that "the space age was increasing the gap between the developed and developing areas at an alarming rate".

An effort was made, therefore, to give space law or initial principles of space law a direction that would also benefit developing countries. A case in point was negotiation of a set of principles relating to remote sensing of the Earth from space, adopted after 13 years of efforts by the legal subcommittee of COPOUS by Resolution A/RES/41/65 of the General Assembly of the United Nations. Here the Committee had to resolve the conflict between the principle of freedom of space activities and the general interest in acquiring global environmental and resource data, on the one hand, and the rights of countries to control access to their natural resources, on the other. Consensus was reached on the principles of a general right to collect data and the right of the sensed states to have immediate access to any data collected over their territories. In the cases of both direct broadcasting and remote sensing, the conflict was intensified in the early stages by fears that the new and somewhat mysterious space technology would revolutionise television broadcasting in the first case and exploitation of natural resources in the second. As the technologies developed and as the practical limits of operational systems became apparent, it became clear that the potential impact of the technologies had been somewhat exaggerated by the agencies that had an interest in promoting them. In the case of remote sensing, the negotiating positions of the parties became more flexible and agreement was reached.

A similar conflict that remained unresolved concerned access of equatorial countries to the geostationary orbit.

A declaration adopted by the COPOUS in 1996, whose lengthy title referred to the need to conduct the exploration and use of Outer Space "for the benefit and in the interest of all states, taking into particular account the needs of developing countries" reflects a further stage of the North South debate on space co-operation. Whether, as some authors believe, it even marks the end of a contentious North-South Debate in this area remains to be seen.

What it certainly does is, to combine the freedom of the exploration and utilisation of outer space with a reminder to space powers to fulfil their obligation to conduct their activities for the benefit of all countries. Space powers should foster international co-operation on an equitable and mutually acceptable basis. Developing countries interested in space activities might thus be motivated to put their energies into a well prepared strategic towards Outer Space. This could make many of them more equal partners in co-operation that the space powers might be ready to accept. Another consequence might be a strategy to pool their resources on a regional basis as even the industrialised countries of Europe had to do.

The mandate of COPOUS to promote international co-operation in the peaceful uses of outer space helps in some measure, to outbalance inadequacies felt in the legal field. In this respect, it is more difficult to point to concrete results, since the Committee itself does not actually carry out space activities. Most space programmes contain some degree of co-operation between countries, the practicalities of which are worked out between the responsible agencies and the technical personnel of the countries involved.

In response to the desires of the developing countries to benefit from space technology, the United Nations, through the Committee, organised two major world conferences on outer space - both in Vienna - in 1968 and in 1982. In response to the first, the United Nations established a space applications programme to provide developing countries with information on how they could use space technology. The 1982 Conference, in which 94 countries and 45 international organisations participated, was dominated by conflict between the developing and the developed countries over rights and obligations with respect to the transfer of technology. While it managed to agree that a major expansion of the Space Applications Programme was desirable, there was no agreement on funding, which continued to remain largely voluntary. Nonetheless, the existing programme does provide for a number of seminars and training courses each year in developing countries and administers a number of fellowships for long-term advanced training in space technology in developed countries. UNISPACE III to be held in Vienna in 1999 will be another attempt in this direction.

There can be no doubt that a substantial body of international space law has been created by the UN Committee on the peaceful uses of outer space, particularly by the work of its legal experts in its relevant subcommittee. This body of law has underpinned a wide array of space law developed by other UN organisations, not least the ITU and it can also be regarded as the groundwork on which regional and sub-regional organisations have drafted various instruments of space law.

On the other hand, space law development has gone, over the past years, through a series of stages that have, as was pointed out earlier, been influenced very clearly by geopolitical developments such as the course of the Cold War or the North-South conflict.

As both these phenomena have either disappeared - as the East-West conflict - or been transformed, as the North-South conflict, the question remain why this has not resulted in a renewed blossoming of space law treaties and a new push to regulate men's conquest of outer space. The last part of this paper will therefore be devoted to discuss possible reasons why even in a new world environment development of space law remains sluggish and slow and early enthusiasms to write or at least codify space rules seems to have completely evaporated.

And indeed there is a clear break between the first decades of space law that saw, after the entry into force of the historic Outer Space Treaty the drafting and adoption of a few more classical legal instruments, such as the 1968 Agreement on the Rescue of Astronauts, the 1972 Convention on International Liability for Damage Caused by Space Objects; the 1976 Convention on Registration of Objects Launched into Outer Space; and the Moon Treaty, which entered into force in 1984, when Austria became the fifth country to ratify it.

Following the Moon Treaty the Committee reverted to the adoption of several sets of principles, the first of which was the „Principles Governing the Use by States of Artificial Earth Satellites for International Direct Television Broadcasting" of December 1982, the last one being the 1996 principles on „Space Benefits“ of which mention was made earlier. Very important set of principles also concern remote sensing of the earth by satellites as well as the use of nuclear power sources in outer space.

While the importance of the adoption of these principles cannot be denied and while they reflect - perhaps with exception of the principles on direct broadcasting which were the only ones adopted by majority vote - a welcome spirit of compromise and understanding they still constitute a significant departure from previous law making efforts.

AS mere principles their legal effect is certainly smaller than that of the previous conventions and while they could be important building blocks of later, more mature space law they certainly reflect a growing resistance of some of the major players in space politics to create too stringent a body of space law. This apparent unwillingness to adopt new space regulations and complete the existing body of space law has become visible once again as first efforts to find legal solutions to the problem of space debris have failed.

One reason for this development is certainly a general public mood that first surfaced in the developed world and then became more and more global to liberalise and deregulate national markets and consequently international economic relations. Such an atmosphere was certainly not conducive to the acceptance of new regulations in space, which at the same time saw the massive entry of particularly aggressive private sector players, motivated by the expectation of rapid growth and major economic opportunity. These new players therefore resisted, as elsewhere, the introduction of a legal framework that they considered to be an artificial barrier to their expansion.

Next to economic considerations, national interest also must have played its part: national space agencies, not least those operating in some of the technologically most advanced countries, apparently saw little merit in accepting new legal obligations of an international character and preferred to cast their international relations in bilateral form. While certainly accepting a responsibility to support efforts of developing countries to become users of space technology most developed countries obviously came to prefer the bilateral approach in their assistance programmes.

At the same time however, technological progress as well as new and multiple uses of Outer Space continue, creating new problems and challenges for which legal solutions are just as important as technical ones.

And while it remains debatable to what extent economic globalisation can safely and successfully continue without some degree of regulation, the global nature of space co-operation certainly requires universally accepted rules to stay on course so as to avoid lawlessness and conflict in Outer Space.

These developments will certainly renew, at some stage in the future, the traditional role of the United Nations and its Outer Space Committee as indispensable instruments and fora for the further development of space law. Their universal nature is also the best guarantee that interests and concerns of all nations can be met and compromise be reached when philosophies, policies and strategies concerning the exploration and uses of Outer Space are opposed.

IISL-97-1.06

The OUTER SPACE TREATY IN PERSPECTIVE

By
Qizhi He
Beijing, China

Abstract

1997 is the 30th anniversary of the Treaty on Principles Governing the Activities of States in the Exploration and Use of Outer Space, including the Moon and other Celestial Bodies, commonly referred to as the Outer Space Treaty. The International Institute of Space Law decided to devote a special session to commemorate the birth of this most important treaty which represents a remarkable culmination to embark on the elaboration of international space law. With the adoption of this treaty, the space activities of states brought about by the great revolution in science and technology have been subjected to a regime of law.

Introduction

The 1967 Outer Space Treaty is regarded as the cornerstone of international space law conventions, or what may be termed the Magna Carta of international space law. It propounds a set of fundamental principles which establish the basic framework for space exploration and utilization. Being the first international convention embodying a number of customary international space law as enunciated in the 1963 Declaration of Legal Principles Governing the Activities of States in the Exploration and Use of Outer Space[1], it prescribed in treaty form their principles adopted in the Declaration, and went further to enrich the law by providing additional important substantive rules.

These basic principles and rules could be summarized as follows:

1. The exploration and use of outer space, including the Moon and other celestial bodies, shall be carried out for the benefit and in the interest of all countries;
2. Outer space shall be free for exploration and use by all states on a basis of equality;
3. Outer space shall not be subject to appropriation by claims of sovereignty, by means of use or occupation, or by any other means;
4. Activities in the exploration and use of outer space must be carried out in accordance with international law, including the Charter of the United Nations in the interest of maintaining peace and security;
5. No nuclear weapons or any other kind of weapons of mass destruction shall be placed in orbits of the earth;
6. The Moon and other celestial bodies shall be used by all state parties to the treaty exclusively for peaceful purposes;
7. Astronauts shall be given every possible assistance;
8. State parties bear international responsibility for national activities in outer space;
9. State parties keep jurisdiction and control over launched objects and personnel recorded in their register;
10. State parties shall avoid harmful contamination of outer space, celestial bodies and the environment of earth, and shall consult with other state parties regarding potential harmful experiments;
11. The UN Secretary-General must be informed about space activities and shall disseminate such information to the public and the international scientific community;
12. All stations, installations etc. on the celestial bodies shall be open to representatives of other parties on a basis of reciprocity;

13. International cooperation and understanding are to be promoted.

The treaty entered into force on October, 1967. Thus, the state parties were contractually obligated to carry out their space activities in accordance with the accepted norms and goals as set out in the treaty, as per its title, is a treaty of principles, capable of broad interpretations, and is considered to form the basis upon which more precise legal instruments could be constructed. Some specific issues call for further elaboration. They include four additional treaties and four sets of principles adopted by the General Assembly[2]. The legal contents of all these instruments are determined by the ideas and principles of the 1967 Outer Space Treaty, which were developed and amplified into more specific provisions, including procedures for resolving disputes which might arise from relevant space activities

Historical Background

Looking back 30 years when the Outer Space Treaty was adopted, at the threshold of the space age, it may be asked what were the circumstances that led to the drafting and adoption of the Outer Space Treaty composed of a set of comprehensive legal rules closed linked with man's journey into outer space, Confronting with the wide range of important issues which were proliferated with the speedy development of space science and technology, states immediately realized the urgent necessity of bringing them within the framework of the law. This is how the movement towards legal regulation of outer space under the aegis of the United Nations has taken place.

The United Nations as a worldwide organization established to maintain peace and security, and entrusted with the task of "encouraging the progressive development of international law and its codification[3]", made quick response and took the initiative for the elaboration of treaty law dealing with space activities. Immediately after the successful launch of the first satellites, the General Assembly set up the Ad Hoc Committee on the Peaceful Uses of Outer Space (COPUOS) in 1958[4], and made it permanent in 1961[5]. COPUOS and its two subcommittees: the Scientific and Technical Subcommittee and Legal Subcommittee, embark on their substantive work in 1962, and have become the main center of international cooperation and coordination in the field of exploration and peaceful uses of outer space.'

One of the tasks entrusted to the committee and the legal subcommittee was "to study legal problem which may arise from the exploration and the use of outer space"[6]. The result of the laborious deliberations were finally embodied in the Outer Space Treaty, which is a monumental work laying down the foundation and the first layers of bricks in the structure of the international law.

The unique characteristics of outer space require a special law-making mechanism which should allow technical experts, government representative and lawyers specializing in this field to interact together to accommodate both political and legal concerns in regulating space activities[7]. The developing events could show that without the timely and appropriate institutional arrangements provided by the United Nations, it would be unlikely that the legal regime provided By the 1967 Outer Space Treaty, could be established in such an efficient and expeditious way within such a short span of time.

International political developments also play an important part which should not be overlooked for the conclusion of the Outer Space Treaty. 1960's witnessed the Sharp confrontation amidst the cold war between the two superpowers which alone posses the satellite launching capabilities, thus placing them in a dominant position in treaty making for outer space. Among the procedures which were not in the framework of COPUOS is the agreement by the USSR and the USA "of their intention not to station in outer space any objects carrying nuclear weapons or any kind of weapons of mass destruction", and called upon all states refrain from such activities[8]. Such a political compromise and arrangement reached outside the United Nations by the two superpowers was a determining factor led to the formulation of key provision of Article 4 of the Outer Space Treaty. Although Article 4 invoked a lot of comments[9], the result of the efforts as a whole were noteworthy. Legal Principles and rules universal in scope were established in the immediate wake of scientific and technological progress, and began to operate for the new dimension. To this, it should be added that states not engaged in space activities in the beginning phase of the space era also played a very active part in the negotiation process by making substantive statements and proposal which find their way into the final instruments. So the Outer Space Treaty was being drafted through the cooperation of the international community as a whole. This fact was all the more instructive since, in practice, outer space was being explored and utilized then by only a very small number of states[10]

Customary Rules of International Space Law

Analysis of state practice in outer space shows that long before the conclusion of the 1967 Outer Space Treaty, important principles had been established as customary international law[11].

Among these practices which were codified into the Outer Space Treaty, the most important are the following: Outer space is open and free for explanation and use by all states; The sovereignty of states does not extend to outer space; Outer space is not subject to national appropriation; and states retain jurisdiction and control over space object launched into outer space.

It should be noted that these principles were grown into being within a very short period of time. Although there were controversies over the problem of the emergence of "instant" customary international law[12], most publicists held that international law, particularly the customary law of outer space, does not require the existence of practice for a long period of time. This was evinced by the dictum of the International Court of Justice in the North Sea Continental Shelf case:

> "The passage of only a short period of time is not necessarily , or , of itself, a bar to the formation of a new customary law."[13]

In the case of the Outer Space Treaty, custom as a source of international law to be codified in the Treaty, shall operate alongside the Treaty provisions and may extend the sphere of the validity of the customary rules of outer space to those states which do not formally accept this Treaty.[14] In this respect, the International Law commission did authoritatively stated:

> "A principle or rule of customary international law may be embodied in a bipartite or multipartite agreement so as to have, within the stated limits, conventional force for the state parties to the agreement so long as the agreement is in force; yet it would continue to be binding as a principle or rule of customary international law for other states."[15]

An example of great significance may be cited here to illustrate the importance of the application of customary rules of outer space to those states which have not accepted the Outer Space Treaty.

In the consideration of legal status of the geostationary orbit in COPUOS, some of the equatorial countries which were not parties of the 1967 Outer Space Treaty stressed the claim that they are not bound to the principles of the Treaty, in particular principles relating to the freedom of exploration of outer space contained in Articles1 and 2 of the Outer Space Treaty, as they are not party members of the Treaty.[16]

However a great majority of states rejected the above argument of the equatorial states on the ground that the fundamental principles and rules of the Outer Space Treaty, as enumerated above represent the existing general customary law which shall bind all members of the international community independent of a formally ratifying or accepting the Treaty.[17]

Thus, the 1967 Outer Space Treaty by codifying customary rules of international space law, regulates the mutual relations of both states which are parties of the Treaty and States which are not, as well as the relations of states which do not participated in the Treaty. In this way, the effectiveness of the fundamental legal rules relating to space activities plays a more important role in the international legal order in outer space.

Future Development

In retrospect, the 1967 Outer Space Treaty together with other related instruments forming the basic parts of space law, have played a significant role to regulate and promote space ventures which have brought enormous benefits to mankind. Now thirty years after the adoption of the Treaty, the question may be raised whether the existing legal regime based on the Outer Space Treaty can cope with anticipated developments in the coming years of the 21st century.

The immediate answer would be the fundamentals of the Treaty would be viable and effective to meet the challenges, but this does not preclude necessary adjustment and changes which should be made to forestall the forthcoming developments.

Space law will grow to cover two principal aspects of space ventures.

In the field of earth-oriented activities, environmental issues, including both earth and space environment, have become a matter of great concern to mankind. An examination of existing law shows that effective protection of the environment requires not only general principles but also more detailed specific legal rules. This indicated the need for further progressive development of space law relating to environment mainly on a treaty basis, since existing treaty law regarding environmental protection is very inadequate , and customary law provided only a few basic legal principles which lay down guidelines to be followed.[18] In such circumstances, and bearing in mind the political and legal realities of the international system, the development of a comprehensive system of space environmental law has to take a realistic approach, dealing with issues step by step based by a comprehensive and in-depth understanding of the environment situation mainly by space technology.

With regard to space -oriented ventures which will extend well into the 21st century, the recent magnificent feat of pathfinder's landing on Mars on July 4, 1997 revives ma's interest and opens a new chapter of man's forays into the universe. The new and successive moves in space would make space lawyers to watch scientific discoveries warily and to be mindful of changing needs in the field of law which would be attendant on the new achievements.

New pages of space law will thus be added to those already written. In the case of mars, the Moon agreement provided its provisions would apply to Mars and other celestial bodies "except in so far as specific legal norms enter into force" for them.[19] This question should be answered so as to ascertain whether the terms of the 1979 Moon Agreement be applied to Mars as one of the other celestial bodies and also to asteroids.

The principle of "the common heritage of mankind"[20] may also be reaffirmed along the lines of the Law of the Sea Convention[21] which was accepted by all states ratifying the Convention.[22] The essential point seems to establish the regime spelled out in para 7 of Article 11 of the Moon Agreement that special consideration should be given to the interest and needs of the developing countries, but also to the efforts of those countries which have contributed either directly or indirectly to the exploration of the Moon, i.e. to the efforts of space faring countries and other countries participating in he activities.

With regard to other space activities in the earlier years of the 21st century, legal provisions perhaps be required for the crew traveled inside space vehicles to space station, and from there to the Moon, as well as inside lunar habitats and for flights to Mars. These rules could be vary if crew members are national or international in an environment equipped with special life-support systems, which are unknown on the earth. Though the undertaken of occasional, short and long duration, multi-person visits to celestial bodies are still far away in the future, they will be realized in the course of the 21st century, with which the legal profession should keep in touch as a natural extension of the existing legal system based on the principles of the Outer Space Treaty.

In addition to treaties in response to anticipated events, there will be in the future, as in the past, accepted international practice which can produce a valued alternative to treaty law, Namely customary international space law.

Moreover, legal rules based on the Outer Space Treaty will have to be tailored to cope with immediate events. This means the attention of the legal profession should be directed to the perfection of regimes both for manned and unmanned flights, the management of space robots and space stations, etc. The priorities will be subject to the necessities of space undertakings.

The prospects are getting better. The diminishing east-west tension and the easing of North-South confrontation will enhance cooperation towards the noble goal of man's ventures into space.

References

[1] G.A. Res. 1962 (XVIII), 13, December, 1963.

[2] The four treaties are: Agreement on the rescue of Astronauts, the return of Astronauts and the return of objects launched into Outer Space of 1968; Convention on international Liability for damage Caused by Space Objects of 1972; Convention on Registration of objects Launched into Outer Space of 1976; Agreement Governing the Activities of states on the Moon and Other Celestial Bodies of 1979. The Four sets of principles are: Principles Governing the Use by States of Artificial Earth Satellites for International Direct Television Broadcasting of 1982; Principles Relating to Remote Sensing of the Earth from Space of 1986; Principles Relevant to the use of Nuclear Power Sources in Outer Space of 1992; Declaration on International Cooperation in the Exploration and Use of Outer Space for the Benefit and in the interest of All States, Taking into Particular Account the Needs of Developing Countries of 1996.

[3] Art.13 of the UN Charter.

[4] G.A. Res.1348 (XIII) of 13 December, 1958.

[5] G.A. Res. 1472 (XIV) of 12 December, 1959.

[6] Ibid.

[7] Cf. N. Jasentuliyana, The Lawmaking Process in the United Nations, in Space Law: Development and Scope, ED. by N. Jasentuliyana, Praeger, 1992, p.35.

[8] G.A. Res.1884 (XVIII), 17 October, 1963.

[9] Criticisms were mainly centered on the Section I of Article 4 which did not mention any weapon other than nuclear weapons of mass destruction, thus allowing the use of other weapons, such as ballistic missiles and rockets in earth orbits. Section 2 of Article 4 was also being criticized on the ground that although the Moon and other celestial bodies shall be used exclusively for "peaceful purposes", yet that term was subjected to varying interpretation. Thus the Outer Space Treaty was described as "an international agreement tailored to the needs and wishes of the US and USSR". See N. M. Matter, Space Policy and Programs Today and Tomorrow; The Vanishing Duopole, Toronto, 1980, p.41. D. Goedhuis also noted the views of some commentators who characterized the Treaty as ":essentially a bilateral agreement between the US and Soviet Union to which 80 states had dutifully accepted." 54 International Law Association Proceedings, the Hague, 1990, p.422

[10] Cf. Manfred Lachs, the Law of Outer Space--An Experience in Contemporary Law-Making, Leiden, 1972, p.141

[11] V. S. Vereshchetin and G. M. Danilenko, Custom as A source of International Law of Outer Space, Journal of Space Law, 1985, vol.13, No.1.

[12] See B. Cheng, United Nations Resolutions on Outer Space: "Instant" International Customary Law? 5 Indian Journal of International Law, 1965, P.36. However other publicists insist customary international law can not come into being "instantly", because custom is based on constant and uniform practice and calls for the passage of at least a certain period of time. supra (11) P.25.

[13] ICJ Report, 1969, p.43.

[14] Up to 1997, there have been 96 states ratified the Outer Space Treaty. See Annual Report, 1997, published by the Standing committee on the Status of International Agreements Relating to Activities in Outer Space of IISL.
[15] Year Book of international Law Commission, 1950, p.368.
[16] See, for example, the statement by the representative of Colombia in COPUOS, UN Doc. A/AC.105/PV.173, p.56 et seq.
[17] See, for example, the statement by the representatives of Czechoslovakia and Italy in the Legal Subcommittee, UN Doc. A//AC.105 C.2/SR.297, 1978, p.4 and p.8 respectively.
[18] Cf. Qizhi He, Space Law and Environment, in Space Law; Development and Space, Ed. by Jasentuliyana, Praeger, 1992, p.169.
[19] Article1 of the Moon Agreement.
[20] Para1, Article11 of the Moon Agreement.
[21] Article 136 et seq. the Law of the Sea Conventions; UN Doc. A/48/950.
[22] In accordance with Article 308, the Convention shall enter into force after 12 months of the deposit of the 60th ratification or accession.

THE DEVELOPMENT OF THE OUTER SPACE TREATIES AND LEGAL PRINCIPLES FROM A THIRD WORLD PERSPECTIVE

by

Nandasiri Jasentuliyana*

I. Introduction

Three days and forty years ago the Space Age was launched. In the years that began the second-half of this century, and even later, when the Americans went to the Moon, space was seen as something esoteric, something beyond the grasp of all but two or three of the most powerful nations of the world. However, space was also seen as yet another environment in which the power-brokers of the world could carry on their race for military superiority. It was also seen as an area of colonial conquest by the powers, for their own use and exploitation. The developing countries, especially, were concerned that the Moon and other celestial bodies would be exploited, to their disadvantage.

The international community therefore realized that it was essential to formulate international rules and regulations for the conduct of human activities in outer space because this new frontier was outside the bounds of existing international law. The responsibility to regulate this new environment fell upon the United Nations, which had been established to "maintain international peace and security" and charged with the task of "encouraging the progressive development of international law and its codification".[1] Thus, it became the focal point for international co-operation in outer space and for the development of international space law.[2]

In 1958, the United Nations General Assembly created the Committee on the Peaceful Uses of Outer Space (COPUOS), which was requested to study and report on, *inter alia*, "the nature of legal problems which [could] arise in the carrying out of programmes to explore outer space".[3] COPUOS was initially set up as an 18-Member State *ad hoc* body. Gradual increases over the years has brought the membership of COPUOS to 61 States. The Committee, in turn, created two Sub-committees, the Legal Subcommittee (LSC) and the Scientific & Technical Subcommittee (S&T), at its second session in 1962.[4]

The first significant step in the development of space law was made in 1963, when the General Assembly adopted the "Declaration of Legal Principles Governing the Activities of States in the Exploration and Use of Outer Space".[5] This important resolution formed the basis of what is sometimes called the "Magna Charta" of outer space law, the Outer Space Treaty of 1967. In the years that followed, COPUOS developed four other multilateral treaties to regulate human activities in outer space. These are the Rescue Agreement of 1968, the Liability Convention of 1972, the Registration Convention of 1976, and the Moon Agreement of 1979. Each of these instruments, with the exception of the Moon Agreement, has been signed by over 100 countries. In addition, the General Assembly has adopted four additional legal principles relating to outer space. These are the Direct Television Broadcasting, Remote Sensing, and Nuclear Power Sources Principles and the Declaration on Outer Space Benefits.[6]

In elaborating and negotiating provisions for all the treaties and principles, the countries of the Third World have played—from the outset—a leading role. Mr Krishna Rao of India, in one of the first UN legal meetings held on space, noted that "outer space was a new field and there were no vested interests to prevent the international community from embarking on a régime of co-operation rather than of conflict."[7]

It is this dual perspective of cooperation and non-conflict, for the benefit of all humanity, that has been at the forefront of thought of the developing countries, from the very beginning of the formation of space law by the United Nations.

Deputy to the Director-General, United Nations Office at Vienna; Director, Office for Outer Space Affairs; President, IISL. The views expressed here are the author's personal views and not that of any organization to which he belongs.

This presentation was made on 7 October 1997.

II. The Treaties and the Role of the Third World

Before going on to a specific discussion on the major contributions made by the Third World to the development of space law, I would like to briefly highlight some aspects of the role they played during the drafting of the space treaties.

In elaborating and negotiating provisions for the Outer Space Treaty, the developing countries played an effective role in its drafting. They especially fought for, with others, in having nuclear weapons banned from space and for the inclusion of such issues as liability of damages caused by space activities.

The demilitarization issue was one of the most controversial during the negotiations for the 1967 Outer Space Treaty. However, only partial demilitarization of space was achieved in the Treaty, and many countries, especially the developing countries were disappointed that the "Magna Charta" of space did not totally demilitarize outer space.[8] India pointed out that the omission of the words "outer space" from the second paragraph of Article IV "was likely to be interpreted to mean that outer space could legitimately be used for military manoeuvres and the like".[9] The expression "outer space, including the moon and other celestial bodies", employed "throughout the treaty, was not used in [the article] and the inference was that the treaty could be interpreted ... as giving legal licence for the use of outer space for non-peaceful purposes."[10]

Developing countries also played a major role in having a liability provision added to the Treaty. This was Article VII of the Treaty, which deals with the liability of launching States, for damage caused to other State Parties to the Treaty. India went further in making its interpretation clearer, when she stated that the term "internationally liable" would only be acceptable to her delegation, if it meant "absolutely liable".[11]

The Treaty, in addition, was one of the first multilateral instruments in which the generally applied principle of sovereignty of States gave way to the modern concept of the internationalization of the Global Commons. Its "province of mankind" principle, though controversial, became the watch-word for international relations, especially in the fields of space law and the law of the sea. The Treaty still provides a strong basis for the regulation of space activities, and its presumed "flaw" of vagueness is now turning out to be a major strength, as space activities are becoming more defined and certain. The Treaty's broad parameters encompass all space exploration and utilization, and they can serve as the foundation on which more detailed rules of space law, covering more specific fields of activity, can be built.

During the negotiations of the Rescue Agreement, many developing countries pressed for compensation for expenses incurred by a State Party that fulfilled its obligations under the Agreement in recovering and returning space objects of another Party, found in its territory.[12] Although the Rescue Agreement does not provide for dispute settlement, many States, especially developing countries, made an attempt to have such a provision included.[13] However, a provision to this effect was later included in the Liability Convention.

When the Legal Subcommittee took up consideration of the issue of liability for damage caused by space objects in 1964, the active participation by several developing countries, ensured that the absolute liability for damage caused by a space object of a launching state, was expressly included in the Convention.[14] Developing countries also successfully lobbied for the inclusion of a dispute settlement provision which was not included in the Rescue Agreement despite the strong efforts of many States.

With regard to the Registration Convention, its weakness lies in the fact that a launching State is only required, "as soon as practicable", to provide information regarding an object it launches into space and certain minimum details of the launched object.[15] Developing countries fought, unsuccessfully, to have Article IV strengthened, so that the launching State would be required to provide greater, and more specific information.[16]

III. The "Common Heritage of Mankind"

As space technology developed, especially in the 1960s, developing countries began to take a more positive and prominent role in the development of space law, so as to ensure that their interests would not be adversely affected. They made a particular impact on space law when, in the late 1960s, their lawyers proposed and developed, for the first time,[17] the concept of the Common Heritage of Mankind (CHM). In June 1967, the Ambassador of Argentina, Aldo Armando Cocca, introduced the concept of the Common Heritage of Mankind in the discussions being held in the COPUOS Legal Subcommittee.[18] Then, a few months later, the Maltese Ambassador to the United Nations, Arvid Pardo, applied the principle to the law of the sea when he stated that the seabed was the "Common Heritage of Mankind".[19] The concept was formalized in a treaty first in the 1979 Moon Agreement,[20] and subsequently in the 1982 Law of the Sea Convention.[21]

The developing countries, in applying the CHM principle to space, call on "... advanced States, which benefit from a de facto quasi-monopoly, to enrich human knowledge by undertaking space activities that are regarded as in the general interest, particularly scientific research ...".[22] In addition, the CHM, as applied to space law has a specific meaning in the "economic sphere, in that it protects the interests of all countries which do not have access to space resources due to a lack of technology; it implies an equitable sharing of what are considered common resources."[23]

However, the concept of the CHM, as applied to the Moon, has been given a very specific meaning in Article 11 of the Moon Agreement in which the article tries to balance both the needs of developing countries, and the interests of those developed countries that contribute to the exploration of the moon.

As Earth-oriented application satellites were developed and became operational in the early 1970s, remote sensing and direct broadcast satellites emerged. These technologies offered new tools for national resource management and national communications, but also threatened to expose countries to foreign eyes and foreign influence in a way they had never been before. Outer space as an open international realm began to pose problems in addition to offering new opportunities. The presence of remote sensing and telecommunications satellites only a few hundred kilometres overhead reinforced the general sense of vulnerability of Third World countries to economic and cultural exploitation, and provoked a strong demand for controls on such space technology.[24]

It is in areas such as these, i.e., direct broadcasting, remote sensing, international cooperation in outer space and satellite communications, that the Third World has created the greatest impact in the law-making process for outer space.

IV. Principles on Direct Broadcasting by Satellite (DBS)

Work on the study of the technical feasibility of communications by DBS began in 1968.[25] The issue was controversial. The opposing views of member States concerning the free flow of information and State sovereignty appeared to be irreconcilable. The divergent views in COPUOS, on particular issues like prior consent, programme content and signal overspill, reflected some of the contested issues in the North-South political debate occurring at that time.

Negotiations continued in COPUOS. By 1982, full agreement on the set of principles had not yet been reached, but it seemed that a basis for consensus would emerge, at least by the next round of negotiations, based on a Canadian-Swedish compromise proposal.[26] More progress was expected to be made in the negotiations. However, pre-empting further discussions in COPUOS, delegations from some developing countries, strongly supported by the former Soviet Union and its allies, who felt that the United States and some western countries were unreasonably dragging their feet, took the set of principles, along with the Canadian-Swedish proposal, for a vote in the General Assembly, where they were adopted by a majority vote. Hence, for the first time in the history of space law-making a set of legal principles was not adopted in COPUOS by consensus and instead adopted by the General Assembly with a majority vote.[27]

V. Principles on Remote Sensing

With regard to the principles of remote sensing, the matter was first formally raised by Argentina when, in 1970, it submitted to the Legal Subcommittee a draft international agreement on activities carried out through remote sensing satellite surveys of earth resources.[28] Of the developing countries, Argentina, Brazil and India played a prominent role in the development of the Principles which were adopted In 1986.

The elements relating to the needs of developing countries, technical assistance to interested States to enable them to interpret remote sensing data themselves, and access to data by sensed States were included in the Remote Sensing Principles because of the insistence of developing countries, as these elements were particularly important in those countries' plans for national development.

Similar to the DBS debate, the member States were divided on fundamental issues of international law. The developing countries were concerned about national sovereignty protection, prior consent and control by the sensed State over the distribution of data retrieved over their territories.[29] On the other hand, the developed countries, particularly the United States, championed freedom of use and non-discrimination of dissemination of space-derived information.[30] The debate on remote sensing, however, took a more positive turn than the DBS principles and the Legal Subcommittee was able to reach a consensus on the draft principles. There were several reasons for such approval. One of the more important reasons was that over the years of negotiation, there were changes both in the perceptions of the impact of remote sensing and in the general atmosphere of international relations.

VI. Outer Space Benefits[31]

Another item largely initiated and developed by some of the developing countries members of COPUOS, was "Consideration of the legal aspects related to the application of the principle that the exploration and utilization of outer space should be carried out for the benefit and in the interests of all States, taking into particular account the needs of developing countries".

The item on "*Outer Space Benefits*",[32] as it was familiarly called, was placed on the agenda of the 28th session of the Legal Subcommittee, in 1989.[33] This item was placed on its agenda because many countries felt that there was a lack of any legal effort to ensure that space exploration and application of space technology benefited all countries. The item, therefore, aimed at the development of a legal regime that would embody and promote the principle of Article I of the Outer Space Treaty, that the exploration and utilization of outer space should be carried out for the benefit and in the interests of all States, taking into particular account the developing countries. Thus, Member States looked at moving beyond Article I of the Outer Space Treaty, possibly to codify the rights and responsibilities of countries with respect to international cooperative space activities, and not let the Article stand, as some States viewed it, merely as an artifact or a moral appeal to the space-faring nations.

For several years, initially, a set of principles co-sponsored by several developing countries had formed the basis for discussions on this agenda item.[34] They were aimed at meeting the concrete needs and expectations of all countries, particularly those of developing countries. The central thrust of the draft Principles was that of the means of access by all countries - especially the developing countries - to the benefits of space technology, and this, ultimately, was a question of the nature of international co-operation among States. In a more general sense, the co-sponsors were saying that they believed that the technological gap between the developed and the developing countries had become too vast and that they wished to reverse the trend, at least to some degree, by the application of these Principles. They also seemed to say that they had lost their confidence in the moral appeals as embodied in the spirit of Article I of the Outer Space Treaty, and that the remedy laid in the establishment of an international legal framework regulating space co-operation and requiring the developed countries to co-operate within specified limits.

Having first refused to discuss the issue at the earlier sessions, at the 1994 session of the Legal Subcommittee, the developed countries indicated their willingness to discuss the matter, though somewhat reluctantly. On the basis of discussions, and further consideration of the matter during the following year, France and Germany submitted a working paper at the 1995 session of the Subcommittee.[35] The paper rested on two basic considerations: (i) States were free to determine all aspects of their cooperation, whether it was bilateral or multilateral or whether it was commercial or non-commercial, including, of course, development cooperation; (ii) States should have the choice of selecting the most efficient and appropriate mode of cooperation in order to allocate resources efficiently.[36]

The paper was divided into three short parts, the first which laid out general elements of international cooperation in the peaceful uses of space, the second which described the modes of such cooperation, and the third which listed possible areas in which this cooperation could be carried out. At the 1996 session of the Legal Subcommittee, the developing countries and France and Germany submitted revised versions of their working papers.[37] The Franco-German paper more-or-less reflected the previous version. However, the paper presented by the developing countries was different, in that it substantially resembled the French and German document. This, perhaps, was the catalyst that provided a breakthrough in the debate on the matter. It indicated a willingness of the developing countries to allay the concerns of the developed countries and strike a compromise to resolve the issue. After a paragraph-by-paragraph discussion of both papers, and intensive informal discussions by the sponsors of the papers, they succeeded in agreeing to a Chairman's consolidated text, with only a few disputed elements that were placed in square brackets. The text was submitted as an informal working paper of the Chairman of the Working Group, and annexed to the report,[38] with the hope that it could be adopted at the next session of the Legal Subcommittee, or at the June 1996 session of COPUOS.

During the course of the Committee session in June 1996, the Chairman of the Working Group on the item conducted informal consultations with members of the Committee which took up the those paragraphs of the text on which the Legal Subcommittee's Working Group could not agree. After exhaustive discussion and negotiation on the remaining paragraphs, the Committee agreed on a new text appropriately amended to reflect the agreement reached.[39] The Declaration was subsequently adopted by the General Assembly, at its fifty-first session, in G.A. Resolution 51/122.

VII. The Geostationary Orbit and the Regulation of Satellite Communications[40]

Another important issue for developing countries which continues to be discussed in the Legal Subcommittee, has been access to the geostationary orbit (GSO) for their communications satellites. The international regulatory regime for satellite communications has long been working on a "first come, first serve" basis which has primarily benefited the industrialized countries. Despite the general legal principle of equal access,[41] a country wishing to put a communications satellite into the geostationary orbit has had to ensure that it did not interfere with any system previously registered with the International Telecommunication Union,[42] essentially placing a burden on the proposed new system. Since the technologically advanced countries were the first ones to set up communication satellite systems, the developing countries felt that the current registration procedures inequitably restricted their access to the geostationary orbit.

The developing countries' concerns over accessibility are based on the natural limitations of the Geostationary Orbit (GSO) and frequency spectrum. Within the geostationary orbit, satellites rely on the radio frequency spectrum for radio communications to transmit or relay information. Thus a satellite in orbit is easily affected by radio frequency interference from other satellites that use the same operating frequency. To avoid interference, limitations need to be imposed on the distance between satellites using a particular frequency. Due to the fact that the frequency spectrum can only be used by a certain number of satellites to avoid problems of congestion and interference between satellite communication systems, access to the radio-frequency spectrum can present a major constraint on the use of the GSO by late-comers. In addition, the technology required to ensure the avoidance of these problems can be very costly.

The existing regulatory regime gave priority to existing satellite systems, and the non-space faring nations, mostly developing countries, saw it as limiting their access to the GSO and the frequency spectrum. Although the developing countries did not have any financial and technical resources to utilize satellite technologies, they wanted to ensure that they would not be precluded from access in the future.

One approach that several developing countries decided to take was that they declared their sovereignty over portions of the GSO. In 1975, Colombia announced a claim of sovereignty over a portion of the geostationary orbit above its territory. In 1976, a meeting of the equatorial States was convened, and representatives from Brazil, Colombia, Congo, Ecuador, Indonesia, Kenya, Uganda and Zaire signed the Bogotá Declaration,[43] proclaiming their sovereignty over their respective portions of the orbit. Needless to say, these claims were rejected by the developed countries and did not receive much support from other developing countries. This issue was first raised in the United Nations General Assembly and was subsequently introduced in COPUOS where it still continues to be discussed in the Legal Subcommittee in the context of the question of the definition and delimitation of outer space.[44] Lack of support from other countries has led the equatorial countries to moderate their views on this matter, and most equatorial countries have moved away from the position stated in the Declaration.

Two opposite view-points are also prevalent on this issue. The first is put forward by developing countries or countries which have not had the financial or technical resources to place a communications satellite in orbit. They say that the geostationary orbit requires a special legal regime to regulate access to and utilization of the orbit. As noted earlier, some countries, particularly those whose national territories lie along the equator, have claimed that they should have special access or a reserved right to this orbit. The opposing view is that the legal regulations of the geostationary orbit are inseparably linked with the ITU which has the sole competence for the coordination and regulation of the radio frequency spectrum. In addition, it has been argued that any type of reservation or an *a priori* claim to the geostationary orbit would amount to an appropriation of outer space which is prohibited under Article II of the Outer Space Treaty.[45]

At the last session of COPUOS and its Legal Subcommittee in 1997, in an attempt to break the deadlock in the discussions, Germany submitted a working paper entitled "Draft resolution: request to the International Telecommunication Union: ensuring equitable access to the geostationary satellite orbit". However, on the basis of exhaustive discussions held at both the Legal Subcommittee and the Committee, the working paper was withdrawn as it became evident that the differing viewpoints remained which did not allow for a consensus compromise to be reached.

The primary - and most successful - thrust of Third World action has been at the International Telecommunication Union (ITU), the specialized United Nations agency that coordinates orbital slots for communication satellites and registers radio frequency assignments to avoid harmful interference between radio signals.[46] Although the ITU does not have authority to enforce coordination of orbit positions or radio signals, it does provide the framework for establishing the procedures for coordination agreed upon at specialized conferences known as World Administrative Radio Conferences (WARCs). It is at the WARCs in particular that Third World countries have sought to change regulation procedures so that they would be guaranteed access to GSO.[47] One major milestone in the effort made by developing countries to bring positive change to ITU regulations was, for example, achieved in the adoption of Article 33 of the 1982 ITU Nairobi Convention.[48] The Article laid down the principle that "radio frequencies and the geostationary satellite orbit are limited natural resources and that they must be used efficiently and economically ... so that countries or groups of countries may have equitable access to both, taking into account the special needs of developing countries and the geographical situation of particular countries".[49]

Another major milestone of the work of the developing countries during the WARCs was the addition to the international regulatory regime for radio communications that provided for an allotment plan for predetermined arcs and bands. At the 1988 WARC where this plan was adopted, the countries present reviewed and revised existing resolutions and recommendations associated with international radio regulations and adopted new Resolutions and Recommendations in the Final Act.[50] These Acts, which entered into force on 16 March 1990, marked an important step for the developing world in regulating the use of the frequency spectrum and the geostationary orbit.

VIII. Conclusions

The United Nations, from the very beginning, has been the primary law-making instrument in the field of outer space, to ensure that the wealth of space is shared by all nations, and not just a few. Space is no longer dominated by a mere handful of countries. Over 150 nations, most of these from the Third World, now directly or indirectly use space technology for national development.

The initial thrust and emphasis of space-law making by the developing countries was two-fold: to ensure that space was not used for military purposes and that the space powers did not extend their colonial ambitions to the new environment.

As space applications began to develop, the Third World countries became concerned that this new technology, over which they had very little control, could be used to dominate them socially and economically. For instance, they felt that direct broadcasting satellites operated by the space powers would amount to cultural imperialism, and remote sensing satellites would result in exploitation of their natural resources. Therefore, their space law-making emphasis grew, to limit the use of this technology.

However, increasing cooperation between the space powers and developing countries have long laid rest to these fears, and the Third World is becoming involved in all aspects of space exploration and use, and increasingly so in the development of space law.

From the view-point of the developing countries, the more recent underlying drive behind their law-making efforts, whether relating to the geostationary orbit or international space cooperation, has been to reduce the ever-growing technological gap between them and the developed countries. By reversing the trend, at least to some degree, they could share in the benefits of space activities.

For those countries that have not implemented their own space activities, the possibility of preclusion from launching satellites and from realizing the benefits of the technology represent a major handicap in their future national development. The countries of the Third World, by actively participating in the law-making process of space law, have sought to mitigate these disadvantages as far as possible, thereby ensuring that their current, and future, interests in space are safeguarded.

Looking to the future, it is certain that the developing countries will continue to place emphasis in space law-making on issues such as sharing of space benefits. However, those countries, coming from the Third World, that now have space capabilities of their own, for instance as satellite owners or with launch capability, will develop a different set of priorities and have different points-of-view, in areas, for example, such as space debris. This could possibly lead to a technological gap within the Third World countries themselves, which will have its own emphasis in the motivation of developing countries in space law-making. An additional point to be considered is that an increase in the complexity of space technology has led to a parallel increase in the complexity of space law-making, for example, in fora such as the ITU. There is a possibility that many developing countries, because of a paucity of technical experts, have been unable to put forward their points-of-view and contribute in the law-making process in these highly specialized fields. Hopefully, this will not happen, and the countries of the Third World are as active in the development of space law in the 21st century, as they were in this century.

ENDNOTES

1. *Charter of the United Nations*, Done at San Francisco on 26 June 1945, entered into force on 24 October 1945, 24 U.S.T. 2225, T.I.A.S. No. 7739, at Art. 13(1)(a).

2. The role of the United Nations in the progressive development of space law began with its resolution 1721 (XVI)A of 20 December 1961, in which the General Assembly stated, for the guidance of States in their exploration and use of outer space, that international law, including the Charter of the United Nations, applied to outer space, and that it was free for exploration and use by all States and not subject to national appropriation.

3. *See*, Resolution 1348 (XIII) of 13 December 1958. The General Assembly, "[c]onsidering that an important contribution (could) be made by the establishment within the framework of the United Nations of an appropriate international body for co-operation in the study of outer space for peaceful purposes", set up the Committee, requesting it to report on, *inter alia*, the following: The activities and resources of the United Nations, of its specialized agencies and of other international bodies relating to the peaceful uses of outer space; ... and, the future organizational arrangements to facilitate international

co-operation in this field within the framework of the United Nations.

4. *See*, the Statement by the Chairman of the Committee, made on 29 March 1962: Doc. A/5109 of 30 March 1962 at para. 4 and Doc. A/5181 of 27 September 1962 at Annex I [hereinafter *Doc. A/5109* and *Doc. A/5181*, resp.]. The S&T held its first session in May/June 1962: *see*, Doc. A/AC.105/5 of 3 July 1962, "Report of the Scientific and Technical Sub-Committee on the Work of its First Session (28 May-13 June 1962)" [hereinafter *S&T 1st*]. The LSC also held its first session in this period: *see*, Doc. A/AC.105/6 of 9 July 1962, "Report of the Legal Sub-Committee on the Work of its First Session (28 May-20 June 1962)" [hereinafter *LSC 1st*].

5. G.A. Res. 1962 (XVIII), adopted on 13 December 1963. The draft proposals for this text submitted to COPUOS by several States are reproduced in the *Official Records of the General Assembly*, 17th Session (1962), Agenda Item 27: *see* Doc. A/5181, Annex III A and E, Doc. A/C.1/879 and Doc. A/C.1/881, for the USSR, UAR, UK and US proposals, respectively. *See also*, LSC 1st, *id.* at para. 11 (1) for the Soviet proposal, Doc. A/AC.105/12 of 6 May 1963, "Report of the Legal Sub-Committee on the Work of its Second Session (16 April - 3 May 1963) to the Committee on the Peaceful Uses of Outer Space" at Annex E, F and G for the UAR, UK and US proposals, resp. [hereinafter *LSC 2nd*]. A discussion of these drafts continued for these two sessions of the LSC, and the General Assembly adopted the 9-paragraph Declaration at its 18th session later in the year.

6. These Principles, in addition to the "Declaration" Principle of 1963 and the five outer space treaties can be found in Doc. A/AC.105/572/Rev.2, *United Nations Treaties and Principles on Outer Space*.

7. UN Doc. A/AC.105/C.2/SR.29-37 at 78.

8. For the contribution that developing countries made to the debate on this question, see, in particular, UN Doc. A/AC.105/L.6 of 14 September 1962, the Working Paper, "Draft Code of International Cooperation in the Peaceful Uses of Outer Space", submitted by the UAR (as Egypt was then called). Also see the statements made by the delegations of Argentina (Doc. A/AC.105/C.2/SR.60), the UAR (Doc. A/AC.105/C.2/SR.62), India, Brazil and Iran (Doc. A/AC.105/C.2/SR.71 and Add. 1) and Tanzania and Ceylon (as Sri Lanka was then called) (Doc. A/PV.1499).

9. UN Doc. A/C.1/PV.1493.

10. Statement of Mr Krishna Rao, A/AC.105/C.2/SR.71 and Add. 1.

11. See, Doc. A/AC.105/C.2/SR. 71 and Add. 1, Statement of Mr Krishna Rao, and A/C.1/PV.1493, statement of Mr Parthasarthi.

12. *See, e.g.*, Argentina's draft proposal, WG.I/24/Rev.1, reproduced in N. Jasentuliyana and R.S.K. Lee (eds.), *Manual on Space Law: Vol. III* (Dobbs Ferry, NY: Oceana Publications, 1981) at 125 [hereinafter *Manual III*]. Article 5 (5) of the Agreement provides that, "Expenses incurred in fulfilling obligations to recover and return a space object or its component parts ... shall be borne by the launching authority.

13. For example, the delegate of Mexico, Mr. Francoz Rigalt, proposed that an arbitration commission should be established to determine, in the event of controversy, whether a launching of a spacecraft, whose crew was to be returned, had been in accordance with the "Declaration" on legal principles on outer space (G.A. Res. 1962 (XVIII) of 13 December 1963).

14. And so it does. *See,* Liability Convention, at Article II which states: "A launching State shall be absolutely liable to pay compensation for damage caused by its space object on the surface of the earth or to aircraft in flight."

15. Article IV.

16. *See for example*, the statements made by Ambassador Cocca in the Legal Subcommittee, Doc. A/AC.105/C.2/SR.201 and SR.203, reproduced in Manual III, *supra* note 12 at 675 and 680.

17. *See,* statement of Ambassador Cocca in Doc. A/AC.105/C.2/SR.250, in N. Jasentuliyana and R.S.K. Lee (eds.) *Manual on Space Law, Vol. IV* (Dobbs Ferry, NY: Oceana Publishers, 1981) at 151 [hereinafter *Manual IV*], where he recalls that the concept, "which had gained ground in other spheres of international law, had been expounded and used for the first time by the Legal Subcommittee".

18. *See*, Doc. A/AC.105/C.2/SR.75 of 13 November 1967.

19. *See,* Doc. A/C.1/PV.1515 of 1 November 1967 and Doc. A/C.1/PV.1516 of 1 November 1967.

20. Article 11(1), which states: "The moon and its natural resources are the common heritage of mankind ...". *But also see, infra*, the Bogotá Declaration of 1976, which was the first official outer space document to apply the CHM Principle.

21. *United Nations Convention on the Law of the Sea*, Opened for signature at Montego Bay on 10 December 1982, entered into force 16 November 1994, (1982) 21 ILM 1261; Cmnd. 8941 [hereinafter *UNCLOS*]. Article 136 states: "The Area and its resources are the common heritage of mankind."

22. *Id.* at 459.

23. *Id.*

24. *See*, I.A. Vlasic, "Remote Sensing of the Earth by Satellites" in N. Jasentuliyana and R.S.K. Lee, *Manual on Space Law, Vol. I* (Dobbs Ferry, NY: Oceana Publications, 1981) at 311 [hereinafter *Manual I*]. *See also*, M. Benkö *et al*, *Space Law in the United Nations* (The Netherlands: Martinus Nijhoff, 1985) 18-29.

25. G.A. Res. 2453 (XXIII) of 20 December 1968 approved the creation, by COPUOS, of the Working Group on Direct Broadcast Satellites.

26. "Elaboration of Draft Principles Governing the Uses by States of Artificial Earth Satellites for Direct Television Broadcasting", A/AC.105/C.2/L.117 of 15 February 1979. Their approach took account of existing technical regulations and balanced the concept of free flow of information with the principle of sovereignty of States; they had tried to allow for potential beneficial contributions, while giving appropriate consideration to the concern about undue unilateralism or abuse; one of the corner-stones of their draft was the principle that international cooperation should form the basis for all direct television broadcasting by satellites; if such a principle were strictly observed, there would remain little cause for international concern about undesirable broadcasting; in addition, true cooperation implied beneficial effects for all States interested in an orderly development of the technology: *see*, the statement by the Swedish delegation, Doc. A/AC.105/127, 133 and 134; A/AC.105/C.2/L.102, reproduced in Manual IV, *supra* note 17, at 279-80.

27. All treaties and principles are adopted by COPUOS by consensus and then recommended to the General Assembly, which normally adopts them, also by consensus. Since all regions of the world are represented in COPUOS, its recommendations are generally accepted by the General Assembly. For a detailed examination of the consensus principle, *see*, N. Jasentuliyana, "A Survey of Space Law as Developed by the United Nations" in N. Jasentuliyana (ed.), *Perspectives on International Law* (The Hague: Kluwer Law International, 1995).

28. *See*, Doc. A/AC.105/C.2/L.73 of 26 June 1970. The paper can be found in Doc. A/AC.105/85 and 133. On 1 February 1974, Brazil submitted a draft of its own (Doc. A/AC.105/122). On 15 October 1974, these two papers were replaced by a joint Argentina-Brazil working paper submitted to the General Assembly: Doc. A/C.1/1047.

29. *See, e.g.*, the statements by the delegates of Brazil and Argentina in the Legal Subcommittee: A/AC.105/C.2/SR.220, and by Mongolia: A/AC.105/C.2/SR.263, reproduced in Manual IV, *supra* note 17 at 402, 407 and 444, respectively. For example, Mr. de Seixas Correa of Brazil stated that international remote sensing implied, in essence, the effective transfer of information on such matters as natural resources between two or more points of the earth's surface. The actual transfer of such information necessarily involved elements relating to the political, military or economic security of the States concerned. In those circumstances, the permanent sovereignty that States exercised over the natural resources of their territory clearly embodied the right to control not only access to information relating to such resources, including data gained through the process of remote sensing, but also the dissemination of the information thus obtained. *Id.* at 403.

30. *See for example*, the American statement in the Legal Subcommittee, Doc. A/AC.105/C.2/SR.222, reproduced in Manual IV, *supra* note 17 at 409. Mr. Stowe of the United States observed that some States had expressed concern regarding remote sensing, and that they feared that the communication of data about their natural resources to third States might jeopardize their economic interests or security. It had therefore been suggested that the international community should establish a system under which the communication of data concerning a particular State would be subject to the prior consent of that State. In the opinion of his delegation, such a system would have many disadvantages; it would in no way protect the interests of States which did not have their own remote sensing system, and it would impede the development of international cooperation. If States which conducted remote sensing activities by satellites were not authorized to share freely the data obtained, they would in the end be the only States to derive genuine benefit from the considerable advantages of remote sensing. In many cases, full benefit could be derived from data

only if they were studied on a regional or even global basis. He also noted that it had been said that, in international law, a State was entitled to exercise permanent control over the dissemination and use of any information concerning the natural resources of that State, no matter where that information was gathered, disseminated or studied. In the opinion of his delegation, that had never been the law in the past and could not become the law then.

31. *See generally*, N. Jasentuliyana, "Article I of the Outer Space Treaty Revisited" (1989) 17 *J. Space L.* 129.

32. The full title of this agenda item was: "Consideration of the legal aspects related to the application of the principle that the exploration and utilization of outer space should be carried out for the benefit and in the interests of all States, taking into particular account the needs of developing countries". For a detailed historical and drafting background of the Outer Space Benefits Principles, *see generally*, *id.*, and N. Jasentuliyana, "Ensuring Equal Access to the Benefits of Space Technologies for all Countries" (1994) 10 *Space Policy* 7 (1994).

33. The Legal Subcommittee, at its 26th and 27th sessions in 1987 and 1988, resp., had considered and finalized the choice of this new item. *See*, U.N. Docs. A/AC.105/385, 411 and 430, being the reports of the 26th to 28th sessions of the Legal Subcommittee.

34. *See*, Report of the 34th session of the Legal Subcommittee, U.N. Doc. A/AC.105/607 of 19 April 1995, for the Working Paper jointly co-sponsored by Brazil, Chile, Colombia, Egypt, Iraq, Mexico, Nigeria, Pakistan, Philippines, Uruguay and Venezuela, "Principles Regarding International Cooperation in the Exploration and Utilization of Outer Space for Peaceful Purposes" (U.N. Doc. A/AC.105/C.2/L.182/Rev.2 of 23 March 1995) [hereinafter *LSC 34th*]. Cuba joined as a co-sponsor of the Paper at the 34th session. In the Working Paper, Principle I, after partly re-stating Article I, paragraph 1, of the Outer Space Treaty, urges all States with space capabilities to promote cooperation with countries with less developed space capabilities. Principle II addresses the issue of ensuring equal access to the applications of space technology. Principle III primarily addresses the question of promoting indigenous capabilities in space science and technology application in developing countries through international cooperative mechanisms and Principle IV refers to the conditions under which such cooperation should be established and implemented. The need to utilize space technology and applications as a vehicle to protect and preserve the Earth and space environments is covered by Principle V. The next Principle deals with the role of the United Nations and its Programme on Space Applications in international space cooperation.

This Working Paper had its genesis in a Group of 77 paper (U.N. Doc. A/AC.105/C.2/L.162 of 1 April 1987) submitted at the 26th session of the Legal Subcommittee: *see*, report of the 26th session, *id.* The first version of Working Paper L.182 was submitted at the 31st session of the Legal Subcommittee: *see*, report of the 31st session, U.N. Doc. A/AC.105/514 at 50. The first revision, L.182/Rev. 1, was submitted in 1993, at the 32nd session (U.N. Doc. A/AC.105/544 at 32). Rev. 2, above, was submitted at the 34th session.

35. The paper, entitled, "Consideration of the legal aspects related to the application of the principle that the exploration and utilization of outer space should be carried out for the benefit and in the interests of all States, taking into particular account the needs of developing countries" (U.N. Doc. A/AC.105/C.2/L.197 of 27 March 1995) is reproduced in LSC 34th, *id.*

36. The statement by the German delegation at the 34th session of the Legal Subcommittee.

37. A/AC.105/C.2/L.182/Rev.3 and A/AC.105/C.2/L.197/Rev.1, respectively. *See*, the Report of the Legal Subcommittee on the Work of its Thirty-Fifth Session (18-28 March 1996), U.N. Doc. A/AC.105/639 of 11 April 1996, at Annex III, Part B and Part C [hereinafter *LSC 35th*].

38. *See* "Declaration on International Cooperation in the Exploration and Use of Outer Space for the Benefit and in the Interests of all States, Taking into Particular Account the Needs of Developing Countries", U.N. Doc. A/AC.105/C.2/L.202 of 27 March 1996, in LSC 35th, *id.* at 36.

39. The text can be found in U.N. Doc. A/AC.105/L.211 of 11 June 1996.

40. *See for example*, "The Physical Nature and Technical Attributes of the Geostationary Orbit" (Doc. A/AC.105/203 of 29 August 1977 and the addenda thereto; and Doc A/AC.105/404 of 13 January 1988); R. Jakhu, "The Legal Status of the Geostationary Orbit" (1982) VII *Annals Air & Space L.*333; P. Abdurrasyid, "The Outer Space Treaty and the Geostationary Orbit" (1987) XII *Annals Air & Space L.* 131.; M.A. Ferrer, "The Use of the Geostationary Orbit" (1977) 20 *Coll. L. on Outer Space* 216; and J.F. Galloway, "Telecommunications, National Sovereignty and the Geostationary Orbit" (1977) 20 *Coll. L. on Outer*

Space 226.

41. The second para. of Art. 1 of the Outer Space Treaty states that, "Outer space, including the moon and other celestial bodies, shall be free for exploration and use by all States without discrimination of any kind, on a basis of equality and in accordance with international law, and there shall be free access to all areas of celestial bodies."

42. Art. 45, para. 1, of the Constitution and Convention of the International Telecommunication Union, Geneva, 1992, reads as follows: "All stations, whatever their purpose, must be established and operated in such a manner as not to cause harmful interference to the radio services or communications of other Members or of recognized operating agencies, or of other duly authorized operating agencies which carry on a radio service, and which operate in accordance with the provisions of the Radio Regulations." The 1992 Geneva Constitution replaced and superseded all previous Conventions of the ITU.

43. *Declaration of the First Meeting of Equatorial Countries*, signed in Bogotá, 3 December 1976. Reproduced in N. Jasentuliyana & R.S.K. Lee (eds.), *Manual on Space Law, Vol. II* (Dobbs Ferry, NY: Oceana Publications, 1979) at 383 *et seq.*, and (1978) 6 *J. of Space L.* 193 *et seq.* In the Declaration, the equatorial countries, stating that the geostationary orbit was not part of outer space, proclaimed their sovereignty over those segments of the orbit that were above their territories.

44. In this regard, the Bogotá Declaration states that, "[t]here is no definition of outer space that is valid and satisfactory for the international community such as might be cited to support the argument that the geostationary orbit is included in outer space." *Id.* at section 4.

45. See, the Outer Space Treaty at Art. II, which states that, "Outer space, including the moon and other celestial bodies, is not subject to national appropriation by claim of sovereignty, by means of use or occupation, or by any other means."

46. See, for example, R. Jakhu, "The Evolution of the ITU's Regulatory Regime Governing Space Radiocommunications Services and the Geostationary-Satellite Orbit" (1983) VIII *Annals Air & Space L.* 381; F. Lyall, "Law of Satellite Communications", in N. Jasentuliyana (ed.), *Space Law: Development and Scope* 114 (Westport, Conn.: Praeger Publishers, 1992); D.M. Leive, *International Telecommunications and International Law: The Regulation of the Radio Spectrum* (Leiden: A.W. Sijthoff, 1970); and, J.G. Savage, *The Politics of International Telecommunications Regulation* (San Francisco and London: Westview Press, 1989).

47. For greater details, see, N. Jasentuliyana, "The International Regulatory Regime for Satellite Communications: The Meaning for Developing Countries" (1994) 2 *Asian Yearbook of Int'l L.* 49.

48. International Telecommunication Convention, Final Protocol, Nairobi, 1982.

49. *Id.* at Article 33 (2), now Article 44 (2) of the Constitution and Convention of the International Telecommunication Union, Geneva, 1992, which supersedes and replaces all previous Constitutions of the ITU.

50. Final Act, adopted by the Second Session of the World Administrative Radio Conference on the Use of the Geostationary Satellite Orbit and the Planning of the Space Services Utilizing It (ORB-88), Geneva 1988.

SOLIDARITY AND HUMANISM IN THE OUTER SPACE TREATY

Aldo Armando Cocca *
The Council of Advanced International Studies
(Argentina)

Background

The creation of the International Astronautical Federation (IAF) in Paris, 1950, offered a new field for the scientific research and technical applications of astronautics. Law was present in the reflections and deliberations.

As Vladimir Kopal rightly said, the congresses of the International Astronautical Federation began including reports on legal issues relating to space flights as early as 1952, Congress of Stuttgart (Prof. A.. Meyer). Reports on legal topics were also presented at the 1954 Congress in Innsbruck (Dr. A.A. Cocca) and at the 1956 Congress in Rome (Dr. A.A. Cocca and Dr. E. Pépin).[1]

In the 50's international academical organizations offered a valuable contributions to the development of the Space Law doctrine: Institut de Droit International, International Law Association, The Hague Academy of International Law.

In addition there are also regional and national institutions devoted to international law, including topics of space law in its programs and meetings; there are as well, centers on Air Law that broaden their researches to Space Law.

Within the framework of the IAF, a Permanent Legal Committee to study the legal problems of space was created on the occasion of the First Colloquium on the Law of Outer Space (The Hague, 1958). At the Second Colloquium (London, 1959) the Permanent Committee became into the International Institute of Space Law (IISL) which obtained its Statute in Stockholm, 1960.

The real enforcement of the doctrine in Space Law is due to the annual Symposia of the IISL, that generates the more complete reference, bibliography, and discussions on Space Law matters. The Institute represents the international academical forum for jurists who, whichever their place of residence or the legal régime to which they belong to, have a common feeling: to create new concepts, new principles and new legal rules.

The Outer Space Treaty on the line of the UN Charter

The principles enshrined in the United Nations Charter needed to be deeply examined and explained in all their scope. The Charter is explicit in establishing that the peoples are the addresses of the benefit previewed by the international organization.

The UN Charter gives, as non international instrument did before, a specific legal value to the expressions: *peoples, generation, mankind, fundamental human rights, dignity and worth of the human person, social progress and better standards of life in*

---

* Honorary Director IISL, Trustee of IAA .

larger freedom.

The progressive development and codification of international law is a mandate received by all nations half a century ago in the opportunity of the enforcement of the UN Charter. We take from the Charter the following concepts:

Peoples:
The first line of the Preamble of the Charter recognizes a new dimension to the international subjectivity of peoples. The **peoples** of the United Nations express themselves in this text, and they do so speaking in first person.

Generation:
A new concept in international law linked with the fundamental human rights which legal content has been analyzed considering necessary its insertion together with the concepts of people and mankind. This concept embodies all human beings without any distinction and for this reason it is a basis for the concept of solidarity. Generation implies all the persons born at about the same time or living in the same period of time.

Mankind:
Without any doubt is in the Space Treaty and in other international instruments derived thereof that it is offered to the juridical science a legal content of the human race, individuals taken collectively, all human beings.

Fundamental human rights:
It is well known that the history of Law is the history of man fighting for his individual rights. These rights were contemplated by the *Ius Gentium* and the *Ius Inter Gentium.* The Universal Declaration on Human Rights, adopted by the UN General Assembly in 1948 represents a main step for recognition and expansion of human rights around the world. But it is in the Outer Space Treaty and in its following international instruments that this aperture consolidates and aims its present dimension. The expression *mankind* as a concept entitling human rights is enshrined in the Space Treaty. In accordance with its Preamble the Parties got inspired by the great prospects that the man's entry into outer space meant for mankind. There is also a recognition of the common interest of *all mankind* in the progress of the exploration and use of outer space for peaceful purposes. The Parties also explain their belief on the benefit for all *peoples,* and remark that this is irrespective the degree of their economic and scientific development.

Article I of the Treaty establishes a new concept: *the province of all mankind* (l'apanage de l'humanité toute entière; incumben a toda la humanidad). This concept implies a recognition of something that always existed but needed of a legal text.

Article V explains the concept of astronaut (cosmonaut) as *envoy of mankind.*
Dignity and worth of the human person:

This principle is enlarged in the Space Treaty from the individual to mankind.

Social progress and better standards of life in larger freedom
Many provisions of the Space Treaty are inspirated in this concept, as it happens with the enshrinement of the principle that all activities must achieve the benefit for all mankind and that space must be used and explored with peaceful purposes, among other.

The step given by the Space Treaty must be considered englobed within the progressive development of international law.

It is opportune to remark that the observance of the Treaty is entrusted to the Secretary General of the United Nations, who has specific functions at this regard.

Codification of Space Law

The last international instrument elaborated by COPUOS was the Agreement Governing the Activities of States on the Moon and other Celestial Bodies, adopted on 5 December 1979, opened for signature on 18 December 1979 and entered into force on 11 July 1984. In the last 18 years the activity of the Legal Sub Committee was elaborating principles related to international Direct Television Broadcasting, Remote Sensing of the Earth from Space, Use of Nuclear Power Sources in Outer Space and International Cooperation in the Exploration and Use of Outer Space for the Benefit and in the Interest of all States, Taking into Particular Account the Needs of Developing Countries.

In the Report of the Legal Sub Committee of COPUOS last session,[2] mention is made about the review of the principles concerning International Direct Television Broadcasting and Remote Sensing of the Earth from Outer Space with a view of possibly transforming those texts into treaties in the future.

We must also recall the importance of the offered cooperation of non governmental organizations like IAF, IISL, International Academy of Astronautics (IAA), International Law Association (ILA), among other and international professional associations in the last years. This cooperation has been reflected in matters as settlement of Space Law Disputes, Space Débris, SETI, in accordance with the interest and concern of the jurists and in view of plenty of doctrine available.

As it could be proved five years ago when celebrating the International Space Year, the growth both in quantity and variety of space activities, had as a consequence the expansion and development of the scope and substance of Space Law.[3] Nevertheless, there are some differences in the interpretation of the terms opportunely adopted, for instance the *common heritage of mankind* principle, enshrined in the Moon Agreement, a concept created in 1954 in order that the exploration and use of outer space be done having mankind as beneficiary.[4]

Peoples have a new legal concept that succeeds interdependency and integration: solidarity. And this concept is a guarantee of future development of Space Law in its own cosmic dimension.

The hope now focus towards to the Third UN Conference on the Exploration and Peaceful Uses of Outer Space. One of the topics relates to the benefits space applications can provide to life on Earth.[5]

Footnotes

1 V. Kopal, Evolution of the Doctrine of Space Law, in *Space Law Development and Scope*, ed. by N. Jasentuliyana, published by Praeger, USA, 1992, p. 25.

2 Report of the Legal Subcommittee on the Work of its Thirty-sixth Session 1-8 April, 1997, A/AC.105/674, 14 April 1997, N° 40, p. 9.

3 N. Jasentuliyana, Introduction, *Space Law, Development and Scope*, cit., p. xvii.

4 A.A. Cocca, Die Rechtliche Natur des Weltraums, *V Internazionalen Kongress*, Innsbruck 1954, Springer Verlag in Wien, 1955, p. 270.

5 Further information in German Journal of Air and Space Law, 2 / 97, Carl Haymanns Verlag, june 1997, p. 235.

Session 2:

CONCEPTS OF SPACE LAW AND THE OUTER SPACE TREATY

Chairmen: G. Catalano Sgrosso, Italy
E. Galloway, USA

Rapporteur: M. L. Longo, Italy

IISL-97-2.01

IMPORTANT CONCEPTS FOR THE INTERNATIONAL LAW OF OUTER SPACE

by

Carl Q. Christol*
Distinguished Professor Emeritus of International Law and Political Science
University of Southern California
Los Angeles, California 90089-0044
Attorney at Law, Member, IISL, IAA, AIAA

Abstract

The province of mankind concept is a classic example of a legal principle in that it is a starting point for legal reasoning. This principle has provided guidance and direction for the international law of outer space. It has fortified the expectation that outer space be explored and used in the interest of and for the benefit of all mankind. It is a fundamental concept of the international law of outer space.

The tensions of the Cold War existed during the negotiation of the 1967 Principles Treaty. Also influencing its content were the goals of the new States with their colonial antecedents. The space treaty was a successful accommodation to these varying outlooks. It captured the optimistic expectations of States which were able to understand that the creation of a viable legal regime was necessary if the resources of space were to serve their general interests.

While the province of mankind principle became a unifying influence, it did not require a State possessing exploitative capabilities to share a specific asset with another State. It was entirely consistent with the *res communis* provisions of Article II of the 1967 Principles Treaty. At the same time the mankind principle encompassed the concept of the sharing of benefits and interests. Such sharing was to be voluntary rather than mandatory.

Introduction

The space age at its outset was not a peaceful one. At the close of World War II two former allies were embroiled in major policy differences. The tensions were so great that it became the era of the Cold War.

There were substantial and unsettling military confrontations. These were accompanied by vigorous and even bombastic diplomatic encounters. These former allies quickly embarked on a space race to perfect the hardware required for successful space ventures. It was their assumption that their national security and exploitative capabilities would be enhanced.

From a Western perceptive the contest was seen as one in which an open and democratic society was under challenge from a closed and totalitarian one. Each power block attracted its supporters and detractors.

At this period the Soviet Union was propounding the theme of "peaceful co-existence." It was composed of carefully selected propositions contained in the U.N. Charter, such as the maintenance of international peace and security, sovereign equality of member states, territorial integrity, equal rights and self-determination of peoples, among others. This theme, from a Western perspective, when measured against the actual international conduct of the Soviet Union, was often perceived as a dangerous deception.

On the other hand, the Western commitment to freedom of speech and press, was interpreted by Soviet officials as facilitating the promulgation and dissemination of dangerous propaganda.

The break up at this time of former colonial empires, with the consequent population explosion of new States, also seriously affected international tranquility. As these fragile entities sought to position themselves in the existing world order, their achievements were, at best, very uneven.

However, on the political front the new States were able to obtain the adoption of resolutions designed to enhance their economic viability. Among these were the 1962 General Assembly Resolution on Permanent Sovereignty Over Natural Resources[1] and the 1974 General Assembly Declaration on the Establishment of a New International Economic Order.[2] The seriousness of purpose, cooperative efforts, and unity of the newly emerging States were clearly evident.

Thus, the 1950s and the 1960s were marked by security concerns, and by economic and political tensions. These were augmented by major environmental concerns respecting the well-being of the human environment, as a result of nuclear testing, and for the marine environment, because of numerous oil spills and the use of the ocean as a dumping ground for all manner of toxic wastes. Such were the "true to life" circumstances confronting responsible statesmen as they, after consulting their own consciences, examined the outlooks of humanity.

It is a sign of humankind's durability that, even in those unsettled times, there could be both strong and unabiding hopes for the future. Thus, with the advent of the space age, there were many who considered that the role of human beings in outer space could be a magnificent opportunity to correct, or at least to modify, some of the Earth's problems. It was perceived that there might be substantial changes in how people worked, lived, and dreamed as they sought to improve the quality of their lives.

Such a higher quality of life would exist, if things went right, for those already blessed with formidable advantages. It would also be enjoyed by the less favored, and in particular for those who composed the underdeveloped world.

As the space age achieved momentum, and even before all of the scientific and technological developments that we know today had been realized, there emerged a classic situation in which expectations were high, and where intransigence, assertiveness, and conflict, which if not having been brought under full control, were, at least, somewhat attenuated. A combination of the above forces, coupled with serious efforts to protect human rights, began to make it more probable that the quality of life available within the advanced countries might also become available to the human beings living in less favorable circumstances.

With the growing acceptance of the commonality of all humanity it followed that every aspect of outer space must be shared among all humans. The emergence of an intellectual framework that gave attention to the needs of all mankind was, in itself, a dramatic commentary of the evolving support for the basic values founded in a sense of community.

These values conditioned the behavior of individuals, of nation-States, and international intergovernmental organizations with the resultant dampening down, if not wholly extinguishing, the remaining vestiges of the Cold War. In this manner it became possible to start the negotiations at the UN which led in 1967 to the Treaty on Principles Governing the Activities of States in the Exploration and Use of Outer Space, Including the Moon and Other Celestial Bodies.[3]

Concepts and Fundamental Concepts

Even to the most casual observer a reading of the 1967 Principles Treaty will bring out its numerous concepts. Fortification respecting its meaning can be obtained through recourse to the seven Resolutions dealing with the peaceful uses of outer space adopted by the UN General Assembly between November 14, 1957 and December 24, 1963.[4] Aside from the common reference in these Resolutions to the peaceful uses of outer space,[5] they contained the central theme that space should be used for the betterment of mankind, that cooperative efforts would contribute to the development of mutual understanding and to the strengthening of friendly relations between nations and peoples, and that international law and the U.N. Charter had application to outer space activities.

Undoubtedly, among the most significant of these fundamental concepts was the requirement that humans in their outer space activities should conform to earth-based laws. Humans assumed they possessed the qualifications, and, even more importantly, the authority to create and administer an effective international legal regime for space and celestial body activities.

Having arrived at this decision, and it was not a difficult one, for, after all, who other than humans should be called upon to regulate the behavior and the products of human ingenuity used for space activities, it became necessary to identify with particularity relevant principles, standards, and rules. Space lawyers were able to make the traditional distinctions with principles conceived of as broad propositions which serve as the starting point for legal reasoning, with rules constituting the sharper and more detailed, consequence-laden depictions of "do's" and "don'ts," and with standards consisting of the intelligent amelioration of the possible harshness of rules so that the fundamental substance of an applicable principle would not be thwarted or disregarded.

Fundamental Concepts Examined

If, as has been suggested, a distinction should be drawn between fundamental concepts and other concepts, it must be asked on what grounds can such a distinction be supported? An easy but not wholly satisfying answer is that some concepts are simply more important or more fundamental than others. When examined more closely it is possible to suggest that certain basic values grounded in long-received traditions of morality should be recognized as fundamental concepts. When such moral precepts are coupled with pragmatic considerations, their joinder produces meaningful law. For example, the U.S. Constitution provides that it is the "supreme law of the land." As a result, subordinate legal prescriptions, inconsistent with the foregoing principle, can be rendered unconstitutional, i.e., without force of law.

Applying this analysis to the concepts contained in the 1967 Principles Treaty it may be posited that the highest legal norm contained in the Treaty is the provision of Article I, paragraph one, that the space environment, namely, outer space, per se, the Moon, and other celestial bodies, including their natural resources, "shall be the province of all mankind." Acceptance of this mandate will assure that the benefits and interests of all countries will receive full protection despite the inequality of their respective space capabilities. Moral and practical considerations are more important than the scientific and technical evolution of political entities.

The importance of such an approach to the new field of human activity in the space environment was highlighted in a report prepared by the staff of the Committee on Aeronautical and Space Sciences of the United States Senate. It expressed the view that the foregoing principle accorded with "a basic tenet of democracy."[6] Account was taken of the prospect that many practical activities in outer space might require more modest capabilities and that almost any country would become involved.

In focusing on the universality of national participation in space activities and in space benefits Article I, paragraph 1 refers to "all" countries and to "all" mankind. This fundamental thesis is confirmed in Article XIV of the Treaty where reference is made to the right of "all" States to become parties. This determination, which was particularly confirmed though the acceptance of the *res communis* international legal principle in Article II, does not allow a single State or a combination of States to establish their sovereignties with respect to the province of all mankind areas or their resources. Article II states that "Outer space, including the moon and other celestial bodies, is not subject to national appropriation by claim of sovereignty, by means of use or occupation, or by any other means."[7]

Particular claims for the recognition of mankind's interest in benefits derived from space activity have been widespread and durable. In this connection an early concern respecting the product of space activity was voiced by President D.D. Eisenhower. In 1958 he called on States "to promote the peaceful use of space and to utilize the new knowledge obtainable from space science and technology for the benefit of mankind."[8]

In support of the 1958 proposal that the U.N. General Assembly examine international cooperation in the field of outer space it was indicated that the General Assembly was "the body most representative of the interests of mankind . . ."[9] and should concern itself with both the peaceful uses of outer space and the establishment of appropriate machinery to deal with the expectation that "outer space will be used solely for the benefit of all mankind."[10] These propositions were repeated in General Assembly Resolutions 1348 of 1958, 1472 of 1959, 1771 of 1961, and 1962 of 1963 where the terminology also included "common interest of mankind" and "betterment of mankind."

The General Assembly resolutions were influenced by the terms of national proposals which had made reference to the concerns of mankind in space activities. Thus, in 1962 the Soviet Union in its Draft Declaration of the Basic Principles Governing the Activities of States Pertaining to the Exploration and Use of Outer Space had proposed that "[t]he exploration and use of outer space shall be carried out for the benefit and in the interests of the whole of mankind."[11] This was reiterated by the Soviet Union in its 1963 Draft Declaration of the Basic Principles Governing the Activities of States in the Exploration and Use of Outer Space where reference was also made to the "common interest of all mankind," to the "betterment of mankind," and "to the benefit and in the interests of the whole of mankind."[12]

In its December 1962 Draft Declaration of Principles Relating to the Exploration and Use of Outer Space the United States referred to "the common interest of all mankind" and the "betterment of mankind and to the benefit of States" in the exploration and use of the space environment.[13]

The Soviet Union in its June 16, 1966 draft treaty on Principles and Activities of States in the Exploration and Use of Outer Space, The Moon, and Other Celestial Bodies referred to its support for the proposition that the exploration and use of outer space "shall be carried out for the benefit and in the interests of all countries."[14] To this was added for the first time the proposal "and shall be the province of all mankind."[15] In its June 16, 1966 proposal, which related only to the Exploration of the Moon and Other Celestial Bodies, the United States retreated from its earlier references to the involvement of mankind in outer space activities. It merely made reference in the preamble to the "interest of all mankind" in the Moon and Other Celestial Bodies.[16]

An interpretation of the motivation lying behind the Soviet choice of the foregoing terminology was provided by Ambassador Morozov in a statement made to the Legal Sub-Committee of COPUOS on July 12, 1966. He observed that his country, in "charting man's course into outer space, had always regarded its achievements in that field of endeavor as belonging to mankind as a whole."[17] Underlying this principle were "the rights and opportunities of States to engage in the exploration and use of outer space but also of all countries and of the international community as a whole. . . ."[18]

The Sub-Committee's working group, following a review of the Soviet text, gave its approval on July 29, 1966.[19] Prior to acceptance by the working group of the foregoing text, the United Arab Republic on July 22, 1966 had endorsed for inclusion in the treaty the "province of mankind" principle.[20] Following further review by the working group on July 29, 1966 it proposed the acceptance of the province of mankind principle.[21]

The province of mankind principle constituted an expression of support for both the protection of the interests and the distribution of the benefits derived from outer space activity to human beings and their governmental institutions, both national and international. Although space activities are not limited to space research and the practical utilization of the benefits derived from such research, key Soviet writers have cast the province of mankind principle in that light. After referring to the foregoing considerations they have indicated that derived benefits will serve "the whole of mankind, and it is for that reason that the exploration and use of outer space are described in the treaty as the 'province of all mankind.'"[22]

When the principles Treaty was reviewed by the Committee on Foreign Relations of the U.S. Senate questions were raised respecting the meaning of "province of mankind." Ambassador Arthur J. Goldberg, who was the principal U.S. negotiator in the last stages of the extended effort to arrive at the agreement, advised the Committee that the term was to be taken in the sense that "exploration shall be carried out for the benefit of mankind,"[23] and should, as the treaty specifically stated, take into account "the benefit and in the interests of all countries, irrespective of their degree of economic or scientific development."[24]

Senator Hickenlooper asked: "Is this in the nature of securing for the non-contributing countries all the benefits of those who put up the money and the expenses of this?"[25] He was concerned that the non-space resource states would have a "free ride"[26] in the distribution of potential benefits. To this Ambassador Goldberg responded that the indicated language constituted a "goal," and that it was not "a free ride, because there are other . . . provisions which are self-executing provisions. This states a general concept that the exploration and use of outer space shall be carried out generally for the benefit of all mankind. . . . [I]t states a . . . worthy purpose."[27]

However, the Senate Committee on Foreign Relations raised the practical question whether the language of the foregoing general principle "might imply a fixed treaty obligation on the part of the United States to share the benefits and results of its space activities, particularly in the communications satellite field."[28] These concerns had previously been mentioned to Ambassador Goldberg by members of the committee.[29] At that time the question was raised whether the United States should append a formal reservation to its ratification whereby it would be specified that the United States had no legal obligation to provide a "free ride" to other States respecting the use of communication satellites.

Both Secretary of State Dean Rusk and Ambassador Goldberg urged the

Committee members not to attach a legal reservation on the ground that this would be an invitation to other States to add reservations with the result that the comprehensive, integrated, and carefully constructed Treaty would become flawed and would lose its intended utility. The merit of this approach was recognized by the members of the Committee. Yet, its members were not entirely persuaded. As a result, in their report to the President it was indicated that "[i]t is the understanding of the Committee on Foreign Relations that nothing in article I, paragraph 1 of the treaty diminishes or alters the right of the United States to determine how it shares the benefits and results of its space activities."[30] A situation in which the committee's cautionary approach might be invoked has never arisen.

The provisions of Article I, paragraph 1 of the Treaty, must be interpreted in the light of terms of its preamble and Articles 3, 10 and 11 which stressed the importance and duty of international cooperation in space activities. The Treaty envisioned progressive scientific and technological developments in the exploration, exploitation, and use of the outer space environment and its natural resources. In light of the preliminary negotiations, the carefully honed terms of the agreement, and the views of both Soviet and American commentators on the province of mankind principle, it may be concluded that it was an open-ended commitment to seek to advance the benefits and interests of all mankind, including the governmental structures within which humans conduct themselves.

The "province of mankind" principle gave direction to the entire Treaty. It proclaimed a goal or a purpose. When the principle is coupled with the duty of international cooperation the combination allows space benefits in the broadest sense to be shared among all who are able to incorporate such discoveries and benefits into their normal lives. It was not designed to effect specific distributions or shares of the product of experiments or practical activities conducted in outer space, or to parcel out orbital positions, or to make allotments to States of radio frequencies used in space communications. The Treaty did not create an international intergovernmental entity charged with the powers to effect such distributions. Absent such a formal institutional presence the province of mankind principle could be implemented only through voluntary sharing processes. This in turn could only result from an understanding of the common interest of all in such sharing.

At the outset the question was asked: Why select the "province of mankind" principle as a fundamental one in the development of the modern international law of outer space? In addition to the earlier responses, it is now possible to add another, and possibly more important that each of the others. This principle can be characterized as having a noble purpose. The principle, though abstract, seeks to serve in the area of space activity the wants, needs, interests, and values of a modern international community. It can serve as a catalyst in support of other worthy purposes.

Other Concepts

Although the "province of mankind" principle has been selected for analysis, the author willingly concedes that there are many other space-law concepts that possess a status sufficient to allow them to be reviewed. Among these are the guarantees of freedom of outer space (non-sovereignty) provisions of Article II, the application of the U.N. Charter and international law pursuant to Article III, and the use exclusively for peaceful purposes of the Moon and other celestial bodies of Article IV. The norm of international cooperation has been mentioned previously.

Also, it would have been possible to examine such provisions as those dealing with rescue and return of astronauts and space objects, international responsibility, damages, registration, and so on. However, none of these is characterized by the quality of novelty. They are, of course, highly important elements of a structured international legal regime designed to normalize and advance wide-ranging space activity.

The Province of Mankind and the Common Heritage of Mankind Principles Distinguished

It may be noted initially that the province of mankind principle and the Common Heritage of Mankind principle enjoy a critically important commonality.[31] Each is a fundamental principle of international space law. Each is founded on a set of common values. Each is goal-oriented in that each is intended to be of service to the well-being of mankind. Each bear the important *imprimatur* of the United Nations with a detailed and well-documented negotiational history. Each has undergone the analysis and commentary of highly qualified students of international space law. The province of mankind principle appears in each of the two agreements.

Nonetheless, it is evident that the two principles carry separate and distinct characteristics. The province of mankind principle is linked to the *res communis* principle which allows for the exploration, use, exploitation, and voluntary sharing of common resources. On the other hand, the Common Heritage of Mankind principle, as contained in the Moon Agreement, may be characterized as a "*res communis* plus" principle in the sense that successful explorers, users, and exploiters of the moon and its natural resources will be obliged to conform to the decisions of the international legal regime identified in Article 11 of that agreement.[32] A broadly worded formula, which is contained in Article 11, paragraph 7.(d) is intended, under the supervision of the new (and yet to be created) international authority to achieve the goal of "equitable sharing." The province of mankind principle does not contemplate the formation of an international intergovernmental body or that there be an obligatory sharing of the tangible acquisitions of Moon and celestial body activity. This simplistic but correct assessment of the two principles makes it evident that the principle of the Common Heritage of Mankind has significantly advanced the concepts which are central to the province of mankind principle.[33]

Conclusion

The 1967 Principles Treaty was written during the Cold War and at a time when the world's maternity ward was filled with newly born States. Following many years of stressful negotiations the diverse negotiators arrived at important accommodations. They were based on the perceived benefits available only through international cooperation.

The new treaty, which has frequently been referred to as the Magna Carta of outer space, contains very important principles. Some are quite abstract and accord well with the precept that a legal principle is a starting point for legal reasoning. Other terms of the agreement are more specific and are based on quite traditional legal concepts. A leading example is the *res communis* principle.

It has been suggested here that the "province of mankind" may qualify as a fundamental concept of the Treaty. This is based on an early, continuing, and general claim that the space environment and its natural resources should serve the common interest of mankind and be employed for the betterment of mankind. This outlook fits into the view that certain basic values, including the ancient theme of sharing, which are grounded in long-held traditions of morality, are entitled to be treated as fundamental. As such they take precedence over other outlooks, frequently set forth in legal rules, which do not meet the critical standards of fundamental legal prescriptions.

Further, the "province of mankind" principle in taking a broad view respecting the exploration, use, and exploitation of the space environment and its natural resources provides encouragement to those who have the practical capabilities to engage in such conduct. In this manner the principle, which does not allow for nor encompass a "free ride" mentality, can give guidance and direction to all of the other key provisions of the Treaty. By identifying a central purpose of the Treaty the "province of mankind" principle can and has served as a meaningful catalyst. It has facilitated innovative space activities. Of equal importance the principle has prevented the erection of artificial barriers to the world-wide dissemination of the scientific and technological benefits of the space age.

NOTES

1. U.N.G.A. Res. 1803 (XVIII), 17 U.N.GAOR, Supp. (No. 17) 15, U.N. Doc. A/5217 (1963), 2 *I.L.M.* (1963).
2. U.N.G.A. Res. 3201 (S-VI), 6 (Special) U.N. GAOR, Supp. (No. 1) 3, U.N. Doc. A/9559) (1974), 13 *I.L.M.* 715 (1974).
3. 18 U.S.T. 2410, T.I.A.S. 6347, 610 U.N.T.S. 205. Done at Washington, London and Moscow, Jan. 27, 1967. Entered into force for the United States, Oct. 10, 1967. At last count it had been ratified by 93 States and signed by an additional 27.
4. These have been collected in C.Q. CHRISTOL, INTERNATIONAL SPACE LAW 441-456 (1966).
5. Unfortunately, in its final form the Treaty provided only that the Moon and Other Celestial Bodies should be used exclusively for peaceful purposes.
6. *Treaty on Principles Governing the Activities of States in the Exploration and Use of Outer Space, Including the Moon and Other Celestial Bodies,* Staff Report, Committee on Aeronautical and Space Sciences,

United States Senate, Committee Print, 90th Cong. 1st Sess., 22 March 1967).

7. *Supra*, note 3. The meaning of "by any other means" has been addressed in my article entitled "Article II of the Principles Treaty Revisited," 9 *Ann. Air & Sp. L.*, 217 (1984). The position was taken that in so far as States may not assert sovereign claims that this denies to States, acting in concert or by way of an international intergovernmental organization, any right to make equivalent claims.

8. *Introduction to Outer Space*. An Explanatory Statement Prepared by the President's Science Advisory Committee, 1 (1958).

9. U.N.G.A., Official Records: Thirteenth Session, Annexes, Agenda Item 60, 4, Sept. 2, 1958.

10. *Ibid.*

11. U.N. Doc. A/AC.105/L2 at 1, September 10, 1962; U.N. Doc. A/5181. Annex III, *supra*, note 4 at 480.

12. U.N. Doc. A/AC.105/C.2/L.6, at 1 and 2, April 16, 1963, *supra*, note 4 at 466.

13. U.N. Doc. A/C.1/881, at 23, December 8, 1962. *Supra*, note 4 at 459.

14. U.N. Doc. A/AC.105/35, Annex I, at 11, September 16, 1966.

15. *Id.* at 12.

16. *Space Treaty Proposals by the United States and U.S.S.R*, Staff Report, Committee on Aeronautical and Space Sciences, United States Senate, Committee Print, 89th Cong., 2d Sess. 16 (1966).

17. U.N. Doc. A/AC.105/C.2/SR.57, at 12, October 20, 1966.

18. *Ibid.*

19. *Supra*, note 14, Annex II, at 4.

20, *Id.*, Annex III, at 6.

21. *Id.* at 12.

22. GENNADY ZHUKOV and YURI KOLOSOV, INTERNATIONAL SPACE LAW 42 (1984).

23. *Treaty on Outer Space, Hearings before the Committee on Foreign Relations*, United States Senate, Executive D, 90th Cong., 1st Sess., 10 (1967).

24. *Supra*, note 3.

25. *Supra*, note 23, at 10.

26. *Ibid.*

27. *Ibid.*

28. *Treaty on Outer Space, Executive Rept. No. 8 of the Senate Committee on Foreign Relations (To accompany Ex. D)* 90th Cong., 1st Sess., 4 (1967).

29. *Supra*, note 23 at 35-37.

30. *Supra*, note 28 at 4.

31. The Common Heritage of Mankind principle was formally acknowledged in the Agreement Governing the Activities of States on the Moon and Other Celestial Bodies of December 5, 1979. U.N. Doc. A/RES/34/68, 14 Dec. 1979; 18 *I.L.M.* 1434 (1979).

32. *Ibid.*

33. The Common Heritage of Mankind Principle has been frequently assessed in the *Proceedings of the International Institute of Space Law.* The author's "The Legal Common Heritage of Mankind: Capturing an Illusive Concept and Applying it to World Needs," appeared in the *Proceedings of the 18th Colloquium on the Law of Outer Space* 393 (1976). Excerpts, including a detailed bibliography, were published in C.Q. CHRISTOL, SPACE LAW, PAST, PRESENT, AND FUTURE, 398-401 (1991). See also, C.Q. Christol, "The Common Heritage of Mankind Provision in the 1979 Agreement Governing the Activities of States on the Moon and Other Celestial Bodies," 14 *The Int'l Law* 429-483 (1980) and the "Agreement Governing the Activities of States on the Moon and Other Celestial Bodies," *Committee Print*, Committee on Commerce, Science, and Transportation, United States Senate, 96th Cong., 2d Sess. (1980). This consists of four parts. Parts 1 and 2 were authored and edited by Eilene Galloway; Parts 3 and 4 were written by U.S. governmental officials employed in a variety of offices and agencies.

IISL-97-IISL.2.02

IS IT NECESSARY TO REDEFINE PRINCIPLES AND CONCEPTS OF THE OUTER SPACE TREATY?

María de las Mercedes Esquivel de Cocca*
(University of Buenos Aires)
Argentina

Introduction

After thirty years of the entering into force of the Outer Space Treaty, some adjustments seem to be necessary, taking into account the development of space activities, consequently the new situations, needs, factors, and resources arisen, and the previewing characteristic that Space Law used to have in its beginnings. At the present we witness how Law hurries up after technology, trying to adjust its provisions to facts already occurred. A deep study towards the real understanding of the principles must be carried out to maintain the enforcement of the *corpus iuris spatialis* and its peculiar spirit. No amendment to the Treaty is necessary at all. A protocol could enshrine those principles that appear to be necessary to rule the new activities and legal relations not previewed in the present international instruments, and may include, as well, an updated interpretation of them adequate to the new times.

Principles in force that should be included in the protocol

Activities in accordance with international law and in the interest of maintaining international peace and security and promoting international cooperation and understanding.

Peace and security are values that depend on order, law and good faith. They are values of common welfare and for that reason must be observed above any particular interest or benefit. The question of space débris is closely linked with this and space contamination should be included within the principles of general benefit, security, peace and observance of international law.

We must remember that in occasion of the Jupiter plutonium powered probe, President George Herbert Bush was sued by several non-profit organizations.[1] The suit was based on the risk the plutonium meant for Earth during the launching and while the probe was in Earth orbit. The same legal and factual reasons move those who protest against the Cassini-mission, also plutonium powered, which ended with the arrests of several dozen people. The Cassini rocket is scheduled to be launched October 13 for a seven-year trip to Saturn. The rocket shall be powered with 72 pounds of plutonium.[2]

---

* Member of IISL

The results of the Cassini mission may be huge for humanity. One cannot imagine how many new data can be achieved. But, bearing in mind what is happening with the Mir mission, what could result from of an accident: the threat of contaminating air space, the Earth's surface, and the sea. In case of an accident the risk may remain for many years. People's health and lives are worthy enough to wait until an absolutely secure technology can be managed. If a collision happens in outer space with another spacecraft there may be loss of human lives and contamination traces spreading in space besides the débris.

Use of space with peaceful purposes

Article IV of the Treaty establishes one of the most important principles in Space Law and in Law in general terms.

The first paragraph prohibits specific acts or attitudes that infringe the concept of peace. But in the second paragraph, the text is more clear and definitive. It states that *the Moon and other celestial bodies shall be used... **exclusively for peaceful purposes***. This sentence is so condensed that it makes all the rest of the article almost redundant.

Astronaut

The Space Treaty was opened to signature the same day that three American astronauts died scorched after the *Apollo I* test failure,[3] and entered into force two years before the arrival of man to the Moon. After thirty years the astronauts still lack international legal protection.

Space transport, close to get to reality, urges the establishment of the legal statute of cosmonauts and commanders in spacecrafts.

Article V of the Space Treaty establishes that State Parties to the Treaty *shall regard astronauts as envoys of mankind in outer space*. This status applicable only when the cosmonaut is in outer space, extends to an unintended landing, accident, distress or emergency on the territory of another State Party or in high seas.

This astronautical solidarity is specially established to rule the need of assistance and rescue in space. We have recently lived the experience of assistance when the crew of Atlantis cooperated with the crew of the Mir station in repairing the module. For the first time a foreign astronaut got out of an American vehicle for an EVA mission.

Cosmonauts are defenseless in space. Radiation, débris, and a hostile environment lurk around this man's health and life. The worker that does more for humankind is the one with less practical and legal protection.

A cosmonaut's responsibility is huge in both professional and human terms. His acts obligate the launching State's international liability, may endanger the mission and also jeopardize the terrestrial environment and human health. In outer space the possibility of mistake can have terrible consequences for his life and for the life of other cosmonauts in orbit. In space he is alone, it is he who makes decisions and puts them in practice.

At the present crews are constituted by more than one cosmonaut, therefore that loneliness is relative but not for the commander, who is in charge of the security of the persons and vehicle and for the prosecution of the mission.

The status of *envoy of humankind* means a great honor for any person, but in the case of cosmonaut should also entail rights. To such a responsibility must encompass parallel rights to get to a balance between duties and rights. The captain of a ship, the commander of an aircraft have their clearly specified duties but are also provided with rights and faculties to achieve a good end for their missions. This should be included in a protocol to the treaty.

The psychological stress that cosmonauts have to support is different than the one suffered by other workers. A sailor may be separated from his family and homeland for several months but he does not work in a place of extremely reduced dimensions without having the possibility of going outside to breath fresh air or of having some fun in a stopover. The cosmonaut works in a hostile environment, his life is in danger during the whole mission and he may get ill and die for having been exposed to radiation, which he cannot avoid.[4]

Space object

To define space objects appears to be urgent due to future space stations, settlements in celestial bodies, EVA activities, etc.

The description made in the Convention on Registration of Objects Launched into Outer Space in Article I, does not define the concept, it just explains what elements should be included within the concept of space object: *launch vehicle and parts thereof.*

We understand that a space object is any device able to transport through outer space persons or things from or to Earth or a celestial body, or to develop any activity in outer space, as well. It is a concept that encompasses either the launching vehicle and the parts thereof.

The importance of reaching a definition of space object resides in the fact that new vehicles have risen in the space scenery: the space planes, that can fly in air space as aircrafts and move through outer space as space vehicles. In this case it is important to refine the concept for the same device can perform air transport and space transport as well.

Human settlements. Notion of humankind

Undreamed possibilities have arisen from space. Who could speak of human civilizations in space in 1967? The project of human settlements on celestial bodies generates the need to reflect on the meaning of humankind concept, to contemplate persons born on other celestial bodies, in space stations and to understand the framework of the relationship that could be established if an extraterrestrial civilization is contacted. Five centuries ago years Francisco de Vitoria did so when America and its inhabitants were discovered.

Common heritage of mankind

Commercial activities have caused legal difficulties in the application of the common heritage of mankind principle. The principle of international cooperation –as a condition *sine qua non* for the lawfulness of space activities– was always a heavy duty for space faring states in the planning and execution of their space enterprises. Vertical cooperation requires not only a great quantity of patience and efficient diplomacy, but also an adequate understanding of the principle of mankind as legal subject and of the fact that said activities, that are performed in an environment owned by mankind, must aim the benefit for all mankind. In this sense, the principle of international cooperation is inseparable from the common heritage of mankind. To harmonize these matters is a hard task, but the fruits derived thereof, deserve the efforts.

A general assumption that the common heritage of humanity and benefit for all humankind principles are source of illegal enrichment of nations —passive in connection with the activities and the necessary investments— has led to the poor ratification of the Moon Agreement and the diminution of advanced proposals, in order to follow the leadership of law in astronautics.

Absolute international responsibility of states

State is not often mentioned in the *corpus iuris spatalis*. When a provision refers to state, it means it is alluding to a duty. Since article II of the Space Treaty State has been deprived of rights in Space Law.

Absolute responsibility is closely linked to the principle of the *restitutio in integrum* established in old Roman Law.

This responsibility is moreover, objective when it is born in damages generated on surface or in air space, and subjective if damage is caused in outer space. How can the responsibility generated in damages caused on a celestial body be defined? What responsibility is applicable to damages caused during transport made with spaceplanes?

Is it necessary to return to the issue of the delimitation of the boundary between space and air spaces? I believe that the answer is and was in the doctrine of the purpose borne in mind when an activity is performed. The teleological theory to determine which legal régime is applicable to a certain situation overcomes the difficulties derived from the delimitation of both spaces. This criterion establishes that if an activity has been performed having in mind a spatial aim, Space Law must rule it in all its aspects. If that activity was conceived to achieve any aeronautical goal, then Air Law rules it in all its aspects.

The criteria established until now have been put in crisis with the event of the space-aircraft.

New principles that should be included in the protocol

Spatial conscience

Future human settlements in space and in celestial bodies put under discussion several issues that may introduce new principles in Space Law. Solidarity is the first manifestation of this phenomenon; assistance in space [5] is due to the feeling of belonging to the same nature, is the result of respect and consideration for the fellow.

The possibility of establishing human permanent settlements in space introduces the question of what civil rights may the settlers have. Shall the concept of nationality be transformed in a concept of race or gender? If States cannot exercise any sovereignty right out from Earth, which shall be the political and civil status of the settlers?

Recent technological and cultural achievements —as Internet, for example— have showed humankind in a clear and indisputable manner that all human beings belong to the same and unique humanity, which, for the meanwhile and up what we know, is placed as civilization in only one planet: Earth. Times shall come when humanities, developed in several celestial bodies and large space stations, will inhabit.

This is not far from the common sense of peoples. Popular literature and cinema preview this phenomenon. This possibility implies a spatial conscience that is being developed and that Law must recall and give legal framework to.

Space Transport

Responsibility in space must be meditated in order to rule future space transport and aeronautical activities closely linked with satellite airnavigation security systems and the utilization of space-aircrafts, within the spirit of the *corpus iuris spatialis.*

Several Air Law institutions must be transformed, as there is no sovereignty in space we cannot talk about the Parties' territory but of State of launching jurisdiction and control. First and second air freedoms should be preserved and adapted to space activities in the following way: 1st freedom: the freedom of moving throughout outer space and in celestial bodies orbit;

2nd freedom: the liberty of descending on celestial bodies or arriving to a space station with no commercial purposes.

The other three air liberties cannot be adapted to space transit or traffic because space settlements may be originated in several states of launching or in other settlements in celestial bodies, thus there cannot be identified any state exercising jurisdiction and control.

Another subject that must be clarified in a future protocol should be duties and rights of the commander and the status of passengers. Should all of them be considered astronauts?

There is no doubt that during the travel the commander should be in charge of the security of the vehicle, persons and things on board. This obligation should end when the vehicle arrives to destination.

Contractual responsibility should be absolute in space transportation, and it should be based on guiltiness of the transporter. There is no reason to change the legal framework of responsibility for damages caused with space objects launched into outer space. The foundation in guiltiness derives from the fact that passenger and shipper assume the space risk when contracting transportation, thus, they are in a different position from the one of the damaged person on the surface, in air space or on the sea.

Conclusions

1. The Outer Space Treaty must not be modified, to preserve its spirit and the integrity of its principles.
2. A protocol should be established to introduce the new necessary principles and to clarify the interpretation of those in force enlightened by the new technological, social and cultural circumstances.

Footnotes

[1] Re: Florida Coalition et all v. George Herbert Bush.
[2] CNN Interactive Quicknews, October 5, 1997, Internet.
[3] See USA Today October 2, 1997, electronic edition by Internet: Air Force pilots flew over Cape Canaveral in the missing-man formation, concluding a service for the three Apollo 1 astronauts killed in a launch pad fire 30 years ago. Virgil "Gus" Grissom, Edward White II and Roger Chaffee died inside their burning capsule during a test on Jan. 27, 1967. NASA has generally not observed the anniversary but has held programs in memory of the space shuttle Challenger's explosion, 11 years ago.
[4] Michael Foale, USA astronaut, got back to Earth after being 144 days in Mir station, separated from his family and planet, so did a Russian astronaut after a similar period after being unable to move his legs for the long permanence in space.
[5] The space shuttle Atlantis headed toward home after a successful six-day visit during which it left behind an American astronaut and a rejuvenated central computer on the Russian space station Mir. The space station, meanwhile, prepared to receive another visitor.

IISL-97-2.04

The Privatisation of the "Province of Mankind"
Time to Reassess Basic Principles of Space Law?

Sylvia Ospina, J.D., LL.M.
Space Law Consultant*

ABSTRACT

Since the Cold War ended, the private sector's involvement and influence in space activities and policies have increased considerably, creating new legal expectations, such as ownership rights to intangible outer space resources, or rights to their exploitation. These endeavours are putting into question basic principles of public international law and space law.

While the Outer Space Treaty states that outer space shall be free for exploration and use by all States, a monetary value is being attached to intangible resources that heretofore have been considered "res communis humanitatis", or the province of mankind.

Should these fundamental tenets be reinterpreted, to allow for the progressive privatisation of outer space, according to the same economic principles that permeate terrestrial activities? Or should space resources remain beyond the scope of the privatisation pendulum, and continue to be governed by the principles of customary international law incorporated in the Outer Space Treaty?

INTRODUCTION

A few years ago, the **"ISMs"** were dominant: communism, commercialism, colonialism, nationalism, capitalism, among others. Now, it is the **"ATIONs"** that prevail: deregulation, commercialisation, globalization, liberalisation, privatisation, and other terms being coined daily.

The **"ATIONs"** imply movement, action, trends that perhaps cannot be stopped. Certainly, space activities, and in particular satellite communications, are evolving at an ever-increasing speed, and are subject to many of the "ATIONs": commercialisation, globalisation, deregulation. These terms are not synonymous, and through time, their meanings have been evolving, and influence our conduct of space activities.

A SYNOPSIS OF THIRTY YEARS OF SPACE ACTIVITIES

Thirty years ago, the major powers, advocates of either communism or of capitalism, agreed on several important principles relating to outer space activities, which were incorporated in the "Outer Space Treaty."/[1]/

One of the most important stated:

> "Outer space, including the moon and other celestial bodies, is not subject to national appropriation by claims of sovereignty, by means of use or occupation, or by any other means."/[2]/

With time, however, the principles of the Outer Space treaty have been subject to reinterpretation by different parties, who allege that the words of the Outer Space Treaty do not really mean what they say. The reinterpretations enable claims similar to those made on terrestrial resources. For example, about 20 years ago, the "Equatorial" countries, basing their arguments on laws of physics, made claims of national sovereignty over certain segments of the geostationary orbit (GSO); these claims evolved into preferential or special rights to the GSO./[3]/ Their claims were rejected by the major space powers, which declared that their sovereignty claims were misplaced. It should be noted that several of the countries which dismissed these claims are now engaging in the auctioning of geostationary orbital positions (slots) and related resources, such as the frequencies to be used by spacecraft, alleging that remuneration for their use merely reflects their most efficient allocation./[4]/

In the past ten years several national events and factors account for the move from government to the private sector to undertake certain space-related activities. The US policies in support of privatisation / commercialisation were embodied in several laws, such as the 1985 U.S. Executive Order on Separate Satellite systems, and the Space Commercialisation Act./[5]/ The creation of private international satellite systems, such as PANAMSAT, and ORION, as separate from the traditional intergovernmental organisations -INTELSAT and INMARSAT- are a reflection of the "commercialisation" and "privatisation" of space communications. Further, the 1986 "Challenger" accident led to curtailment of non-military payloads on the space shuttle, and to the "commercialisation" of many launches./[6]/ Hence, the incipient "commercialisation" and "privatisation" of certain space activities gained in importance./[7]/

By the late 1980s, many corporations, with the approval of their governments, became increasing involved in all phases of space activities. With the impending demise of communism, the major space powers seemed to realise that their public sector did not have the financial resources to pursue many space activities, but hoped that the private sector would be able to find them.

The demise of "Communism" led to the further shifting of activities from government to the private sector. By the early 1990s, economic considerations seemed to become paramount, with the result that the private sector began performing functions that heretofore had been considered the purview of governments. Private corporations and financial institutions began assuming a leading role, while governments were being relegated to a secondary role in many of these activities, especially in their financing.

PROMOTING PRIVATISATION

In the US's satellite telecommunications sector, these policy and economic changes became more evident with the implementation of the "separate satellite systems", and the planning of new satellite systems in non-geostationary orbits (NGSOs)/[8]/, to be operational by the end of this century. The NGSOs will revolutionise the way we communicate, the way we relate to each other; they will also change the regulatory environment, and concepts of responsibility of entities engaged in these activities./[9]/ The end of this century will see outer space teeming with numerous private satellite systems in a variety of orbits, comprised of scores of spacecraft, balloons and platforms./[10]/

These activities are a reflection of one of the most important policy changes that has occurred in the last decade: the separation of regulatory responsibilities from daily operations of telecommunication systems. Regulatory entities have been created, while operations are slowly being "demonopolised", and transferred from the public sector to private companies./[11]/ As the private sector's involvement and influence increase, some corporations and individuals seem to be of the view that the ITU should take second place to the World Trade Organisation (WTO) in the allocation of frequencies and regulation of telecommunications systems, even though telecommunications services only recently became subject to the GATT (General Agreement on Trade and Tariffs)./[12]/

The views reflected in the WTO agreements signal important changes in the regulation and provision of services in the telecommunications sector. Wireless communications are expanding and their importance is increasingly linked to development, with radio frequencies viewed as potential "gold mines in the sky Hence the "auctioning" of radio frequencies, a new means of generating revenues for the governments, while creating expectations of (sovereign?) property rights over the frequencies and orbital slots. This practise is seen as acceptable within the framework of corporate / economic development, especially as industries and corporations becomes "globalised". Many strong advocates of the commercialisation and privatisation/[13]/ of outer space activities and resources (including radio frequencies), however, are not familiar with their international regulation; thus, they do not consider that auctioning these resources may be in violation of any international legal instrument, such as the Outer Space Treaty, and potentially the ITU Convention as well.

As the "ATIONs" gain ground, and "ISMs" become concepts of another era, underlying legal principles are also subject to reinterpretation. This is reminiscent of George Orwell's "Animal Farm". /[14]/ This fairy tale is an indictment of several "ISMs": communism, totalitarianism, authoritarianism; i.e., political systems that preclude expressions of dissent or disagreement with the decision-makers. During the first stages of the animals' rebellion, several tenets of "Animalism" were agreed to, one of the main ones stating that "all animals are equal." /[15]/ With time, however, a few other words were added to this simple phrase: "There was...now a single Commandment:

ALL ANIMALS ARE EQUAL BUT SOME ANIMALS ARE MORE EQUAL THAN OTHERS. /[16]/

Similar revisions seem to be occurring with outer space activities: some States claim that they place a greater value on outer space, and thus are able to make better use of these resources than others. Admittedly, some States have greater access than others to these resources, but this does not confer on them the right to unilaterally decide what is "best" for the others. Nor does the size of the state bear any relation to these claims. (New Zealand was one of the first countries to auction parts of its spectrum, while Tonga set the precedent of charging fees for the use of orbital positions. Since then, auctioning of orbital slots has gained in importance as a means of creating revenues for the government. It also creates expectations of property rights in the slots.)/[17]/

While the World Trade Organisation (WTO) and the ITU recognise that the radio frequency spectrum is a scarce resource, their views of how this resource should be allocated, by whom it should be managed and used, are quite divergent. While making the "best" or "most efficient" use of these resources is a laudable goal, the question remains: who will define and determine what is meant by "most efficient"? The party willing to pay the highest price for the resource, even though it may not be able to use it?

Economics may be one criterion, but it should not be the dominant, or sole basis for re-allocating, or assigning certain frequencies or orbital positions./[18]/ While the ITU is increasingly subject to criticism -- that its definitions and means of allocating these resources are not optimal, that they lead to non-use, or less efficient use – so far there is no indication that "privatising" these endeavours or resources, or making them available through auctions, will lead to their more efficient use. Rather, access to these resources, based on economic or financial considerations or power alone, will lead to their (even more) skewed distribution, and their concentration in a few "mega-corporations". Ironically, this outcome would be contrary to one of this decade's key goals: competition and "market forces" replacing monopolistic control.

DÉJA VU?

With the implementation of privately owned and controlled non-geostationary satellite and other communication systems, national Administrations are being pressured to reconsider their concept of national sovereignty, to review their licensing procedures, to modify their long-standing policies on control and use of radio frequencies, and to contract with corporations for their future satellite communications. In return, the corporations promise to bring fruits of development and of better communications to all countries./[19]/
In this respect, however, the Outer Space Treaty, other UN Resolutions, the Preamble of the INTELSAT Agreement, even the ITU Constitution and Convention, should be re-examined, as these, and nearly every other document related to development make reference to the benefits that communications will bring to developing countries. Many of the benefits of space communications which were being promised thirty years ago are being reiterated by the corporations which plan to provide Satellite Personal Communication Services (S-PCS), or Global Mobile Personal Communication Services (GMPCS) via NGSOs. The principal difference is that private parties, rather than governmental organisations, are making the promises to other private parties, who in turn, are to convince Administrations and end users of the benefits of their new means of communication. The next decade will see the implementation of many new communication systems, (but not necessarily of the promises), whose principal beneficiaries will be the corporations of developed countries.

RESPONSIBILITY and LIABILITY ISSUES

Privatisation of outer space resources, their increasing control by only a few corporations, raises the issue of their accountability under the terms of the Outer Space Treaty. While it is often considered the "Magna Carta" of space activities, it is being subject to re-interpretation, to accommodate to changing times and pressures.

Administrations are being relegated to secondary roles in many respects, but a question remains: are private corporations prepared to take on full responsibility for the consequences of their endeavours, from which they hope to reap great financial benefits, most of which are increasingly beyond the control of governments? It should be recalled that under the terms of the Outer Space Treaty, it is still the Parties, or Governments that are internationally liable for damage to another State Party to the Treaty or to its natural or juridical persons by such object or its component parts..../[20]/, since State Parties to the [Outer Space] Treaty ...retain jurisdiction and control over such object[s].../[21]/

As with the rulers of "Animal Farm", one of the drawbacks to transferring power, or decision-making to a few, such as with the "privatisation" of certain activities, is a shift of accountability, even responsibility. Private investors are accountable primarily to their shareholders, whose main interest is financial profit. However, the public has little control over private investors, few means of holding them responsible or accountable for their actions. Government officials, on the other hand, may be held accountable to the persons who elected them, and thus removed.

Further, with the "globalization" of multi-national corporations, responsibility and accountability become even more tenuous or nebulous, as their "headquarters" become more difficult to define or locate. In the end, it may be that those "animals that are more equal than others", those who are the most persuasive, will manage to convince the others that private corporations alone are the best judges of what is best for the communications world.

The result could be a return to a monopoly situation, a new kind of colonialism, but with a variation. Rather than governments, a few multi-national monopolies so large as to be uncontrollable will have total domination of the means and content of communications. They will be in control of the hardware and software necessary to operate any communication system. As in prior colonial times, there will be total dependence on the part of the less economically affluent societies, on the largesse of corporations, not of other governments.

Some economists and political leaders would have us believe that the healthy growth of most enterprises will benefit from the globalization of their theories and practises, that national borders are obstacles to the economic benefits to be reaped from their activities. Globalization has led to increasing economic interdependence of countries, but the financial health of a few corporations seems to obscure the potential negative consequences of this trend. Ironically, globalised communications, and particularly satellite communications which should eradicate the differences and boundaries between countries, may have just the opposite effect.

CONCLUSION

In the last few years, "ISMS" have given way to "ATIONS", raising the question as to the future management (in the broadest sense of the term) of outer space activities and resources. The fundamental role of government is also put into question: is there still a need, let alone room, for it?

Privatisation of outer space resources, their increasing control by only a few corporations, also raises the issue of their accountability under the terms of the Outer Space Treaty.

Perhaps the basic principles of Article I of the Outer Space Treaty need to be restated, and reaffirmed:

The exploration and use of outer space...shall be carried out for the benefit and the interests of all countries, [not only for the benefit of a few corporations] … and shall [remain] the province of mankind."/[22]/

[1] Treaty on Principles Governing the Activities of States in the Exploration and Use of Outer Space, including the Moon and Other Celestial Bodies. United Nations Publication. [Cited as the Outer Space Treaty hereinafter.]

[2] Article II, Outer Space Treaty.

[3] *See* The Bogota Declaration of 1976. *See* also the UNCOPUOS Reports to the UN General Assembly, where the issue of these claims and principles to govern the use of the geostationary orbit (GSO) have been debated at nearly every session of the UNCOPUOS. Even in that forum, the importance of a new (commercial) envirnoment is recongised in the *Declaration on International Cooperation in the Exploration and Use of Outer Space for the Benefit and in the Interests of All States, Taking into Particular Account the Needs of Developing Countries.* UN Document A/AC.105/L.211, 11 June 1996. The International Telecommunication Union (ITU) Convention also reflects the evolution of these claims, and their acceptance, or at least acquiescence by the international community. (Article 33 of the 1982 Nairobi Convention, now Article 44 of the Convention, Geneva 1992 .Whether the ITU or the UN should take the lead in resolving htese issues has also been subject to debate for years.

[4] So far, the frequencies that have been auctioned have been those allocated for national terrestrial services. However, since the US has done away with the difference between domestic and international satellite systems, the frequencies and orbital slots that heretofore were for domestic communications systems, may now be used for these, as well as for transborder and international services. See the US FCC's "DISCO 1" January 1996 Report and Order.

[5] *See* President Reagan's 1985 Executive Order allowing for the creation of international satellite systems separate from INTELSAT. *See* also the U.S. 1984 Commercialization of Space Act, allowing for the provision of launch services by private corporations, and subsequently by joint ventures in partnership with corporations of the former Soviet Union.

[6] The US Commercialisation of Space Act was drafted in the early 1980s, enacted in 1989, and subesequently amended on various occasions. An examination of these changes is beyond the scope of this paper.

[7] The author is grateful to Dr. Eilene Galloway for supplying several key NASA definitions:

> ***Commercialisation:*** the employment of private sector financial and other resources to provide goods and services which can meet Government requirements as well as those of a substantial number of other customers. Because the Government is neither the sole nor the predominant customer, it neither dictates specific design or operational parameters, nor bears more than its proportionate share of development or overhead costs.
>
> ***Privatisation*:** Generally the transfer to the private sector of responsibility for providing ongoing, necessary services or functions currently being provided by the Government for itself and others through in-house labor and property resources. When the Government transfers to the private sector the responsibility for services or functions performed by the Government for the sole benefit of the Government, the transfer is referred to *contracting out, a subset of privatisation.*

[8] Among the NGSO systems are (in alphabetical order): Globalstar, ICO, IRIDIUM, ODYSSEY, ORBCOMM, Skybridge, Teledesic. Other proposed systems would use balloons ("SkyStation"), or stratospheric platforms to provide a variety of communications.

[9] Carl Pope, Executive Director of the Sierra Club warns against the nefarious effects of the "enhanced power that globalisation [is] giving to multinational corporations." Economic profits are paramount, while human considerations diminish in importance. He states that "[t]he challenge to building a decent world order is to ...find solutions that guarantee that the fruits of human invention and labor are used to improve the lives of both present and future generations." *"Beware Globalisation That Leaves People Behind."* International Herald Tribune, p. 6, 22-23 November 1997.

[10] The ITU's listing of proposed satellite systems has increased exponentially in the

last few years, resulting in many Administrations advocating the implementation of measures, such as "due diligence" tests which would prevent or at least reduce the number of filings of "paper" satellite systems. Resolution 18 of WRC '95 Final Acts was addressed and amended at the 1997 World Radio Conference (WRC).

[11] In most countries, there are now regulatory bodies, separate from the major operator or provider of basic telecommunications. The new regulatory entities enjoy varying degrees of success, independence from the Ministry, and adequate financing.

[12] The World Trade Organisation(WTO) concluded an agreement on General Basic Telecommunications in February 1997; WTO Members agreed to open, non-discriminatory, transparent regulations and treatment in licensing procedures, and access to the radio frequencies which are essential for satellite communications services.

[13] *See* note 7, *supra*, for definition of these terms.

[14] George Orwell, "Animal Farm", Copyright 1945; Penguin Group, publishers of Penguin Books (1989).

[15] Ibid., p. 15 This is the seventh of the "The Seven Commandments", "...the unalterable law by which all the Animals on Animal Farm must live for ever after."

[16] Ibid; p. 90. (Emphasis in original.)

[17] These issues are not a merely "domestic" questions. With the globalisation of many economic enterprises, as well as of corporations, practises which are deemed successful in one country are adopted by others. For example, Mexico is considering auctioning parts of "its" spectrum, while the trade press talks of the orbital slots "owned" by Mexico . *See Space News,* p.1., October 1997.

Auctions of resources that until now are deemed to be the "province of mankind" under the Outer Space Treaty could set unwelcome precedents: if they can be auctioned off, and become the "property" of private parties, why not do the same with other elements that are part of the "common heritage of mankind," suçh as carving up the high seas, and assigning the right to use certain shipping lanes in the or air lanes in the skies to specific coprorations? The high seas and air space are as intangible as satellite uplinks and downlinks, and their exclusive use may increase the profits (or lessen the costs) of members of the transportation industry.

[18] The ITU Radio Regulations provide definitions of these terms of art: *Allocation* (of a frequency band to particular services), *allotment* (of a radio frequency or radio frequency channel, for use by one or more administrations for a radiocommunicaiton service), and *assignment* (authorisation given by an Administration to use a radio frequency under specified conditions. ITU Radio Regulations, Chapter S1, Article S1, Section II, S.1.16-1.18. [Italics in original].

[19] In another context, Pope, *supra, note 9,* points out that while globalisation has enhanced the power of multi-national corporations, the benefits of this economic strategy have not all been positive. Unemployment is rampant, living standards have fallen, the environment adversely affected, particularly in developing countries.

[20] Outer Space Treaty, Article VII. There seems to be little concern on the part of the proponents of the NGSOs of the impact on the outer space environment that thousands of satellites and hundreds of launches are likely to have. The concept of responsibility and liability in Art. VII does not seem to include responsibility for these potentially ill effects.

[21] Ibid, Article VIII.

[22] Outer Space Treaty, Article I . Proposed amendments or changes in wording are indicated by the square brackets [ ].

IISL-97-2.05

UN GENERAL ASSEMBLY RESOLUTIONS AND OUTER SPACE LAW

Andrei D. Terekhov*
Doctor of Law, Moscow
Member of the Board of Directors of the International Institute of Space Law
Member of the International Academy of Astronautics
Senior Legal Officer, Office of Legal Affairs, United Nations

Introduction

The leading role in developing outer space law is played by treaties, while custom acquires an increasingly meaningful role as the mankind expands its activities in outer space and more States are becoming involved in the exploration and use of this environment[1].

From the beginning of the space age, the United Nations has become a focal point in the international cooperation in the field of outer space, including developing a legal framework for orderly activities in the "sixth ocean".[2] In 1966, the General Assembly of the United Nations commended the Treaty on the Principles Governing Activities of States in the Exploration and Use of Outer Space, including the Moon and Other Celestial Bodies (the Outer Space Treaty)[3], which was destined to become a universally recognized cornerstone of international space law.

The starting point for elaborating the Outer Space Treaty was General Assembly resolution 1962 (XVIII) entitled "Declaration of Legal Principles Governing the Activities of States in the Exploration and Use of Outer Space" adopted on 13 December 1963. Since that time the Assembly has adopted, and continues to adopt, numerous resolutions, including declarations of principles, on various outer space matters.

The purpose of this paper is to attempt to examine the role played by the UN General Assembly resolutions in the development and codification of international space law.

A. Legal status of General Assembly resolutions: brief general overview

As Paul Szasz noted, "it is common practice to start, and sometimes to end, any listing of various forms of international law by referring to the four sub-paragraphs of Article 38(1) of the Statute of the International Court of Justice".[4] Indeed, according to that Article, there are three sources of international law: international conventions, international custom, and the general principles of law recognized by civilized nations. The Article also refers to judicial decisions and the teachings of the most highly qualified publicists as "subsidiary means for the determination of rules of law".

A norm of international law is a legally binding rule of behaviour of States and other subjects of international law in their international relations[5]. In accordance with the Charter of the United Nations, the General Assembly makes recommendations on the issues within its competence. Thus, as a general rule, Assembly's resolutions are not obligatory and do not create norms of international law.

A wide variety of opinions exists, both among States[6] and in the doctrine, concerning the legal role and force of at least some categories of General Assembly resolutions. The matter of contention appears to be whether or not certain Assembly resolutions should be viewed as sources of international law in addition to the three traditional sources (treaty, custom, general principles).

Many distinguished authors wrote on the subject of the legal character and force of General Assembly resolutions. Professor Blaine Sloan provided an almost ten pages-long bibliography list which, in his words, "gives a broad sampling of opinion on General Assembly resolutions"[7]. Professor Sloan's article contains a comprehensive review of various views on the subject, and, although this author does not share all the conclusions of the above article, there is no need to duplicate that thorough effort.

This article is not intended to address all aspects of this continuing debate or review general aspects of the problem. Rather the objective of this author is to review specifically some aspects of the role that General Assembly resolutions have played and are playing in the development of the law of outer space. However, for purposes of this paper, it is necessary to briefly summarize the existing views on the topic in order to put in a correct perspective this author's comments on the role of General Assembly resolutions in the development of international space law.

While, as stated above, substantive differences

 The views expressed herein are those of the author and do not necessarily reflect the views of the United Nations.

exist, it appears possible to identify areas of converging views on the subject of the legal force of General Assembly resolutions. Everybody seems to agree that an overwhelming majority of such resolutions are recommendations as provided in relevant Articles of the UN Charter. Those resolutions are not binding. However, they have moral and political weight, and cannot be simply ignored especially if adopted by consensus or without voting. Everybody also seems to agree that there are General Assembly resolutions that are binding: they concern the internal functioning of the United Nations, administrative, budgetary and some other questions. It is a third category or group of General Assembly resolutions which are evaluated differently both by States and scholars. Those resolutions are, in the most cases, adopted in the form of "declarations" which have "a special place among the General Assembly resolutions".[8] Without going into details, according to one view, all those resolutions (declarations) are merely recommendations. According to the second view, some of those resolutions, sometimes referred to as "soft law"[9], establish obligatory rules binding on States or rather help to establish existence and contents of such rules. According to the third view, while not exactly binding, those resolutions, contribute to the establishment of legal rules which may become obligatory in due course.

The United Nations Office of Legal Affairs has written that "...there is probably no difference between a 'recommendation' and or a 'declaration' in the United Nations practice as far as strict legal principle is concerned". This opinion, provided to the General Assembly some 35 years ago, stated that:

> "A 'declaration' or a 'recommendation' is adopted by a resolution of a United Nations organ. As such it cannot be made binding upon Member States, in the sense that a treaty or a convention is binding upon parties to it, purely by the device of terming it a 'declaration' rather than a 'recommendation'... However, in view of the greater solemnity and significance of 'declaration' it may be considered to import, on behalf of the organ adopting it, a strong expectation that Members of the international community will abide by it. Consequently, insofar as the expectation is gradually justified by State practice, a declaration may by custom become recognized as laying down rules binding upon states".[10]

In 1981 the same Office advised that:

> "In the practice of the United Nations a declaration is a formal and solemn instrument suitable for those occasions when principles considered to be of special importance are being enunciated. Apart from the solemnity and formality associated with a declaration there is legally no distinction between a declaration and a recommendation which is less formal."[11]

The above views continues to be fully valid nowadays.

General Assembly resolutions, both declarations and, as will be shown, "conventional" resolutions, play an important role in the development of the law of outer space. Several areas in this respect may be distinguished as described below.

B. Declarations of legal principles adopted by the General Assembly

Before examining the five outer space declaration, a potential terminological misunderstanding should be clarified. Four of the five declarations have the term "principles" in their titles. It is important to remember that these principles are not those meant in Article 38 of the Statute of the International Court of Justice as one of the sources of international law (the general principles of law recognized by civilized nations).

As indicated by Professor Bin Cheng over 30 years ago in connection with the 1963 Declaration:

> "Although the title... is 'Declaration of Legal Principles Governing the Activities of States in the Exploration and Use of Outer Space', this does not mean that the principles are necessarily binding. For example, in innumerable treaties, States have developed a vast number of optional rules of conduct in the field of international economic relations and international air transport. These optional rules are legal in character, but their binding force is potential (*in posse*) rather than actual (*in esse*)... [The 1963 Declaration], to the extent to which it is not restating existing rules of international law, merely expresses non-binding standards of international law governing the activities of States in the exploration of outer space".[12]

The above evaluation is equally applicable to outer space declarations adopted after the 1963 one.

The following parts of this section briefly review the five outer space declarations elaborated in the United Nations Committee on the Peaceful Uses of Outer Space (COPUOS) and adopted by the General Assembly, and also examines certain related legal issues.

1. Declaration of Legal Principles Governing the Activities of States in the Exploration and Use of Outer Space

As indicated above, the starting point for elaborating the Outer Space Treaty was General Assembly resolution 1962 (XVIII) of 13 December 1963, "Declaration of Legal Principles Governing the Activities of States in the Exploration and Use of Outer Space".

This Declaration was the first comprehensive[13] international intergovernmental document reflecting the position of the world community as a whole concerning general principles by which States should be guided in the exploration and use of outer space.

Those principles stipulate[14] that:

- outer space should be explored and used for the benefit and in the interests of all mankind;
- outer space is free for exploration and use by all States;
- outer space is not subject to national appropriation;
- activities in outer space should be carried out in accordance with international law;
- States bear international responsibility for national activities in outer space;
- In the exploration and use of outer space, States should be guided by the principle of co-operation and mutual assistance;
- States should retain jurisdiction and control over space objects which they registered;
- States should be internationally liable for damage caused by their space objects;
- States should regard astronauts as envoys of mankind in outer space, and should render them all possible assistance.

Although legally not binding at the time of adoption, this unanimously approved document "solemnly declare[d]" the above principles all of which subsequently had became obligations under the Outer Space Treaty.

Manfred Lachs who was Chairman of the Legal Subcommittee of COPUOS at the time of the adoption of the 1963 Declaration, having reviewed the circumstances of the approval of the document, came to the conclusion that "it is difficult to regard the 1963 Declaration as a mere recommendation: it was an instrument which has been accepted as law".[15]

The importance of the 1963 Declaration has not disappeared with the conclusion of the Outer Space Treaty[16]; it continues to serve as universally accepted guidelines, especially for those States which, for certain reasons, have not so far acceded to that Treaty.[17]

In connection with the 1963 Declaration, as a basis for the Outer Space Treaty, it is pertinent to recall that, at the same session, the General Assembly adopted another important resolution the substance of which had also been subsequently incorporated in the Outer Space Treaty.

This resolution, 1884(XVIII) of 17 October 1963 entitled "Question of General and Complete Disarmament", welcomed the expressions by the USSR and the USA "of their intention not to station in outer space any objects carrying nuclear weapons or other kinds of weapons of mass destruction". In addition, that resolution "solemnly call[ed] upon all states:

> (a) To refrain from placing in orbit around the earth any objects carrying nuclear weapons or any other kinds of weapons of mass destruction, installing such weapons on celestial bodies, or stationing such weapons in outer space in any other manner;
>
> (b) To refrain from causing, encouraging or in any other way participating in the conduct of the foregoing activities".

This resolution may be viewed as a symbolic "ratification" by the international community of an agreement reached by the USSR and the USA. The Secretary-General of the United Nations, addressing the Assembly because of the significance of the occasion, stated that the adoption of the resolution implied acceptance by the UN of the political and moral responsibility for its implementation.[18] As a general obligation, this provision subsequently formed a part of Article IV of the Outer Space Treaty.

Both above resolutions - 1962(XVII) and 1884(XVII) - are "recalled" in the Preamble of the Outer Space Treaty.[19]

Of all General Assembly resolutions relating to outer space, the 1963 Declaration is the one which has come closest, in the opinion of this author, to stating law. Its unanimous adoption, the intention of the Assembly to make it a "solemn" document reflecting its significance, a strong expectation that members of the international community will abide by it,[20] subsequent confirmation of all its substantive provisions in the Outer Space Treaty, actual practice of States acting in conformity with the principles proclaimed in the Declaration - the totality of those circumstances can be viewed as evidence that at least most fundamental principles of the Declaration are acquiring (and perhaps have already acquired) the character of customary norms of international law.[21]

Indeed, as a theoretical example concerning just one principle of the declaration, if, a State which is **not** a Party to the Outer Space Treaty (which prohibits national appropriation of outer space and celestial bodies) were to attempt to-day to "nationally appropriate" part of, say, the lunar surface, there is little doubt that such action would be viewed by international community as violating **universally recognized** international law.[22]

2. <u>Principles Governing the Use by States of Artificial Earth Satellites for International Direct Television Broadcasting</u>

On 10 December 1982 General Assembly adopted resolution 27/92 "Principles Governing the Use by States of Artificial Earth Satellites for International Direct Television Broadcasting" (the DTBS[23] Principles). This action by the Assembly concluded a decade of intensive negotiations conducted in COPUOS and its Legal Subcommittee.[24]

The DTBS Principles contain ten sections dealing with "Purposes and objectives", "Applicability of international law", "Rights and benefits", "International co-operation", "Peaceful settlement of disputes", "State responsibility", "Duty and right to consult", "Copyright and neighbouring rights", "Notification to the United Nations", and "Consultations and agreements between States".

The DTBS Principles are the only outer space principles adopted by General Assembly by voting.[25] While the detailed examination of the negotiating process and of provisions of this document goes beyond the scope of this paper, it is pertinent, however, to recall that the main disagreement with regard to the DTBS Principles was whether or not establishment of international direct television broadcasting service should require a prior consent of a State which would be receiving those foreign broadcasts. This central difference between proponents and opponents of the notion of free flow of information had not been overcome, and eventually led to voting whereby some Western and developing States either voted against or abstained. The final principle of the document, "Consultations and agreements between States", envisages that DTBS service "shall only be established... on the basis of agreements and/or arrangements". In other words, under the DTBS Principles, consent of a receiving State is required.

Explaining the US vote on the DTBS Principles, the American delegate, basically reflecting sentiments of other countries which did not support the resolution, stated, inter alia, that:

> "His delegation recognized that the concept of direct television broadcasting by means of artificial earth satellites caused much wider concern than other forms of broadcasting, even among Governments which did not have a philosophy of state control of all information flows. It strongly believed however that, in a United Nations set of principles, such legitimate concerns should not be dealt with by suggesting that every State, including totalitarian States, should be afforded a non-conditional veto over that form of broadcasting. In the view of his delegation, any principle requiring that broadcaster must obtain the consent of a foreign Government would violate United States obligations towards both the broadcasters and the intended audiences; it would also violate article 19 of the Universal Declaration of Human Rights on the right to freedom of expression".[26]

The DTBS Principles resolution is obviously a recommendation, and thus is not legally binding. As far as its moral and political force is concerned, it is undermined by the fact that the declaration was adopted by voting and that a few major nations actively involved in the exploration and use of outer space, including first of all the US, did not support it. As Eilene Galloway indicated in connection with the DTBS voting, "this type of decision is not effective in gaining compliance as compared with consensus".[27]

It is only natural that General Assembly resolutions supported by all States acquire more significance than those to which objections had been raised, and compliance with their provisions has a much better chance to become a reality than in the case of a resolution adopted by voting.

3. Principles Relating to Remote Sensing of the Earth from Outer Space

On 3 December 1986 the General Assembly adopted resolution 41/65 "Principles Relating to Remote Sensing of the Earth from Outer Space" (the RS Principles). Unanimous approval of this document was a final act of over a decade of complex negotiations.[28]

This declaration contains fifteen principles relating to: definitions (Principle I); carrying remote sensing activities for the benefit of all States, and taking into particular consideration the needs of developing countries (Principle II); conducting those activities in accordance with international law (Principle III); conducting remote sensing activities in accordance with principles contained in Article I of Outer Space Treaty; on the basis of respect for the principle of full and permanent sovereignty of all States and peoples over their own wealth and natural resources, and with due regard to the rights and interests of other States and entities; remote sensing "shall not be conducted in a manner detrimental to the legitimate rights and interests of the sensed State" (Principle IV); promoting international cooperation in remote sensing field (Principle V); maximizing the availability of remote sensing benefits (Principle VI); providing technical assistance (Principle VII); promoting international cooperation by the UN and relevant UN agencies (Principle VIII); informing the UN Secretary-General and States of remote sensing activities (Principle IX); protection of Earth's natural environment (Principle X); protection of mankind from natural disasters (Principle XI); access of the sensed State to remote sensing data and information concerning its territory (Principle XII); consultations (Principle XIII); international responsibility (Principle XIV); and settlement of disputes (Principle XV).

The RS Principles have been serving, already for a decade, as general guidelines for activities in the field of remote sensing from outer space. While not binding, they are respected by those conducting outer space activities. This author is not aware of any disputes among States concerning the application of the RS Principles which fact proves that the document is a balanced one and, so far, has adequately met the expectations and served the interests of both sensing and sensed States.

Professor Christol remarked that "the principal customary rule found in the [RS] Principles is the right of States, international intergovernmental organizations and non-governmental entities to engage in such sensing activities".[29]

4. Principles Relevant to the Use of Nuclear Power Sources in Outer Space

On 14 December 1992, the General Assembly adopted, without a vote, resolution 47/68 "Principles Relevant to the Use of Nuclear Power Sources in Outer Space" (the NPS Principles).[30] COPUOS and its two subcommittees worked on this subject since the late 70s after the "COSMOS-954" incident.[31]

The NPS Principles consist of a preambular section and the following eleven principles: "Applicability of international law", "Use of terms", "Guidelines and criteria for safe use" (which has a brief opening part and three sections: "General goals for radiation protection and nuclear safety", "Nuclear reactors" and "Radioisotope generators"), "Safety assessment", "Notification of re-entry", "Consultations", "Assistance to States", "Responsibility", "Liability and compensation", "Settlement of disputes" and "Review and revision".

The NPS Principles contain rather detailed guidelines on a very complex subject of the use of nuclear power sources in outer space. While not binding, the document carries a heavy moral and political weight, in particular because in relates to the use of nuclear energy - a sensitive public subject in many countries.

States, including those actively involved in launching space objects, appear to follow the recommendations contained in the NPS Principles. In November 1996, the Russian Federation, acting "in accordance with the Principles Relevant to the Use of Nuclear Power Sources in Outer Space", dispatched a communication to the Secretary-General of the United Nations informing him of the planned launching "Mars-96" probe carrying "radionuclide heat source based on plutonium-238".[32] Subsequently, when the mission malfunctioned, Russia sent another notification to the UN[33].

The General Assembly played an important role in the elaboration of the NPS Principles not only because the Assembly had instructed COPUOS to consider this problem and the Principles were eventually adopted as GA resolution, but also because certain NPS guidelines were first approved by the Assembly even before the NPS Principles as a whole were finalized.

General Assembly resolution 33/16 of 10 November 1978, in operative paragraph 9, requested "launching States to inform States concerned in the event that a space object with nuclear power sources on board is malfunctioning with a risk of re-entry of radio-active materials to the earth". Subsequently this provision has been further developed and specified, and eventually became Principle 5, Notification of re-entry, of the NPS Principles.

General Assembly resolution 42/68 of 2 December 1987, in operative paragraph 11, endorsed "the agreements reached in the Scientific and Technical Sub-Committee with respect to the use of nuclear power sources in outer space". Those agreements[34] were the recommendations formulated by the technical experts with the view to ensuring safe use of NPS in outer space, which recommendations had been subsequently reflected in the NPS Principles.

Thus, Member States used General Assembly resolutions as a mechanism for putting into effect certain recommendations concerning the use of NPS in outer space before the conclusion of the elaboration of the NPS Principles. These actions were explained perhaps by the importance which the international community attached to the need for attaining safety in this area of human activities in outer space.

5. Declaration on International Cooperation in the Exploration and Use of Outer Space for the Benefit and in the Interest of All States, Taking into Particular Account the Needs of Developing Countries

On 13 December 1996 General Assembly adopted, without a vote, resolution 51/123 "Declaration on International Cooperation in the Exploration and Use of Outer Space for the Benefit and in the Interest of All States, Taking into Particular Account the Needs of Developing Countries" (the OSB[35] Declaration). This subject has been considered in COPUOS and its Legal Subcommittee since 1989.[36]

The OSB Declaration consists of eight paragraphs envisaging that: (1) international cooperation in outer space field shall be conducted in accordance with international law, for the benefit and in the interest of all States, it shall be the province of all mankind, and particular account should be taken of the needs of developing countries; (2) States are free to determine all aspects of their participation in such cooperation, and contractual terms in space cooperative ventures should be fair and reasonable; (3) all States should contribute to promoting space cooperation giving particular attention to the benefit and the interests of developing countries; (4) cooperation should be conducted in the modes that are considered most effective and appropriate by the countries concerned; (5) cooperation should aim, inter alia, at the goals of promoting the development of space science and technology and of it applications, fostering the development of space capabilities in interested States, and facilitating the exchange of expertise and technology among States; (6) national and international agencies, institutions, organizations and countries should consider the appropriate use of space applications and the potential of international cooperation for development goals; (7) COPUOS role in the exchange of information on international space cooperation should be strengthened; and (8) all States should be encouraged to contribute to the UN Space Applications Programme and to other initiatives in outer space field.

Too little time has elapsed to attempt to make conclusions as to how States are following the recommendations contained in the OSB Declaration. However, the fact that the Declaration has been elaborated in COPUOS on the basis of consensus and adopted by the Assembly without a vote permits the expectation that the Declaration will indeed be taken into account by States, and will make a contribution in the further development of international cooperation in the exploration and peaceful uses of outer space.

6. References to existing law in outer space declarations

All outer space declarations adopted by the General Assembly, except the 1963 one, contain references to existing law.

For example, Principle IX of the RS Principles envisages that "a State carrying out a programme of remote sensing shall inform the Secretary-General of the United Nations... in accordance with article IV of the [Registration Convention] and article XI of the [Outer Space Treaty]". Principles 8 and 9 respectively of the NPS Principles stipulate that "States shall bear international responsibility for national activities involving the use of nuclear power sources in outer space" and "shall be internationally liable for damage caused" by space object carrying an NPS on board "in accordance with article VI" and "in accordance with article VII" of the Outer Space Treaty. The OSB Declaration states that "international cooperation in the exploration and use of outer space for peaceful purposes... shall be conducted in accordance with the provisions of international law, including the Charter of the United Nations and the [Outer Space] Treaty".

As stated by Erik Suy, the former Legal Counsel of the UN,

> "The General Assembly's authority is limited to the adoption of resolutions. These are mere recommendations having no legally binding force for member states. Solemn declarations adopted either unanimously or by consensus have no different status, although their moral and political impact will be an important factor in guiding national policies. Declarations frequently contain references to existing rules of international law. They do not create, but merely restate and endorse them. Other principles contained in such declarations may appear to be new statements of legal rules. But the mere fact that they are adopted does not confer on them any specific or automatic authority". (Emphasis added)[37]

Whether or not all relevant actions recommended by the above outer space declarations are indeed obligations established by the agreements to which those Declarations refer, is an interesting legal subject. It appears that, for example, Principle 8, Responsibility, of the NPS Principles envisions something that is not required by Article VI of the Outer Space Treaty although this Principle 8 opens with "In accordance with article VI" of that Treaty.[38] Generally, it appears that this subject merits additional study.

7. "Shall/should" issue in outer space declarations

At certain stage of elaborating practically all existing outer space declaration a question was discussed as to whether the document would use "shall" or "should" as its grammatical construction.

In this respect the view was expressed that usually "shall" is used in treaties which contain legally binding provisions, while "should" is more appropriate for non-binding documents like GA resolutions. Often until the very last moment the negotiating text of a declaration either contained both expressions in square brackets or had an asterisk indicating that the choice of "shall" or "should" will be made later.

As far as final texts of those declarations are concerned, the 1963 Declaration solemnly declared that in the exploration and use of outer space States "should" be guided by the principles contained in the Declaration. However, the principles themselves use not "should", but rather "shall" or simply the present tense.

The DTBS Principles use mostly "should" with occasional "shall" and a couple of present tenses.

Both the RS and the NPS Principles use "shall".

The OSB Declaration uses mostly "should" with a couple of "shall"s and one present tense.

The eventual selection of "shall" or "should" by the drafters of the above documents did not, however, affect their legal character - they are still recommendations. Moreover, the fact that, for example, the RS Principles contain "shall" and the OSB Declaration uses mostly "should" is not perceived as an indication that the former makes stronger recommendations than the latter. In view of the foregoing, it appears that the "shall/should" controversy has basically lost its relevance at least as far as outer space declarations are concerned.[39]

8. Declaration *versus* treaty

After reviewing specific General Assembly declarations on outer space, and certain other related issues, one may ask a question as to why space law-making in the United Nations has resulted in both treaties and resolutions. In other words and given the fact that, notwithstanding all the existing differences of opinion concerning the legal force of GA resolutions, it is indisputable that treaties establish unequivocally binding obligations for parties, and, from this perspective, it may have seem preferable to establish in treaties, rather than in GA resolutions, all guidelines for activities in outer space.

The simple answer to this question is of course that Member States have decided that certain guidelines for outer space activities must be obligatory and, accordingly, must take the legal form of binding agreements, and that certain other guidelines for such activities must not be obligatory and, accordingly, must take the form of non-binding resolutions. However, which reasons guide States in making the choice is a complex question,[40] especially because at least some of those reasons may be not necessarily always stated openly.

There may be several explanations of the fact that space law-making in the United Nations has produced both binding treaties and non-binding declarations. First, outer space activities are a relatively new domain, and some States perhaps feel that it is premature to establish certain rules in this area immediately in mandatory manner - in treaties - while no sufficient experience has been accumulated. "The form of United Nations principles is particularly convenient when not all members of the international community are convinced about the usefulness and viability of immediate treaty regulation".[41]

Second, and even more important, is perhaps the fact that space science and technology are rapidly developing, and there may be a certain degree of reluctance to adopt mandatory treaty rules for new evolving areas of human activities in outer space, taking into account that, once established in a treaty, such obligations are rather difficult to change: amending a treaty may take a long time and require consent of other parties.

Third, an advantage of a General Assembly declaration is that, although non-binding, it is addressed to all States, while a treaty commits only its parties.

Finally, "one of the reasons for preparing non-binding normative instruments is that it is much easier to do so than to formulate, adopt, and bring into force agreements that are intended to have binding effect".[42]

Naturally each particular State may have additional specific reasons for preferring to have rules for certain outer space activities established in the form of a non-binding document rather than in a mandatory treaty.

C. Annual General Assembly resolutions on outer space

Since the late 50s, each regular session the General Assembly considers an agenda item entitled "International cooperation in the peaceful use of outer space".[43] Under this item, the Assembly reviews the work of COPUOS and its two subcommittees, and determines their programme of work for the future. Detailed examination of all aspects of those resolutions goes beyond the scope of this paper. The following aspects, however, appears to be of interest in connection with the subject under review.

The above annual resolutions on outer space (sometimes they are referred to as "omnibus" resolutions) are recommendations. Their provisions, such as, for example, a traditional invitation to States to become parties to outer space treaties, are not binding.

However, a major part of each of those resolutions determines agenda items to be considered at the following year's sessions of COPUOS and its Scientific and Technical, and Legal Subcommittees. Thus, as far as legal topics are concerned, it is the General Assembly who decides the perspective of the UN work in the field of progressive development and codification of outer space law.

It is well established in the practice of the United Nations that a request by a principal or parent body to its subsidiary organ that an item be placed on the latter's agenda is considered to be a directive to be followed by the subsidiary body concerned.[44] Accordingly, a decision of the General Assembly to the effect that COPUOS or its subsidiary bodies should consider a given subject is binding. This does not of course mean that such a decision by the Assembly obliges delegations to take a certain defined position on the **substance** of the subject assigned by the Assembly to COPUOS or its Legal Subcommittee. A decision that a given topic should be considered merely means that it must be included in the agenda of the relevant organ, and, by implication, that States should implement Assembly's decision by making bona fide efforts to examine the subject.

Some of the formulations of items on the agenda of the Legal Subcommittee, both in the past and currently, do not directly envisage the elaboration of a treaty or a declaration of principles.[45] Whether a legal document should be elaborated under such agenda items or the work on such item should be limited to the exchange of views only, has been a perennial matter of disagreement among various delegation. It would appear that no such obligation exists, although nothing prevents the Subcommittee to initiate such a process of elaboration if it believes that action appropriate.

In addition to adopting a programme of work on legal subjects, annual resolutions are sometimes used as a mechanism for putting into effect certain specific recommendations concerning activities in outer space (see section on the NPS Principles above).

D. General Assembly resolutions concerning outer space agreements

1. Resolutions commending outer space agreements

COPUOS has successfully elaborated five international agreements.[46] The final action in the process of elaborating each of those instruments was the adoption of General Assembly resolution commending the agreement and requesting either Depository Governments (in case of the Outer Space Treaty and the Rescue Agreement) or the Secretary-General of the UN (in case of the Liability Convention, the Registration Convention and the Moon Agreement) to open the instrument for signature.

Thus, those General Assembly resolutions play the role of a final "seal of approval" of outer space agreements. Besides, in some cases those resolutions contained provisions which put the action (commending the agreement) in a specific historic perspective, and gave concrete guidance to COPUOS concerning its future work. For example, resolution 2222(XXI) of 19 December 1966 not only commended the Outer Space Treaty, but also instructed COPUOS to continue its work on agreements concerning space liability and assistance to astronauts, as well as to begin the study of questions relative to the definition of outer space.

Finally, General Assembly resolutions commending outer space agreements may, in some cases, serve as an additional means of interpretation of certain provisions of those agreements.

2. Review and revision clause of outer space agreements: General Assembly role

Three of the five outer space agreements elaborated in COPUOS (the Liability Convention, the Registration Convention and the Moon Agreement) contain identical provisions envisaging that:

> "Ten years after the entry into force of this Convention [Agreement], the question of the review of this Convention [Agreement] shall be included in the provisional agenda of the United Nations General Assembly in order to consider, in the light of past application of the Convention [Agreement], whether it requires revision".[47]

At its sessions in 1982, 1986 and 1994 respectively, the General Assembly reviewed the three above agreements and reflected the results of those reviews in its resolutions.[48] In all three cases the Assembly did not come to the conclusion that revision of each of the three instruments was necessary. However, the fact itself that drafters of the three agreements had assigned to the Assembly such an important role is noteworthy and raises a few interesting legal questions.

For example, what the next step should be in a hypothetical situation whereby the General Assembly would adopt a resolution to the effect that a revision of an outer space agreement is necessary, and this resolution is adopted by a majority vote with States Parties to that agreement voting against.[49]

Whatever is the answer to that question, it is clear that the General Assembly had been assigned an important role in "deciding the fate" of the above space agreements.

3. General Assembly's "promotion" of outer space agreements

Over the years a more or less established structure or pattern of General Assembly resolutions adopted under the above-mentioned agenda item has developed, and some of their provisions are reproduced every year without any changes.

For example, for many years the second operative paragraph of annual GA resolutions contains basically identical provision inviting States that have not yet become parties to the international treaties governing the uses of outer space to give consideration to ratifying or acceding to those treaties[50]. With a very few exceptions, those resolutions have been adopted without a vote, and when voting did take place, disagreements which necessitated it had nothing to do with the paragraph in question.

One may ask a question whether as result of this many years of reiterating this invitation, which obviously has become an established practice in General Assembly work on outer space, this provision has become a binding one in the sense that States are obligated by it to consider acceding to outer space agreements[51].

In the opinion of this author, the above provision is not intended to acquire and, therefore, has not acquired any binding character, and, therefore, continues to serve merely as a reminder to States to consider acceding to the five outer space treaties. In this respect, General Assembly resolutions play the role of a mechanism "promoting" outer space treaties and thus consolidating the rule of law in the exploration and use of outer space.

Concluding remarks

Most of the above-described types of General Assembly resolutions on outer space matters are recommendations. While some of the provisions contained in some declarations are obligatory, they are/became binding not because they were adopted by the Assembly, but rather because they reflect already existing (by the time of the adoption of the declaration) either conventional norms of international law or customary norms (such as, for example, the right to conduct remote sensing of foreign territories). On the other hand, declarations help in the process of establishment of such customary rules and also help to interpret those rules.

The 1963 Declaration of Legal Principles Governing the Activities of States on the Exploration and Use of Outer Space was not binding at the time of its adoption.[52] However, its unanimous adoption and other factors described above, including subsequent State practice, allows to view the provisions of that declaration as currently reflecting customary rules of international law.

Other resolutions, even if they are recommendations and are not binding, should not, however, be discarded as having no significance. Those resolutions guide States in situations where specific treaty norms have not yet been adopted or are too general, and, thus, contribute to ensuring orderly and dispute-free interaction of States in various areas of human activities in outer space. Besides, sometimes those resolutions "prepare the soil" for subsequent elaboration of outer space agreements.

One of the important functions of the General Assembly is the determination of the programme of legal work of COPUOS and its Legal Subcommittee.

The above report demonstrates that the General Assembly of the United Nations plays a significant role in the progressive development and codification of outer space law, as well as in guiding activities in the "sixth ocean"

NOTES

1. See **Vereshchetin, V.S.** and **Danilenko, G.M.**, "Custom as a Source of International Law of Outer Space", Journal of Space Law, Volume 13, Number 1, 1985, pages 22-35.

2. As **N.C. Goldman** noted, "In international space law, the United Nations has been a dominant force". (**Goldman, N.C.**, "American Space Law: International and Domestic", 2nd edition, 1996, page 23).

3. Treaty on Principles Governing the Activities of States in the Exploration and Use of Outer Space, including the Moon and Other Celestial Bodies, 27 January 1967, 18 U.S.T. 2410, T.I.A.S. 6347, 610 U.N.T.S. 205 (entered into force: 10 October 1967).

4. **Szasz, P.**, "General Law-making Processes", in "United Nations Legal Order", edited by **O. Schachter** and **C.C. Joyner**, 1995, page 38.

5. "Dictionary of international law", Editor-in-Chief: **B.M. Klimenko**, Moscow, 1986, page 234 (in Russian).

6. Judge **Schwebel** noted that "the views of states are profoundly divided on the issues of the impact of General Assembly resolutions on the content of international law" (**Schwebel, S.M.**, "The Effects of Resolutions of the U.N. General Assembly on Customary International Law", American Society of International Law, Proceedings of the 73rd Annual Meeting (1979), page 307).

7. **Sloan, B.**, "General Assembly resolutions revisited (forty years later)", The British Year Book of International Law (1987), Volume 58, pages 39-150 (bibliography is given on pages 142-150). Professor Sloan's bibliography list contains only one (although very important) reference to the works of former Soviet scholars - that of Gregory I. Tunkin. For more information on this subject, see **Osakwe, C.**, "Contemporary Soviet Doctrine on the Sources of General International Law", Proceedings of the American Society of International Law, 73rd Annual Meeting (1979), pages 310-324.

8. **Kopal, V.**, "The Role of United Nations Declarations of Principles in the Progressive Development of Space Law", Journal of Space Law, Volume 16, Number 1, 1988, page 19.

9. **Paul Szasz** indicated that "the term 'soft law', in the context currently in use, is meant to refer to norms that states actually do follow or at least subscribe to". (Op. cit., supra note 4, page 39, see also pages 45-47). On this subject, also see **Ferrazzani, M.**, "Soft Law in Space Activities", in "Outlook on Space Law over the Next 30 Years. Essays published for the 30th Anniversary if the Outer Space Treaty", Editor-in-Chief: **Gabriel Lafferranderie**, 1997, pages 429-447.

10. UN doc. E/CN.4/L.610 of 2 April 1962, cited in **Schwebel, S.M.**, "The Effect of Resolutions of the U.N. General Assembly in Customary International Law", American Society of International Law, Proceedings of the 73rd Annual Meeting (1979), page 304.

11. United Nations Juridical Yearbook 1981, page 149.

12. **Cheng, B.**, "United Nations Resolutions on Outer Space": 'Instant' International Customary Law", 5 Indian Journal Of International Law 23 (1965), reproduced in "International Law: Teaching and Practice", Edited by **Bin Cheng**, London, 1982, page 40.

13. General Assembly resolution 1721(XVI)A of 20 December 1961 "commend[ed] to States for their guidance in the exploration and use of outer space the following principles:

(a) International law, including the Charter of the United Nations, applies to outer space and celestial bodies;

(b) Outer space and celestial bodies are free for exploration and use by all States in conformity with international law and are not subject to national appropriation".

14. The provided enumeration of principles is not the exact quotation, but rather conveys their basic contents.

15. **Lachs, M.**, "The Law of Outer Space. An Experience in Contemporary Law-Making", Sithoff Leiden 1972, page 138.

16. As **C. Wilfred Jenks** observed back in 1965, "The authority of the Declaration of Legal Principles may be expected to grow with the passage of years. While it is somewhat less than a treaty it must already be regarded as rather more than a

statement of custom. It represents the Twelve Tables of the Law of Space..." (**Jenks, C.W.**, "Space Law", 1965, page 186).

17. See "International Law", Edited by **Y.M. Kolosov** and **V.I. Kuznetsov**, Moscow, 1994, page 537 (in Russian).

18. The United Nations and Disarmament, 1945-1970. United Nations, New York, page 177.

19. A third General Assembly resolution referred to in the Preamble of the Outer Space Treaty is resolution 110(II) of 3 November 1947 condemning war propaganda.

20. For analysis of some of the statements at the adoption of the 1963 Declaration, see **Cheng, B.**, supra note 12, pages 35-36.

21. "Dictionary of International Space Law", Editor-in-Chief: **V.S. Vereshchetin**, Moscow, 1992, page 32 (in Russian).

22. Professor **Sloan** noted that "norms once formulated in a resolution may be taken over in a convention, and resolutions in tandem with conventions may also facilitate the transmission of treaty norms into customary law... The interplay between the two may assist in the generation of customary norms of both general application and certain effect". (**Sloan, B.**, Op. cit., page 67).

23. DTBS (direct television broadcasting satellites) - this abbreviation was used by negotiators in COPUOS.

24. See **Christol, C.Q.**, "The Modern International Law of Outer Space", 1982, pages 605-719.

25. The resolution was adopted by 107 votes to 13, with 13 abstentions. (UN doc. A/37/PV.100 of 17 December 1982, pages 36-37).

26. UN doc. A/SPC/37/SR.34 of 26 November 1982, page 11.

27. **Galloway, E.**, "Expanding Space Law into the 21st Century", 35th Proc., Washington, D.C., AIAA, 1993, page 53.

28. For a review of the RS Principles see **Christol, C.Q.**, "Remote Sensing and International Space Law", Journal of Space Law, Volume 16, Nos. 1 @ 2, 1988, pages 21-44; **Kopal, V.**, "Principles Relating to Remote Sensing of the Earth from Outer Space: a Significant Outcome of International Co-operation in the Progressive Development of Space Law", Proceedings of the 30th Colloquium on the Law of Outer Space (hereafter - {Colloquium number} Proc.), Brighton, AIAA, 1988, pages 322-329.

29. **Christol, C.Q.**, "Remote Sensing and International Space Law", Journal of Space Law, Volume 16, Nos. 1 @ 2, 1988, page 40.

30. For history of the NPS negotiations and evaluation of the declaration, see **Jasentuliyana, N.**, "Multilateral Negotiations on the Use of NPS in Outer Space", Annals of Air and Space Law, Volume XIV, 1989, pages 297-337; **Kopal, V.**, "The Use of Nuclear Power Sources in Outer Space: A New Set of United Nations Principles?", 34th Proc., Montreal, AIAA, 1992, pages 124-131; **Jasentuliyana, N.**, "An Assessment of the United Nations Principles on the Use of Nuclear Power Sources in Outer Space", 36th Proc., Graz, AIAA, 1994, pages 312-321; **Terekhov, A.D.**, "Review and Revision of the Principles Relevant to the Use of Nuclear Power Sources in Outer Space", 36th Proc., Graz, AIAA, 1994, pages 336-348.

31. The Soviet satellite COSMOS-954, equipped with a small nuclear reactor, was launched on 18 January 1977. As reported in the press, on 6 January 1978, "sudden depressurization" of the satellite occurred as a result of which it began to "descend uncontrollably". On 24 January 1978, COSMOS-954 re-entered the atmosphere over Canada and scattered radioactive debris on its northern territories. Fortunately, the area of impact was scarcely populated and nobody was hurt. Canada, assisted by the United States, has conducted successful search, recovery and clean-up operations. As a result of the incident, Canada presented to the Soviet Union a claim for compensation for damage and subsequently received 3 million Canadian dollars "in full and final settlement". (See 20 I.L.M. 689 (1981), see also **Terekhov A.D.**, "International Liability for Damage Caused by Space Objects with Nuclear Power Sources on Board", 35th Proc., Washington, D.C., AIAA, 1993, pages 151-162). Besides, Canada, supported by other States, initiated the elaboration of NPS Principles in COPUOS (see **Jasentuliyana, N.**, "Multilateral Negotiations on the Use of Nuclear Power Sources in Outer Space", McGill Annals of Air and Space Law, Volume XIV, 1989, pages 297-337).

32. UN doc. A/AC.105/647 of 20 November 1996.

33. See UN doc. A/AC.105/648 of 21 November 1996.

34. See UN doc. A/AC.105/386 of 3 March 1987, paragraphs 55-63.

35. OSB (outer space benefits) - this abbreviation was used by negotiators in COPUOS.

36. See **Jasentuliyana, N.**, "Article I of the Outer Space Treaty Revisited", Journal of Space Law, Volume 17, 1989, pages 129-144; **Jasentuliyana, N.**, "Ensuring Equal Access to the Benefits of Space Technologies for all Countries", Space Policy, Volume 10, 1994, pages 7-18; **Benkö, M.**, **Schrogl K.-U.**, "The 1996 UN Declaration on 'Space Benefits' Ending the North-South Debate on Space Cooperation", 39th Proc., Beijing, AIAA, 1997, pages 183-186.

37. **Suy, E.**, "Innovations in International Law-Making Processes", in "The International Law and Policy of Human Welfare", edited by Macdonald, Johnston and Morris, 1978, page 190.

38. On this subject see **Terekhov, A.D.**, "International Responsibility for Using Nuclear Power Sources in Outer Space - Reflections on the Text Adopted by COPUOS", 34th Proc., Montreal, AIAA, 1992, pages 147-149.

39. However, some commentators believe that at least one of the declarations - the NPS Principles - should be revised, and that, in particular, the word "shall" should be replaced with the word "should". (See **Bennett, G.L.**, "A Technical Review of the U.N. Principles on the Use of Nuclear Power Sources in Outer Space", 38th Proc., Oslo, AIAA, 1996, pages 274-284; **Bennett, G.L.**, "Summary of the US Ad Hoc Working Group Meeting on Revising the U.N. Principles on the Use of Nuclear Power Sources in Outer Space", 39th Proc., Beijing, AIAA, 1997, pages 246-256).

40. Some of the reasons, provided by States, for choosing a non-binding declaration over a binding treaty in the case of the 1963 Declaration were examined by **Bin Cheng**, see supra note 12, pages 29-35.

41. **Kopal, V.**, Op. cit., supra note 8, page 20.

42. **Szasz, P.**, supra note 4, page 96.

43. For the list of resolutions adopted by the General Assembly under this and certain related items up to 1991, see "Space Activities of the United Nations and International Organizations", United Nations, New York, 1992 (UN doc. A/AC.105/521), Annex II, pages 281-286.

44. This practice was specifically reaffirmed, for example, at the session of the UN Disarmament Commission in 1986 when a proposal was made not to include in the agenda of that body an item recommended by the General Assembly. (See UN doc. A/CN.10/PV.102 of 10 May 1986).

45. For example, currently the formulation of one of the agenda items of the Legal Subcommittee reads as follows: "Matters relating to the definition and delimitation of outer space and to the character and utilization of the geostationary orbit, including consideration of ways and means to ensure the rational and equitable use of the geostationary orbit without prejudice to the role of the International Telecommunication Union".

46. Treaty on Principles Governing the Activities of States in the Exploration and Use of Outer Space, Including the Moon and Other Celestial Bodies, of 27 January 1967, 18 U.S.T. 2410, T.I.A.S. 6347, 610 U.N.T.S. 205 (entered into force: 10 October 1967); Agreement on the Rescue of Astronauts, the Return of Astronauts and the Return of Objects Launched into Outer Space, of 22 April 1968, 19 U.S.T. 2389, T.I.A.S. 6599, 672 U.N.T.S. 119 (entered into force: 3 December 1968); Convention on International Liability for Damage Caused by Space Objects, of 29 March 1972, 24 U.S.T. 2389, T.I.A.S. 7762, 961 U.N.T.S. 187 (entered into force on 1 September 1972); Convention on Registration of Objects Launched into Outer Space, of 14 January 1975, 28 U.S.T. 695, T.I.A.S. 8480, 1023 U.N.T.S. 15 (entered into force on 15 September 1976); Agreement Governing the Activities of States on the Moon and Other Celestial Bodies, of 18 December 1979, UN General Assembly resolution 34/68 of 5 December 1979, annex, 18 I.L.M. 1434 (entered into force on 11 July 1984).

47. See Article XXVI of the Liability Convention, Article X of the Registration Convention and Article 18 of the Moon Agreement.

48. See General Assembly resolutions 37/91 of 10 December 1982, 41/66 of 3 December 1986 and 49/34 of 9 December 1994.

49. On this subject see **Terekhov, A.D.**, "Review Clause of Outer Space Treaties: Reflections on the Forthcoming Review of the Moon Agreement", 33rd Proc., Dresden, AIAA, 1991, pages 356-361.

50. Currently the five outer space agreements in question have the following status: the 1967 Outer Space Treaty - 93 ratifications and 27 signatures; the 1968 Rescue Agreement - 83 and 25; the 1972 Liability Convention - 76 and 25; the 1975 Registration Convention - 39 and 4; 1979 Moon Agreement - 9 and 5 (see UN doc. A/AC.105/C.2/1996/CRP.2, this data is also cited in UN doc. A/AC.105/639 of 11 April 1996, Annex III.G.)

51. On the repetition of provisions in resolutions as a factor relevant to their weight and effect, see **Sloan, B.**, op. cit., pages 132-133.

52. It is pertinent to recall in this context that Professor **Bin Cheng**, in his often cited article "United Nations Resolutions on Outer Space: 'Instant' International Customary Law?", did not suggest that the 1963 Declaration was an "instant international custom". He generally concluded that "international customary law requires only one single constitutive element, namely the *opinio juris* of States" which may "grow overnight", but observed that the 1963 Declaration was not a "law-finding" resolution and was "not legally binding on any Member State of the United Nations *qua* Assembly resolution[s]".(See supra note 12, section "Conclusions").

IISL-97-2.06

OUTER SPACE AS A GLOBAL COMMON

Vladimír Kopal
Professor of International Law
Doctor of Sciences
Prague, Czech Republic

Introduction

During the lifetime of our generations, new areas were opened for the performance of activities of human kind which became important theatres of the newly developing international relations. This evolution has been effected under the conditions prevailing in the world of today which are characterized, from the viewpoint of international order, by the non-existence of a cetralized power structure which distinguishes the world community from the internal legal systems of individual states. International law has been described as "a horizontal system of law" which "operates in a different manner from a centralized one and is based on principles of reciprocity and consensus rather than on command, obedience and enforcement." [1] Moreover, states as members of the present international community are unequal in strength, though all of them are considered as sovereign and equal in law, and "there are always one or two states which are so strong that other states are usually too weak or too timid or too disunited to impose sanctions against them." [2]

What has just been said about international community and its members in general, is particularly valid with regard to their participation in the development of activities in newly opened areas that, however, have become areas of a global concern in which all states, in spite of their unequal capacities, wish to play an adequate role.

The present areas of a global concern are: Antarctica; outer space, including the moon and other celestial bodies; and oceans, particularly the high seas, and the sea-bed and ocean floor. In addition, the Earth's environment belongs more and more to "the global commons", as it became evident at the United Nations Conference on Environment and Development /UNCED/ held in Rio de Janeiro in 1992, and during the following development. [3] All these areas which are of common concern to all nations have offered them vast opportunities but, at the same time, many new responsibilities have emerged. The opportunities and problems, however, are not identical in each of these areas and the world community therefore did not decide to cope with them jointly by an attempt to establish one single legal regime which would be valid for each of them. Instead, specific legal systems have been developed for each particular area, though the necessity to do so emerged during the same historical period. In the process of establishing the individual regimes for "the global commons" and in the current results of these efforts, it is possible to identify a number of similarities. But it is also necessary to observe some significant differences which do not allow mechanical transfer and application of the solutions to the issues relating to one specific area to the others.

Fundamental Significance of the OST For the Legal Regime of Outer Space

The 1967 Outer Space Treaty /OST/ [4] laid down the foundations of the legal regime for human activities in those parts of the universe surrounding our planet which have had a significant impact on our lives. Moreover, unless otherwise agreed upon in the future, the principles established by the OST shall also apply to missions extending human activities deeper into our solar system and beyond. In the legal field, it is generally accepted that the OST created a solid basis for the progressive development of space law and belongs to the most important law-making treaties of the whole system of contemporary international law. [5]

Through the OST an attempt was undertaken to find a balanced compromise between the common interests of all nations, the aims of mankind as a whole, and the interests of individual states as members of the world community and traditional subjects of international law. This compromise is particularly evident from the principles inserted in the first three articles of the Treaty. It should be emphasized that the architects of the OST avoided to make an explicit and perfect definition of the legal status of the new area. Instead, they agreed on the purpose and orientation of space activities by saying that "the exploration and use of outer space, including the moon and other celestial bodies, shall be carried out for the benefit and in the interests of all countries, irrespective of their degree of economic or scientific development, and shall be the province of all mankind." Though the provenience of the phrase "province of all mankind" can be revealed and can be also ascertained at which stage of the negotiations relating to the OST this phrase appeared, it is not quite clear why this particular language was chosen by the drafters of the OST. However, since any treaty must be interpreted in accordance with the ordinary meaning given to its terms in their context and in the light of its object and purpose, a comparative analysis of different language versions of the OST should be also made in order to clarify the precise meaning of this phrase.

From among the definitions of the term "province" spelled out in a traditional dictionary of English language, the nearest definition to our subject seems to be that which refers to its abstract sense: "a department of knowledge or activity". [6] In the Russian language the phrase "dostoyanie vsego chelovechestva" is used and it means 'indivisible property /or wealth/ of all mankind'. In French "l'apanage de l'humanité tout entière" means in the figurative sense 'appurtenance' or 'privilege' of all mankind. A similar idea is spelled out in the Spanish text, according to which "e incumben a toda la humanidad". It should be noted in this context that in all these language versions this phrase differs from the term "common heritage of mankind" which has a specific meaning and distinct consequences. At the same time it must be stressed that the phrase "and shall be the province of all mankind" as well as the whole leadding principle of the OST does not refer to outer space itself /including the moon and other celestial bodies/ but to its "exploration and use" which shall be carried out for the benefit and in the interests of all countries. [7]

The second and third paragraphs of the same Article I declare two important freedoms which characterize the legal regime of outer space, namely the freedom of outer space, including the moon and other celestial bodies, for exploration and use by all states without discrimination of any kind, on a basis of equality and in accordance with international law, and the freedom of scientific investigation in this area which shall be facilitated and encouraged by states through international cooperation.

It may be said in this respect that in establishing the legal regime for outer space, the OST followed the example of the legal regime of the high seas which crystallized during centuries of struggles and has been characterized by a series of "freedoms of the seas". [8] And similar to the high seas, outer space, including the moon and other celestial bodies /according to Article II of the OST/, "is not subject to national appropriation by claim of sovereignty, by means of use or occupation, or by any other means". This principle is wide enough to cover not only outer space as a whole but also any part thereof, and national appropriation cannot be effected by any means whatsoever. A weak spot of these provisions, and after all of the whole OST, remains in the silence of the Treaty about the delimitation of outer space which shall be protected against any attempts of national appropriation and other violations of the legal order valid in this area.

Article III of the OST also has a fundamental meaning for the legal regime of outer space and activities developed in this area. In this provision the imperative of legality of activities in the exploration and use of outer space - "in accordance with international law, including the Charter of the United Nations " - is spelled out. Moreover, the necessity of keeping the peaceful character of such activities is emphasized "in the interest of maintaining internatioal peace and security and promoting international cooperation and understanding". This provision should be read in conjunction with the preambular paragraph of the Treaty in which the States Parties to this Treaty express their desire "to contribute to broad international cooperation in the scientific as well as the legal aspects of the exploration and use of outer space for peaceful purposes".

The general principle which became later on known as "maintaining outer space for peaceful purposes", as spelled out in Article III of the OST, was accompanied by a number of measures for demilitarization of outer space which are stipulated in Article IV. States Parties to the Treaty have undertaken "not to place in orbit around the Earth any objects carrying nuclear weapons or any other kinds of weapons of mass destruction, install such weapons on celestial bodies, or station such weapons in outer space in any other manner." While this provision concerns only the weapons of mass destruction, in the second paragraph of the same article still more far-reaching limitations of military activities have been enshrined; however, they have concerned only the moon and other celestial bodies, not outer space itself. According to this provision, "the Moon and other celestial bodies shall be used by all States Parties to the Treaty exclusively for peaceful purposes" and this general clause was accompanied by a number of specific prohibitions on different kinds of military activities.

Other principles of the OST are also significant for the characteristics of the legal regime of outer space which can be designated as "a global common". These include: assistance to astronauts, which are compared to "envoys of mankind in outer space", in the event of accident, distress, or emergency landing; international responsibility for national activities in outer space, whether they are carried out by governmental agencies or by non-governmental entities, and for ensuring that national activities are carried out in conformity with the OST; liability for damage caused by such activities to other States Parties to the Treaty or to their natural or juridical persons; registration of space objects and jurisdiction and control over such objects based on such registration; cooperation and mutual assistance in the exploration and use of outer space with a specific role for the United Nations and its Secretary-General in the development of such cooperation. These principles mentioned here and in the preceding paragraphs are the essential elements of the 1967 Outer Space Treaty.

Finally, one more significant feature which characterizes the area of outer space must be mentioned. While it was possible to create specialized organizations for other areas of international cooperation, such as the International Atomic Energy Agency /IAEA/, the International Maritime Organization /IMO/ and the International Sea-Bed Authority, as well as the United Nations Environment Programme as a less formal body of the UN system, in the field of space activities no specialized organization of the UN system emerged. Instead, the relevant functions have been dispersed among several bodies and organizations with the focal role of the COPUOS - a special committee entrusted with international cooperation within the United Nations itself. Practice has confirmed, more or less, the reasonableness of this solution, but a proliferation of bodies engaged with specific functions in the vast field of space activities requires good coordination and mutual cooperation in performing their specific programmes.

It may be said that the existing structure of international cooperation in space activities remains in harmony with the OST which did not provide for establishing a specialized agency within the United Nations system that would deal with international cooperation in all relevant space matters. And it is obvious that a more expanded structure in the form of such specialized agency will not emerge in the near future. There are, however, some issues of a global impact which deserve a deeper interest from the world community, but nobody - except some non-governmental organizations - takes adequate care of them. These issues include: control and improvement of the Earth's environment by means of space technology, protection of space environment, exploration and possible use of natural resources from celestial bodies, energy from outer space, search for extra-terrestrial life and eventual communication with extraterrestrial intelligences. Though some of them are often labelled as remote, these problems knock at our door and should be eventually considered in the interest of all humanity.

The OST and the Moon Agreement

The architects of the OST did not intend to work out a comprehensive legal instrument that would forecast and govern all possible aspects of the then ongoing and expected space activities. This is evident from the title of the Treaty which was concluded only on principles governing such activities. And, of course, this character of the OST is also evident from its juridical content. The Treaty thus left the door open for a further development of space law by additional international agreements and sets of principles which were elaborated step-by-step and adopted during the three decades following the entry into force of the OST. All of them recall the OST as the basic legal document and reflect the desirability of maintaining a harmony between the concepts of the 1967 Outer Space Treaty and the subsequent instruments relating to outer space. [9]

However, one of these instruments - the 1979 Moon Agreement - went relatively farther in the attempt to define and develop the provisions of the OST relating to the moon and other celestial bodies than the other instruments with "regard to further progress in the exploration and use of outer space". [10] This tendency is particularly visible in Article 11 of the Moon Agreement in which the moon and its natural resources are declared as "the common heritage of mankind". In Paragraph 5 of the same article, States Parties to this agreement undertake "to establish an international regime, including appropriate procedures, to govern the exploitation of the natural resources of the moon as such exploitation is about to become feasible".

Furthermore, in Paragraph 7 of Article 11, the main purposes of this international regime which is to be established are enumerated, including "an equitable sharing by all States Parties in the benefits derived from those resources, whereby the interests and needs of the developing countries, as well as the efforts of those countries which have contributed either directly or indirectly to the exploration of the moon, shall be given special consideration."

These provisions of the Moon Agreement reflected similar endeavours that resulted in the provisions of Part XI of the 1982 United Nations Convention on the Law of the Sea and its relevant annexes with regard to the area of the sea-bed and ocean floor beyond the limits of national jurisdiction. Though the 1979 Moon Agreement was finalized on the basis of consensus, and was adopted and commended by the UN General Assembly, a sharp opposition against this instrument emerged, mostly due to its provisions including the common heritage principle. As is known, only a few states to date have become parties to the Moon Agreement.[11]

It must be observed, however, that in comparison with the Sea Law Convention, which includes detailed rules and provides for the establishment of a full-scale international organization - the Sea-Bed Authority - and also for a system of dispute settlement with a special international sea law court, the 1979 Moon Agreement is a modest instrument. It has not yet established an international regime to govern the exploitation of the natural resources of the moon. States Parties to the Agreement only recognize that such a regime will be established as its need becomes really impending. Its creation would depend on the adoption of amendments to the Agreement which would enter into force only for States Parties accepting such amendments. Moreover, the undertaking to establish such a regime for the moon refers only to the <u>exploitation</u> of the natural resources of the moon. Other provisions of the Moon Agreement explicitly guarantee the right to exploration and use of the moon as well as the freedom of scientific investigation of the moon, the right to collect and remove from the moon samples of its mineral and other substances, and the right to use these substances for the support of space missions.[12]

It should be also recalled that the ratifications of the 1982 Sea Law Convention, in which a detailed implementation of the common heritage principle, including a complex system of prospecting, exploration and exploitation with a central role of the International Sea-Bed Authority and its Enterprise, has been incorporated, also proceeded very slowly for years. But the main obstacles that hindered this process were removed in 1994 by an Agreement Relating to the Implementation of Part XI of the Convention in which the ways and means of how to carry the controversial part of the Sea Law Convention into effect were found. Since then most countries of the world either adopted the 1982 Convention and the 1994 Agreement or are about to do so in the near future.[13]

The method by which obstacles were removed to enable a wide acceptance of the 1982 Sea Law Convention might be considered as an example of how to proceed with the issue pertaining to the 1979 Moon Agreement and an attempt to reach an agreement on the implementation of Article 11 of this Agreement might be initiated by informal consultations some time in the future. Such an attempt would be in accord with the repeated appeals of the UN General Assembly addressed to states that have not yet become parties to the international treaties governing the uses of outer space to give consideration to ratifying or acceding to those treaties. The list of instruments referred to in these appeals also includes the 1979 Moon Agreement. [14]

Let us also note in this connection that an agreement on a new item for the agenda of the COPUOS Legal Subcommittee, which is called "Review of the status of the five international treaties governing outer space", has been recently reached. According to the sponsor of this item, its consideration should enable the Legal Subcommittee "to propose mechanisms towards achieving the fullest adherence to the five outer space treaties". [15] As to the fifth treaty - the 1979 Moon Agreement - this aim is also important with regard to its possible application "to other celestial bodies within the solar system, other than the earth", unless "specific legal norms enter into force with respect to any of these celestial bodies". [16]

Space Law As a System

The 1967 Outer Space Treaty laid down the cornerstones for the whole building of international space law. In spite of its general character, its main principles are valid and useful. This conclusion also relates to those provisions which established the legal regime of outer space, including the moon and other celestial bodies, as a "global common". As such, the OST does not need any revision. At the same time, however, some provisions of the OST, and also those of the other United Nations space treaties, need clarification and their application should be adapted in the light of new phenomena and issues. Basic principles of the OST should be supplemented by further instruments.

Taken as a whole the present law of outer space must be considered as a legal system which forms a part of contemporary international law. However, unlike the whole system of international law which "may now properly be regarded as a complete system",[17] space law is not yet a complete system; more than any other branch of international law, space law has a progressively developing character. Notwithstanding that, the principles and norms included in the United Nations space treaties and other legally relevant documents have established a specific political and legal status of outer space and provided a special body of rules, the purpose of which is to govern different categories or all activities in this vast area. They represent a legal complex which is based on the 1967 Outer Space Treaty, to which other parts of the whole are linked, being at the same time mutually interrelated.

Space Law, as established by the United Nations, became an important part of international law and a significant contribution to the rule of law in international relations. And its progressive development should continue during the years to come.

References

1. Cf. Peter Malanczuk, Akehurst's Modern Introduction to International Law, Seventh revised edition, Routledge, London and New York, 1997, p. 6.

2. Ibidem, p. 5.

3. For an overall image and assessment of the UNCED see, e.g., Environmental Policy and Law, IOS Press, Vol. 22, Number 4, August 1992.

4. See Treaty on Principles Governing the Activities of States in the Exploration and Use of Outer Space, Including the Moon and Other Celestial Bodies, in United Nations Treaties and Principles on Outer Space, UN doc. A/AC.105/572/Rev.2, United Nations, 1997, pp. 4-9. This document is published in all official languages of the United Nations.

5. In this connection, the assessments by Dr. Ram Jakhu, Director of the Master of Space Studies Program of the International Space University, Strasbourg, France, as expressed at a symposium held to celebrate the 30th anniversary of the OST on the occasion of the 36th session of the COPUOS Legal Subcommittee in Vienna, 1 April 1997, should be recalled: "I would like to join those scholars from all of the world who have almost unanimously been declaring the 1967 Treaty a big success in creating an appropriate order in outer space. I will say it has been perhaps the finest achievement of diplomacy and statesmanship of those involved in its negotiation and drafting, especially during the height of cold war when the world was deeply polarised in two diagonally opposed political and economic systems. In those very tense days, when the human civilisation was almost at the

brink of extinction, the 1963 Resolution of the UN General Assembly on outer space which eventually was transformed into the 1967 Treaty, showed a ray of hope that the human beings are capable of finding peaceful solutions to global problems." /Cf. Jakhu, Ram, Application and Implementation of the 1967 Outer Space Treaty, in Proceedings of the 1997 IISL/ECSL Symposium, UN doc. A/AC.105/C.2/1997/CRP. 6 of 8 April 1997, p.1 of Jakhu's paper.

6. Cf. Webster's New Collegiate Dictionary 150th Anniversary Edition, p. 921.

7. This conclusion is abundantly clear from all the above-mentioned language versions of the OST, in particular from its French text which reads as follows:

"L'exploration et l'utilisation de l'espace extra-atmosphérique, y compris la Lune et les autres corps célestes, doivent se faire pour le bien et dans l'intérêt de tous les pays, quel que soit le stade de leur développement économique ou scientifique; elles sont l'apanage de l'humanité tout entière." /Underlined by the author of this paper./

8. For the evolution of the legal regime of the high seas and the victory of the freedom of the seas cf. C. John Colombos, The International Law of the Sea, Fifth revised edition, Longmans 1962, p. 44 ff.

9. For a recent assessment of this process see N. Jasentuliyana, A survey of Space Law as Developed by the United Nations, in Perspectives on International Law, Ed. by Nandasiri Jasentuliyana, Foreword by Boutros Boutros-Ghali, Kluwer Law International, 1995, p. 349 ff.

10. See the text of the 1979 Agreement Governing the Activities of States on the Moon and Other Celestial Bodies, in United Nations Treaties and Principles on Outer Space, pp. 28-36.

11. The following states became parties to the 1979 Moon Agreement: Australia, Austria, Chile, Mexico, Morocco, Netherlands, Pakistan, Philippines and Uruguay, while France, Guatemala, India, Peru and Romania signed the Agreement but have not ratified it yet. /Cf. Status of International Agreements Relating to Activities in Outer Space, As of 1 April 1997, UN doc. A/AC.105/C.2/1997/CRP.5./

12. See Article 11, Para. 4 and Article 6, Paragraphs 1 and 2 of the 1979 Moon Agreement.

13. For the status of the United Nations Convention on the Law of the Sea, which entered into force on 16 November 1994, and the status of the Agreement adopted by the UN General Assembly on 28th July 1994, as at 31 March 1997, see Law of the Sea, Bulletin No. 33, United Nations, New York, 1997.

14. Cf., e.g., the General Assembly resolution 51/123 of 13 December 1996, Paragraph 2 and Note 4.

15. Cf. Report of the Legal Subcommittee on the work of its Thirty-Sixth Session /1 - 8 April 1997/ UN doc. A/AC.105/674, 14 April 1997, p. 10 and 22-23.

16. See Article 1, Paragraph 1 of the 1979 Moon Agreement.

17. Cf. Oppenheim's International Law, Ninth Edition, Vol. I Peace, Introduction and Part 1, Ed. by Sir Robert Jennings and Sir Arthur Watts, Longman, England, 1992, p.12.

IISL-97-2.07

THE DARK SIDE OF THE MOON
The Status of the Moon: Public Concepts and Private Enterprise

Frans G. von der Dunk*

"There's no 'dark side of the moon', really;
Matter of fact, it's all dark..."
J. Driscoll, 1973

Abstract

Thirty years after the entry into force of the Outer Space Treaty most of its principles and concepts have not only found general acceptance as such, but also undergone considerable further development and elaboration. One of the most notable exceptions was the definition of the status of the moon (and other celestial bodies). This was partly due to the circumstances surrounding the drafting, conclusion, entry into force and then ultimate neglect by most states of the Moon Agreement, which was supposed to progressively develop and elaborate this issue. Now that an increasing interest may be discerned in returning to the moon - with the United States as most outspoken example - for its own sake, or as a jumping board for other celestial bodies, it seems worthwhile to revisit the issue of the status of the moon.

The concurrent trend towards increasing privatization of space and space-related activities provides further justification for such an analysis. For example, for any private entity potentially involved in the return to the moon one of the decisive elements for risking huge investments is the measure in which such investments can be guaranteed. This applies to any factual or legal obstacle to freely dispose as such of any physical objects owned or leased by the entity in question. Likewise, it applies to the factual and legal protection of their non-physical rights in objects such as intellectual property rights.

In other words: which particular legal regime applies to these issues is of supreme importance for private enterprise. And which legal regime *can* apply, is in its turn dependent upon international space law, to some extent aided by principles of general public international law, as the most fundamental pertinent body of legal rules. How can laws dealing with the pertinent issue, providing for rights and obligations of private entities such as companies, apply when international space law is a body of law of almost purely public character, with rights and duties exclusively for states?

Here, the issue of the status of outer space and in particular of the moon comes in. Both the international legal regime applicable to outer space as an area, and the international legal regime applicable specifically to the sub-areas of the moon and other celestial bodies do exclude national appropriation and the exercise of sovereignty on a territorial basis. Yet, complications regarding the question of the international legal status of the moon have to be kept in mind. To some extent territorial sovereignty as a mechanism for providing legal regimes still provides a relevant point of departure, and other mechanisms exist as well which may be relevant in this context, such as

*Co-Director, International Institute of Air and Space Law, Leiden University, The Netherlands; Member IISL; Board Member ECSL; Member AIAA; Member ILA; Corresponding Member CIDA-E.

jurisdiction over nationals and the issue of registration of space objects. In summary, as to the legal regime(s) applicable to the moon, for private enterprise a fragmented picture arises, with many gaps and many overlaps.
The conclusion therefore seems to impose itself that a rather dark side of the moon would lie in its lack of a generally acknowledged and agreed legal status on the international legal level. This results in obvious shortcomings on the level of national legislation, as the hitherto most appropriate tool for dealing with private enterprise. Neither does it look like the situation, either in general or for specific purposes such as intellectual property rights, will soon turn decisively for the better.

1. Introduction

Since a few years the moon and other celestial bodies have made some sort of a comeback in the attention of mankind - or at least of the space lawyers community. Following the end of the Apollo programme, discussions amongst the latter had for some time dealt with the future exploitation of especially the moon. The legal problems regarding such exploitation could not be solved, however. At the same time, this applied to the technical problems, which made solution of the legal ones less necessary. Consequently, the moon (and other celestial bodies) became an area in which only a limited number of people maintained interest of a moreover largely theoretical character.
Then, the plans proposed a few years ago by the United States to use the moon as a basis for travel to Mars rekindled general interest. The recent adventures of the United States Pathfinder and Sojourner on the red planet constituted world news. Increasing interest in the moon and celestial bodies was also evidenced by the devotion of a full session of the Thirty-Ninth Colloquium on the Law of Outer Space of the International Institute of Space Law, held in Beijing October 1996, on the subject of property rights in this unique environment.[1]

Legal issues pertaining to the moon and other celestial bodies, moreover, did not remain confined to theoretical debates amongst a few legal expert specialists. A claim by a German pensioner that he held a private title of ownership regarding the moon, and that as a consequence the German government should initiate international action against a Californian businessman auctioning plots of land on the moon, received widespread attention in the media.[2] Likewise, even more recently it has been rumoured that three Yemenites filed a lawsuit locally against NASA for infringing - by means of Pathfinder's and Sojourner's operations - their purported ownership of Mars.[3]

2. The Moon and the Law

Such instances serve to highlight the resurging interest of private persons and private enterprise in the moon and the other celestial bodies. At the same time, it increases the relevance of clarifying to the extent still necessary the public concepts which are ultimately underlying any solution in practice to such problems.
Obviously, legal analysis of the international legal status of the moon should start with the Outer Space Treaty as the often-quoted 'Magna Charta' for outer space and space activities.[4] Before its entry into force in 1967 (or at the earliest until the United Nations Resolutions from 1957 onwards[5] provided for any statement on the law relating to outer space), no law, man-made as that is, could be applicable to outer space.[6]
As a result, international legal rules, like those requiring conscious and substantive acts such as occupation and effective control[7] for sovereignty and ownership, public as well as private (the latter moreover being dependent on the former), of an 'area' did not apply or even exist yet. Thus, claims basing themselves on the situation in 1756 or even of 3,000 years ago were and remain void.

3. The Moon and the Outer Space Treaty

With respect to the Outer Space Treaty, Article II as the most fundamental legal provision specifies the particular application of these very general principles regarding sovereignty and ownership to outer space. It provides that "outer space, including the moon and other celestial bodies, is not subject to national appropriation by claim of sovereignty".

This clause is widely perceived to exclude the applicability of territorial sovereignty to outer space or any particular part thereof.[8] In other words: outer space does not form part of any state's territory, as legally defined for purposes of the scope of its sovereign authority. Neither *can* it ever become part of such a national territory: outer space is not *res nullius* or *terra nullius*, and is not susceptible to legal occupation, conquest or cession.[9] This obviously also applies to the moon, being part of outer space.

4. The Moon as *Terra Communis*

Under present international legal doctrine, this would still leave open two options as to the status of outer space, including the moon. As to the first such option, already in the times of Hugo Grotius it had been recognized that certain geographical areas were in a very principled sense outside the reach of any state's territorial sovereignty.

Following from the foregoing, outer space indeed would qualify as such a *terra communis* or *res extra commercium*, a geographically defined area where freedom ruled in principle just like the high seas.[10] Only the states of the world acting collectively can provide for legal conditions to any activity in this area. No individual state could call the tune to which other states or their entities would have to dance, not even for a part of that area such as the moon. *Vice versa*, each state (or its entities) could equally profit from that fundamental freedom, without hindrance from any particular rival state.

The application of this notion to outer space is further supported *inter alia* by such provisions in the Outer Space Treaty as the ones regarding the freedom of exploration and use of, and of scientific investigation in outer space.[11] It also arises out of the general character of the Outer Space Treaty as providing the legal framework for all activities in outer space.[12] The Outer Space Treaty itself provides for the application of international law in general to outer space,[13] as well as for the most important restrictions on the fundamental freedom of space activities.[14]

It thereby makes clear that, indeed, only the community of states can establish the legal regime for outer space *in principalem*, while at the same time, to the extent such a regime is *not* in place, the freedom of space activities remains. Individual states furthermore in consequence are directly held accountable for their activities (or those of their entities) towards the other states by means of the principles of international responsibility and international liability.[15]

5. The Moon and the Common Heritage of Mankind-Principle

Relatively recently, however, a second theoretical option for defining the status of an area like outer space, of specific importance in the context of the moon, has entered the international legal discussion: that of the common heritage of mankind. Its application to specific (categories of) geographical areas, and its exact contents and consequences remain the topic of intensive debate.[16] The principle as such however may be said to have achieved a measure of acceptance since a few decades.

It was most intensively discussed with respect to the status of the ocean floor in the framework of the United Nations Conference on the Law of the Sea, taking place from 1974 till 1982.[17] The core issue in the eyes of the proponents of applicability of the common heritage of mankind-principle to the ocean floor amounted to one crucial step beyond the recognition of the *terra communis*-status which the opponents clung to. The 'classical' *terra communis* went with the presumption of complete freedom of activities unless the contrary could be proven.[18]

Those pronouncing the ocean floor the common heritage of mankind on the contrary essentially presumed that any substantial - especially commercial - activities required the consent of the community of states. Consequently, they proposed that an international body should be established to preserve these rights of the world community, and act as a sort of caretaker.[19] Individual states (or their private entities) should only be allowed to undertake commercial activities as long as this international caretaker would see to it that the community of states, especially the developing countries, would actually and materially benefit from those activities.

6. The Moon, Outer Space and the Ocean Floor

Coming back to the area of outer space in general, several traces of this common heritage of mankind-principle found their way into the Outer Space Treaty. The "common interest of all mankind" and the "benefit of all peoples" are major leading principles in guiding the exploration and use of outer space.[20] Furthermore, in its very first sentence, the Outer Space Treaty provides that the exploration and use of outer space "shall be carried out for the benefit and in the interests of all countries".[21] Finally, the hitherto unknown phrase "province of all mankind", as 'defining' the 'status' of exploration and use, is also introduced.[22]

While this phrase indeed seems to echo the common heritage of mankind-principle, most authors as well as the most important spacefaring states agree that its use denies rather than confirms any perceived status of outer space as common heritage of mankind.[23]

At the United Nations Conference on the Law of the Sea, application of the common heritage of mankind-principle to the ocean floor took the form of a rather explicit arrangement regarding any prospective commercial activities in that area.[24] An international body, the Seabed Authority, was foreseen to license such exploitation activities. It should, moreover, license them only under conditions which would allow the other states of the world community - especially the developing ones - to materially profit from any particular licensed activity as well. An international enterprise was to actually undertake exploitation activities of the ocean floor on their behalf. Such bodies, however, of course were not established by the Outer Space Treaty, and, at that point, not even foreseen by the states involved.[25]

Consequently, the provision regarding the "benefit and (...) interest of all countries" should only be interpreted in a 'negative' way. As long as a particular activity in outer space did not (significantly) harm another state, it would be allowable under the fundamental freedom of space activity. No 'positive' benefits accruing to other states were required to make any exploitation of outer space legal.[26] The only difference of the status of outer space with the traditional formulation of *terra communis* would consequently be that, this time, the obligation not to cause significant harm was explicitly included. This, however, should be considered as being of marginal importance from a conceptual point of view. The status of outer space should therefore be generally equated to *terra communis*.

7. The Status of the Moon

These conclusions also apply to the moon, as one specific area within the larger area of outer space. The Outer Space Treaty does make an important distinction between the moon (and other celestial bodies) on the one hand, and outer space in general on the other hand, concerning the stricter regime established in respect of the former when it comes to military or similar activities.[27] This however seems to be of little consequence for any analysis of the legal status of moon or other celestial bodies.

More interesting is the provision that "all stations, installations, equipment and space vehicles on the moon and other celestial bodies shall be open to representatives of other States Parties to the Treaty".[28] The impact of this clause is somewhat mitigated by conditions, the most important one that being of reciprocity. Nevertheless, in principle an obligatory openness to prying eyes results. To that extent, full and uninhibited enjoyment of the ownership of hardware, software and know-how, whether by public or by private entities, is also at issue. In the last instance, however, this interesting clause does not of itself result in the moon being the common heritage of mankind.

8. The Moon and the Moon Agreement

Of course, the Outer Space Treaty is not the only treaty to take into account when trying to establish the international legal status of the moon. Precisely because of the special physical character of the moon (and the same obviously applies to other celestial bodies) when compared to outer space as an area, the Moon Agreement was drafted to deal specifically with these celestial bodies.[29] However, while the Outer Space Treaty was ratified by over 90 states, including all those relevant in terms of space activities, the Moon Agreement has only been ratified by nine states, while being signed but not ratified by five more.[30]

Yet, the Moon Agreement is in force, since five ratifications sufficed for that purpose.[31] Consequently, also for those states signatory though no party, the existence of the Moon Agreement under international legal doctrine would result in an obligation not to jeopardize its object and purpose.[32] Especially the presence of two such important spacefaring nations as France and India amongst the signatories makes it worthwhile to have a closer look into the question, whether the text of the Moon Agreement would lead to different conclusions regarding the international legal status of the moon.

9. The Moon Agreement and the Outer Space Treaty

A number of provisions in the Moon Agreement directly or indirectly affect the international status of the moon. The Moon Agreement in many of these provisions essentially follows the Outer Space Treaty as far as the *terra communis*-character of the moon is concerned. Repeatedly, direct reference is made to the Outer Space Treaty as such.[33] The Moon Agreement clearly attempts to itself establish the conditions under which the free exploration or use of the moon, in conformity with the Outer Space Treaty, may be conducted.[34] It does not allow for national appropriation of (a part of) the moon, nor does it allow any individual state to call the tune in that respect.[35]

Furthermore, mention is made of the "corresponding interests of all other States Parties" in "promoting international cooperation and mutual understanding", which activities on the moon shall take due account of.[36] This echoes the Outer Space Treaty's provision made in Article IX. In view of the circumscribed number of parties to the Moon Agreement, this provision as such is of limited application however. In other words, the confinement to other parties, of the duty to respect corresponding interests, takes on much more significance here. This, however, at the most provides further proof for the contention that the moon is not generally considered the common heritage of mankind: "common" here effectively refers only to the nine states parties, not to "mankind" as a whole.

10. The Moon: Exploration and Use

Consequently, it is of much greater significance that the Moon Agreement repeats the first part of Article I of the Outer Space Treaty.[37] Actually, the phrase "province of all mankind", as applicable to exploration and use, in the Moon Agreement is moved to the first part of the sentence, thus giving it greater emphasis.

While repeating the partial explanation of the notion by means of "the benefit and (...) interest of all countries", the Moon Agreement then adds further precision: "due regard shall be paid to the interests of present and future generations as well as to the need to promote higher standards of living and conditions of economic and social progress and development".[38] The first part foreshadows the recent discussions regarding 'sustainable development'; while the second part tastes of the inter-state solidarity which had been largely responsible for the development of the common heritage of mankind-doctrine.

Nevertheless, it can not be concluded therefrom that the moon, or its exploration and use, should be considered the common heritage of mankind by the states parties to the Moon Agreement. No specific instruments or mechanisms are defined by these provisions to ensure that all states, especially the developing ones, benefit from any material activities on the moon. Rather than an absolute obligation to achieve a particular result, it constitutes an obligation to undertake a certain effort, an obligation of good faith which may be overruled by other circumstances.

Related provisions, such as the one providing for guidance of states in their exploration and use "by the principle of cooperation and mutual assistance", or the one providing for obligations to inform other states in relevant cases "to the greatest extent feasible and practicable" confirm this analysis.[39] Also, the freedom of scientific investigation as a form of exploration, already provided for by the Outer Space Treaty in its Article I, is reaffirmed with respect to the specific case of the moon (and the other celestial bodies).[40] Finally, no convincing reason has been put forward for using two different phrases when supposedly reference is to be had to the same principle.

All in all, it is therefore once more submitted that the exploration and use of the moon, as province of all mankind, is essentially *res communis* rather than common heritage of mankind even under the Moon Agreement.

11. The Moon: Exploitation and Status

Upon closer scrutiny, however, next an essential difference arises between the Moon Agreement and the Outer Space Treaty. The latter, as established, deals with the whole area of outer space, alternatively by dealing with specific sorts of activities taking place therein. This concerns exploration (including for this purpose scientific investigation) and use; leaving aside the special issue of military activities, these two categories together more or less comprise all activities envisaged in outer space.[41] The rules specifically devised for one or both of these categories (usually exploration and use are dealt with in combination), as implied above, do further substantiate the conclusion that outer space as an area in the legal sense of the word constituted *res* or *terra communis*.

The Moon Agreement, however, dealing with tangible *res*, read *terra*, such as the moon and the other celestial bodies, had to envisage a third category of activities: that of physical exploitation. Article 11 forms the core provision in this respect, since it provides that "the moon and its natural resources are the common heritage of mankind".[42] Thus, it creates a dichotomy between the status of the exploration and use, and by analogy of scientific investigation on the one hand, and the exploitation of the natural resources and the status of the moon as a whole on the other hand. The former remains the "province of all mankind", previously argued to be a kind of *res communis*-status with the addition of explicit provisions regarding the obligation not to cause significant harm.

12. The Moon as Common Heritage of Mankind

The exploitation of the moon however is explicitly defined as the common heritage of mankind. The essence thereof is then further elaborated upon: "neither the surface nor the subsurface of the moon, nor any part thereof or natural resources in place, shall become the property of any State" or other entity undertaking activities on the moon.[43] An "international regime, including appropriate procedures" is furthermore to be established, as soon as relevant, "to govern the exploitation of the natural resources of the moon".[44] This international regime finally should *inter alia* guarantee "an equitable sharing by all States Parties in the benefits derived from those resources, whereby the interests and needs of the developing countries, as well as the efforts of those countries which have contributed either directly or indirectly to the exploration of the moon, shall be given special consideration".[45]

The application of the common heritage of mankind-principle to the exploitation of the moon is thus clear and unambiguous. The inclusion of the moon *as such* in this principle, however, is emptied of all meaning beyond such exploitation in view of the other categories of activities envisaged. Neither exploration nor use, nor specifically scientific investigation could be considered common heritage of mankind even under the terms of the Moon Agreement. The freedom of exploration and use of the moon is reconfirmed also by Article 11 itself, while the freedom of scientific investigation is also reconfirmed.[46]

It should be concluded therefore, that even under the Moon Agreement the moon does not have a status as either comprehensively *res communis* or comprehensively common heritage of mankind. While exploration, use and scientific investigation fall under the former categorization, exploitation of natural resources falls under the latter one. To that extent, the Moon Agreement clearly differs from the provisions of the Outer Space Treaty relevant for the moon and other celestial bodies.

This, however, obviously applies only to those states party to the Moon Agreement, and, arguably, to those signatory to it. Even for them, no unequivocal definition of the moon as common heritage of mankind can be deduced; only for exploitation the application of the principle would be beyond doubt.

13. The Moon and Private Enterprise: National (Space) Legislations

The present international rules concerning space activities are primarily directed to states, and subsidiarily to public international organizations. This normative system is also applicable to private commercial activities, of course. However, at least presently private enterprise is not directly bound by those rights and obligations.[47] International space law in this regard does itself determine *how* private entities, through national legislation, should be bound to the substance of international space law.

States, after all, are held internationally responsible and liable for space activities, even if these space activities are partly or wholly privately conducted.[48] One should leave aside here the discussion on *which* categories of private space activities will incur the responsibility (and liability) of *which* particular state, as the "appropriate State" whose "national activities" they constitute. In the abstract, it is obvious that states will likely provide for national (space) legislations with a view to regulating those activities they might be held responsible (and liable) for.

14. The Moon and Three Types of Jurisdiction

From the point of view of private enterprise, for example on the issue of intellectual property rights, furthermore, it is to be noted that generally speaking states have three options for the purpose of legislation.[49] First, there is the option of territorial jurisdiction, which is the type exercised most commonly and comprehensively for such purposes.[50] Territorial sovereignty, however, is not applicable to outer space.[51] If based on the territoriality-principle, legislation of a state can only apply to persons or occurrences on its territory. It does therefore present a rather

flawed tool for the purpose of, for example, protecting intellectual property rights on the moon.
Secondly, there is what I label 'national jurisdiction': jurisdiction of a state over persons of its nationality undertaking certain activities.[52]
In view of the fact that especially intellectual property rights legislation is usually based on territorial jurisdiction rather than on national jurisdiction, as well as from a practical point of view, this type of jurisdiction is (also) of little help in this regard.
That means, that the third type of jurisdiction becomes especially important. This concerns the quasi-territorial jurisdiction over space objects, by registration thereof and thus the grant of a quasi-nationality to it, as provided for by Article VIII of the Outer Space Treaty.[53]
It provides the most efficient means to extend the reach of juridical authority into space for the purpose of regulating important aspects of private activities, such as especially those of intellectual property. For that reason, for example the United States amended its patent rights legislation, in principle applicable to American territory, to inventions made on board of space objects with United States registration.[54]
It should however be noted, that such extension of jurisdiction by definition can only apply to space objects, and possibly to activities directly linked to particular space objects. Once for instance more permanent habitats on the moon will become feasible, the inherent limitations of this third type of jurisdiction will immediately become clear.

15. The Dark Side of the Moon

Summing up, the moon might present an interesting prospective opportunity for private commercial activities as long as, *inter alia*, the applicable legal system is structured in a suitable manner for private enterprise to pursue its aims. Private enterprise is most obliged in this respect (within the parameters provided by law as far as justified from the perspective of the general public at large) by a uniform, transparent and logical legal system as regards its activities.
From this perspective, the moon actually and substantially has a dark side when it comes to undertaking activities there. Firstly, at the international level nothing akin to an international regime can be detected. More importantly, even an accepted basis for such a regime is absent: the status of the moon as *terra communis* is challenged by both those adhering to application of the common heritage of mankind-principle to the moon under the Moon Agreement, and those desiring to read more into the province of all mankind-principle than such a *terra communis*-status. On the other hand, any applicability of the common heritage of mankind-principle to the moon is both denied by a large majority of states and legal experts, and not even unequivocally established by the Moon Agreement itself.
The most relevant provision in space law from the perspective of private enterprise is the one resulting in non-application of territorial sovereignty to outer space. It leads to the second major problem: in the absence of any international regime providing for uniform rules, national regimes should do the job of duly protecting private enterprise's justified interests in space. However, they are fundamentally handicapped when trying to substitute an international regime. National space legislations as such, of course, already carry within them the risk of complications, confusions and absence of uniformity. But this is strongly aggravated by the fact that national law can apply in any comprehensive sense to outer space objects only. However, registration, which is required to achieve this result, depends on a conscious act and does not exist *ipso facto*.
As a consequence of the last aspect therefore, finally, the actual situation is even worse. Only a few states have actually and consciously taken action to provide for extension of scope of any national laws to space objects. The problems arising with respect to the European module of the international space station are a

case in point.[55]

Private enterprise, in conclusion, will confront a large measure of confusion as regards the legal parameters for any prospective commercial operations on the moon. While some entrepreneurs might come to the conclusion that, as a result, the moon presents interesting opportunities as a *de facto* free-for-all, both from the perspective of private enterprise and from that of the interest of the public at large - for instance in the safety of space activities - this bodes ill for the future.

References:

1. Session 1, "The Legal Status of Property Rights on the Moon and Other Celestial Bodies", Thirty-Ninth Colloquium on the Law of Outer Space, International Institute of Space Law (IISL), Beijing, 8 October 1996; see e.g. Second Announcement 47th International Astronautical Congress, 29, 34.

2. The German, Mr. Martin Jürgens, claimed that Frederick II the Great, King of Prussia, had bequeathed one of his ancestors with ownership of the moon in return for services rendered back in 1756, and that he had inherited such ownership. The Californian, Mr. Dennis Hope, had registered his claim to ownership of the moon locally at some point in the early eighties, and since then had begun selling plots (at a cost of some US$ 16 per square meter). Amongst his clients is rumoured to be famous US actor Burt Reynolds. See e.g. P.M. Krämer, *Wem gehört nun den Mond? [Who does the Moon belong to after all?]*, Kölner Universitäts-Journal 3/4-1996, 40-1; G. Gál, *Acquisition of Property in the Legal Regime of Celestial Bodies*, paper presented to the Thirty-Ninth Colloquium on the Law of Outer Space, Beijing, October 1996.

3. Mr. Adam Ismail, Mr. Mustafa Khalil and Mr. Abdullah al-Umari are said to have claimed inheritance of the red planet from their ancestors 3,000 years ago, as well as to be able to prove this by means of legal documents; as per news flash filed by The Associated Press.

4. Treaty on Principles Governing the Activities of States in the Exploration and Use of Outer Space, including the Moon and Other Celestial Bodies (hereafter Outer Space Treaty), adopted 19 December 1966, opened for signature 27 January 1967, entered into force 10 October 1967; 18 UST 2410; TIAS 6347; 610 UNTS 205.

5. E.g. UNGA Resolution 1148(XII) of 14 November 1957, UNGA Resolution 1348(XIII) of 13 December 1958, but especially UNGA Resolution 1721(XVI) of 20 December 1961 and the Declaration of Legal Principles Governing the Activities of States in the Exploration and Use of Outer Space, UNGA Resolution 1962(XVIII) of 13 December 1963; texts e.g. in N.M. Matte, Aerospace Law (1969), Annexes I, II, IV and VII resp.

6. See e.g. M. Lachs, The Law of Outer Space (1972), 12-4.

7. See e.g. I. Brownlie, Principles of Public International Law (3rd ed.)(1979), 117, 134-5, 141 ff.

8. See extensively e.g. Lachs, 42-8; S. Gorove, Developments in Space Law (1991), 25-6.

9. See for *res nullius* e.g. Brownlie, 109, 180-1.

10. See on *terra communis* and *res communis* e.g. Brownlie, 181, 266-70; N. Singh, *Introduction to International Law of the Sea and International Space Law*, in International Law: Achievements and Prospects (Ed. M. Bedjaoui)(1991), 825 ff.; V.F. Wodié, *The High Seas*, in International Law: Achievements and Prospects (Ed. M. Bedjaoui)(1991), 887 ff.; also Art. 2, Convention on the High Seas, Geneva, signed 29 April 1958, entered into force 30 September 1962; 450 UNTS 82; TIAS

5200; 13 UST 2312; Artt. 87, 89, United Nations Convention on the Law of the Sea (hereafter LOS Convention), Montego Bay, signed 30 April 1982, opened for signature 10 December 1982, entered into force 16 November 1994; 21 ILM 1261 (1982).

11. See Art. I, Outer Space Treaty.

12. Cf. e.g. C.Q. Christol, The Modern International Law of Outer Space (1982), 12, 20; S. Gorove, *Sources and Principles of Space Law*, in Space Law - Development and Scope (Ed. N. Jasentulyana)(1992), 46-7.

13. See Art. III, Outer Space Treaty.

14. One important example thereof concerns the limitations to military and/or non-peaceful uses of outer space, as provided by Art. IV, Outer Space Treaty.

15. Artt. VI resp. VII, Outer Space Treaty.

16. Cf. e.g. A.A. Cocca, *Property Rights on the Moon and Celestial Bodies*, paper presented to the Thirty-Ninth Colloquium on the Law of Outer Space, Beijing, October 1996.

17. See also Artt. 136, 137, LOS Convention.

18. Cf. e.g. Art. 87(1), LOS Convention, providing the general rule of freedom, and e.g. Artt. 87(2), 88, 89, as providing exceptions thereto.

19. Cf. esp. Art. 137(2), LOS Convention.

20. Preamble, Outer Space Treaty.

21. Art. I, Outer Space Treaty.

22. Art. I, Outer Space Treaty.

23. Cf. e.g. Christol, 252, ff.

24. See Artt. 150 ff., LOS Convention.

25. This arose only at the discussions regarding the Moon Agreement to be established; see e.g. Christol, 286 ff.

26. Art. IX, Outer Space Treaty, effectively had exactly that effect, by providing *inter alia* for duties of a due care-character in respect of activities in outer space, and consultation.

27. See Art. IV, Outer Space Treaty.

28. Art. XII, Outer Space Treaty.

29. Agreement Governing the Activities of States on the Moon and Other Celestial Bodies (hereafter Moon Agreement), adopted 5 December 1979, opened for signature 18 December 1979, entered into force 11 July 1984; 18 ILM 1434 (1979); 1363 UNTS 3.

30. Australia, Austria, Chile, Mexico, Morocco, The Netherlands, Pakistan, The Philippines and Uruguay are parties; France, Guatemala, India, Peru and Rumania signatories to the Moon Agreement. See United Nations Treaties and Principles on Outer Space (1996), A/AC.105/572/Rev.1; 60-70, status as of March 1996.

31. See Art. 19(3), Moon Agreement.

32. Cf. e.g. Art. 18, Vienna Convention on the Law of Treaties, signed 23 May 1969, entered into force 27 January 1980; 8 ILM 679 (1969).

33. See Preamble, Artt. 9(2), 10(1), 14(2), 16, Moon Agreement.

34. Cf. e.g. Artt. 2, 3, Moon Agreement.

35. See Art. 11(2), Moon Agreement.

36. Art. 2, Moon Agreement.

37. See Art. 4(1), Moon Agreement.

38. Art. 4(1), Moon Agreement.

39. Resp. Art. 4(2) and Art. 5, Moon Agreement; note the qualification of any substantial duty by the use of the terms "guided" resp. "feasible and practicable".

40. See Art. 6, Moon Agreement.

41. Cf. e.g. Artt. I, III, IX, Outer Space Treaty.

42. Art. 11(1), Moon Agreement.

43. Art. 11(3), Moon Agreement.

44. Art. 11(5), Moon Agreement. See also Art. 18. This obviously echoes the relevant provisions of the LOS Convention.

45. Art. 11(7(d)), Moon Agreement. Again, this echoes clauses of the LOS Convention.

46. Art. 11(4), resp. Artt. 6 and (implicitly) 11(7(d)), Moon Agreement.

47. See e.g. P.L. Meredith & G.S. Robinson, Space Law: A Case Study for the Practitioner (1992), 58, 67.

48. See Artt. VI, VII, Outer Space Treaty. See further e.g. Wassenbergh, Principles of Outer Space Law in Hindsight (1991), 23; B. Cheng, *The Commercial Development of Space: the Need for New Treaties*, 19 Journal of Space Law (1991), 36-40; H. Qizhi, *Certain Legal Aspects of Commercialization of Space Activities*, 15 Annals of Air and Space Law (1990), 337; H.L. van Traa-Engelman, Commercial Utilization of Outer Space (1993), 281-2; Meredith & Robinson, 67.

49. See e.g. Cheng, 37.

50. See e.g. F.A. Mann, *The Doctrine of International Jurisdiction Revisited After 20 Years*, in Further Studies in International Law (1990), 4-10; D.W. Bowett, *Jurisdiction: Changing Patterns of Authority over Activities and Resources*, 53 British Yearbook of International Law (1982), 4-10.

51. See Art. II, Outer Space Treaty. See further *supra*, discussion at para. 3.

52. Cf. e.g. Cheng, 37 ff.; Gorove, Developments, 20; Bowett, 4-10.

53. Of course, the word 'nationality' appears nowhere in Art. VIII or in any other legal text of the *corpus juris spatialis internationalis* in relation to registration. However, by analogy to the law of the sea and of the air, and in view of the resulting jurisdiction, for all practical purposes the registration of space objects should be deemed to result in a nationality thereof; cf. respectively Art. 5(1), Convention on the High Seas, and Art. 91(1), LOS Convention; and Art. 17, Convention on International Civil Aviation, Chicago, signed 7 December 1944, entered into force 4 April 1947; 15 UNTS 296; TIAS 1591. See further Christol, 213-4; R. Oosterlinck, *Private Law Concepts in Space Law*, in Legal Aspects of Space Commercialization (Ed. K. Tatsuzawa)(1991), esp. 52, i.a. quoting G.P. Sloup; also on ships and aircraft Brownlie, 242-3, 254-5, 319-20, 424-7.

54. See U.S. Patents in Space Act, Public Law 101-580, of 16 November 1990, 35 U.S.C. 105. Further e.g. G. Catalano-Sgrosso, *Copyright and Intellectual Property in Outer Space - Legal Protection of Discoveries and Inventions Made in Conditions of Microgravity*, in Proceedings of the Thirty-Seventh Colloquium on the Law of Outer Space (1995), 139.

55. See discussion e.g. at A. Farand, *The Astronaut in the Space Station Era*, in Outlook on Space Law over the Next 30 Years (Eds. G. Lafferranderie & D. Crowther)(1997), 153-4, 158-9; Catalano Sgrosso, 138-9.

Thirty Years of Functionalism

Dr. Gyula Gál

Recently an excellent monograph was published bearing the following meaningful title: „The never-ending Dispute: Delimitation of Air Space and Outer Space." (1)

Praising the Space Treaty of 1967 the reasonable question arises, how did live this fundamental source of space law three decades without an amendment or specific agreement on spatial definition of Outer Space. The title given to my paper „Thirty Years of Functionalism" demonstrates that my modest observations attempt to examine whether this theory could stand the test in the light of positive law and international practice.

Before the Treaty

All movements connected with human activities in the presatellite age were regulated according to the legal nature of the space where these activities occurred. Forty years ago (October 4, 1957) with Sputnik-1 a revolutionary new kind of human activity, movement of a man-made, artificial satellite appeared — following the rules of celestial mechanics. It is worth remembering the basic scientific facts underlying all space activities.

Objects launched at a velocity less than 7.9 km/sec will follow a ballistic trajectory and fall back to the Earth.In case of Sputnik-1 the first time in history a launch vehicle gave the necessary energy to an object resulting a velocity exceeding 7.9 km/sec but less than 11 km/sec. This carries satellites on an elliptical orbit, one of the focuses of which coincides with the centre of the Earth (Kepler's first law). Meanwhile the line connecting the satellite and Earth (radius vector) sweeps out equal areas in equal times. The velocity of orbiting satellites therefore will be the higher when nearest the Earth (perigee) and the lower when farthest from it (apogee — Kepler's second law). The minimum velocity for an object to enter a parabolic trajectory i.e. escape velocity from the Earth's surface is about 11.2 km/sec from the Moon 2.4 km/sec. Above this velocity the object will follow a hyperbolic path. (2)

The path followed by a celestial object (natural or artificial) that is moving in a gravitational field is the orbit. Movement on this path an orbital movement. (Fig. 1-2.) This kind of movement connected with human activity meant a real challenge for traditional legal thinking. First of all because it was an inertial movement which could not have respect for state sovereignties theoretically extended in vertical space ad infinitum.

The answer to this problem seemed to be obvious: space law will be the body of rules governing movements in a spatially delimited outer (cosmic) space, while air law remains the legal order of an other space fixed vertically by this delimitation: the air space in the sense of Article I. of the Chicago Convention. Accordingly air law and space law would cover the space above the Earth's surface split into two slices by different legal regimes as the legal status of territorial sea differs from that of the free open see.

Those who found this solution unreasonable held that this new legal domain, other than all earlier norms of human conduct can not be associated with a limited space, but only with the character of the activity under regulation.

Thirty-seven years ago I joined this theory. The conclusions of my treaties on Air Space and Outer Space were the following:
1. An infinite air space is a conceptual impossibility. 2. The facts of astronautics cannot be squeezed into the conventional framework of international air law. 3. The synthesis of territorial air space and free outer space can only be assured by an international agreement which can break away from the conventional forms and lay emphasis on the ***activity element*** of rockets, artificial satellites and planets launched by the states. (3) This is the main idea of the functionalist approach I have been representing since the very beginning before and after the conclusion of the Space Treaty.

The functionalist element determining space activity i.e. the basis of space law to my mind has been the orbital character of the activity, namely the orbital movement involved or intended.

Nevertheless before the Treaty the „never-ending dispute" on delimitation was going on. In my book published in 1964 I could already list not less than 49 proposals from 12 km (Beresford) to 384000 km (Rinck) *and ad infinitum*. (4)

A very realistic position was taken by the U.N. Ad Hoc Committee on the Peaceful Uses of Outer Space. Its report of July 14, 1959 stated:

> It was generally believed that the determination of precise limits for air space and outer space did not present a legal problem calling for primary consideration at this moment. The Committee noted that the solution of the problems which it had identified as susceptible of primary treatment *was not dependent upon the establishment of such limits.* (III.A.28) (5)

It was true also for the fundamental principle of space law: freedom of outer space declared and not constituted by GA. Res. 1721/XVI — December 20, 1961. (6) This meant freedom of space (orbital) activity generally accepted as customary law rule — without any delimitation of air space and outer space.

The Treaty

Fortunately the creation of the Space Treaty itself was not made dependent of the establishment of delimitation, respectively on definition of outer space. The Treaty refers not less than 37 times to „outer space" without attempting to give an authentic definition to it. No wonder, the Scientific and Technical Subcommittee of CUPUOS stated at the same time (1967) that „it is not possible at the present time to identify scientific or technical criteria which would permit a precise and lasting definition of outer space".

As a consequence no right and obligation regulated by the Treaty has been restricted to a definite area or altitude where the legally relevant act occurs. I venture the statement: the rules of the Treaty bear fundamentally and necessarily functional character. Its carrying out necessitates such an approach. In functional sense space activity is an orbital activity. The Treaty regulates consequently activities carried out by orbital movement of means of space exploration and uses of outer space.

At the same time I have to admit that the Treaty contains rules of seemingly spatial character. E.g. in Article V „astronauts as envoys of mankind *in outer space*", VI „responsibility for national activities *in outer space*" or VIII „...the state of registry retains jurisdiction and control over objects launched *while in outer space*". Any movement in outer space is necessarily orbital and any activity on the Moon or other celestial body is the result of an orbital movement. In this context spatial elements of the wording of the Treaty may be harmonised with the functional approach.

Other main sources of international space law following the Treaty are not inconsistent with this interpretation. In this respect the Convention of Registration deserves attention. Space objects are to be registered in a national register when it is launched ***into Earth orbit or beyond*** (Article II). The state of the registry has to furnish information to the Secretary General of the United Nations, among others ***basic orbital parameters*** including nodal period, inclination, apogee and perigee. (Article IV.1) Elements of spatial delimitation do not appear in this system of international registration, and suborbital launchings are clearly excluded from the obligation of registry. Though the term ***space object*** has no real definition in the convention (Article I.b) it is obvious, that objects can be qualified as space object by a ***functional attribute namely orbital movement.***

National air laws

Failing spatial delimitation in space law the question reasonably arises: what was the reaction of municipal air laws, if any, to this insufficiency of space law regulation? To cast a glance at sea law analogies in this respect may be thought-provoking for the space lawyer. Security by self-defence is elementary postulate both for littoral and subjacent states.

In the presatellite age this demand was firmly linked with spatiality i.e. this principle was the essence of territorial sovereignty. Security meant effective control over definite land and sea areas.

In sea law the principle of sovereignty of the littoral state over territorial sea developed contemporaneously with the doctrine of freedom of open seas (7) As formulated by ***Bynkershoek***: the littoral state could dominate only such width of coastal waters as lay within the range of cannon shot from shore batteries. (8) This „cannon-shot rule" became amalgamated with a three-miles limit which in the nineteenth century received widespread adoption in theory and state practice. It never became a general rule of international law. Municipal sea law legislations determined the width of territorial seas adopting the three-miles or a wider limit corresponding to the security demand of the state concerned.

At the very beginning of air law literature for drawing the upper limit of state air space („territorial air") some authors supported the principle of effective control (***Bluntschli*** (9), ***Rivier*** (10)). This theory even after the Paris and Chicago Conventions could refer to such authorities as ***H. Kelsen***: „It stands for reason, that a state can enforce the provisions of this Convention or its own legal order... within that part of the air space over which it has effective control", (11) or ***A. Verdross***: „The Convention in question does not know any limit nevertheless we must suppose the existence of such a limit... the entire air column capable of being ruled (beherrschbar) above the state territory". (12)

As far as I know in positive air law only the Bolivian Air Code of October 24, 1930 accepted this idea in this form: state sovereignty extend to the column of air which covers the surface of the national territory within the limits of the frontiers the height being determined by the range of defensive methods of the country. (13)

Needless to say that this modern form of „cannon-shot rule" apparently would mean that the upper limit of air sovereignty in case of each country would depend on its technical level of defence capability. Consequently free outer space would have no general but singular limits of big variety according to the conditions of the subjacent states.

In the presatellite age there was no need to be concerned with the ultimate consequences of the concepts of „completeness" and „exclusiveness" of sovereignty in vertical direction. Theoretical extension „usque ad coelum" and downwards to the centre of the Earth presented no difficulty.

of astronautics the international lawyers will once more face the question which they believe to have banned from science for ever, and have definitely solved in favour of the boundlessness in height of air sovereignty." (14)

After launching of Sputnik-1 international law really was faced this question. Ten years elapsed before signing the Treaty and thirty years after. The dispute in the theory has been going on between spatial and functional approach now for more than forty years.

Delimitation of air space and outer space in absence of international agreement could be expected logically from national legislations. However, the link between security demand and upper limit of air sovereignty had no technical reality anymore. In the age of ICBM-s self-defence would be impossible within a delimited space above the territory of subjacent states as it could be effective against an attack from the open sea outside territorial waters of littoral states. (15)

Spatiality in this respect is not feasible anymore. To my knowledge not a single municipal air law in the thirty years of the Treaty undertook a spatial delimitation of air sovereignty. Complete and exclusive sovereignty of states over their air space (Chicago Conv. Art. 1) and freedom of outer space (S. T. Art. 1) are living together.

Hungarian air law is but an exception standing beside the rule. The new air code „Act Nr. 97 of 1995 on the Air Traffic" declares that the Hungarian Republic has complete and exclusive sovereignty over Hungarian airspace i.e. „the part of airspace above the territory surrounded by confines of the state to the altitude where air traffic is physically possible" (§ 4.1.) It is not meaningless that the first draft of the Act would have accepted the theory of effective control stating that „the Hungarian airspace ends at the altitude of air navigation and the effective range of antiaircraft defence." This new air law, having been valid since July 1, 1996 could be hardly interpreted otherwise than acceptance of the aerodynamic theory of delimitation. Under the present-day conditions — at least above Hungary — outer space would begin somewhere in an altitude of 35-40 km.

Anyhow this definition of Hungarian airspace is not bound on a numerically fixed upper limit, and — concerning security do not exclude making use of the right of self-defence beyond this sovereign air space under Article 51 of the UN Charter.

Municipal space law

In the last years space activities has been carried out more and more by private entities. This development demanded space legislation especially for space-faring nations. According to the Treaty „Space activities of non-governmental entities shall require authorisation and continuing supervision by the appropriate State Party to the Treaty". (S. T. Art. VI.) Municipal space acts intend to fulfil this obligation last not least the possibility of shifting damages upon the non-governmental entity, in case the authorising state will be internationally liable for damage caused by the private space activity. (S. T. Art. VII.)

The system of regulation in this municipal space laws is clearly functional.

The *Outer Space Act of the United Kingdom* (16) of 1983 sets as its aim to secure compliance with the international obligations of the United Kingdom with respect to the launching and operation of space objects and carrying on other space activities *in outer space* by persons connected with this country.

The act applies to a) launching or procuring the launch of space object, b) operating a space object, c) *any activity in outer space*. As definition of outer space it contains only the statement that it „includes the Moon or other celestial bodies. To space object: „it includes the component parts of a space object, its launch vehicle and the component parts of that." The two unsatisfactory definitions follow the wording of the Liability Convention (Art.

I.d) and the Convention on Registration (Art. I.b)

The above phrase *in outer space* seems to refer to spatially determined activities being subjects of authorization. The systematic interpretation of the Act, however, excludes such a conclusion.

According to Sec, 5.1 namely the licence shall describe the activities authorised by it. In particular this licence may contain the basic *orbital parameters* including nodal period, inclination, apogee and perigee, possibly the advance approval for deviation from the *orbital parameters* and „disposal of the payload *in outer space* on the termination of operations under the licence" (Sec.5/2 b,d,g)

The Secretary of State maintains a register of space objects corresponding to international obligations of the United Kingdom. This is a clear reference to the Convention on Registration obliging launching states to furnish to the U.N. Register „basic orbital parameters" of the space objects.

The British space act accordingly concerns activities carried on by *orbital movement* — without any reference to spatial determination.

The *Swedish act (Act on Space Activities 1982:963)* (17) seems to follow spatial concept stating: „This Act applies to *activities in outer space* (space activities)". To activities carried on entirely in outer space in addition are included the launching of space objects into outer space and all measures to manoeuvre or in any other way affect objects launched into outer space. The decree on Space Activities (1982:1069) is clear in respect of activities: ruled by the Act. Sec. 4 provides the National Board for Space Activities shall keep a register of the space objects for which Sweden is a contracting State in accordance with Article 1 of the Convention on Registration. The data to be registered are accordingly among others basic orbital parameters. Exclusion of suborbital undertakings from the notion of space activities is clear from Sec.1 „Merely receiving signals or information in some other form from object in outer space is not designated as space activities according to this Act, *nor is launching of sounding rockets designated as space activities.*

The *Law of the Russian Federation on Space Activities* of August 20, 1993 represents an essentially different system. Definition of space activity (18) covers an extraordinary wide range of uses of space devices: space communications, remote sensing, manned space flights, space research, manufacture of products in outer space and „other kinds of activities performed with the aid of space technologies."

The *South African Space Affairs Act (No. 84 of 1983)* (19) covers a similarly wide range of activities. Space Activities are defined as activities directly contributing to the launching of spacecraft and the *operation of such crafts in outer space.* From this definition logically follows that the act applies to launching with orbital aim and resulting operation by orbital movement. Similarly to the Russian Act the South African Space Act includes to space acts as „space related activities" all activities supporting or sharing mutual technologies with space activities.

The *National Aeronautics and Space Act of the United States 1958 as amended 1983* (20) regulates both legal domains together. Aims and definitions correspond to this combined system.

The act intends to provide for research into problems of flight *within and outside the Earth's atmosphere.* Sec 103 accordingly defines *aeronautical and space activities*:

A/ research into and the solution of, problems of flight *within and outside the Earth's atmosphere*
B/ the development, construction, testing and operation for research purposes of aeronautical and space vehicles
C/ the operation of a space transportation system including the space shuttle, upper stages, space platforms and related equipment, and

D/ such other activities as may be required for the exploration of space.

Aeronautical and space vehicles are: aircraft, missiles, satellites and other space vehicles manned and unmanned together with related equipment, devices, components and parts.

Concerning differentiation between aeronautical and space activities in the Act no elements are to be found, which would refer to any spatial criterion. The same can be stated also for other basic sources of municipal space law of the United States such as Communication Satellite Act of 1962 and Commercial Space Act of 1984.

As it appears to me from the foregoing can be stated, *municipal space laws do not contain elements referring to delimitation. Facts provided by them are bound to launching into orbit respectively operation in orbit.* State authorization and supervision is by no means restricted to activities carried out in a space above certain limits. Consequently neither municipal air laws nor space laws of the past thirty years support the spatial approach against functionalism.

It should be considered that state practice of thirty years demonstrates: effectiveness of the Treaty never was made dependent on any limits between air space and free outer space by states. On the contrary the Treaty has been living without any delimitation.

This is why I venture the conclusion that thirty years of the Treaty demonstrate the necessary application of the functional approach without express acceptance.

De lege ferenda

Professor *A. Górbiel* named me in 1980 „the most unequivocal adherent of the functional theory in the literature of socialist countries." (21) Not only at colloquia of our Institute but also at Space Law Seminars of INTERCOSMOS my papers on various subjects were built upon this footing indeed. (22)

Nevertheless, I admit the practicability and usefulness of delimitation by an international agreement. At any rate in addition to functional basis of the system of space law.

May I refer as a possible solution to the Soviet proposals of 1983 and 1987. The Soviet Union in a working paper of April 14, 1983 suggested the conclusion of an agreement on delimitation: „The boundary between air space and outer space shall be established by agreement among states at an altitude not exceeding 110 km above sea level."

The „compromise proposal" of 1987 added a new element to above formula: „an object launched into outer space shall be considered as being in outer space at all stages of its flight after launch at which its altitude above sea level is 110 km or more" (23)

A functionalist can be fully satisfied with this formula except one point: the Russian proposal seems to exclude attempted launches in case the object after lift-off do not gain the altitude of 110 km. Theoretically this could be interpreted in the way that e.g. launching of an object carrying nuclear weapons would not constitute a delict of international law, breach of Space Treaty Article IV — if it fails and the carrier rocket crashes from an altitude of 50 km. The solution is only that all launches aiming orbital movement should be governed by space law from the very moment of the lift-off.

The altitude of 110 km otherwise corresponds about to the lowest possible perigee of satellites, according to some authors being excepted as customary-law rule.

The „compromise proposal" is not less than an attempt to harmonize two opposite concepts. To my humble opinion an international agreement following this line would successfully end the „never-ending dispute".

Footnotes

1. By *R. F. A. Goedhart*: Forum for Air and Space Law. Ed. by M. Benkő in cooperation with Willem de Graaf. Vol. 4. 1995

2. Collins Dictionary of Astronomy. Ed. by *V. Illingworth*, Galsgow 1994, pp. 87-88, 316-318. Űrhajózási lexikon (Encyclopedia of Astronautics) Budapest 1981 pp. 345-346. Űrtan (Space Science) — SH-Atlasz — Springer, Budapest 1966 pp. 41-44

3. *G. Gál*: Légitér és világűr (Air Space and Outer Space) Jogtudományi közlöny, Budapest 1960, pp. 565-575. English transl.
in Legal Problems of Space Exploration. A Symposium U.S. Senate, Washington 1961, pp. 1141-1157

4. *G. Gál*: Világűrjog (Space Law) Budapest 1964, pp. 146-150

5. U.S. Senate Symposium p. 1270

6. „Outer space and celestial bodies are free for exploration and use for all states…"

7. *J. G. Starke*: Introduction to International Law, London 1977, pp. 228-229

8. *Bynkershoek*: De dominio maris dissertatio. 1702 „Terrae potestas finitur ubi finitur armorum vis."

9. *Bluntschli*: Das moderne Völkerrecht der zivilisierten staaten 2. Aufl. 1878

10. *Rivier*: Principes du droit des gens, 1896

11. *H. Kelsen*: General Theory of Law and State. Cambridge 1946, p. 217

12. *A. Verdross*: Völkerrecht, Wien 1955, pp. 198-199

13. Quoted by *A. G. Haley*: Survey of Legal Opinion on Extraterrestrial Jurisdiction. III Coll. IISL 1960 Proceedings, p. 69

14. *V. Mandl*: Weltraumrecht, Mannheim-Berlin-Leipzig 1932, p. 31

15. *R. F. A. Goedhart* — otherwise not supporting functionalism — states: „Considering the real danger that ballistic missiles apparantly pose, it is actually not feasible to choose any upper limit of state territory meeting all security requirements of the target states under threat of attack." op. cit. p. 112

16. *K.-H. Böckstiegel — M. Benkő* Ed. Space Law Basic Legal Documents Vol. 2/1, E. I.

17. Ibid. Vol. 2/1, E. II.

18. *F. G. von der Dunk*: Two New National Space Laws: Russia and South Africa. In Proceedings XXXVIII. Coll. 1995

19. Ibid.

20. Note 16, Vol. 2/1, E. III.

21. *A. Górbiel*: Legal Definition of Outer Space, Lódz 1980, p. 48

22. *G. Gál*: Space Law. Budapest—Leyden—Dobbs Ferry 1969, pp. 105-117. The Question of Delimitation—After Twenty Years.
XXII Coll. IISL 1979 Proceedings pp. 125-129, — Fundamental Links and Conflicts Between Legal Rules of Air and Space Flights
XXVI. Coll., 1983. pp. 77-80. — Territorial Sea and Airspace — Passage to Free Spaces? XXVIII. Coll., 1985, pp. 134-139. — Space Objects „While in Outer Space" XXXVII Coll., 1994, pp. 84-86, — Air Law and Space Law—Coincidences and Controversies. INTERCOSMOS Seminar Prague 1982 — Doklad pp. 185-193. — Problems of Delimitation and Geostationary Orbit at the Twenty-third Session of the Legal Subcommittee, Seminar Varna 1984 — Doklad pp. 151-175. etc.

23. U. N. Doc. A/AC.105/C.2/L.139, April 4, 1983 and U. N. Doc. A/AC.105/ L.168, June 5, 1987. See *A. D. Terekhov*: Passage of Space Objects Through Foreign Airspace. XXXII Coll. IISL Proceedings pp. 52-53

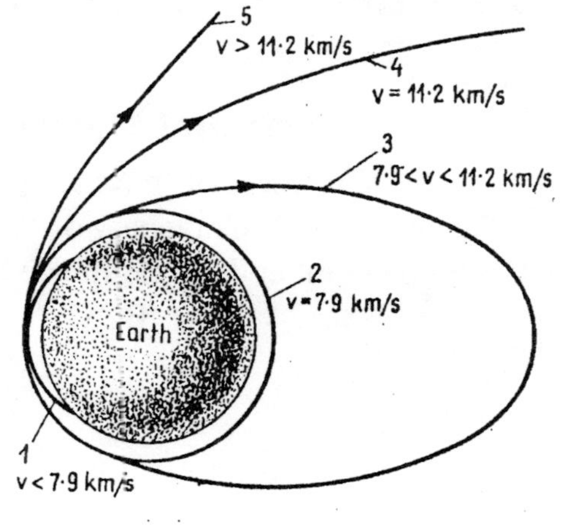

Fig. 1.— 1 trajectory of an ICBM, 2 circular orbit, 3 elliptic orbit, 4 parabolic orbit, 5 hyperbolic orbit

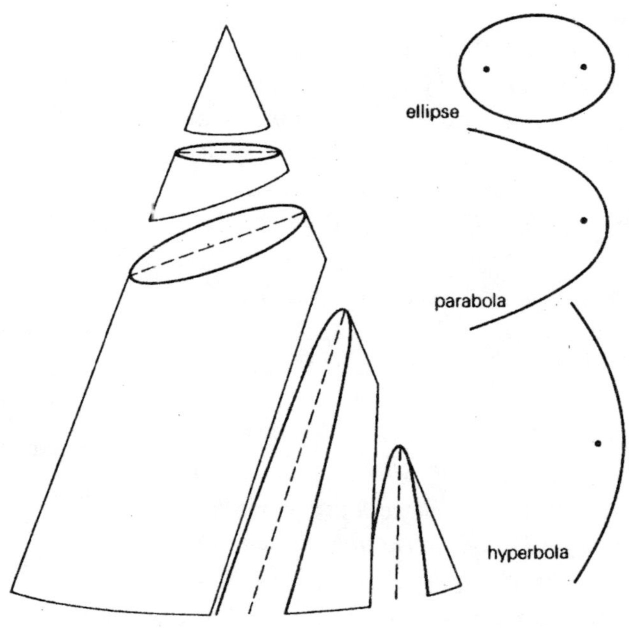

Fig. 2.— Conic sections.

Remarks on the Responsibility and Liability for Damages Caused by Private Activity in Outer Space.

Armel Kerrest
Professor of International Law
Faculty of Law, University of Western Brittany.
Centre de Droit et d'Economie de la Mer (CEDEM)
IISL ECSL *

Abstract:

The question of liability for damages caused specifically by spacecraft is well documented. The victims are relatively well protected by article The Outer Space Treaty and the Liability Convention. However, nowadays, the main issues of debate concern potential damages of other kinds caused by general space activities and not only by possible "traffic accidents".

Whilst the launching State is absolutely liable for damages caused by space objects, the issue of liability and responsibility for other damage is still unclear under the Outer Space Treaty.(1967). The definition of direct damage is found in OST article VII. The OST establishes international responsibility of the "appropriate State" which bears international responsibility for national activities in outer space. The interpretation of this disposition is still open for debate.

Unlike liability under article VII, responsibility under article VI is a common international responsibility. There must be present damage, and a violation of international law in the sense of there being a wrongful act of the State concerned. Thus, the main issue is the extent of the States' control obligation over private activities under international law.

The other question which is a major issue, is the interpretation of the words: "national activities in outer space". What about activities partly conducted in space and partly on earth which is the case in nearly every space activity?If a wide meaning is applied, it will impose an obligation to control, and responsibility for such activities on the State. (Cf. article III OST). Bearing in mind the current extent of private activities in outer space, such an interpretation would have dramatic consequences.

The question of proving the extent of the international rule remains unanswered, as are questions relating to the lack of a judicial control in these circumstances.

Introduction

The new trend towards a privatisation of space activities is a major challenge to space law. Even if, at the time of the outer space treaty (1967) and of the Liability Convention (1972), [1] private activities were considered and

---

.* Email: Armel.Kerrest@univ-brest.fr

---

1 Treaty on Principles Governing the Activities of States in the Exploration and Use of Outer Space, Including the Moon and Other Celestial Bodies, opened for signature January 27, 1967, 610 U.N.T.S. 205 18 UST. 2410 (1967) 6 ILM 386 (hereinafter referred to as OST).
Convention on International Liability for Damage Caused by Space Objects Oct 9 1973 961

authorised, they were rare. Nowadays, we may see an explosion in the number of projects initiated by private companies: Private telecommunication satellites (television and telephone), but also others which were probably unimaginable only a few years ago such as private launches[2], and even private mining of asteroids.[3].

With the rise of private activities in the international domain, questions of the control of such activity arise, especially relating to issues of liability and responsibility. The situation on the high sea is not very encouraging as it is very difficult to control such activities. The solution is, of course, international legislation through co-operation. Given the structure of international society, this process is a very long and difficult one which has involves issues relating to the sacrosanct principle of sovereignty of states. Another possible solution is to limit the extend of the international domain by giving certain sovereign rights over those zones. This approach was used for issues relating to the conservation and utilisation of natural living resources of the sea (creation of the EEZ). In outer space we are now at the crossroads. We can choose between a state of anarchy and the maintenance of a special control exercised by states over space activities.

U.N.T.S. 2389 (hereinafter referred to as "Liability Convention")

2. The Sea Launch project (See our " The Sea Launch project, Launching spacecraft from the sea and the Outer Space Treaty : this colloquium at 1997 Turin IISL 3.15) and more generally: A. Kerrest "The Launch of Spacecraft from the Sea" in: "An Outlook on Outer Space Law in the Coming Thirty Years". Lafferranderie et Crowther edit. (Kluwer 1997).

3 See New Scientist editorial: A giant leap for lawyers: the race is on to grab a piece of celestial body and the article of Kurt Kleiner on the same subject on p.18 September 20, 1997.

Preliminary remark: Liability and Responsibility

Like some other languages, French uses only a word for both these notions: "responsabilité". For a French speaking lawyer, it is very interesting, if difficult, to analyse the distinction made in English between these two principles. In fact the French word "Responsabilité" has two meanings. When the little Prince of Antoine de Saint Exupery writes that he is "responsable de sa rose" (responsible for his rose), it does not only mean that the Prince is liable but that he is responsible i.e. that he has to take care of his rose.

If we examine the English use of both words, the distinction is less clear. In the specialised fields of international domains, space law and the law of the sea, this distinction seems to be used in a way which may please a Cartesian reason. Professor Stephen Gorove is one of the rare authors who deal with this distinction in a way appropriate to space law:

> *"It should be stressed that in the field of international space law two closely connected terms have been used: "liability" and "responsibility". Neither of these terms has been defined in space law but the term "liability" has been used to set the launching state's liability for damage caused by space objects, whereas the word "responsibility" has been used to mandate international responsibility by the appropriate State party for national activities in outer space.*
>
> *Even from this brief reference to the use of the two terms, it appears that in connection with "liabilities" we are dealing with legal consequences (mostly in terms of damages) arising from a particular behaviour. In contrast, it seems that when we speak of "responsibilities", we are dealing primarily with obligations imposed on people and institutions who are supposed*

to carry out certain activities or are accountable in given situations, though not necessarily in the form of compensation for damages. Thus to some extent the two concepts are interrelated. Normally, liability deals with compensation for damages resulting from loss of life, personal injury loss or damages to property, whereas responsibility may not always include compensation for damages but could include criminal accountability. [4]

In the Law of the Sea Convention[5] this distinction is also made but the French word is not the same, the terms "the state has the responsibility are translated by : "l'Etat doit contrôler". In the OST both terms are used. Basically, responsibility is connected with the obligation to control and thus with a fault or a wrongful or unlawful act. Liability may be a consequence of a fault but may also be related to an act without any fault, it is then "objective liability" i.e. strict or absolute liability. On the other hand, an claim relating to responsibility must be based on a fault or breach of the law. In the case of "objective liability" there is no fault, and even no wrongful act; it is a liability imposed without any sense of moral culpability. For reasons, mainly relating to the wish to offer protection to a possible victim, it has been decided to make somebody liable in compensation irrespective of their own actions. The choice of the liable entity depends on its efficiency to provide compensation, to avoid damage or even on the possible profit generated by the activity. If the owner of a oil tanker is held liable by the Brussels convention (1969)[6] it is both because he is supposed to be able to pay or to take insurance cover and because he is responsible for the good maintenance of the ship.

We have to keep this in mind when we turn to space activities. Just like the owner of the ship, the launching state may very well have nothing to do with the control of the satellite. In such cases, the need to protect the victims is paramount. It is very useful in case of ultra-hazardous activities where the damage may be huge and where, given the technical complexity of the processes, it is extremely difficult to find and prove fault.

Both liability and responsibility are of course interrelated. Being responsible will, in the main, impose a duty to compensate the victim who has suffered damage due to a lack of control or mismanagement. On the other hand, a person who is liable will certainly wish to control the activity in question to be sure of avoid incurring heavy losses.

I The liability of private actors in outer space.

Private companies are not subjects of international law and, as such, they cannot be parties to cases before international tribunals or courts. Only states and, in some cases International Organisations, are granted such status. A possible solution is to bring legal action on a national level. There, municipal law applies which can in certain instances incorporate international law. This does not seem to be the case with the OST which has no direct effect on citizens. When there is a national space law, it may establish strict or absolute liability on operators as a burden to private actors acting in outer space, and insurance may be compulsory. In this case, liability does not derive from the liability

4 S. Gorove, «Liability in Space Law: An Overview» dans *Developments in Space Law – Utrecht Studies in Air and Space Law*,-Dordrecht, Martinus Nijhoff, 1991 at pp. 224-225.

5 United Nations Convention on the Law of the Sea open for signature on 10 December 1982 in Montego Bay Jamaica (hereinafter referred to as "Montego Bay Convention")

6 International Convention on Civil Liability for Oil Pollution Damage Brussels 1969 (in force 19 June 1975) (9 ILM 45) hereinafter referred to as "Brussels Convention")

convention which does not apply to private persons, but from national space law itself.

In the Liability convention, only the launching state is taken into consideration. This state may very well transfer the burden of the liability to private companies or to insurance. It may also, to ease insurance, put a ceiling on the liability. This is the case for French Guyana Kourou launches. The private companies acting there, like Arianespace, are liable and must take out insurance to a certain level (here 400 Mfr). If the damage is higher than such a level, France must assume the extra cost without any ceiling in accordance with the liability convention.

If there is no special national space law, no lex specialis, the lex generalis applies. Most of the time it will be ordinary liability i.e. fault liability with all the difficulties already mentioned. In some cases the municipal law uses strict liability in case of ultrahazardous activities. All this depends on the municipal law. It would be a very difficult task for the victim to obtain compensation. Because of the nature of space activity, the risk is really international. The victims have to deal with the variety in the law, the different tribunals, the difficulty of obtaining compensation from a judge, and the issues of legal costs. Moreover it may be necessary to indulge in some "forum shopping". By choosing a nationality, the private company can choose the most favourable liability law.

II The absolute liability of states.

The principle of liability

Even if the provisions of OST and of the Liability convention are not quite clear as far as private activity is concerned, it may be pointed out that the states are responsible and liable for any space activity of entities under their control. The OST, article VII and the Liability Convention put that burden on the "launching state". It is an absolute liability which does not include any moral consideration. The only point to consider is the damage on the one hand and the activity on the other hand.

Because of our Judeo-Christian philosophical background, we are very much linking liability to a fault even if this fault is only to conduct a ultra hazardous activity. We do not readily accept liability being imposed on someone who is not directly involved in the activity.

We have to remind ourselves of the very basis of the rule. At the beginning of the space era, states acting in outer space wanted to obtain a freedom of "navigation" and in general a freedom of use in a place which was often claimed as a part of state territory. States "sovereignty expanded "ad infinitum". To avoid any criticism and to make this change acceptable for the benefit of two states only, they decided to guarantee compensation to third party victims of space activity through an absolute liability rule. As is always the case for liability without fault, imputation had to be decided. Given their special role in launching objects into outer space, the launching states where chosen as liable states. This choice is the best for many reasons. A launch in outer space is impossible to hide, those states were very powerful, rich, and, as such, able to compensate for any possible damage. Moreover, the burden is shared with the other states taking part in the launch, especially the states which procure the launch.

Thus an important distinction was made between the actors of space adventure on the one side, and terrestrial third parties, potential victims, on the other. This distinction is still made today: absolute liability protects "innocent" victims. In the case of other space actors, a fault has to be proven.

The extend of the liability

The liability convention states that : *"A launching State shall be absolutely liable to pay compensation for damage caused by its space object on the surface of the earth or to aircraft flight".*

The definition of "space object" is often discussed. In some cases this discussion takes place with a view to lowering liability. In fact, the word "object" is the widest possible. It includes any kind of "objects", whatever their size, whatever their capacity to move by themselves etc. The term itself is so wide that we have to define : "man launched space object".

The main issue is much more the causal relation, the interpretation of "by" (damage *by* the space object). The liability in the case of direct damaging contact is obvious. But it must be highlighted that no text limits liability to the fall to earth of the space object. It is difficult to imagine possible damages which have not yet have taken place. The damage may be caused by the space object without its falling to earth. Let us imagine certain situations: a spacecraft equipped with mirrors to illuminate a town, goes out of control and illuminates a telescope for some months making any research impossible. Damage may also be caused by data from a remote sensing satellite. Other damage may occur following improper work by a GPS satellite used for Air Navigation Control. Given the lack of judicial decisions on these issues, it is very difficult to speak with certainty as to whether there is a sufficient causal link between the activity in question and the damage incurred. Some of these problems may be considered not as direct but as indirect damage.

Taking into consideration the very basis of space liability, it seems possible to make a distinction between damage caused to third parties and damage caused to people participating in the space activity. In the case of a failure in the GPS, the aircraft company using this system is participating in the activity which includes the space activity. No absolute liability should be involved and the issue must be resolved by reference to the contract. As for the victims in the aircraft itself, they may be considered as third parties but the damage should perhaps not be considered as a direct damage. On the other hand, the scientists working at the above mentioned observatory whose activity is disturbed, are third parties. They suffered a direct damage by a space object.

III The responsibility of states for space activities.

The "national activities"

If the damage cannot be considered as being directly caused "by" a space object, but by the space activity or by an activity partly using outer space, the victims are not left without any possible action. Responsibility of the state must be taken into consideration. The OST article VI states : " *States Parties to the Treaty shall bear international responsibility for national activities in outer space...whether such activities are carried on by governmental agencies or by non-governmental entities*". This is a unique provision in international law. In general international law, states are only responsible for their own activities. In the case of private activity, a state is only responsible if it fails to control such activities. In the well known US-Canada Trail Smelter case[7], the Arbitral Tribunal decided that Canada was responsible for a lack of efficient control of the smelter's activity. Canada was not responsible for the pollution as such. On the other hand, the OST article VI provides that the state is internationally responsible for any activity in outer space, either for a state activity conducted by its own agencies or for any private company which may be considered as a "national space activity"; in fact an activity conducted by a company enjoying the nationality of that state. The appropriate state is not only responsible for a lack of control, as would be the case without the OST provisions, but it is also responsible for any national activity taking

[7] US-Canada Trail Smelter Case 11 March 1941 Arbitral Award 3 U.N.R.I.A.A. 1911

place in outer space, such activity being considered as its own activity. This is much more onerous. The obligation of authorisation and control which is provided for in the same article is only a definition as to the obligation of the state. It does not modify the principle of assimilation of private activities with public ones. The consequences of this are rather important : in the case of a violation by the private entity of any international regulation or principle, the state should be responsible without having the possibility to avoid liability by proving ignorance of such a violation, nor even by showing it had made its best effort to control the activity.

This action and thus this violation are considered as its own.

The existence of a fault.

International responsibility is not absolute liability, by nature responsibility is linked with fault. Concerning states, we prefer to use a notion free from any moral background: the breach of the law. Thus the extend of states' responsibility is a matter to be decided by the applicable law. The law to be taken into consideration is international law as a whole as defined in article 38 of the statute of the ICJ, not forgetting principles of international law and of space law, including for instance principles of peaceful use, non appropriation and benefit of all people.

The appropriate responsible state

Responsibility is no more directed to the launching state like it is for liability. Responsibility as seen before is much more linked with issues relating to control. It must take into consideration the real control behind the operation. The responsible state is the "appropriate state", a very vague provision which is connected with the notion of "national activity". The state is responsible for any national activity in outer space. The question is then to define "national activities". Are they only activities conducted by nationals ?

If we take into consideration national space laws, we can see that they apply very broadly. The US Commercial Space Launch Act refers obviously to activities conducted on or from the territory of the United States. It refers also not only to citizens of the United States but also to foreign companies controlled by US citizens. This seems to indicate that "national space activities" are to be considered broadly.

Conclusion: remarks *De lege ferenda*

The Achilles' heel of this system is that it is wholly state oriented. Both responsibility and liability are imputable to a state. It is often an advantage as far as technical and financial capacity are required. But concerning the question of the settlement of dispute, it may be more interesting for the victims to be able to sue a private company before a national judge. The international settlement of disputes is a political issue. The State of the victim who has suffered loss brings its own action and is free to act. The liable or responsible state can, in fact, use its sovereignty to jeopardise the solution of the dispute refusing the control of any international judge.

In a country where a comprehensive national space law is in force, which is only the case in the United States, it will be possible to act before a national judge on the basis of this national law. This action may be pursued against the state or against the company depending on the national rules. Every country conducting space activity should enact a national space law dealing with liability, responsibility and control. The US Commercial Space Launch Act may serve as a useful pattern. An international convention unifying

8 Luigi Condorelli: *La réparation des dommages catastrophiques causés par les activités spatiales"* in Travaux des journées d'études juridiques Jean Dabin Brulant Bruxelles 1990 ff 263-292

the main rules relating to liability for damage caused by private entities should be used to avoid too much diversity in levels of protection offered against an activity which is by nature very international.

Above all, it is necessary to maintain the present system and its burden of absolute liability and responsibility on states. It will act as an ultimate safety net. We need only look to the US CSLA to see how useful it is. The main aim of this law is to regulate the relations between the US Government and the licensed private entity, taking always into consideration the international obligations of the state on the one hand and the utility for the US of a private activity in outer space on the other.

There are certainly room for improvements on the question of settlement of disputes; an international body should be designated as the appropriate forum and, at the very least, decisions of the Claims Commission provided for in article XIV of the liability convention should be made compulsory. Such action was not acceptable at the time of the cold war, but the time is now appropriate for change.

To schematise the issue of compensation for damage caused by space activities:

I *On a national level (private entity):*

- 1 **The Law**: National, either special or general.
- 2 **The liability:** Either fault, strict or absolute liability as determined by the national law - possible obligation of insurance, possible ceiling.
- 3 **The judge** : National.
- 4 **The damage**: Any damage caused by any activity, (either “caused by the space object”» or by any activity) (either on Earth or in space).

II *On an international level: Liability under OST article VII or liability convention (states)*:

- 1 **The Law**: OST or Liability Convention as lex specialis.
- 2 **The liability:** Absolute liability without any ceiling for damage on earth, fault for damage "elsewhere than on the surface of the earth".
- 3 **The judge:** "Claims Commission" and generally International settlement of dispute, no compulsory judge.
- 4 **The damage:** Only the damage "caused by the space object".

III *On an international level: international responsibility (states):*

- 1 **The Law:** General international law.
- 2 **The liability:** International responsibility for breach of the international law, restitutio in integrum or adequate compensation.
- 3 **The judge:** International settlement of dispute, no compulsory judge.
- 4 **The damage:** Any damage caused by any activity, (either "caused by the space object" or by the activity) (either on Earth or in space).

IISL-97-2.11

EXOBIOLOGY AND THE OUTER SPACE TREATY: FROM PLANETARY PROTECTION TO THE SEARCH FOR EXTRATERRESTRIAL LIFE

By Patricia M. Sterns* and
Leslie I. Tennen**
LAW OFFICES OF STERNS AND TENNEN✪
Attorneys and Counselors at Law
Phoenix, Arizona USA

ABSTRACT

The Outer Space Treaty established the legal obligation of states to prevent both forward and back contamination of the Earth, the Moon and other celestial bodies. These policies originated in the planetary quarantine requirements developed by the international scientific community for interplanetary missions. Subsequent instruments, especially the Moon Treaty, have elaborated on the obligations of states in relation to exobiological matters, including the requirement of public disclosure of the discovery of evidence of organic life. The recent announcement of possible fossilized traces of Martian forms of life in meteorite ALH84001 has focused renewed attention on planetary protection policies.

This article discusses the essential purposes and development of environmental protection provisions in the law of outer space. Interplanetary missions and the lunar quarantine procedures utilized in the Apollo program are examined as examples of the application of these obligations. Suggestions are proposed for specific mechanisms to promote and enhance the effectiveness of policies to preserve pristine environments and protect the integrity of scientific investigation in the exploration of our celestial neighbors.

INTRODUCTION

The fundamental questions of science and philosophy concern the origin and existence of life: How did life on Earth begin, and are we alone in the universe? Perhaps clues to these mysteries may be found on our celestial neighbors, as the late 1990's has witnessed a resurgence in interplanetary exploration. Cross-contamination by human intervention to the

* Member IAA, IISL, ASIL, ABA, IBA, ILA
Senior Member AIAA

** Commissioner, Arizona Space Commission
Member IAA, IISL, IEAS, ASIL, ILA
Senior Member AIAA

✪ Member IAF, IISL

The opinions expressed in this article are those of the authors only and should not be attributed to any organization with which they may be affiliated environments of Earth and celestial bodies by the introduction of biological materials could have unforeseen and catastrophic consequences. Such materials could mutate in unpredictable ways, or cause the mutation of any extant forms of life. It is axiomatic that the maintenance of integrity of scientific investigation mandates that measures be taken to prevent the biological contamination of the environment of Earth and celestial bodies. Accordingly, both the international scientific and legal communities have adopted measures for this purpose.

HISTORICAL DEVELOPMENT OF THE PLANETARY PROTECTION POLICY

The issue of planetary protection has received serious attention since the mid 1950's.[1] Following extensive study, the Committee on Space Research (COSPAR) of the International Council of Scientific Unions called for the imposition of international controls to prevent back contamination, that is, the contamination of Earth's environment by the return of extraterrestrial materials.[2] Further, in 1964, COSPAR approved recommended planetary quarantine requirements (PQR) regarding forward contamination, which declared that the probability of contamination:

> of a single viable organism aboard any spacecraft intended for planetary landing or atmospheric penetration would be less than 1 x 10^{-4}, and a probability limit for accidental planetary impact by unsterilized flyby or orbiting spacecraft of 3 x 10^{-5} or less.[3]

The probability of contamination (P(c)) for any mission was determined by application of the following formula:[4]

$$P(c) = m(i)(o) \cdot P(vt) \cdot P(uv) \cdot P(a) \cdot P(sa) \cdot P(r) \cdot P(g)$$

where:

m(i)(o) = initial microbial burden at launch, after decontamination

P(vt) = probability of surviving space vacuum-temperature

P(uv) = probability of surviving ultra-violet space radiation

P(a) = probability of arriving at celestial body

P(sa) = probability of surviving atmospheric entry

P(r) = probability of release

P(g) = probability of growth

The COSPAR guidelines obligated states to take active measures to reduce the initial microbial burden of an interplanetary craft at launch. At a minimum, fly-by craft were subjected to clean room assembly, while landing craft, such as Viking, were sterilized by heat, gas or radiation. Mission profiles were developed which minimized the risk of unintentional and accidental contact between orbiting vehicles and celestial bodies. The PQR were applied to all interplanetary missions, and any deviation from the policy constituted a specific exception.

The United States implemented the COSPAR requirements by the issuance of a series

---

1. *See generally* C.R. PHILLIPS, THE PLANETARY QUARANTINE PROGRAM: ORIGINS AND ACHIEVEMENTS 10 (1975), NASA Pub. No. SP-4902, U.S. GPO Stock No. 3300-00578; M. WERBER, OBJECTIVES AND MODELS OF THE PLANETARY QUARANTINE PROGRAM 4 (1975), NASA Pub. No. SP-344, U.S. GPO Stock No. 3300-00588.

2. *See generally* J.R. BAGBY, JR., BACK CONTAMINATION LESSONS LEARNED DURING THE APOLLO QUARANTINE PROGRAM, JPL Contract No. 560226 (1975).

3. COSPAR Res. 26, COSPAR INFOR. BULL. at Annex 4 (1964), Fifth International Space Science Symposium, Florence, Italy.

4. Phillips, *supra* note 1, at 38.

of NASA Management Instructions and NASA Policy Directives.[5] In 1978, the Space Science Board of the National Research Council established specific values for the probability of growth (P(g)) factor, which, in their opinion, eliminated the need to employ any decontamination techniques whatsoever in complying with the COSPAR planetary quarantine requirements for exploratory craft to Jupiter, Saturn, Uranus or Neptune.[6] Further revisions to the planetary protection policy in the 1980's provided that planetary protection constraints *may* be imposed, depending upon the nature of the mission and the target body or bodies to be explored. For certain missions and/or target bodies, including the Moon, the policy did not require any planetary protection techniques to be utilized, nor was any specific documentation required. For other target bodies, the classification for planetary protection purposes was to be determined on a case by case basis.

The planetary protection policies were revised most recently in 1994, in relation to exploratory missions to Mars. Specifically, the policy ties the utilization of decontamination and cleanliness controls to mission objectives. That is, craft landing on Mars which carry life detection instruments are subject to Viking level sterilization. However, landing craft without such life detection instruments, such as Pathfinder, are subject to less stringent decontamination techniques.[7]

The issue of back contamination of the environment of Earth mandated the imposition of strict quarantine controls for the Apollo program. The astronauts of the first landing missions, together with the samples of lunar materials returned to Earth, were subject to a rigorous quarantine and examination for several weeks. However, the investigations were focused primarily on whether exposure to lunar materials had any discernable effect on terran life, and not the detection of any evidence of life or the precursors thereof which may have been present. Furthermore, the lunar quarantine experiences were marred by several lapses in procedure, where release of any extraterrestrial life forms present could have occurred.[8]

The discovery of possible evidence of Martian life in meteorite ALH84001,[9] and the current series of missions to the red planet, have renewed concerns over the issue of back contamination. Even prior to the ALH84001 discovery, the scientific community had begun to formulate plans and procedures to ensure that the environment of Earth is not contaminated by organisms which might be present in returned Martian samples, and has stressed that strict quarantine of the material be implemented.[10]

---

5. *See, e.g., Outbound Spacecraft Basic Policy Relating to Lunar and Planetary Quarantine Control,* NASA Policy Directive 8020.7 (1967); *Outbound Planetary Biological and Organic Contamination Control,* NASA Policy Directive 8020.10A (1972); *Quarantine Provisions for Unmanned Extraterrestrial Missions,* NHB 8020.12A (1976); *Biological Contamination Control for Outbound and Inbound Planetary Spacecraft,* NMI 8020.7A (1988).

6. Committee on Planetary Biology and Chemical Evolution, Space Science Board, NATIONAL RESEARCH COUNCIL, RECOMMENDATIONS ON QUARANTINE POLICY FOR MARS, JUPITER, SATURN, URANUS, NEPTUNE AND TITAN 27-28 (Appendix C) (1978).

7. *See* AN EXOBIOLOGICAL STRATEGY FOR MARS EXPLORATION 49 (1995), NASA Pub. No. SP-530.

8. *See generally* Bagby, *supra* note 2.

9. *See* McKay, Gibson, Thomas-Keprta, Vali, Romanek, Clemett, Chillier, Maechling & Zare, *Search for Past Life on Mars: Possible Relic Biogenic Activity in Martian Meteorite ALH84001,* 273 SCIENCE 924 (1996).

10. *See* DeVincenzi, *Planetary Protection Issues and the Future Exploration of Mars,* in 12 ADV. S. RES., No. 4, 121 (1992); D.L. DEVINCENZI, H.P. KLEIN & J.R. BAGBY, JR., PLANETARY PROTECTION ISSUES AND FUTURE MARS MISSIONS, NASA Conf. Pub. 10086 (1991).

The formulation of planetary protection policies, especially in the context of a Mars sample return mission, will attract public involvement and scrutiny, and thereby involves issues of both science and politics.[11] A substantial grey area exists between the two, in which political issues overlap onto areas of scientific uncertainty. Although all available scientific data must be considered, a residual level:

> [of] uncertainty means that decisionmakers cannot determine policy on purely scientific grounds. At this point uncertainty itself becomes an aspect of the factual picture, and the question of what level of risk is acceptable in light of the uncertainty becomes a question of value, requiring political determination. . . . Failure to recognize the trans-scientific character of such questions too often lends 'scientific' credibility and authority as well as an air of 'factuality' to assertions or determinations that are at least as dependent on value choices as they are on 'scientific fact.'[12]

Viewed in this light, it may be questioned whether the relaxation in the planetary protection policies over the past three decades is prudent. There is no doubt that the utilization of decontamination and cleanliness controls adds to the cost and complexity of missions, and that the scientific data on which the policies have been based are constantly subject to review and re-evaluation. The advances in scientific knowledge on which the reductions in the policy requirements have been predicated generally centered on an assessment of the probability of growth for terrestrial organisms in the anticipated Martian environment. Nevertheless, these revisions in policy have occurred in a virtual scientific vacuum, especially in relation to Mars, as no new data had been collected *in situ* in the twenty years between Viking and Pathfinder. Moreover, Pathfinder does not carry life detection instruments, and the examinations conducted by the Viking landers were inconclusive.[13]

The risk of harm presented by biological contamination necessarily is unknown and difficult to quantify, and should not be determined solely with reference to the current state of knowledge. At best, our understanding of celestial environments is incomplete. Moreover, it would be premature to conclude that life does not exist elsewhere in the solar system merely because the scientific community has not been able to identify any alien organisms from the investigations conducted to date. Life on Earth has demonstrated that it is adaptable and resilient, and that once it takes hold, it is imbued with a tenacious will to continue to exist.[14] Therefore, prudence would dictate that planetary protection policies continue to be developed and enforced pending future scientific investigations.[15]

---

11. *See* Race, *Planetary Protection: Legal Ambiguity and the Decision Making Process for Mars Sample Return*, 18 ADV. S. RES., No. 1/2, 345 (1996).

12. Allen, *The Current Federal Regulatory Framework for Release of Genetically Altered Organisms into the Environment*, 42 FLORIDA L. REV. 531, 538-39 (1990).

13. *See generally* ORBITING QUARANTINE FACILITY: THE ANTAEUS REPORT 39-40, (D. DeVincenzi & J.R. Bagby, Jr., eds. 1981) NASA SP-454.

14. *See* Sterns & Tennen, *Preserving Pristine Celestial Environments: The Planetary Protection Policy*, 77 SCIENCE AND TECHNOLOGY SERIES, SPACE SAFETY & RESCUE 1988-1989 399 (1990); *see also* Phillips, *supra* note 1, at 36-7 (discussing the discovery of an alpha hemolytic *Streptococcus mitis* bacterium in a camera from Surveyor 3 returned from the lunar surface by Apollo 12).

15. *See* DeVincenzi, *supra* note 10; DeVincenzi, Klein & Bagby, *supra* note 10; Sterns & Tennen, *Legal Aspects of Planetary Protection for Mars Missions*, 15 ADV. S. RES., No. 3, 281 (1995).

EXOBIOLOGY CONSIDERATIONS IN THE OUTER SPACE TREATY

The conclusion of the Outer Space Treaty[16] in 1967 was a remarkable achievement by the international legal community. Born of political necessity, the instrument reflects both the ideological tensions of the cold war, as well as the most positive of mankind's aspirations. The Outer Space Treaty contains numerous provisions which relate directly or indirectly to matters of exobiology, particularly concerning forward and back contamination.

Article IX of the Outer Space Treaty requires that:

> States Parties to the Treaty shall pursue studies of outer space, including the moon and other celestial bodies, and conduct exploration of them so as to avoid their harmful contamination and also adverse changes in the environment of the Earth resulting from the introduction of extraterrestrial matter and, where necessary, shall adopt appropriate measures for this purpose.

This provision applies both to forward as well as back contamination by biological materials. A strict construction of this provision may obligate States parties only to *avoid* harmful contamination or adverse changes in the environment.[17] Nevertheless, the introduction of biological materials to a new environment should be considered as potentially harmful, at least until demonstrated otherwise. Although interpretive questions remain concerning what constitutes "harmful contamination" or "adverse changes" to the environment, this article of the Outer Space Treaty comprised the primary statement of international policy to protect and preserve the environmental integrity of Earth and space during the initial period of interplanetary exploration.

The protection and preservation of the natural environments of space and celestial bodies properly is considered as an extension of the principle of the common province of mankind as referenced in article I of the Outer Space Treaty. That is, the interests to be promoted by ensuring the continued existence of pristine environments is common to all mankind. No state, political philosophy or generation can claim the right to jeopardize or destroy a natural environment. In the event biological contamination of a natural environment were to occur, the rights of all states to use and explore that environment would be jeopardized.

The principle of non-appropriation, as expressed in article II of the Outer Space Treaty, prohibits claims of national sovereignty in space by claim of appropriation, by means of use or occupation, or by any other means. A substantial alteration of a celestial environ would deny other entities the right to use and explore the pristine balance previously existing on that body. Thus, it could be argued that appropriation had occurred by virtue of the modification of the environment.[18] Such biological contamination also could be considered as an interference with the activities of other states parties.[19]

---

16. Treaty on Principles Governing the Activities of States in the Exploration and Use of Outer Space, Including the Moon and Other Celestial Bodies, *opened for signature* January 27, 1967, 18 U.S.T. 2410, T.I.A.S. No. 6347, 610 U.N.T.S. 205, *text reproduced in* UNITED NATIONS TREATIES AND PRINCIPLES ON OUTER SPACE 4 (1996)[hereinafter referred to as the "Outer Space Treaty"].

17. *See* Almond, *A Draft Convention for Protecting the Environment of Outer Space*, in PROCEEDINGS OF THE 23RD COLLOQUIUM ON THE LAW OF OUTER SPACE 97 (1981).

18. *See* Sterns & Tennen, *Current U.S. Attitude Concerning Protection Of The Outer Space Environment*, in PROCEEDINGS OF THE 27TH COLLOQUIUM ON THE LAW OF OUTER SPACE 398 (1985).

19. *See* Outer Space Treaty, *supra* note 16, at art. IX; *see also* Agreement Governing the Activities of States on the Moon and Other

The issue of back contamination may be governed, to some extent, by Article VIII of the Outer Space Treaty. This provision imposes international liability on states for damages to another state party, or its natural or juridical persons, caused by a space object or its component parts.[20] This provision could become applicable where damage is caused by a craft, due to its exposure to or release of harmful extraterrestrial biological contaminants. Additionally, the Liability Convention obligates states to examine the possibility of rendering appropriate and rapid assistance upon request where damage caused by a "space object presents a large scale danger to human life or interferes with the living conditions of the population or the functioning of vital centers."[21] Although numerous interpretive and factual issues may be present, the release of exobiological materials could present such a large scale danger to life or interference with living conditions.

States are obligated to provide assistance to astronauts as a result of accident or other conditions of distress pursuant to article V of the Outer Space Treaty. Furthermore, astronauts making an emergency landing in the territory of another, together with objects and components, are to be returned to the launching state.[22] The Return and Rescue Agreement obligates states to recover objects or component parts if requested by the launching state.[23] There is a question, however, whether the independent duties to return personnel or space objects to the launching state would be applicable if the object or astronauts had become exposed to or infected with some type of harmful exobiological contamination, particularly if the mere handling and transportation of the objects or persons would pose a hazard to the state party. A distinction may be drawn between astronauts and objects in this regard. Astronauts are declared to be envoys of mankind,[24] thereby constituting a special and protected class of personnel. Moreover, humanitarian considerations would dictate that nations render aid to and return contaminated personnel.

The national registries of objects launched into space referred to in the Outer Space Treaty,[25] and as required by the Registration Convention,[26] may assist in the imposition of international responsibility and liability by aiding in the identification of the responsible state of registry. The Registration Convention provides that where a state party is unable to identify the state of registry of a space object which "caused damage to it or to any of its natural or juridical persons,

---

Celestial Bodies, *entered into force* July 11, 1984, art. 8, ¶ 3, 1363 U.N.T.S. 3, *text reproduced in* UNITED NATIONS TREATIES AND PRINCIPLES ON OUTER SPACE 28 (1996), *and* 18 I.L.M. 1434 (1979)[hereinafter referred to as the "Moon Treaty"].

20. *See also* Convention on International Liability for Damages Caused by Space Objects, *opened for signature* March 29, 1972, arts. II, III, 24 U.S.T. 2389, T.I.A.S. No. 7762, 961 U.N.T.S. 187, *text reproduced in* UNITED NATIONS TREATIES AND PRINCIPLES ON OUTER SPACE 14 (1996).

21. *Id.* at art. XXI.

22. Outer Space Treaty, *supra* note 16, at art. VIII.

23. Agreement on the Rescue of Astronauts, the Return of Astronauts, and the Return of Objects Launched Into Outer Space, *opened for signature* April 22, 1968, art. V, ¶ 2, 19 U.S.T. 7570, T.I.A.S. No. 6599, 672 U.N.T.S. 119, *text reproduced in* UNITED NATIONS TREATIES AND PRINCIPLES ON OUTER SPACE 10 (1996).

24. *Id.* at Preamble; *see also* Outer Space Treaty, *supra* note 16, at art. V; Moon Treaty, *supra* note 19, at art. 10, ¶ 1.

25. Outer Space Treaty, *supra* note 16, at arts. V, VIII.

26. *See* Convention on Registration of Objects Launched Into Outer Space, *opened for signature* January 14, 1975, 28 U.S.T. 695, T.I.A.S. No. 8480, 1023 U.N.T.S. 15, *text reproduced in* UNITED NATIONS TREATIES AND PRINCIPLES ON OUTER SPACE 23 (1996).

or which may be of a hazardous or deleterious nature," other states parties shall render assistance to the greatest extent feasible when requested.[27] This provision could encompass the hazards presented by contamination from exobiological materials. Such language is not restricted to circumstances where damage has already occurred, but could include situations which pose a risk of harm.

The provisions of the Outer Space Treaty with direct application to issues of exobiology have been supplemented by the Moon Treaty. Article 7, paragraph 1, of the Moon Treaty provides as follows:

> In exploring and using the moon, States Parties shall take measures to prevent the disruption of the existing balance of its environment, whether by introducing adverse changes in that environment, by its harmful contamination through the introduction of extra-environmental matter or otherwise. States Parties shall also take measures to avoid harmfully affecting the environment of the earth through the introduction of extraterrestrial matter or otherwise.

This article of the Moon Treaty is more specific and comprehensive than article IX of the Outer Space Treaty. Pursuant to the Moon Treaty, states are required to *prevent* the disruption of natural celestial environments. Furthermore, the Moon Treaty, in Article 7, paragraph 2, imposes an affirmative obligation on states to report to the Secretary-General the measures taken to comply with the Treaty. The expression of this obligation implies that it is incumbent upon states to take precautions for *all* missions to prevent forward and back contamination.

Article 7, paragraph 3, of the Moon Treaty, provides for the establishment of areas with special scientific interest as "international scientific preserves." Article 5, paragraph 3, further provides that states shall inform the Secretary-General of the United Nations of the discovery of any indication of organic life on celestial bodies. Presumably, areas containing evidence of organic life will be designated as "international scientific preserves," and subject to more rigorous standards of environmental protection to be developed.[28]

The Outer Space Treaty requires states to immediately inform the Secretary-General or other states parties of any phenomena they discover in space, including the Moon and other celestial bodies, which could constitute a danger to the life or health of astronauts.[29] The Moon Treaty expands this obligation, and requires disclosure "of any phenomena they discover in outer space, including the moon, which could endanger human life or health, *as well as any indication of organic life*."[30] The Moon Treaty does not limit disclosure to other states parties or the Secretary-General, but requires that the information be disseminated to the public and the international scientific community. Nevertheless, there is a considerable degree of discretion allowed to the discovering state in making the disclosure.

The Future of Space Law and Exobiology: The Detection of Extraterrestrial Intelligence

The current exobiological considerations in the law of outer space primarily relate to matters of environmental protection and the prevention of forward and back contamination. These policies help to ensure scientific integrity in the search for life in the universe, as well as to protect indigenous life on Earth from possible extraterrestrial contaminants. We must be

27. *Id.* at art. VI.

28. *See* DeVincenzi, Klein & Bagby, *supra* note 10, at 18.

29. Outer Space Treaty, *supra* note 16, at art. V, ¶ 3.

30. Moon Treaty, *supra* note 19, at art. 5, ¶ 3 (emphasis added).

prepared, however, for the possibility that a discovery will be made not just of extraterrestrial life, but of intelligent extraterrestrial life. The *corpus juris spatialis* can and should foresee the interaction of humankind with such beings.

The detection of an ETI signal will require mankind to consider the circumstances and implications from a planetary environmental and sociological perspective.[31] Any ETI will be beings "with their own understanding of a kind of 'rules of behaviour' and thus, be legal subjects."[32] Furthermore, in regard to contact between two intelligent races, a basic understanding of "mutual rules will lead to a 'code of conduct.'"[33] This is the starting point for metalaw and a fundamental link for exobiological interaction between humankind and extraterrestrial life.

Haley's classic principle of metalaw states that:

> We must do unto others as they would have done unto them. To treat others as we would desire to be treated might well mean their destruction. We must treat them as they desire to be treated. This is the simply expressed but vastly significant premise of metalaw.[34]

This principle could present a significant dilemma if the ETI were to desire us to act in a manner toward them which would be repugnant to our moral code or sensibilities, scientific or otherwise. A further dilemma could arise in circumstances where a race is unaware that its acts are harmful to the other intelligent life.[35]

An ETI's ability to think and reason will be derived from its indigenous environment, its unique history, and its biology. We must consider the possibility that in logic, as in physics, everything may be relative.[36] The rules relating to exobiology will need to be expanded *vis-a-vis* the political, psychological and sociological implications for both intelligent beings. The ultimate philosophical question and driving force, however, may be the respective positions of humankind and an ETI relative to the origin of life and their respective views of self, prior to discovery and verification of the other's existence.[37] Upon discovery of an ETI, our legal and scientific obligations will clearly cross over into another realm of consideration, where any remaining distinction between science and politics disappears. At this juncture, exobiological concerns will become the primary focus for survival of both species.

Conclusions and Recommendations

The legal protection of natural celestial environments must be supported and supplemented by effective policies and procedures of the international scientific community. The planetary protection policies developed by

---

31. *See* E. Fasan, Metalaw: Relations with Alien Intelligences (1970); Kopal, *International Law Implications of the Detection of Extra-terrestrial Intelligent Signals*, in Proceedings of the 29th Colloquium on the Law of Outer Space 118 (1987); Reijnen, *Extraterrestrial Intelligence and Earthians*, in Proceedings of the 18th Colloquium on the Law of Outer Space 126 (1976).

32. Fasan, *Legal Consequences of SETI Detection*, IAF Paper No. IAA-95-IAA.9.2.04 (1995).

33. *Id.*

34. A. Haley, Space Law and Government 395 (1963).

35. *See* Sterns, *SETI and Space Law: Jurisprudential and Philosophical Considerations for Humankind in Relation to Extraterrestrial Life*, IAF Paper No. IAA-96-IAA.9.2.08 (1996).

36. *See* Körner, *Laws of Thought*, in 4 The Encyclopedia of Philosophy 416-17 (P. Edwards ed. 1972); Von Muldau, *Has the Fact that Humans are Alone on the World (or Not) Begun to Influence Ethics?*, IAF Paper No. IAA-96-IAA.9.2.09 (1996).

37. Sterns, *supra* note 35.

COSPAR and other entities have been important contributions to this process. However, extreme caution must be exercised in the consideration of reductions in decontamination and sterilization requirements, particularly where scientific knowledge is incomplete. Samples of Martian materials returned for study, for example, must be presumed to be hazardous until thorough examination and analysis establishes otherwise.

In relation to the legal obligations of states, recognition should be given to the fact that harmful or potentially harmful exobiological contamination of the environment of the Earth, by its very nature, would present exigent circumstances. Thus, the law of outer space should contain a clear and express requirement to disclose and provide detailed information concerning the nature, scope, extent and location of the contamination and the risk of harm. In the face of possible global catastrophe from extra-terrestrial contamination, states cannot be allowed to withhold information which could be crucial in counter-acting the threat of damage.

The law of outer space also should obligate states to report to the Secretary-General of the United Nations of the measures taken by any mission or program to ensure the preservation of natural environments of space and the Earth prior to launch of the spacecraft.[38] The requirements of reporting phenomena which may endanger human life[39] should include natural phenomena in space or on celestial bodies which may present a risk of danger or contamination, as well as phenomena caused by man's activities which may pose such a risk. Finally, the obligation to disclose the discovery of organic life found on the Moon or elsewhere[40] should be clarified and strengthened by requiring the public release and disclosure of both the fact and content of a detected signal or other discovery of an ETI within a specified period of time following verification.

This paper is dedicated to my Father, Lawrence P. "Waco" Sterns, a true aviation pioneer who helped pilot the way to the stars. PMS and LIT

38. *See* Moon Treaty, *supra* note 19, at art. 7, ¶ 2 (requiring advance disclosure, to the maximum extent feasible, of the intended use or placement of radioactive materials on the surface or subsurface of the Moon); *see also* Moore & Leaphart, *Manipulation and Modification of the Outer Space Environment: International Legal Considerations*, in PROCEEDINGS OF THE 25TH COLLOQUIUM ON THE LAW OF OUTER SPACE 115, 119 (1983).

39. *See* Moon Treaty, *supra* note 19, at art. 5, ¶ 3.

40. *Id.*

THE JUDICIAL SETTLEMENT OF DISPUTES ARISING OUT OF SPACE ACTIVITIES : RETURNING TO AN OLD PROPOSAL *

by

Prof.Dr. Nicholas M. Poulantzas **

There is no doubt that liability and the settlement of disputes arising out of space activities are some of the most challenging areas of space law. International instruments of space law, and in particular the "Treaty on Principles Governing the Activities of States in the Exploration and Use of Outer Space, Including the Moon and Other Celestial Bodies" (which entered into force on October 10, 1967) thirty years ago, and the "Convention on International Liability for Damage Caused by Space Objects", effective on October 9, 1973, are already outdated and lack precision.

Thus, neither Article VII of the previously mentioned Outer Space Treaty, nor Articles XIV ff. of the above mentioned Liability Convention, which provide for the establishment of a Claims Commission, if a settlement through diplomatic negotiations has failed, do seem in fact to cover the gap of dispute settlement. The lack of such a mechanism would be certainly more felt, if essential rights of the space powers concerning outer space and celestial bodies might be at stake.

Now, the conclusion of a new agreement setting up a specialized International Court of Justice for Outer Space Matters would be a completely unrealistic idea and proposal. It is true that there is some tendency in international law today to establish specialized international tribunals, as for instance, the International Criminal Tribunal for the Former Yugoslavia (ICTFY), which, as is well known, also has its seat at The Hague, like the International Court of Justice.

Another important example of a Specialized International Court of Justice is the International Tribunal for the Law of the Sea, which was set up by Part XV of the United Nations Convention on the Law of the Sea (1982), which entered into force on November 16, 1994.[1] The Tribunal started already its work in Hamburg on October 1, 1996, following the election of the 21 members of the Tribunal on August 1, 1996.[2]

However, neither the international community nor the United Nations would be willing to set up a specialized International Court of Justice for Outer Space Matters on

---

** Professor of Admiralty (Shipping) Law, and of the International Law of the Sea at the Department of Maritime Studies of the University of Piraeus in Greece.

account of the rarity of the occurrence of international space disputes, of the expences involved, etc. Indeed, the Cosmos 954 incident, when a Russian naval reconnaissance satellite, carrying a nuclear reactor with a highly radioactive material for providing power to its remote sensing and communication systems crashed over a remote area of Canada on January 27,1978, was the first ever and the last until now known case involving international liability of a state for its outer space activities. The "Challenger" disaster did not involve the international liability of the launching state, which was the United States.[3]

The final settlement of the Cosmos 954 case, when the Soviet Union -at that time- agreed to pay Canada three million Canadian dollars in full and final settlement for all claims arising out of the disintegration of its satellite, did not permit the recourse to any Claims Commission, or to any judicial authority to deal with the problems of substance or procedure connected with the satisfaction of such a claim.

Therefore, with a view to solving this problem, i.e., the judicial settlement of disputes arising out of space activities, we suggest to return to a proposal advanced more than thirty years ago by an international and space law lawyer, wh ich has endured the test of time. **Dr.Dion.M.Poulantzas** proposed during the meeting of 1965 of the International Institute of Space Law that the Chambers of the International Court of Justice be used for the settlement of disputes arising out of space activities.[4]

While at that time the proposal seemed somewhat theoretical, this is not the case anymore. The legal advisers of several states have realized the practical advantage of having recourse to a Chamber of the I.C.J. rather than to the full Court. The first ever case to come before a Chamber of the I.C.J. was that involving the delimitation of the maritime boundary in the Gulf of Maine Area between the United States and Canada.[5]

Thus, by a special agreement ("compromis") notified on November 25,1981, and filed in the Registry of the I.C.J. the same day, the Governments of the United States and Canada submitted to a Chamber of the International Court of Justice their difference over the delimitation of the maritime boundary in the Gulf of Maine Area. By virtue of the Special Agreement, dated March 29,1979, and modified subsequently, the two Governments agreed to submit their differences regarding this issue to a Chamber of the I.C.J., to be set up pursuant to Article 26, para.2, and Article 31 of the Statute of the Court.[6]

The question submitted to a Chamber of the I.C.J. regarded the course of the single maritime boundary that divides the continental shelf and fisheries zone of the two interested states, namely, the United States and Canada, in the Gulf of Maine Area. By virtue of Articles 26 to 31 of the Statute of the I.C.J., the Court by an Order of January 20,1982, formed a Chamber to deal with this case, according to Articles 70 to 73 of the Rules of the Court.[7]

The Judgement of the I.C.J., sitting as a Chamber of five Judges -including the ad hoc Judge for Canada- was pronounced on October 12,1984.[8] The proceedings took a rather long time not only because it was the first time that the I.C.J. sat as a Chamber and there was not enough experience in this procedure,[9] but also because of the fact that it was the first time that the Judges of the Court had to draw a single maritime boundary line between two different jurisdictions in an area hotly disputed by both states because of its fishing grounds, rich especially in lobster and scallops.[10] However, the line which was drawn by the Chamber of the I.C.J. -"The Hague line" as it has been called- has ever since almost always been respected by fishermen of both states.[11]

Following that case, several other cases were brought by agreement ("compromis") between the parties before a Chamber of the I.C.J. Most of these cases regarded the delimitation of maritime or other areas between adjacent states.[12]

As it was noted by the previously mentioned author Dr. Dion.M.Poulantzas, the referral of a dispute arising out of outer space activities - by agreement of the parties- to a Chamber of the I.C.J. presents many advantages. He also mentioned the following reasons which call for the choice by the parties of a Chamber of the I.C.J. rather than referring the case to the full Court, to a Claims Commission, or to an Arbitration Tribunal.[13]

1. The flexibility of a Chamber of the I.C.J., which is usually composed of only three to five Judges, as compared to fifteen Judges of the full Court.

2. A Chamber, because of its flexibility, can sit with the consent of the parties at a different location than at the seat of the Court at The Hague. Accordingly, a Chamber can play in fact -because of its flexibility and specialization, especially in territorial disputes- the role of a kind of a fact-finding committee as well, which at the same time may deliver a binding judgment upon the parties. Therefore, the Chambers may be more aware of the particular circumstances of a case through an on-the-spot investigation.

3. The deliberations and proceedings in a Chamber are usually much shorter than in the full Court. This is important in cases where the peace between the parties is at stake, requiring a speedy solution.[14]

4. A definite also advantage of the Chambers over the full Court is that the Chamber's procedure seems more acceptable to those states, which previously seemed more distrustful of the I.C.J. on account of the different -alleged or real- ideology of the majority of the members of the Court. Moreover, the presence of an ad hoc Judge, having the nationality of the parties is far more important within a small Chamber than in the full Court.

5. Finally, the authority of the judgments delivered by the Chambers is equal to the decisions of the full Court. Under Article 27 of the Statute of the I.C.J. the judgments of the Chambers are regarded as rendered by the full Court.[15] This has also been demonstrated by the more recent decisions of the I.C.J. sitting as a Chamber.

It should be also noted that one of the major legal instruments of recent years, the United Nations Convention on the Law of the Sea (1982), which, as already mentioned, entered into force on November 16, 1994, in Part XI regarding the novel notion of "The Area" in the International Law of the Sea, includes several provisions dealing with the establishment of a Sea-Bed Disputes Chamber of the International Tribunal for the Law of the Sea.[16] Moreover, the Statute of the International Tribunal for the Law of the Sea includes several provisions regarding the establishment not only of a Sea-Bed Disputes Chamber, but also of an *ad hoc* Chamber of the Sea-Bed Disputes Chamber, composed of three or more of its elected members to deal with particular categories of disputes.[17]

Concluding, one should express the hope that specialized jurisdictions today, like the Chambers of the International Court of Justice, or of the International Tribunal for the Law of the Sea, which present some of the advantages of an arbitration tribunal or a Conciliation Commission, because of the small number of adjudicators and the institution of the *ad hoc* Judge, with the increased authority of the International Court of Justice, or, in the course of time, of the International Tribunal for the Law of the Sea, will be used for the settlement of disputes, which might arise out of space activities as well.

FOOTNOTES

1. See N.M.Poulantzas, "The New International Law of the Sea and the Legal Status of the Aegean Sea", *Revue hellénique de droit international (RHDI)*, Athens, 1991, pp.251-272.

2. See N.M.Poulantzas, *The Law of the Sea* (student edition in Greek language), Piraeus, 1997.

3. For a discussion of the legal issues involved in "The Challenger" disaster of February 1986, see Dion.M.Poulantzas and Nicholas M.Poulantzas, "Intellectual Property Rights Within the Framework of International Organizations and National Agencies", in *Revue de droit international*, Geneva, 1988, No.3, pp.191-210.

4. See Dion.M.Poulantzas, "The Chambers of the International Court of Justice and Their Role in the Settlement of Disputes Arising Out of Space Activities", *Proceedings of the I.I.S.L.*, 1965, p.127 ff. Reprinted in the *Revue hellénique de droit international*, Athens, 1965, p.150 ff. See also the same author, "Einige Betrachtungen über die Beilegung von durch Tätigkeiten im Weltraum entstehenden Streitfällen", in *Zeitschrift für Luftrecht und Weltraumrechtsfragen*, Cologne, 1965, p.319 ff.

5. See Nicholas M.Poulantzas, "The Chambers of the International Court of Justice and the Judicial Settlement of Disputes: The Delimitation of the Maritime Boundary in the Gulf of Maine Area Case", in *Revue de droit international*, Geneva, 1985, No.4, p.323 ff.

6. See "Special Agreement Between the Government of Canada and the Government of the United States to Submit to a Chamber of the International Court of Justice the Delimitation of the Maritime Boundary in the Gulf of Maine Area".

7. See *Case Concerning the Delimitation of the Maritime Boundary in the Gulf of Maine Area (Canada/United States of America)*, Order of January 20, 1982, *Constitution of Chamber*, pp.3-13.

8. See *Case Concerning Delimitation of the Maritime Boundary in the Gulf of Maine Area (Canada/United States of America)*, *I.C.J.*, *Recueil* 1984.

9. See *Case Concerning Delimitation of the Maritime Boundary in the Gulf of Maine Area (Canada/United States of America)*, Order of 1 February 1982; Order of 28 July 1982; Order of 5 November 1982; Order of 27 July 1983; etc.

10. Following that case, several other cases for drawing a single maritime -or other- boundary were brought before the International Court of Justice or other jurisdictions. See, *e.g.*, *the Libya-Malta Continental Shelf Case*, decided by the I.C.J. in 1985; *the Guinea/Guinea-Bissau case*, decided by a Court of Arbitration in 1985; *the Guinea-Bissau/Senegal case*, decided

by an Arbitration Tribunal in 1989; *the St.Pierre-Miquelon Maritime Areas Dispute* between Canada and France, decided by the I.C.J. in 1992; *the Land, Island and Maritime Frontier Dispute* between El Salvador and Honduras, decided by the I.C.J. in 1992; and *the Maritime Delimitation (Greenland/Jan Mayen) case* between Denmark and Norway, decided by the I.C.J. in 1993.

There are also some other cases still pending before the I.C.J. See, *e.g.*, *the Guinea-Bissau/Senegal case*; *the Timor Gap Dispute* between Portugal and Australia; and *the Maritime Delimitation case* between Qatar and Bahrain.

11. See also *Nicholas M. Poulantzas*, "Recent Developments Regarding the Right of Hot Pursuit in the International Law of the Sea", *Revue de droit international*, Geneva, 1997, No.2, at p.204, and note 61, at p.220.

12. See, *e.g.*, the following cases: *Frontier Dispute case* (*Burkina Faso v. Mali*, decided by a Chamber of the I.C.J. in 1985; *the Land, Island and Maritime Frontier Dispute* (El Salvador v. Honduras), decided by a Chamber of the I.C.J. in 1987; and *the Elettronica Sicula S.p.A. (ELSI) case*, decided by a Chamber of the I.C.J. in 1989.

13. See *Dion. M. Poulantzas*, "The Chambers of the International Court of Justice and Their Role in the Settlement of Disputes Arising Out of Space Activities", *op.cit.*, p.127 ff.

14. This is, for instance, the case in the Falkland/Malvinas islands crisis, when a speedy solution of the dispute between the Governments of the United Kingdom and Argentina, at that time, could have averted the armed conflict between the two nations.

15. Thus, the reservations expressed in the past by the late Judge *A. Alvarez* have not proven justified (*Le droit international nouveau*, Paris, 1959, p.599). Judge *Alvarez*, namely, contended that "the judgments by the Chambers in the eyes of public opinion have only less value since they do not express the opinion of all the judges and consequently they cannot create a judicial precedent like the decisions rendered by all members of the Court" (translation by the present writer).

16. See Articles 186-191 of the Convention, as well as Part XV of the Convention, which deals with the "Settlement of Disputes".

17. See Article 188 of the U.N. Law of the Sea Convention (1982), and Articles 14-15 of the Statute of the International Tribunal for the Law of the Sea.

------------.------------

Session 3:
APPLICATIONS & IMPLEMENTATION OF THE OUTER SPACE TREATY

Chairmen: S. E. Doyle, USA
G. Lafferranderie, France

Rapporteur: F. G. Von der Dunk, Netherlands

SPACE LAW AT UNISPACE III (1999) AND BEYOND

Dr. Marietta Benkö, Attorney-at-Law, Cologne
Dr. Kai-Uwe Schrogl, German Space Agency DARA, Bonn

Abstract

At its 1997 session, the United Nations Committee on the Peaceful Uses of Outer Space (UNCOPUOS) adopted the agenda of the third UN space conference (UNISPACE III), which will take place in 1999 at Vienna. Space law is explicitly mentioned as a point of discussion under the agenda item „Promotion of international cooperation". This paper introduces the concept of UNISPACE III, analyses the agenda in view of identifying subjects relevant to space law and proposes ways to take up these subjects in order to prepare, in the most efficient way, the future development of space law, which will have to be accomplished in the Legal Subcommittee (LSC) of UNCOPUOS.

1. The concept of UNISPACE III

When, in 1993, India came up with the idea to hold a third UNISPACE after 1968 and 1982, nobody imagined that such an event would take place already before the end of the decade. It was then fiercely discussed if such a conference would make sense or if it would be politicised. Also it was unclear if India was simply aiming for the opportunity to take again the chairmanship (as it was the case in the two conferences before) or if Austria simply wanted to host a big event in Vienna (as it also was the case in the two conferences before). The debate, however, quickly got more rational and delegations in UNCOPUOS got more and more convinced that there had considerable substance emerged in the field of space activities since UNISPACE II, which could be discussed on a global level.

In that spirit, the UN General Assembly decided in 1996 to hold UNISPACE III as a UNCOPUOS meeting open to all States Members of the UN (and not only the 61 member States of UNCOPUOS).[1] Since this will mean that the Chairman of UNCOPUOS will also chair UNISPACE - currently the renowned Indian expert U.R. Rao - and since the location of the conference will, for cost-saving reasons, be at the place, where the UN Office for Outer Space Affairs is located - since 1994 at Vienna - a satisfactory solution for all interests could be found. UNCOPUOS is now acting as Preparatory Committee and ist Scientific and Technical Subcommittee (STSC) as Advisory Committee for UNISPACE III. In that capacity, the agenda of UNISPACE III was prepared at the 1997 session of the STSC in February and adopted at UNCOPUOS in June of that year.[2] The primary objectives of UNISPACE III will be „(a) to promote effective means of using space technology to assist in the solution of problems of regional or global significance and (b) to strengthen the capabilities of Member States, in particular developing countries,

[1] UNGA Res. 51/123 of 13 December 1996, para. 28.

[2] Report of the Scientific and Technical Subcommittee on the work of its thirty-fourth session, UN Doc. A/AC.105/672 of 10 March 1997, Annex II. The following quotation are taken from that Annex and are indicated in the text by their respective number of paragraph.

to use the applications of space research for economic, social and cultural development." (para. 18). By that, UNISPACE should give a strong impetus for more intensive use of space technologies in the Member States and also promote these technologies in the UN system as well.[3]

This stress on concrete applications and distinctive refusal to discuss abstract problems of political nature is also reflected in the agenda, which consists of five „substantive items:
(a) Status of the scientific knowledge of Earth and its environment;
(b) Status and applications of space science and technology;
(c) Information needs and the global approach;
(d) Promotion of international cooperation;
(e) Economic and societal benefits." (para. 21).

Although (d) and (e) seem to have a great weight in the agenda, they are in fact marginalized, when it is taken into account that (a) to (c) cover almost 30 neatly described topics and that (d) only is described in two short subparagraphs. The first deals with a review of existing and new mechanisms for international cooperation in space (para. 22), the second (still para. 22) reads: *„There would be a review of the current status of law of outer space, including ways and means of promoting wider adherence to the existing international space treaties and principles."*

2. Agenda items relevant to space law

Therefore, the „substantive item": „Promotion of international cooperation" will be the main focus for the discussion of space law. The scope of the subject is phrased as broadly as possible, so that an imaginative debate is to be expected. Space law development will, however, not only turn around itself. It will also be of relevance in other fields, in particular substantive items (b) and (c) where telecommunications, navigation and earth observation are discussed, where environmental law will meet space law, and where intellectual property rights get into the spotlight in various fields. In addition to that, it was decided that UNISPACE III will have „additional components". This means that Workshops and Seminars (some of them open to the public) could be held before or during that intergovernmental conference. The field of space law is explicitely mentioned and the International Institute of Space Law (IISL) is named as the international organization which could act as the organizer of such an event.

3. The political setting for discussing space law at UNISPACE III

UNISPACE I and II, which took place in 1968 and 1982 respectively, were overlayed by political conflicts between East and West and North and South. These systemic and distributive conflicts made it difficult to focus on the benefits of space activities. Instead, the promotion of space activities could not develop as intensive as it was possible due to competing ideas of how to organize cooperation. In particular the conflict between North and South about the distribution of resources (natural resources like the Geostationary Satellite Orbit as well as technologies and funds) made it difficult to find modes of cooperation, which were of mutual benefit to all parties involved. Since the beginning of the 90s a new approach on global cooperation including new ways and means for development cooperation (demand pull with own investments instead of technology push without sustained foundation) emerged. This was also facilitated by the fact that a number of developing countries

---

[3] Numerous UN bodies are involved in the use of space technologies. They coordinate in the „Inter-agency meeting on outer space affairs". The report of the last meeting, which took place in May 1997 is contained in UN Doc. A/AC.105/676 of 2 June 1997. The programme of work for 1997 and 1998 of the relevant bodies is contained in UN Doc. A/AC.105/675 of 12 May 1997.

have become space powers themselves and that important conflicts over resources have been solved to the satisfaction of the developing countries (in particular the use of the Geostationary Satellite Orbit through the 1985/88 WARC-ORB of the International Telecommunication Union).

In UNCOPUOS this new trend found its manifestation in the unanimous adoption of the „Space Benefits Declaration“ in 1996[4], which stresses the freedom of cooperation and the need for effectiveness in international cooperation, denouncing at the same time forced cooperation and dirigist approaches for reaching re-distribution. Space law purists might have been disappointed by the „Space Benefits Declaration“ in that it is less a legal text dealing with a specific legal problem (like, for example, the use of nuclear power sources, NPS) but more a political text. But in view of UNISPACE III, the Declaration came exactly at the right time, in order to make clear that no distributive or systemic conflict in the field of space applications and cooperation is alive. To the contrary, all States share the high expectations in a liberal system based on mutual repsonsibility and effectiveness. UNISPACE III will therefore be saved from „politicised“ discussions and can concentrate on finding the best way of letting all people benefit from space applications.

The substantive preparation of UNISPACE III has already begun. STSC and UNCOPUOS will concentrate from 1998 on preparing the report of UNISPACE III. This report will be composed of numerous pieces drafted in different fora and circles. For example, the subject of Basic Space Science (astronomy and planetary exploration) has been dealt with in that way, by discussing a draft text for the UNISPACE III report already at the 7th UN/ESA Workshop on Basic Space Science, which took place at Honduras in June 1997.[5] In the field of space law, the LSC of UNCOPUOS will certainly take the lead. Non-governmental fora like IISL can, however, have an impact as well.

4. The chance to reinvigorate the debate in the Legal Subcommittee of UNCOPUOS

The current agenda of the LSC is in a deplorable state. The item on delimitation and GSO has been treated for decades without any result. Even the most benevolent approach with regard to the GSO was doomed by the intransigence of the sponsors of outdated concepts.[6] The agenda item on the question of review and possible revision of the NPS principles is suspended, and „Space benefits“ has disappeared after the adoption of the Declaration in 1996 by the UN General Assembly. Nothing else had been left, if the LSC had not put its act together and agreed on a new agenda item. Upon an initiative by Mexico[7], the LSC will begin to discuss „Review of the status of the five international treaties governing outer space“ from 1998 on („Review item“). This agenda item will be structured along a three-year workplan - an immense organizational improvement, in order to avoid remnants of agenda item like delimitation/GSO. The main focus will be to improve the state of ratification of the treaties. The General Secretary of the UN has already asked the member States to

[4] UNGA Res. 51/122 of 13 December 1997. See *Marietta Benkö/Kai-Uwe Schrogl*, History and impact of the 1996 UN Declaration on „Space Benefits“, in: Space Policy (13,2) 1997, 139-143 (presented before as IISL-96-IISL.3.12).

[5] See the paper presented at that workshop by *Kai-Uwe Schrogl*, Basic space science at UNISPACE III 1999. It can soon be found on the Web Site of this series of workshops at: http://ecf.hq.eso.org/~ralbrech/un/un-homepage.html

[6] See the Working-paper presented by Germany at the 1997 session of the LSC with a draft resolution „Request to the International Telecommunication Union and ist member States: Ensuring equitable access to the geostationary satellite orbit/radio frequency spectrum resources“, UN Doc. A/AC.105/C.2/L.207/Rev.2 of 7 April 1997.

[7] See the Working-paper presented by Mexico at the 1997 session of the LSC, UN Doc. A/AC.105/C.2/L.206/Rev.1 of 4 April 1997.

report on their views regarding the status of the treaties.[8]

This agenda item exactly prepares one of the main thrusts of discussion on space law at UNISPACE III as quoted above: „...including ways and means of promoting wider adherence to the existing international space treaties and principles." It would, however, be a less imaginative debate, if only the sheer facts of the status of ratifications would be discussed in the LSC. In order to create more orientation, States will certainly take the chance to describe, where room for improvement in the treaties lies - without negating the treaties themselves. By that, all the currently open questions could be discussed in relation to possible shortcomings of the existing body of space law. Among them are:[9]

- strengthening of the provisions of the Registration Convention thus providing the States and the general public with more relevant information on space activities,
- improving the notification of NPS and possibly transfering the Principles into treaty law,
- finding a way to incorporate further provisions on space debris mitigation in the already existing body of space law.

While the three-year work-plan of the „Review item" will certainly not lead to any concrete measure in the context of filling a legal vacuum, the main task will be to identify subjects which can in the future be dealt with under a work-plan of their own in the LSC. UNISPACE III will have an important task to act as a clearing-house for such subjects. There, all the problems can be discussed not only on a theoretical level; it will be possible to reflect them on the spot with the technological reality in that the close interaction with the technical discussions under the other agenda items have to be sought. It will then be an important and difficult task to select the most pressing problems and focus them in such a way that discussing them in the LSC should reach tangible results. The LSC will not be able to deal with all subjects at one time. Political or technological priorities of the States will also determine the selection of items for further consideration. All this will act as the background for the discussion of space law at UNISPACE III.

5. Summary

Space Law is integrated into the agenda of UNISPACE III 1999. This provides for the possibility to discuss all relevant matters concerning the current state and future development of space law at this conference, which might have great impact on space activities during the next decade. The positive political setting for discussing space law without the straight-jackets of ideological conflicts has been reached by the end of the Cold War and in paeticular by the „Space Benefits Declaration" of 1996. The debate at UNISPACE will be prepared by the deliberations on the „Review item" in the LSC of UNCOPUOS but might also get impetus through initiatives by non-governmental organizations like the IISL. UNISPACE will have the task, not to discuss as many legal problems as possible, but to select legal problems, check them with the technological reality presented at UNISPACE, and pre-structure them for focused discussions based on work-plans in the LSC. In that way, UNISPACE III can give direction to the fruitful further development of space law.

---

[8] OOSA/97/004, CU97/149 of 14 July 1997.

[9] An excellent overview presents the book edited by *Gabriel Lafferranderie/Daphné Crowther*, Outlook on Space Law over the Next 30 Years, The Hague/London/Boston (Kluwer) 1997.

IISL-97-3.03

The Outer Space Treaty and the International Organisations conducting Space Activities

G. Lafferranderie
ESA Legal Adviser

Introduction

In considering the 1967 "Outer Space Treaty" ("the Treaty") and the place it accords to international organisations, I shall also be looking at the other Agreements established by the United Nations Committee on the peaceful uses of outer space (UNCOPUOS) and its Legal Subcommittee, namely, the Agreement on astronauts, the Convention on international liability for damage, the Convention on registration and the Agreement on the moon (note 1). The various Principles enshrined in Resolutions adopted by the United Nations General Assembly (on broadcasting, remote sensing, nuclear power sources, and the recent Declaration on the international cooperation for the benefit of mankind) have not been taken into account for the purposes of this article.

This body of law when setting out the principles governing the activities of States whilst underlining the need of international cooperation, adopts initially a restrictive position on international intergovernmental organisations. It is, however, inherently putting itself in difficulty, in that it cannot avoid emphasising international cooperation and the need to encourage such cooperation, though without going so far as to give full recognition to the international organisation.

The first mention of international organisation appears in Article VI. In discussing this point, it was impossible to be unaware of the differences among legal writers as to whether or not international organisations should be recognised as having legal personality. History was to favour a solution based effectively on the terms of reference of those organisations, which will continue to be -perhaps in a different form- undoubtedly among the major players in the conduct of space activities.

I propose to trace the course of this evolution, from the first efforts of the Member States of ESRO[1] and ELDO [2] (now ESA) through to the latest developments, notably in the context of the Intergovernmental Agreement on the international space station (IGA). What is becoming of the division that was introduced into space law at its inception? Should the entire subject be reviewed in the light of the general trend towards greater international cooperation involving more players in the conduct of space activities?

CHAPTER 1

1 ESRO: European Space Research Organisation set up by the Convention opened for signature on 14 June 1962

2 ELDO: European Organisation for the Development and Construction of Space Vehicle Launchers set up by the Convention opened for signature until 30 April 1962

Principles: the legal precedence of States and the joint responsibility of States and international organisations in the conduct of space activities

The principle of international cooperation to further the objectives pursued by States in their space activities was a basis in COPUOS debates from the outset (stated in the UNGA resolutions) I am referring to Western Europe. No country in Western Europe had the necessary resources to proceed alone. International cooperation was essential and ELDO and ESRO were consequently established being coordinated on the political plane by the "European Space Conference" (ESC) (note 2), a body without legal personality. The ESC in turn set up a Committee charged with monitoring the work of the United Nations[3] and I am sure many of us will still remember the Chairman of that Committee, Mr A. Vranken, who played a prominent role in negotiating the Convention on liability and the New Delhi "compromise".

(a) The question first arose at the time of the Declaration of legal principles of 13 December 1963 and an initial answer is to be found at the end of § 5 which states that "when activities are carried on in outer space by an international organisation, responsibility for compliance with the principles set forth in this Declaration shall be borne by the international organisation and by the States participating in it". That line was to be taken again in the 1967 Treaty. Discussions on the subject eventually led to a distinction being drawn between the situation of States and that of international organisations (see note 3) and it was decided not to treat them in the same way, notably because international organisations could not exercise some of the powers exercised by States, a problem that still arises in practice. Article VI of the Treaty provides that States alone are to bear international responsibility for national activities in outer space but it acknowledges the reality of the international organisation in terms of the responsibility it shares with States:

"When activities are carried on in outer space, including the moon and other celestial bodies, by an international organization, responsibility for compliance with this Treaty shall be borne both by the international organization and by the States Parties to the Treaty participating in such organization."

Article XIII of the Treaty affirms that States are to have precedence, even when activities are conducted within the framework of an international organisation:

"The provisions of this Treaty shall apply to the activities of States Parties to the Treaty in the exploration and use of outer space, including the moon and other celestial bodies, whether such activities are carried on by a single State Party to the Treaty or jointly with other States, including cases where they are carried on within the framework of international intergovernmental organizations.

Any practical questions arising in connection with activities carried on by international intergovernmental organizations in the exploration and use of outer space, including the moon and other celestial bodies, shall be resolved by the States Parties to the Treaty either with the appropriate international organization or with one or more States members of that international organization, which are Parties to this Treaty."

The Treaty, like the other Agreements, was open for signature only to States (Article XIV.1).

It should be noted that this applies only to international intergovernmental organisations, not to institutions established under private law, which - as we know - now play a considerable part in the conduct of space activities. In this case, the principle of Article VI of the Treaty applies.

However, this approach left some questions open (for example, the international organisation will not be bound by the international Agreement since it is not a signatory or party to it). The beginnings of a change are already observable in the 1968 Agreement on the rescue and return of

3 ESRO and ELDO obtained observer status in COPUOS on 12 September 1972.

astronauts, which contains the following special provision (Article 6):

"For the purposes of this Agreement, the term "launching authority" shall refer to the State responsible for launching, or, where an international intergovernmental organization is responsible for launching, that organization, provided that organization declares its acceptance of the rights and obligations provided for in this Agreement and a majority of the States members of that organization are Contracting Parties to this Agreement and to the Treaty on Principles Governing the Activities of States in the Exploration and Use of Outer Space, including the Moon and Other Celestial Bodies."

This provision implies tacit recognition that the international organisation has legal personality. Of course, this was a limited field (the astronauts rescue), of interest to humanity at large and calling for general cooperation.

(b) Further thought was given to the matter when the draft Convention on international liability for damage came to be examined in a completely different context, at a time when the first launches were beginning to take place, conducted by or on behalf of international organisations. The organisation itself had to reflect on its position. In Western Europe, the task of considering the subject was taken up by the above mentioned ESC Working Group charged with monitoring the work of the United Nations and the results were embodied in a draft Article submitted by the delegations of the States belonging to ESRO and ELDO. That Article served as a basis for Article XXII of the Convention on liability signed on 29 March 1972, which provides that:

"1. In this Convention, with the exception of Articles XXIV to XXVII, references to States shall be deemed to apply to any international intergovernmental organization which conducts space activities if the organization declares its acceptance of the rights and obligations provided for in this Convention and if a majority of the States members of the organization are States Parties to this Convention and to the Treaty on Principles Governing the Activities of States in the Exploration and Use of Outer Space, including the Moon and Other Celestial Bodies. "

This Article doesn't stop here and continues with provisions being guidelines for the States organization or what they have to do. It continues:

"2. States members of any such organization which are States Parties to this Convention shall take all appropriate steps to ensure that the organization makes a declaration in accordance with the preceding paragraph.

3. If an international intergovernmental organization is liable for damage by virtue of the provisions of this Convention, that organization and those of its members which are States Parties to this Convention shall be jointly and severally liable; provided, however, that:

(a) any claim for compensation in respect of such damage shall be first presented to the organization;

(b) only where the organization has not paid, within a period of six months, any sum agreed or determined to be due as compensation for such damage, may the claimant State invoke the liability of the members which are States Parties to this Convention for the payment of that sum.

4. Any claim, pursuant to the provisions of this Convention, for compensation in respect of damage caused to an organization which has made a declaration in accordance with paragraph 1 of this Article shall be presented by a State member of the organization which is a State Party to this Convention."

That Article marks an important stage in the process of according the international organisation the same treatment as States. Instead of "the provisions of this Convention shall be deemed to apply to any international organisation in the same way as to a State...", it adopted the following form of words, which was to become standard: "In this Convention, references to States shall be deemed to apply to any international intergovernmental organisation ...".

As we know, the same clause was to be included in the Convention on registration (Article VII) and the Agreement on the moon (Article 16).

States and international organisations are not yet on an equal footing, as this presents certain legal difficulties, but we are in practice very close to that situation.

(c) The declaration of acceptance presupposes that a majority of the States members of the organisation are already Parties to the Treaty and to the Agreement in question. However, the organisation may not propose amendments to the Agreement (Article XXV of the Convention on liability) and nothing is said about what happens if the organisation withdraws its declaration of acceptance (Articles XXV and XXVII). The solution that has been adopted is therefore not an entirely satisfactory one from the international organisations' point of view and, taking into account the place occupied by international organisations, there is a case for resuming the discussions on the subject. As regards the manner in which the provision is applied, it is interesting to turn to the international organisation itself (at that time: ESRO, ELDO, Intelsat and Intersputnik) and in particular ESRO/ELDO, which merged in 1975 to form ESA.

CHAPTER II

The declaration of acceptance: procedures and implications

(a) How are the terms of the "standard" Article to be interpreted in practice? What does "declare" mean? The texts were silent on the subject, that is to say they left it to the organisation itself to find an answer. At the European Space Agency, the subject was discussed in the International Relations Advisory Group (IRAG), now the International Relations Committee (IRC) and a draft declaration was prepared, submitted to the ESA Council which adopted it unanimously, and finally lodged with the depositaries of the Agreement in question. In due course, the Director General of ESA forwarded declarations of acceptance of the Agreement on the rescue of astronauts, the Convention on liability and the Convention on registration, respectively, the other condition, namely that a majority of the Member States of ESA be Parties to the Agreement or Convention, being fulfilled in each case. The Director General of ESA also informed the Secretary-General of the United Nations of the establishment of the registry of space objects.

The standard text of the declaration of acceptance (note 4) is relatively simple, comprising:

- a preamble recalling the purpose of the European Space Agency, noting the provisions of the Agreement or Convention in question, and considering that a majority of the Member States are parties to the 1967 Treaty;

- a Declaration that the Organisation "accepts the rights and obligations" and "considers that the references made to the States Parties to the Convention (or Agreement) apply to it with effect from the date of this Declaration". In the case of the liability convention it makes no mention of recognition or of the fact that the decision of the Claims Commission is binding.

The declaration is a unilateral act which is binding upon all the Member States of ESA and which is consequently adopted by unanimous decision of those States. It should be noted that, because of the peculiar legal situation obtaining at that time, the declaration is subscribed, in accordance with an approved form of words, by "the European Space Research Organisation, established by a Convention opened for signature at Paris on 14 June 1962, conducting its activities as from 31 May 1975

under the name of European Space Agency".[4]
The Agency was the first intergovernmental space organisation to follow this practice in its declarations of acceptance (Eutelsat has since followed suit) and its declarations have been published in the ST/SG/SER-E)115 series.

(b) Other practical questions remained to be settled within the organisation.

(i) International liability for damage

The European Space Agency conducts both mandatory activities, in which all the Member States are required to participate, and optional ones, undertaken only by States wishing to participate, which contribute in accordance with a scale which they themselves decide. How can it be ensured that Member States not participating in an optional programme will not have to bear part of the cost of compensation, if damage is caused by a space object developed under the programme in question? What if a participating State withdraws from the programme before the damage occurs? What if a State participating in the optional programme in question is not a party to the Space Treaty? And what about the fact that, although the Agency is concerned solely with development and has no responsibility for commercial launches - which are carried out by the Arianespace company -, they take place on Agency property, namely the Ariane launch site in the French Department of Guiana? So the Agency is a "launching State"(note 5).

The principles, the fundamental answers, are to be found in a Resolution adopted by the ESA Council on 19 December 1977 (note 6) and in a series of Agreements concluded between the Agency and the French Government on access to and use of the Guiana Space Centre (CSG) and the Ariane launch site, as well as in the Convention with Arianespace and the latter's launch contracts with its clients, not to mention the Arrangements or Declarations relating to the optional programmes (note 7).

(ii) Registration of space objects

The Convention on registration of space objects contains the same Article on acceptance by an international organisation. However, the application of that Article by the organisation raises questions concerning the difference between a State and an international organisation in respect of certain powers which the international organisation does not have.

Before accepting that Convention, ESRO had asked the UN Legal Adviser about keeping the registry of space objects and about the concept of a "State of registry". In his reply, dated 19 September 1975, the UN Legal Adviser said that an international organisation was regarded as a launching State and could therefore register space objects. He acknowledged that the question of jurisdiction and control over space objects and personnel on board was not addressed in the Convention itself: "It would seem reasonable ... that the Organisation when considering the registration of the space object, should consider also the question as to which of its Member States could most appropriately exercise jurisdiction over the space object and its personnel... When an international intergovernmental Organisation launches a space object jointly with one or more States, the provisions of para. 2 of Article II are relevant" (note 8).
It remains to determine what constitutes a space object and a joint launch. The first case in point arose with the 1973 Spacelab Intergovernmental Agreement between the US Government and the Member States of ESRO participating in the Spacelab programme. Was Spacelab a space object eligible for registration? (Note 9) The American authorities thought not, because Spacelab was not self-contained but remained attached to the Shuttle. Non-American personnel and the results of the experiments, including those financed by Europe, were consequently

4 The ESRO Convention served as a legal basis during the period of de facto application of the Convention of the European Space Agency, which was opened for signature on 30 May 1975, and Article XIX of the ESA Convention governed the continuity of rights and obligations from its entry into force on 30 October 1980.

governed by US law. The Europeans took the opposite view, contending that the concept of a "space object" had not been defined and there did not appear to be any justification for holding that all activities on board should automatically be governed by US law, especially as the same Agreement took a different line on the subject of damage that might be caused by Spacelab on its return!

The subject came up again in connection with the international space station and the 1988 IGA and the same solution was adopted in a new version, which is to be signed soon. The answer (see Article 5) is based on the concept of jurisdiction and control and on the fact that Europe's contribution, the Columbus module now known as the COF, is regarded as a space object and therefore eligible for registration despite the fact that it is attached to the station. The same reasoning applies to the Canadian robotic arm. It is true that we are now moving towards a wide definition of space objects but this was the only way of protecting the legitimate interests of the various partners and reaching a generally agreed solution.

This broad approach is reflected, for example, in the provisions on ownership of assets, intellectual property, the implementation of the Convention on liability, patents, criminal jurisdiction, etc. As regards the personnel involved, the safety of the on-board team will be the determining factor and here too solutions will have to be found that are in keeping with the international nature of the station. The IGA is thus a major extension of the Space Treaty and, as we know, responsibility for implementing it has been assigned to national bodies (NASA, RKA, CSA) and to ESA, with the result that the various MOUs and Arrangements have become a rich source for the development of space law.

The European Space Agency therefore maintains a registry of space objects launched by the Agency or on its behalf (satellites, the third stage of Ariane), though there are for the present no entries under the heading "State of jurisdiction".

(iii) The crew - There are three sources for the provisions governing the international crew and, in particular, the role of the station commander: the IGA, the MOUs between the Agencies involved in the venture, and the Code of conduct drawn up by the Agencies and approved by the partners (a draft has still to be drawn up). As an extension of the Agreement on rescue, consideration is now being given to life and work on a space station, today in outer space, tomorrow on the moon. Add to this the provisions on criminal liability, and we have the beginnings of a legal status for astronauts in space (note 10).

Conclusion

The evolution of space activities, space law and the role of the international organisations

To conclude, in view of the ever increasing international cooperation and ever greater resources required to conduct space activities (mission to Mars for instance) and the demand for their effects to be deployed to the greatest possible advantage, the international organisation, possibly in a more advanced form [5](note 11), is becoming an essential player, not yet in the field of primary law but certainly in the context of secondary law. I offer the following examples:

- observation of the Earth's resources, and the MOUs concluded by ESA on the collection and processing of satellite data (ERS-1/ERS-2, etc.), in particular, as well

5 It should be mentioned that there is a move to privatise certain intergovernmental space organisations (Eutelsat, Inmarsat, Intelsat) and we therefore need to consider ways of ensuring continuity in the application of the principles of space law, of Article VI of the Treaty in particular and of the relevant provisions of other Agreements, Liability and Registration Conventions.

as in the areas of meteorology and satellite navigation (GNSS)[6];

- monitoring space debris, and the leading role played by ESA in this area, in establishing the Inter-Agency Space Debris Coordination Committee (IADC);

- scientific research (Hubble, Giotto, etc.) and astronomy, including ground stations.

I should not wish to close this brief review without mentioning the teaching of space law. It is taught in our universities, of course, but its essential elements are also becoming part of the intellectual baggage of the citizens of tomorrow. The Agency supports the activities of the European Centre for Space Law, which has been instrumental in reviving Europe's interest in this branch of the law (note 12). Space law is constantly expanding, on the firm basis of the Treaty, and the international organisation's contribution to this development is far from negligible.

6 Various organisations are concerned: R&O, exploitation and operators (private?), plus the regulatory aspects (ICAO). How to ensure the ??? between the Space Law and the law of space activities?

Notes

(1) Declaration of legal principles governing the activities of States in the exploration and use of outer space - UNGA Res. 1962 (XVIII) - 13 December 1963.

Treaty on principles governing the activities of States in the exploration and use of outer space, including the Moon and other celestial bodies, opened for signature on 27 January 1967, entered into force on 10 October 1967.

Agreement on the rescue of astronauts, the return of astronauts and the return of objects launched into outer space, opened for signature on 22 April 1968, entered into force on 3 December 1968.

Convention on international liability for damage caused by space objects, opened for signature on 29 March 1972, entered into force on 1 September 1972.

Convention on registration of objects launched into outer space, opened for signature on 14 January 1975, entered into force on 19 September 1976.

Agreement governing the activities of States on the Moon and other celestial bodies, opened for signature on 18 December 1979, entered into force on 11 July 1984.
See UN Treaties and Principles on Outer Space - A/AC.105/572 1994.

(2) M. Bourély, La Conférence spatiale européenne, A. Colin, Dossiers U2 - Paris, 1970.

(3) Marco G. Markoff, Traité de droit international public de l'espace, Pedone Paris, 1973.

N. Jasentuliyana, Manual on Space Law, Oceana publications, 1979.

Y. Kolossov, International Space Law, Praeger, 1984.

(4) G. Lafferranderie, Responsabilité juridique internationale et activités de lancement d'objets spatiaux au CSG - Bulletin ESA, novembre 1994.

H. Kaltenecker and J. Arets: La position et la responsabilité des Organisations internationales de l'espace dans la Convention relative à la responsabilité concernant les dommages causés par le lancement d'objets dans l'espace extra-atmosphérique - XII IISL Colloquium, 1969.

(5) Declaration of acceptance (attached).

(6) G. Lafferranderie, L'application par l'Agence spatiale européenne de la Convention sur l'immatriculation des objets lancés dans l'espace extra atmosphérique - Annals of Air and Space Law, Vol XI, 1986, 226-235.

(7) ESA Council Resolution, December 1977 - cf. ESA Basic Texts, Vol. I (attached).

(8) CSG and ELA Agreements between the French Government and the Agency - Articles 13 and 14, respectively, ESA Basic Texts.

A further problem arises should the Agency's facilities at the CSG be used when a space object belonging to a non-Member State is launched from an aircraft or from the CSG.

The Agency agrees to indemnify the French Government against any claim for damage, prejudice or loss sustained by an ESA Member State, non-Member State or nationals of such States as a result of carrying out an ESA programme at the CSG, save where the damage is caused by gross negligence, a wilful act or deliberate omission on the part of a servant or agent of the French Government. The Agency bears sole liability where its activities do not involve use of CSG facilities.

(9) M. Bourély, Legal problems related to Spacelab Flight, IISL Colloquium, 1978.

G.B. Sloup, Legal regime of international space flights: criminal jurisdiction and command authority aboard the space Shuttle/Spacelab, 21 Colloquium IISL, 1979.

Gorove, Space Shuttle and the law, 1980, University Mississippi Law Center.

(10) G. Lafferranderie - Pour une Charte de l'astronaute, 12 Ann. Air & Space Law 263, 1987.

M. Bourély - Towards a Convention on the legal Status of Manned International Space Flights - IISL, 1980.

Draft Convention on manned space flights, a joint project by Prof. Bockstiegel, Prof. Vereschetin and Prof. Gorove - in Journal of Space Law and Manned Space Flight, Prof. Bockstiegel - Ed. Cologne, 1993.

(11) Outlook on Space Law over the Next 30 years, Kluwer 1997. Ed. G. Lafferranderie and D. Crowther.

(12) G. Lafferranderie and P.H. Tuinder - The role of ESA in the evolution of Space Law. Journal of Space Law, vol. 22, n° 1 and 2, 97-113.

PROBLEMS AND REALITIES IN APPLYING THE PROVISIONS OF THE OUTER SPACE TREATY TO INTELLECTUAL PROPERTY ISSUES

by Bradford Lee SMITH* and Elisabetta MAZZOLI**

*European Patent Attorney, Senior Patent Counsel, Alcatel Alsthom Intellectual Property Dept., Paris- France

**Docteur en Droit, LUISS Guido Carli, Università di Roma, Italia

[1]Abstract

Several potential conflicts may be discerned between basic tenants of the Outer Space Treaty (OST) and modern patent practice. The temporary, territorially- limited monopoly granted to a patent proprietor by a national patent office shall be difficult to reconcile with the "Space Benefits clause" of OST art.1, particularly in the case of essential patents (e.g. Williams, assigned to HAC). Extension of national territory to Outer Space (e.g. US Space Bill) would seem at odds with the "non-appropriation clause" of OST art. 2, as would patent claims whose scope is expressed in terms of geographical regions of Outer Space (e.g. Horstein et al., assigned to TRW).

The first problem encountered in applying the provisions of the OST to Intellectual Property (IP) law and practice is that the different national patent legislations have no mechanisms to allow for any effects of international law within the delivery procedure up until grant.

Furthermore, the rights granted to the patentee are uniform throughout the territory of validity of the patent, including any artificial extensions of national territory on orbit and beyond.

In such a situation, there is no provision which foresees the introduction of the notions of the OST before a judge in a case of alleged patent infringement in Outer Space. Anyway, judges who are competent in patent cases are by and large surely ignorant of the OST provisions and their possible impact on the judgement concerning inventions made or used in Outer Space.

These problems may be stated in a more general manner. In sum, there is no truly competent body for granting or denying patents in accordance with the provisions of the OST. Further, there is no competent judicial body for judging whether the use of a space-related patent is consistent with the OST provisions.

The authors propose that the UN legal subcommittee of the COPUOS should take the initiative in indicating how the provisions of the OST should be implemented with respect to IPR legislation and its use before the courts.

The recent UN resolution (Feb., 1997) is a first step in this direction. However, in order to be efficiently implemented into national legislations, an agreement similar to the GATT/TRIPS treaty, and its implementation mechanisms could be proposed. The provisions of the OST will only be effectively implemented with respect to IP legislation and practice when the different national legislations are modified to take the OST provisions into account, and a competent judicial body is designated for enforcement of the OST.

Basic Principles

In order to examine conflicts which may arise between application of existing Intellectual Property (IP) law and the basic tenants of the Outer Space Treaty (OST), we shall first recall the principles as expressed in the various legal instruments, as well as some specific US IP-related legislation and the IP clauses of the InterGovernmental Agreement (IGA) for the International Space Station Alpha (ISSα).

The basic principles underlying national IP laws may be illustrated by one of the most venerable texts on the matter, the United States Constitution (1787). Article 1, section 8, paragraph 8 states :

> "Congress shall have power ...to promote the progress of science and useful arts by securing for limited times to authors and inventors the exclusive right to their respective writings and discoveries." (emphasis added)

The first national patent law was enacted in the US in 1790, followed shortly thereafter by a similar law enacted in France. All modern national patent laws are based on the same basic principles, giving rise to codification of Intellectual Property Rights (IPR) in the respective countries.

A general definition of IPR emerges from the mosaic of national laws. IPR is a legal right, which is obtained, exercised, interpreted, and judged according to nationally enacted legislation and ensuing case law. It is the right to forbid third party exploitation, or to allow such exploitation by license on terms dictated by the registered IPR owner or his designated successor.

The scope of the protection of IPR is defined by the filed instruments, for example by the claims of a patent. The geographical scope of the protection is that of the territory of the State which has registered the IPR. And the IPR has a limited lifetime, for example twenty years after the filing date for patents.

We shall now recall some basic principles of the Outer Space Treaty for comparison.

OST Article 1, Paragraph 1 states, in what has been called the "Space Benefits" clause :

> "The exploration and use of outer space ... shall be carried out for the benefit and in the interest of all countries, irrespective of their degree of economic or scientific development, and shall be the province of all mankind."
>
> Outer space... shall be free for exploration and use by all states without discrimination of any kind, on a basis of equality and in accordance with international law, ..."
>
> There shall be freedom of scientific investigation in outer space ... and States shall facilitate and encourage international co-operation in such investigation."

Article II continues :

> "Outer space ... is not subject to national appropriation by claim of sovereignty, by means of use or occupation, or by any other means."

Until this point, the only potential conflict which could be pointed out is that between a monopoly granted to an inventor, on the one hand, and the sharing implied by the Space Benefits clause of the OST on the other.

However it is also important to note that the US has taken the lead in extending the applicability of its national IPR legislation into orbit and beyond, through the "US Space Bill", and the NASA act.

Unilateral Territorial Extension for IPR

President Bush, when commenting on his signature of the US Space Bill, [2] explained how this bill would promote progress of science and useful arts by securing a temporary monopoly for US commercial entities which would apply also to space in the same manner that they would receive if their activities were conducted on earth (i.e. in the US).

The motivation for the US Space Bill as expressed by President Bush, was to obtain legal certainty that US patent law protection would be extended to Outer Space for US entities under certain broad conditions. The US Space Bill adds an article [3] to US patent law, which reads in part :

> " Any invention made, used or sold in outer space on a space object or component thereof under the jurisdiction or control of the United States shall be considered to be made, used or sold within the United States for the purposes of this title, except with respect to any space object or component thereof that is specifically identified and otherwise provided for by an international agreement to which the United States is a party, or ... carried on the registry of a foreign state in accordance with the Convention of Registration of Objects Launched into Outer Space."...

The US Space Bill thus follows the "flagship principle" [4] as applied to vessels on the high seas, or aircraft flying over international waters. In a logical extension of the Flagship concept, US legislation enacted in the "NASA act" [5] concerning US Government Interests in Patents issued from space activities :

> (k) "...Any object intended for launch, launched, or asssembled in outer space shall be considered a vehicle for purpose of section 272 of title 35, United States Code." (temporary presence).

Patchwork Territorial Solution for ISS α

This "flagship principle" is pursued to its logical limits in the provisions governing applicable law (including, but not limited to IPR law) in the Inter Governmental Agreement (IGA) for the International Space Station Alpha (ISSα) , which states [6] :

> "... for purposes of intellectual property law, an activity occurring in or on a Space Station flight element shall be deemed to have occurred only in the territory of the Partner State of that element's registry, except that for ESA-registered elements any European Partner State may deem that the activity to have occurred within its territory.
>
> For avoidance of doubt, participation by a Partner State, its Cooperating Agency, or its related entities in an activity occurring in or on any other Partner's Space Station flight element shall not in and of itself alter or affect the jurisdiction over such activity provided for in the previous sentence."

Forum Shopping via UN Registry

The Flagship principle as embodied by the IGA solves the forum shopping dilemma for this project by agreement between the partners. However, the Flagship principle in maritime law has lead to convenience registry. Could this also happen for registry of space objects under the most favorable regime ?

---

[2] S459, Nov. 16, 1990, publ. in BNA vol. 41, pp. 90-93 (11/22/90).

[3] 35 U.S.C. 105

[4] cf. Senate Report on S 459, publ. in BNA vol. 41, p. 91 (11/22/90) : "Extraterritorial application of the patent laws".

[5] U.S.C. 42, § 2457(k), in J.M. Samuels, Patent Trademark & Copyright Laws, 481 1997.

[6] IGA art, 21, para. 2 (1988).

US patent law and IGA make explicit reference to the State of Registry to determine the applicable law. However, for space activities undertaken by entities in the private sector there remains the possibility of a willful choice of the State of Registry.

U.N. Registry : In Which Country ?

Several legal consequences may govern the choice of a registry state for space objects : Ownership, liability, jurisdiction and control, hence applicable law.

The 1975 U.N. Registry Convention[7] provides that the "Launching State" will register each object launched in a national register, and inform the U.N. Secretary General. However the "Launching State" has multiple definitions in art. 1 of the 1975 treaty :

- the State that launches ;
- the State that procures the launching ;
- the State from whose territory an object is launched ; or
- the State from whose facility an object is launched .

Example : Sea Launch Project

A current example of the Registry problem is given by the Sea Launch [8] Project . Which is the "Appropriate State" for U.N. Registry in a consortium comprising a US firm (Boeing, the leader), a Nowegian firm (Kvaerner) for platform and boat construction, a Ukranian firm (Yuzhnoye) for two-stage rockets, and a Russian firm (Energia) for the third rocket stage ? Considering also that the Sea Launch firm itself is registered for tax purposes in the Cayman Islands, a British Crown Colony, and that the Sea Launch vessel has its permanent berth in Long Beach, California, and is flying a Liberian Flag ? Furthermore, the Sea Launch may launch from any territorial waters (with permission), or international waters if more convenient.

As it turns out, the Sea Launch consortium has opted for the US as a Launching State, perhaps as it is financially preferable to do so in light of the bilateral agreements concerning the former USSR launch offers on the civilian launch market. As a consequence, US IPR law also applies to Sea Launch activities.

Which is the Applicable IPR Law in Outer Space ?

In view of the US Space Bill, the IGA, and provisions on UN Registry for determining the "nationality" of a space object, it appears that the US law is the law which is most often appplicable. Indeed, in the case where US law is not applicable, it would seem that there is no clear answer, as no other country has passed space-specific IPR legislation.

However US patent law has no explicit mechanisms to take into account the provisions of the Outer Space Treaty (OST) in the examination procedure leading to the grant of a patent. This means that the U.S.P.T.O. can (and does) grant patents in conflict with the basic principles of the OST.

Possible Conflicts with OST

Is there conflict between 1) IPR use in outer space ; 2) OST principles in articles 1 & 2 ; 3) basic principles of IPR ?

There would seem to be a fundamental conflict between a temporary monopoly granted to the owner of IPR, and the broad general terms of the OST Space Benefits clause. In some cases, IPR may also conflict with the non-appropriation clause of OST (Art. 2). We will illustrate such possible conflict by reference to a couple of real cases in the following paragraphs, which may also lead us to ask if there may be a contradiction between the use of IPR in space, and the basic underlying principle of IPR itself, i.e. to

---

[7] A/AC.105/572/Rev.1

[8] For a thorough discussion, cf. Kerrest, A., "The Launch of Spacecraft from the Sea", in Outlook on Space Law over the Next 30 Years, pp. 217-233, Kluwer Law International, Netherlands, 1997.

promote the progress of science and the useful arts.

Our first example concerns a US patent granted to TRW in July 1995 [9]. This first issue has been followed by similar patents issued from the same parent application via the US procedure "Continuation in Part" (CIP). The most recent of this series, which also has the broadest claims, issued in September 1996[10].

The main claim of the first ('726) patent has the following features :

•Launch of a constellation of satellites to between 5,600 and 10,000 nautical miles above the earth;

•At least one satellite to have a reduced antenna field of view, FOV, less than full earth coverage;

•The satellites to be oriented in a plurality of predetermined orbital planes;

•Receiving radio frequency signals by at least one satellite from a plurality of mobile handsets with omni-directional antennae;

•Overlapping of a portion of the coverage region of a departing satellite with a portion of the coverage region of an arriving satellite;

•Predetermined criteria for the assignment of calls to or from users within the coverage overlap region from a departing satellite to an arriving satellite (call hand-over).

The main claim of this patent may be interpreted as reserving an orbital "shell" surrounding the earth between the altitudes of 5600 and 10,000 nautical miles, for virtually all conceivable practical applications in the field of satellite-based communications to mobile handsets.

As soon as the patent was granted, a foreign (non-US) competitor, ICO Global Communications Ltd., was warned that its planned satellite system was considered to be a potential infringement of that patent.

[9] US-A-5,433,726 to Horstein et al., 1995.

[10] US-A-5,551,624 to Horstein et al., 1996.

The warning was followed by a lawsuit in a California court to enforce the monopoly, and purporting to exclude that foreign competitor from implementing a global mobile satellite communications system having satellites in the forbidden (appropriated) altitude range mentioned in the main claim.

This action based on legitimately obtained US IPR seems to be in direct conflict with both the non-appropriation (Art.II) and the Space Benefits (Art. I) provisions of the OST.

The TRW-ICO case was recently dismissed in the first instance, according to press reports, by the conclusion that no infringement has yet occured, as although the satellites are now under construction, they are not in and of themselves infringing *per se.* In a more recent development, Odyssey (TRW and Teleglobe Canada), is apparently having financial difficulties and industry speculations are that TRW will never realise a system according to its patent. According to the terms of the GATT/ TRIPS agreement, this could be considered sufficient and valid grounds for obtaining a compulsory license in many countries, but not the US.

TRW Vice President and Managing Director of Odyssey Bruce Gerding was quoted in the press [11] during Telecom Geneva as having boasted :

> "We have built a fence around our real estate", said Gerding. The patents now allow "us to build that fence higher, thicker and stronger. We will rigorously protect our intellectual property rights."

This belligerent, "no-trespassing" attitude of the patent proprietor betrays a direct conflict with the non-appropriation principle of the OST.

Our second example concerns the famous Hughes Aircraft Company (HAC) Williams patent [12], concerning a method for obtaining and maintaining satellite attitude on orbit. Proper attitude is necessary in

[11] "Satellites of Regulation", Network Europe, Nov.-Dec. 1995.

[12] US-A-3,758,051, 1973.

order to allow the satellite to properly aim its directional antennas in order to fulfill communications missions, and in some platform architectures, to orient the solar energy collectors to supply electrical energy to the payload.

Once again, shortly after this patent was granted in 1973, HAC attacked the US Government for infringement in the same year, followed by multiple attacks on foreign entities. In this case, litigation dragged on for years, practically during the whole term of the patent, and by the time the first judgement was handed down in 1983, there were 108 alledgedly infringing craft, although when the litigation started there were only 14 satellites incriminated.[13] HAC had not sought an injunction or a temporary restraining order to prevent the launches of the additional satellites, and was then encouraged to assert its patent aginst other, non-US entities by the successful judgement against the US government and the substantial (multibillion dollar) damages awarded in the first instance [14].

However, it would have theoretically been possible, especially concerning foreign entities, to request the court to halt the alleged infringing activities pending final judgement of the case. If this had happened, in view of the necessity of a method for maintaining the satellite attitude in order to fulfill communications missions, this patent could be seen to hinder, rather than to promote, the progress of science and the useful arts, at least during its term of enforcement.

These two examples serve to illustrate that in practical situations in the use of IPR, conflicts may arise with the basic tenants of the OST, and even with the fundamental underlying principles of IPR itself.

Existing attempts to obtain legal certainty for IPR in space activities through extension of territorial sovereignty via the flagship principle (e.g. US Space Bill, IGA) present only ad-hoc, piecemeal solutions concerning the forum for IPR enforcement. However these efforts do not attempt to address the fundamental underlying conflicts illustrated above.

Current Situation and Tendancies

Space activities are attracting multi-billion dollar investments from the private sector. The use of IPR in space activities is becoming more and more agressive, with attempts tending towards the total exclusion of competitors from certain activities or certain regions of space (or both). The founding principles of IPR and the OST seem to have been forgotten by the main players. This may be simply because the main players are no longer those States who signed the Outer Space Treaty thirty years ago, but rather private enterprises hoping to cash in on the enormous revenues foreseen in consumer mass markets. In the opinion of the authors, it is high time to take corrective action before it is too late.

Possible Corrective Actions : Harmonisation

In order to resolve apparent or potential conflicts with the fundamental principles which may arise from the use of IPR in space activities, one possible corrective action could be an attempt towards harmonisation of applicable law. Harmonisation would require that there be at least one other national law in addition to the US law, which is for now the only explicitly applicable law for IPR in outer space.

Shouldn't some of the other space-faring nations consider legislation similar to that of the US Space Bill ?

In Europe, a unique opportunity for such legislation is presented by the

---

[13] For discussions of the Hughes case, cf. BNA vol. 52, pp. 250-252, idem vol. 46, pp. 428-430, idem vol. 36, pp. 555-556, idem vol. 26, pp. 491-492.

[14] For discussion of the dispute on the evaluation of damages in the Hughes case, see Christol, C. "Damages and Intellectual Property " An Up-Date on Hughes Aircraft Company vs. USA", in Proc. 39th Colloquium on the Law of Outer Space, pp.210-214 ,AIAA, Washington D.C. 1996.

European Commision's Green Paper on the Community Patent [15]. Interested parties are invited to comment on this proposition for a patent legislation having a uniform effect in the countries of the European Community, up until 7 November 1997.

However, even if other space powers eventually undertake appropriate legislation to make their national IPR laws applicable to space activities, we will still be confronted with a patchwork of virtual territorial considerations in order to determine which is the applicable law in a given case, and forum shopping is likely to become a favorite pastime under such a regime.

Possible Corrective Actions : Globalisation

A preferable solution would be "Globalisation" of the jurisdiction in outer space activities, i.e. a single, worldwide IP legislation for space activities. This could be imagined as a treaty under the auspices of the UNCOPUOS. We could recommend to **establish space and its accesses (launch sites, vehicles) as a single territory with a single, uniform law**.

A "Space Patent" could be imagined as a new "Country" designation on a PCT (Patent Cooperation Treaty) application, to be examined and granted under the auspices of the WIPO (World Intellectual Property Organisation).

Such a space patent, once granted, should be administered and interpreted by **a single, universal enforcement body** such as an international court of law or an international arbitration authority. Perhaps such an arbitration authority could be created under the auspices of the World Trade Organisation (WTO), aided by the World Intellectual Property Organisation (WIPO) for the IPR aspects. WIPO has already established a board of arbitration for Intellectual Property matters, and it should not be difficult for them to acquire the necessary competence to act on space matters.

[15] COM(97)314 final, 24.6.97

This board could be empowered to arbitrate on matters such as space patent validity and compatibility with international law, alleged infringement, conditions of licensing to third parties, etc.

To aid the Board in its considerations, we propose that a code of conduct be elaborated for the use of such patents. This code of conduct should reflect the basic principles and rely on them for its terms. We are committed to promoting the progress of science and the useful arts, while rewarding inventors for their efforts, and procuring Space Benefits for all mankind. Any appropriation (or monopoly) of any region of space for any use should be formally precluded. Patentability, if incompatible with the basic principles, should be excluded. Such exclusion would be most effective if operated in the patent examination phase.

As for licensing conditions to third parties, in order to fulfill the principles of the OST, licenses should be made on a non-exclusive, non-discriminatory basis, on fair and reasonable terms and conditions. Such fair and reasonable conditions may depend on the economic and scientific development of the Licensee, as reflected in the UN Resolution on International Cooperation in the Exploration and Use of Outer Space for the Benefit and in the Interests of All States, Taking into Particular Account the Needs of Developing Countries, [16] which states in particular :

> in Para. 2 : " States are free to determine all aspects of their participation in international cooperation ... on an equitable and mutually acceptable basis. Contractual terms in such cooperative ventures shold be fair and reasonable and they should be in full compliance with the legitimate rights and interests of the parties concerned, as, for example, with intellectual property rights. "
>
> And in Para. 5, last alinea : "International coooperation, while taking into particular account the

[16] A/RES/51/122, 4 Feb. 1997

needs of developing countries, should aim, *inter alia*, at the following goals ...

... Facilitating the exchange of expertise and technology among States on a mutually acceptable basis. "

We urge the necessity of a globalisation of applicable law, which could be obtained through the efforts of an international treaty organisation such as UNCOPUOS.

Possible Corrective Actions : Enforcement

The efficacity of the Outer Space Treaty, thirty years after, is cast under doubt by the lack of interpretation and enforcement. In the opinion of the authors, if the OST cannot be enforced, it cannot be applied and respected. One way of obtaining enforcement would be to introduce litigation for violation of OST provisions.

Litigation could be introduced by any interested party before the US Federal District court for invalidity of a US patent granted in violation of OST provisions. This would help define the limits within which the USPTO should grant patents on vast regions of outer space, for the exclusive use of patentees for certain lucrative commercial applications.

Also, any Party to the OST could sue the US government for non-respect of its treaty obligations, as the Signatory State is responsable for the acts of its nationals (i.e. U.S.P.T.O.. or TRW). However any Signatory State contemplating such action will surely be aware of the considerable trade sanctions which may be applied, under the super 301, for such untowardly behavior.

If no Action is Taken, OST will Fade Away

As exposed above, the problems arising in space activities in the field of IPR put the OST to the test of its applicability and efficacity. A thirty-year old legal instrument which has never been interpreted nor used in the courts runs the risk of becoming obsolete and useless. If not used, the OST provisions will succomb to economic forces, which are too strong and inevitable to be resisted by the lofty principles of the Outer Space Treaty.

Save the Outer Space Treaty

If the OST is to survive into the 21st century, which promises to be really the golden age of commercial space endeavors, then a few essential accessories need to be added by OST defendors. We need binding legal interpretation of the obligations of the OST for the signatories, and clear mechanisms for applying this to the activities of nationals of the signatory states. In addition, we need to create an efficient enforcement means, which protect the legitimate interests of inventors and patent proprietors, to the extent which they are not in conflict with the basic principles of the OST and the founding principles of IPR, i.e. promotion of the progress of science and the useful arts, the sharing of Space Benefits, and the non-appropriation of Outer Space.

IISL-97-3.07

THE DEVELOPMENT OF ARTICLE IX OF THE 1967 SPACE TREATY

by

Professor Dr. Maureen Williams *
University of Buenos Aires

1. Introduction

At the time of the conclusion of the 1967 Space Treaty, or Treaty on General Principles, as it is currently known, a number of fundamental rules to govern activities in these areas were adopted which, thirty years on, continue to be valid. Stemming from these principles, more detailed agreements were subsequently concluded in order to give a more precise legal framework to certain matters which, as a result of advances in science and technology, needed further clarification. Such is the case of the 1968 Astronauts Agreement, the 1972 Liability Convention, the 1974 Registration Convention and the 1979 Moon Agreement, all of which have clarified, respectively, the meaning of articles V, VI and VII, VIII, and II, of the 1967 Space Treaty. In addition to these treaties which now have the hierarchy of positive international law, much effort has been directed to other fields of the law of outer space which called out for precision. In this respect, mention should be made, inter alia, of the Draft Convention on Manned Space Flight prepared by a team headed by Professors Böckstiegel, Gorove and Vereshchetin and which was thoroughly analysed in 1992 at a Colloquium held in Cologne. Reference should also be made to the ILA Convention on Dispute Settlement (Paris 1984) presently under review by the Space Law Committee of the International Law Association, and to the Buenos Aires International Instrument on Damage caused by Space Debris, adopted in 1994 by the 66th Conference of the said Association. This Instrument, which the Space Law Committee of the ILA keeps under permanent study so as to keep pace with technological developments, was presented to the Legal Subcommittee of COPUOS as well as to the Full Committee by Professor Böckstiegel in 1995 (1), and explained further at the 1995 and 1996 Sessions.

---

* **Rapporteur of the Space Law Committee of the International Law Association. The author is grateful to the University of Buenos Aires for the support given for research on this topic.**

2. The background of Article IX of the Space Treaty

Indeed, article IX of the 1967 Treaty, dealing with damage to the environment as a result of space activities, is far from satisfactory nowadays. Suffice it to say that the stage of mere exploration of outer space is definitely over. We are now well within the stage of exploitation. The use of the geostationary orbit for telecommunication satellites is a glaring example in this sense. Consequently, the need to have more precise rules on the matter is clearly manifest. And yet, the topic has not been included on the agenda of the Legal Subcommittee of COPUOS so far.

I have referred to certain legal aspects of this subject in recent years and different occasions, such as the International Colloquium on Environmental Aspects of Activities in Outer Space (Cologne 1988), the IISL Colloquium in Oslo (1995), the 66th Conference of the International Law Association (1994), the Journal of Space Law (Jan.1995) and the Reports that preceded the adoption of the 1994 ILA Buenos Aires Instrument on Space Debris (Queensland 1990 and Cairo 1992). I would now like to pause on some other issues and developments related to space debris which appear important when considering the development of article IX of the Space Treaty.

De lege ferenda, there is a great deal to be said on this topic. In addition, it is essential to have in mind that a number of rules of international customary law in connection with the protection of the environment have become such over the last thirty years. Furthermore, international cooperation in this field is no longer viewed as a mere expression of ideals but, rather, as a general obligation and a pre-requisite for carrying out activities in outer space. International cooperation, however, has not yet achieved the status of international customary law but it is certainly moving in that direction.

One of the first landmarks which went a long way in elucidating the question was, no doubt, the Colloquium organised by the University of Cologne in May 1988. This meeting was possibly the first of a truly interdisciplinary character (2). The 63rd Conference of the ILA followed shortly in Warsaw and entrusted the Space Law Committee with the task of elaborating principles and guidelines on debris and pollution resulting from activities in outer space and to start work on a draft convention on the subject (3).

Two months later, in October 1988, a Conference was held in Asunción del Paraguay , organised by the Ibero-American Institute of Air and Space Law, which briefly touched upon the issue during its Space Law Session. This meeting declared, in no uncertain terms, that some kind of obligation ought to be imposed on states and international organisations, when launching a satellite into orbit, to arrange for removal of the satellite once its useful life had ended. The obligation to remove debris is still an outstanding question, rather vague in the context of the outer space treaties in force (4) and which the ILA Instrument, with the idea of starting at a low level of compulsion and then slowly moving up the scale, has left within the area of international cooperation (namely, as "the general obligation to cooperate", as worded in article 3 of this Instrument).

2.1 Unilateral removal of space debris

Some authors have engaged in research and discussion concerning the possibility of unilateral removal of inactive satellites. The majority coincide in the sense that no unilateral action is permissible in accordance with international law. As Perek acutely observes, satellites were considered so valuable at the time of the drafting of the 1967 Space Treaty that, according to article VIII, they had to be returned to the state of registry when found outside the territory of the launching state (5).

This discussion has led to interesting reasoning: for example, Ferrer's position on the "droit de route" or "right to trajectory", a thesis which this Argentine professor has held ever since the IISL Constance Colloquium in 1970, and elaborated further at subsequent international meetings and in his writings (6). According to Ferrer, when a satellite is describing a given route or orbit, and provided the launching and orbit have been duly notified to the UN Secretary-General and other formalities duly observed, the state of registration has a right "erga omnes" to maintain this satellite in its orbit, even when it becomes inactive, for whatever reason. Should another State-Party to the Treaty wish to use the orbital slots occupied by that inactive satellite, by no means can it take the law in its own hands and proceed to its removal.The right procedure, according to Ferrer, would be to hold consultations during which the state demanding the removal of the satellite should prove to the state of registration that that satellite is no longer complying with the principles laid down in the Space Treaty, that is to say, that the activity is not being carried out for the benefit and in the interest of all countries.

The same principle is, of course, applicable where parts of satellites or particles therefrom (second generation debris) are concerned. The problem is aggravated in that small particles are almost impossible to be detected from the Earth and that in the geostationary orbit they travel at speeds that are likely to cause severe damage in case of collision with active satellites.

Therefore, the outlook today is considerably more dangerous and complex. It thus follows that not only old law should be adjusted but also new law progressively developed. Hence the raison d'être of the ILA Instrument and, particularly, of "a general obligation to cooperate", as indicated above. It is not unreasonable to expect that in due course, and with the accelerated growth of space activities, this initial step should lead to the adoption of more stringent measures.

Malanczuk has also tackled the issue of unilateral removal of space debris (7) and concluded, like Ferrer and most of the doctrine today, that such a right does not exist, except with the prior consent of the state of registration or as a result of an international agreement envisaging such action.
To be realistic the establishment of an obligation to remove, in such blunt terms, within the ILA Instrument, would have endangered the support from the international community, a goal which the Instrument has always had as one of its main objectives. It is a true fact that states are reluctant to accept obligations when they are not quite sure what these obligations will ultimately result in. Therefore, the ILA

Space text chose to start at a low level of compulsion, as priorly remarked, by laying down a "general obligation to cooperate" drafted along the lines of Principle 11 of the Declaration of Principles for the Protection of the Environment" (Ottawa, February 1989), and which would cover other situations as well.

2.2. A glance in retrospect

An important precedent to the 1988 Meetings described above was an ILA Regional Seminar held in Buenos Aires in December 1987 where one of the Special Working Groups, chaired by the present writer and where Professor Cocca was Rapporteur, studied the matter and offered its conclusions to the then forthcoming Warsaw Conference (8).

The Buenos Aires Regional Seminar of the ILA concluded that environmental risks arising from space activities may have their effects not only in outer space proper but also on planet Earth itself. Thus, a uniform set of rules should be devised to be applicable both to the Earth and Outer Space environments.

Indeed, this conclusion is consistent with, and was in clear anticipation of, article 1 (d) of the 1994 ILA International Instrument which states that for the purposes of that Instrument "environment" includes both the outer space and earth environments within or beyond national jurisdiction (9).

The Buenos Aires Regional Seminar also recommended that a protocol to the 1967 Space Treaty or some other kind of separate international instrument should be elaborated with a view to giving a more precise meaning to the general principles on the protection of the environment enshrined in article IX of that Treaty as well as in other agreements in force dealing with environmental risks (10).

These meetings, and other precedents such as the Darmstadt First European Conference on Space Debris (Darmstadt 1993), the Brighton IISL Colloquium (1987) and the 62th Conference of the ILA (Seoul 1986) clearly revealed the need for seriously addressing the problem of space debris and its legal aspects in the light of the new circumstances which the space scientists had explained only too well during the Cologne Colloquium and other meetings. (A Second European Conference took place in Darmstadt in 1997, to which reference will be made later).

2.3 Recent Developments

As a result of of the above-described currents of opinion, attention was directed to article IX of the Space Treaty and opinions were initially divided, ranging from those which advocated the need for revising the whole Space Treaty to more moderate stances supporting the view that a separate protocol or instrument should be drafted to supplement, with more detailed and stricter provisions, the call for international cooperation embodied in article IX of the Treaty. And within these two extreme positions a variety of intermediate shades was to be found.

One of the strongest criticisms aimed at Article IX was, in fact, that it did not go beyond international cooperation and that, if mankind was to survive, we could not afford to run further risks. The expresion "having reason to believe" was equally questioned by the doctrine given its highly subjective and discretional nature: if a state "had reason to believe" that its activity would cause no harm it would have duly conplied with its obligations under Article IX of the 1967 Space Treaty. This, in practice, implied a severe risk of causing irreversible damage to the environment.

On the other hand and particularly during the seventies, some writers believed that, because of the very broad interpretation that could be given to the definition of "damage" established in article I (a) of the Liability Convention (i.e, loss of life, personal injury or other impairment of health; or loss of or damage to property of states or of persons, natural or juridical , or property of international intergovernmental organisations), nothing much could be done to improve the legal situation. Should, for example, environmental damage occur during refuelling operations of a spacecraft, this assumption would unquestionably come under the terms of article I of that Convention and thus be subjected to a regime of absolute liability in accordance with that Convention and the 1967 Space Treaty. Yet, almost nobody in those days questioned the advisability of having a better-defined set of rules on this topic.

The ILA Instrument on Space Debris has taken the Liability Convention definition almost literally (article 1 (c)) adding, at the end, the words "or any adverse modification of the environment in areas within or beyond national jurisdiction or control", with the obvious thought of including the global commons. Moreover, in the interest of accuracy, it was considered necessary to have more detailed provisions within this Instrument such as, for example, those contained in article 3 concerning its scope of application. Thus, the Instrument is applicable to space debris which causes or is likely to cause direct or indirect, instant or delayed damage to the environment, or to persons or objects.

3. <u>The world of today</u>

In the view of Perek, article IX of the 1967 Treaty can hardly be interpreted as applying to space debris which is of terrestrial origin. Rather, this scientist -who is one of the scientific consultants of the ILA Space Law Committee- sees that article as referring to contamination of the environment by extraterrestrial matter. After mentioning the definition of "orbital debris" contained in the IAA Position Paper (11)
Perek concludes that it coincides, in substance, with the one adopted by the Buenos Aires Instrument (12). With his usual clearness, Perek identifies major reasons for having legal rules on space debris. Apart from the fact that they constitute around 95% of all objects in space and represent a hazard to active satellites, they a have a legal status of their own, different from that of active spacecraft, and a different terminology should be used as well. As Perek puts it, the term "space object" should not apply to space debris. It is not difficult to conclude that the author in question fully recognises the need for having a separate convention on space debris without having to think of revising the outer space treaties presently in force (13).

The Second European Conference on Space Debris was held in Darmstadt on 17-19 March 1997. Results clearly indicate that one of its essential objectives had been reached, namely that awareness of this question had increased significantly.

As in the previous Conference (Darmstadt 1993), the meeting was of an interdisciplinary nature. Likewise, there was an interesting contribution from space lawyers. Kopal, for example, strongly supported the idea of having a legal document on space debris. In his opinion, the term "damage" as defined in article I (a) of the Liability Convention, did not include damage caused by space activities to the space environment. Insofar as dispute settlement related to space debris is concerned, Kopal considers that the present rules of international law are much too general and leave many a question unanswered. Kopal strongly supports the idea that the Buenos Aires ILA Instrument should include an obligation to remove "non-functional space objects" from the most saturated areas of outer space (14).

Benkö and Schrogl, for their part, focused on space debris, in the United Nations, addressing aspects of law and policy. These authors hold the view that solutions should be sought on the basis of the existing law, supplemented by comprehensive regulations, and maintaining the existing space treaties intact (15).

Among the scientific experts voicing their thoughts at Darmstadt 1997, the chairman of the Scientific and Legal Subcommittee of COPUOS and Consultant of the ILA Space Law Committee, Dietrich Rex, made very accurate comments on the threat of overcrowding, both as regards tracked space objects and untrackable population even though the risk of collision was still low. However, impacts on satellites by man-made objects smaller than 1cm were frequent today (16). Interesting are Rex's observations on the way space debris got on to the agenda of the Scientific and Technical Subcommittee, first under the item "other matters" and since 1994 as a separate topic on that agenda.

Following this precedent and having in mind that in 1996 during the 36th Session of the Legal Subcommittee the legal aspects of space debris was listed as a possible item for inclusion on the agenda, and that, among the five topics considered for inclusion, three of them were closely connected to space debris: the above-mentioned (legal aspects), the review of the existing norms of international law applicable to space debris, and the comparative review of the principles of international space law and international environmental law, it is reasonable to expect its addition to the agenda of the Legal Subcommittee in a not too distant future.

All these topics were the object of thorough discussion at the different stages leading to the adoption of the ILA Buenos Aires Instrument, as may be seen in the Conference Reports.
And, as remarked by Kopal, "the ILA Buenos Aires International Instrument on the Protection of the Environment from Damage caused by Sapce Debris offers a good example of how a legal document on the issue of space debris could be drafted"(17).

At the 36th Session of the Legal Subcommittee (March-April 1997) mentioned above it was recommended that further informal
consultations on specific proposals for new agenda items should continue in 1998. Shortly after that, in June 1997,

the Chairman of the ILA Space Law Committee -Prof. Böckstiegel- , in his statement to the Committee on the Peaceful Uses of Outer Space (40th Session, Vienna) reiterated his comments on the ILA Buenos Aires Instrument providing the meeting with a progress report on the matter and noting with satisfaction the work on the topic carried out so far by COPUOS. However, he did not fail to notice a certain reluctance on the part of the Legal Subcommittee to take up the subject in spite of the firm position of some delegations to the contrary. After stressing the importance of exchanging information between the two Subcommittees Professor Böckstiegel observed that it was important for the Scientific and Technical Subcommittee to
discuss the possibility of having an international instrument on space debris and, in this sense, the expertise of the Legal Subcommittee would be important (18).

4. Conclusion

When the idea of including space debris on the Legal Subcommittee's agenda finally materialises, the work which led to the adoption of the 1994 ILA International Instrument, enriched with the suggestions and contributions from the experts in the legal and scientific fields, will surely mean a useful and realistic contribution to the elucidation of the legal rules which should govern such a crucial matter.

As indicated by the different stances referred to in this paper, there is almost consensus among lawyers and scientists on the increasing danger of damage to be caused by space debris at the turn of this century. The need to develop the general principles enshrined in article IX of the 1967 Treaty is therefore fully justified. Hence, every effort tending to encourage a political decision to include the legal aspects of space debris on the agenda of the Legal Subcommittee of COPUOS should be hailed as a step forward for the benefit and in the interest of the international community as a whole.

NOTES

1. See Böckstiegel's statements to the Legal Subcommittee of COPUOS and to the Full Committee in 1995, 1996 and 1997. Also Kopal's presentation to the Legal Subcommittee (March-April 1995).

2. See ENVIRONMENTAL ASPECTS OF ACTIVITIES IN OUTER SPACE, Carl Heymanns Verlag, 1988.

3. See Report of the 63rd Conference of the International Law Association, Warsaw 1988, pp.842-844.

4. See P.Malanczuk, REVIEW OF THE REGULATORY REGIME GOVERNING THE SPACE ENVIRONMENT, published In the Proceedings of the 38th Colloquium on the Law of Outer Space, AIAA, pp.355-382.

5. See L. Perek, OUTER SPACE TREATY IN PERSPECTIVE, 40th Colloquium on the Law, Torino 1997. Manuscript by courtesy of the author.

6. See Ferrer, EL DERECHO A LA TRAYECTORIA, Proceedings of the XIII Colloquium on the Law of Outer Space, Constance 1970,
pp. 160-164. Also, same author's book DERECHO ESPACIAL, published by PLUS ULTRA, Buenos Aires 1976.

7. See Malanczuk, op.cit. in note 4, at pp. 377-8.

8. See BUENOS AIRES REGIONAL SEMINAR OF THE INTERNATIONAL LAW ASSOCIATION, Buenos Aires 2-4 December 1987, published in a special number of the REVISTA DEL COLEGIO DE ABOGADOS DE LA CIUDAD DE NUENOS AIRES, March 1988, pp. 133-162.

9. See REPORT OF THE 66TH CONFERENCE OF THE INTERNATIONAL LAW ASSOCIATION, Buenos Aires 1994, p. 9 et seq.

10. See op.cit. in note 8, p.133.

11. See ACTA ASTRONAUTICA, Vol. 31, p.169, October 1993, quoted by Perek in op.cit. in note 5.

12. Ibid., page 2 of manucript .

13. ibid., p.5

14. See Kopal, PRESENT INTERNATIONAL LAW PRINCIPLES APPLICABLE TO SPACE DEBRIS AND THE NEED FOR THEIR SUPPLEMENT, in Proceedings of the Second European Conference on Space Debris, Darmstadt, 17-19 March 1997.

16.See Benkö and Schrogl, SPACE DEBRIS AND THE UNITED NATIONS: ASPECTS OF LAW AND POLICY, in Proceedings of the Second Conference... etc, Darmstadt, 17-19 March 1997.

17. See Kopal, SPACE DEBRIS: A REVIEW OF THE CURRENT REGULATORY STRUCTURE, in Proceedings of the 39th Colloquium on the Law of Outer Space, AIAA, Beijing 1996, pp. 343-350, at p.349.

18. See Böckstiegel's statement to the 40th Session of the COPUOS, Vienna, June 1997.

Buenos Aires, October 1997

IISL-97-3.08

THIS TREATY NEEDS A LAWSUIT*

D. O'Donnell
G. Robinson III
G. Robinson IV
United Societies in Space, Inc.

Abstract

The Outer Space Treaty of 1967 has been discussed for 30 years but the treaty burden of benefit sharing has not been litigated. Instead, it has been politically removed from the protagonist southern states by UNGA resolution. However, the treaty was not amended so a juridical decision by a competent court could re-instate the forced transfer of space resource benefits requested by the southern states. Because of this future open door and because the financial stakes are getting higher, this treaty needs a lawsuit. The naked possibility that an injunction against the use of orbits or other space resources could obtain by court order in the future, or damages for trespass in the past could be calculated, requires judicial action now. This is viewed as curative and preemptive: a proper lawsuit may be the only way to define the treaty burden adequately. Strategies and problems are discussed.

Introduction

The treaty burden of benefit sharing has been restated in UN General Assembly (UNGA) Resolution 51/122, adopted December 13, 1996, to mean that nations should freely engage in international cooperation and that space powers should not forget to integrate the developing nations into space exploration. During the decade leading up to this political emasculation of the entire concept, northern and developed and spacefaring nations refused to consider any agenda that could lead to a legal commitment on benefit sharing. Only philosophical discussion was allowed. Since UNCOPUOS is conducted on a basis of 100 percent consensus voting, this refusal has forced the protagonist under-developed nations to concede their case:

> "It marks the end of a North South Debate which has focused on the introduction of forced cooperation and transfer of resources. By providing

---

Author Declan J. O'Donnell, JD, is a member of the IISL, President of United Societies in Space, President of World-Space Bar Association, and principal attorney in Declan Joseph O'Donnell PC in Denver, Colorado. Author George S. Robinson III is retired as associate general counsel at the Smithsonian Institution in Washington, DC., and is now in private practice with author George S. "Fox" Robinson IV in Fairfax, Virginia.

> an authoritative interpretation of the cooperation principle of Art. 1 of the Outer Space Treaty the Declaration prevents further confrontation on the general political level."[1]

This political solution culminates a deep seated and heart felt campaign at the United Nations by a group of 77 nations (called G77). They have relied on Article 1 of the 1967 treaty to mean what it says and say what it means: "taking into account the particular needs of developing countries" indicated the need for unequal treatment for states in unequal situation.[2] The developing countries viewed outer space benefit sharing as a key part of their conception of "New International Order" involving a whole world economy and benefits from the satellite communications system. One draft paper by nine such states in 1991 "was rich in New International Order language aiming at forced cooperation and an automated transfer of financial and technological resources from North to South.[3]

The 1979 Moon Treaty had provided hope to the South because it expanded the concept of benefit sharing to individuals and companies, as well as clarified in much detail that the "common heritage of mankind" would be required as space treaty law. In 1982 this phrase was defined by the UN to require administration of the resources by all nations and equal distribution of benefits to all nations.[4] It is no wonder that G77 expected northern nations to respect their New International Order of mandatory transfers to them of money and resources in respect to space activities.

UNGA Resolution 51/122 (1996) may represent a political solution to benefit sharing, but it does not represent a legal solution. The treaties are not amended by this resolution, nor are they formally interpreted. The issue is technically unresolved.

If the southern states appear to be asking for too much, it may also be observed that the northern states may be offering too little. The satellite information and telecommunications industry has been relying on a rather novel theory of benefit sharing. It relies on the idea that its work product is valuable to the people who use it so that a benefit is made available to mankind and that complies with the treaty principle.[5] This self-serving declaration is true, but it means little to G77 citizens who can not afford it.

Somewhere in between these extremes there may be a fair and legal solution to the administration of a truly valuable principle. That depends on three procedural points that need to be fleshed-out before a court could deal with the substance:

1. Whether or not the treaty principles are enforceable or are they mere policy. The prevailing view appears to be that this principle is not legally enforceable.[6]
2. Whether or not a judicial forum can be found that can deal with such an issue. It has been asserted that no such forum is available.[7] The International Court of Justice has no case in controversy, but it could render an advisory.[8] This may not be

adequate.

3. Whether or not parties will consent to service of process.[9]

This paper will not deal with the substantive arguments regarding benefit sharing, except in order to frame the issue. The need for litigation is now clear because UNGA Resolution 51/122 (1996) was over reaching politically. The prospect of maintaining litigation in a procedural sense is our focus and it is the next logical step in the North South Debate.

Litigation Strategies

For the southern states there is one overriding strategy for litigation: political solutions are not available at the UN. The going to court scenario is forced as a consideration. The bottom of it rests on the legal meaning of a treaty and courts are cautioned that nations must respect treaties as binding agreements to be performed in good faith.[10] A competent court will not be able to refuse legal discussion of the issue in favor of merely philosophical discussions as forced against the South by UNCOPUOS during the last decade. All facts, circumstances, and UN Treaties will be briefed and argued towards a more fair hearing with due process.

This strategy is timely because the Deep Sea Bed Conservation Treaty is gaining momentum.[11] Not only does this agreement relate to analogous kinds of world resources, it also provides specific and enforceable procedures for the protection of deep sea bed resources as the common heritage of mankind. Not surprisingly, there has been published some basis for the juridical analogy of sea bed assets to outer space resources.[12] The temptation to connect the two by court order is compelling, if not obvious.

The South would be able to represent all nations in this proposed litigation. The treaty burden would fall on the relatively few nations and companies that gain revenues from space resources. However, the treaty benefit is clearly stated as applicable to all. The Moon Treaty of 1979 mentions "equitable sharing" in addition to the "common heritage of mankind" so a degree of apportionment is possible.[13] There are many ways to accommodate our space constitution level principal legal directive but the 1996 UN Resolution probably is not one of those methods.

For the North the strategy is closure. It is not tolerable to avoid this issue for 30 years while wrenching resources from mankind. The low earth orbits (LEO) are saturated with telecommunication satellites and more are expected by the year 2000 AD. Constellations of small satellites in LEO may bring the total number of satellites to well over 2,000 by that time. The amount of money earned by revenues from space by private industry surpassed government space expenditures world wide in 1996, according to *SPACE NEWS*. Revenues for the global space industry are at $77 billion USD per year. This will be $100 billion USD by the year 2000 AD.[14] It is too dangerous for private investors to build their castles in the sky out of shifting sands on earth. Litigation is probably the only way to clarify this issue and set it in concrete.

Enforceability of the UN Benefit Sharing Burden

Article 1 of the Outer Space Treaty and the UNGA resolutions that preceded it were obviously assertions of principles only. No standards for enforcement were provided. Based on this there has been a high degree of open and free cooperation internationally under the aegis of the UN Office of Outer Space Affairs.[15] The lack of enforceability has not thwarted healthy cooperation among diverse states and under-developed countries have experienced much benefit. (There is no criticism to be implied in this paper of the status quo).

However, treaties are to be considered binding and the advent of the Moon Agreement in 1979 and the Sea Bed Treaty in 1982 and the enormous strides by the environmental movement over the past 30 years need to be considered. Public interest litigation in the outer space arena is being prepared by United Societies in Space Inc. for the benefit of all people and as trustee for all nations. It asserts that this principle may now have legal significance because standards have evolved. There is no question about the specific, detailed, and comprehensive standards now a part of international conservation law in the Sea Bed Treaty. If the common heritage of mankind as defined there was equated through the wording of the Moon Treaty of 1979 into Article 1 of the 1967 treaty, then Sea Bed procedures would be relevant. If relevant and if enforceable, this would impose all of the following onto the use of space resources:[16]

1. Neither the area nor its space resources could be appropriated by anyone, except as follows:
2. All of the nations in the world must have an active sharing in the management of the area and its space resources, and,
3. All benefits reaped from these space resources must be shared by, and actually distributed to, all such nations, and,
4. The area and its resources must be preserved for future generations of mankind.[16]

These directives are legally cognizable, albeit some creative mechanisms would be required to implement them. Courts normally engage in such activity and the North and the South may soon be required to propose to a proper court how that should be accomplished. Enforceability of benefit sharing as a treaty law is now predictable, perhaps because in part the political process has failed.

What Forum is Available

The International Court of Justice would represent a good forum for resolution of this issue. It is an instrumentality of the UN and its judges have the appropriate background. However, there is no case pending and nations would have to consent to jurisdiction in that court for such a case. There are 185 UN member nations and it is unlikely they will all submit voluntarily to such litigation. The likely alternative is declaratory relief at the request of the UNGA or any other office of the UN.[17]

No declaratory relief of any kind is acceptable. Political and/or administrative solution have not gotten to the bottom and no unenforceable judgment could do much more. It is time for a lawsuit, not just for another opinion.

Each nation has its own court system available to it. Argentina could file any time to test this principle in its own courts. So could the USA and so could representatives of any nation or combinations thereof, like G77. The costs of administering such a large suit would be very substantial and the losers may have to pay the litigation expenses of the winners. One reason to doubt that nations will rush forward to sue themselves and each other is that it has never happened before.

Another reason is that there are conflicts of interest inherent in the subject matter. For example, an attorney representing the USA would have to be on both sides of the controversy: It earns revenue subject to the burden and is also one of the class of beneficiaries of that burden. No ethics code of any bar association would permit this double dealing by its lawyers.

Without reciting all of the possible forum considerations, one traditional and likely avenue of litigation is planned: a Rule 23 class action in a US Federal District Court. United Societies in Space Inc. would bear the cost of maintaining such an action and it is proposing to file it soon. The basic research has been accomplished and the strategies have been evaluated.[18] Much of this analysis has centered on the Antarctic Treaty and the benefit sharing analogies to be made therein.[19] Clearly the oil and gas under Antarctica are in no immediate danger of being appropriated by anyone.

The local US Forum is believed appropriate here because the telecommunication industry is the target defendant. Most of that legal, financial, and regulatory activity is located in America.[20] Also, most of the plaintiffs and defendants can be located there and the legal community is well versed in these kinds of suits. Adequate representation is available.

The class action under Rule 23 gathers together a class of plaintiffs called all nations in the world. There are 269 of these according to *Town & Country* magazine (but eight do not order Coca-Cola and, therefore, may not be real). The defendants can be defined as nations and enterprises that receive revenues from the use of LEO satellites. It too is a class.

Personal Jurisdiction and Service of Process

The Rule 23 procedure includes a preliminary hearing to determine the suitability of the class as named.[21] Notice is the focus rather than service of papers for legal process. No individual is a party but the class itself is the party. This rule may represent the only way to get started on this mission. Of course, there is an "opt out" clause so any company or nation who would prefer to go it alone could be served and joined as a formal party to the pending suit, or deferred for litigation after judgment is rendered. Because of the economy and the need for clarification already stated above, it is expected that each class will remain in the notice group.[22]

The court traditionally focuses primarily on whether or not the class is well represented. This anticipates not only class attorneys and space lawyers but, also, a healthy array of *amicus curie* briefs and direct representation of the major actors by competent counsel.

The class action is designed to provide equitable remedies against the defendant class, such as an accounting and an injunction against future use of orbits.[23] However, as a matter of practice, ancillary damages may be assessed and settlements that exchange full releases for money are commonly approved by the court. No dismissal or settlement is permitted without specific court approval following a hearing on that subject.[24]

The monetary damages could be sizeable. Even a modest award of money, even a nominal $1 USD per orbit per satellite, could be substantial. There have been hundreds of satellites in various orbits for 20 years. The space treaties do not contain any statute of limitation and the operators could not claim lack of notice of this burden. All of their nations signed the 1967 treaty.

The likely defendants are Comsat, Pan Am Sat, and their sponsoring nations. Comsat is not an agency of the US government and could be served directly with legal process. It would bring in its sponsors by interpleader in such a case. The class action rule is preferred, however, in order to avoid selective litigation within the industry.

Conclusion

The Outer Space Treaty needs a lawsuit in order to test and measure benefit sharing as a burden. US Federal District Court litigation under Rule 23 class action procedures is recommended. United Societies in Space Inc. of Colorado is considering public interest litigation to that end.

References

1. Benko M and Schrogl K, "History and Impact of the 1996 UN Declaration on Space Benefits," *Space Policy Journal,"* Vol. 13:2 (May 1997), p. 139 at p. 143.

2. Jasentuliyana N, "Ensuring Equal Access to the Benefits of Space Technologies for all Countries," *Space Policy Journal,* Vol. 10:1 (Feb. 1994), p. 7 at p. 11.

3. op cite Note No. 1 at p. 140.

4. UN Agreement on the Law of the Sea, 1982, signed by the USA in 1995.

5. Doyle SE, "Benefits to Society from Space Exploration and Use," *Acta Astronautica*, Vol. 19:9 (1989), p. 749; Doyle SE, "Space Law and the Geostationary Orbit: The I.T.U.'s WARC-ORB 85-88 Concluded," *Journal of Space Law,* Vol. 17:1 (1989), p. 13; and Doyle SE, "Equitable Aspects of Access to and Use of the Geostationary Satellite Orbit," IAF, IAA-87-642, Brighton, UK, 1987.

6. Jasentuliyana N, "Review of Recent Discussions Relating to Aspects of Article 1 of the Outer Space Treaty," IISL 26 (1983), p. 7 at p. 9.

7. van Traa-Engelman H, "Commercialization of Space Activities," *Space Policy Journal,* Vol. 12:2 (May 1996), p. 119 at p. 126.

8. UN Charter, 1947, Chapter XIV, Article 96, para. 1 and 2 providing for advisory opinions at the request of various UN offices and instrumentalities.

9. Nations may not be sued without their consent to waiver of sovereign immunity.

10. Vienna Convention on the Law of Treaties, 1969, Article 26: "Pacta Sunt Servanda," meaning that every treaty in force is binding upon the parties to it and must be performed in good faith.

11. op cite Note No. 4 above.

12. O'Donnell DJ and Harris PR, "Is it Time to Replace the Moon Agreement?" *ABA Journal of Air & Space Law,* (1994), p. 3.

13. The Moon Treaty, 1979, *in passim*. For proposed sharing formulations see: Dekanozov, "Juridical Nature and Status of the Resources of the Moon and Other Celestial Bodies," IISL-23 (1980), p.5; Christol C, "The Modern International Law of Outer Space," IISL-25 (1982), p. 350; Fason E, "Some Legal Problems Regarding the Moon," IISL-23 (1980), p. 9; Sterns PM and Tennen LI, "Utilization of Extraterrestrial Resources: Law, Science, and Policy," IISL-35 (1992), p. 499; Christol C, "The Common Heritage of Mankind Provision in the 1979 (Moon Treaty)," *International Lawyer* 14 (1980), p. 434; Menter M, "Commercial Space Activities under the Moon Treaty," *Syracuse Journal of International Law and Communications,* 7 (1979-80), p. 13; Gorove S, "The Concept of Common Heritage of Mankind: A Political, Moral, or Legal Innovation," *San Diego Law Rev.,* Vol. 9 (1982), p. 402; Filiate AR, "The Commercial Space Launch Act: America's Response to the Moon Treaty," *Fordham International Law Journal,* Vol. 10 (1987), p. 763; Qiwu Z, "Some Reflections on the Most Important Principle of Outer Space Law: The Common Interests of All Mankind," IISL-32 (1989), p. 25; Cocca AA, "Space Law and the Rights of Mankind," IISL-33 (1990), p. 278; Nakamura M, "Community Interests of International Society, Especially in the Field of Space Law," IISL-33 (1990), p. 230;

Dula A, "Free Enterprise and the Proposed Moon Treaty," *Houston Journal of International Law,* Vol. 2 (1979), p. 3; Finch ER, "1979 United Nations Moon Agreement Encourages Lunar Mining and Space Development," IISL-22 (1980), p. 120; and O'Donnell DJ, "Benefit Sharing: The Municipal Model," IISL-39 (1996).

14. KPMG Peat Marwick, "State of the Space Industry: 1997 Outlook," *Space Publications* (1997).

15. op cite Note No. 2, pp. 15 and 16.

16. Op cite. Note No. 4, Sea Bed Treaty provisions contained therein.

17. op cite Note No. 8.

18. Robinson G (Fox), "Do the Space Treaties Need a Lawsuit?" *Space Governance Journal*, Vol. 4:2 (July 1997), p. 116.

19. The Antarctic Treaty, 12 U.S.T. 794; T.I.A.S. 4780; 40 U.N.T.S. 71 (Dec. 1959), entered into force in 1961.

20. Rothblatt M, "Lex Americana," *Journal of Space Law,* Vol. 23:2 (1995), p. 123.

21. Federal Rules of Civil Procedure (US), 23(c)(1), "Determination by Order Whether Class Actions to be Maintained; Notice, Judgment, Actions Conducted Partially as Class Actions."

22. F.R.C.P. 23(c)(2).

23. F.R.C.P. 23(b)(2).

24. F.R.C.P. 23(e).

IISL-97-3.09

APPLICATION AND IMPLEMENTATION, THE LAW OF OUTER SPACE.

Prepared by Dr. Harry H. Almond, Jr., Professor, the Edmund A. Walsh School of Foreign Affairs, Georgetown University. Member, New York Bar; of Gray's Inn, London, Barrister-at-Law.

ABSTRACT

The law of outer space is the enigmatic symbol of globally shared controls operating to attain common objectives of states: the law we are seeking is regulatory law - law that regulates and ensures the safety of hazardous activities and promotes the equitable access and use of the resources and energy of space. Security of states - security in the comprehensive sense that includes but extends beyond the military security that each state claims - is the primary objective at the formative stage of the public order in space. Security in the comprehensive system raise the ambiguous and continuing problems of permissible uses of force, of the overall social use of coercion, of the maintenance of international peace and security, and so on, all of which are beyond this paper.

The law of outer space relates to the activities of human beings, and to their attempts, currently through cooperative efforts building upon their technologies, to control or subdue obstacles to their activities in space. The nature of space technologies is such that as they are refined they impose the controls of natural science upon the efforts to exploit space, and these in turn promote cooperative effort under a principle of scientific economy and burden reduction. It is therefore not surprising that space law and its principles are similar to the terrestrial law: principles relating to responsibility, duties and liability, for example, vary primarily in the operative facts not in the law to be construed and applied.[1] We are of course ready to recognize that major factual elements subjected to law may, in combination designed for the application of law, lead to changes or variations in the applicable law. This law offers us the potential for establishing among states a universal law, comparable to the law of the sea, applicable in a framework of public international law to all states. Law for outer space, traceable to customary international law and to the major, law-making treaties reveals that the law in action as distinguished from the law that is on the books is the continuing outcome of a complex, and dynamic, process of claim and counterclaims among states: that law is the outcome of what states will tolerate from each other, derived from the claims process itself.

Theories of law provide guidelines to enable us to go about shaping future law from the decision flow among states. Theories generally provide us with the supportive principles, the means to draw inferences from those principles, and they gradually shape the essential doctrine and necessary concepts. Theories for the law of outer space must therefore provide us with a theory of global public and legal order. A universal law for outer space applied to operate in a framework shaped by a variety of operating theories of law is meaningless and likely to lead to instability and insecurity.

Operational components of law, like enforcement, guide the conceptual law to make it effective in action and implementation.

I

The 1967 Principles Treaty for outer space is the foremost instrument, and the primary source of doctrine, for space - so much so that courts can draw from those principles to establish specific rules from the customary behavior of states. In outer space, in particular, our concern is that of law affected by the rapidly advancing technologies critical for exploiting and subduing the threats of space. Moreover, control over the use of technologies must be distinguished in point of time and mode of application from control over the research and development essential to refining the technology as an instrument in itself.[2] This law making process has already begun to take on substantial proportions supported by both public and private law.

Among the foremost activities in space calling for regulatory controls are two: the launching of space objects, and the occupation, maneuvering, orbiting and movement through space. There is in addition an increasing interest in meeting the challenge of space debris. To establish controls and regulation over new activities, in the absence of established law or rule, it is necessary to turn to the balancing processes and balancing of policy under the standards of reasonableness. McDougal and his Associates identify this operational balancing as it is applied in the common and customary law as follows:

> And for all types of controversies the one test that is invariably applied by decision makers is that simple and ubiquitous, but indispensable, standard of what, considering all relevant policies and all variables in context is reasonable as between the parties.[3]

All lawyers through their practice and through the cases with which they are familiar, determined in their own jurisdictions are familiar with the linkage or impacts of liability to the dangerous nature of objects, and to control under law to protect the public at large and the community that peoples have formed from those who are in charge of the activities involved. Hence a primary problem in a major number of claims is the problem of assessing liability and establishing a decision framework for the allocation of responsibility for harm and for the payment of compensation or provision of other relief for harm found to have occurred.

Control over a rapidly advancing technology for space is the most urgent challenge. This involves control under law at all of the points where technology has an impact upon the activities or relations of states. This may mean that additional principles will be added to the 1967 Treaty, but added in such a way as to encourage universal support, at least equivalent to that of the Treaty. In general we seek social order support to our instruments - law - formulated for the purposes of social control. None can escape the importance of this problem, because we will need a regulatory regime constructed of sound state practices, patterns of behavior, expectations of peaceful arenas, and so on: outer space is both a locus of threats to those on earth if turned to uses in armed conflict, and a promise of great significance reaching into limitless energy and resources, and into the possibility of habitation elsewhere in space in the more remote future in the event the earth is destroyed by a major asteroid, comet, wayward planet, or otherwise.[4]

We must as soon as possible consider the changing conceptual and operational framework arising from the growing activity of states and their citizens in outer space.[5] The conceptual element will help us to reach a consensus on the operational factors. With a strengthened consensus of the fundamental notions and principles of the applicable law, we can turn, during state practice to the realities and at any given time to the "givens" of state behavior in space and to the realistic assessment or appraisal of state activities in space to our own problem of designing in an action-oriented framework a regime for the regulation and control over the programs, implemented plans and policies, institutions, procedures and practices that make up our thinking about law as a means of social control.

The givens of state behavior, reflected in their conduct and symbolized in the regulating or governing principles relating to their conduct, include those situations in which they are compelled to resort to self defense, to actions for the purpose of security or defense, and to the appropriate devices to promote the human dignity of their citizens and those under their jurisdiction.[6] A part of the task will include a continuing assessment of the custom, practice, and ultimately a critical assessment of customary international law. In short the conceptual and operational elements overlap and interact, so that consideration of one without the other is likely to be feckless.

The context of analysis is strategic in nature because our goals are controls that will shape the public order and its objectives. Strategic instruments of policy are aimed at long term goals that states seek to achieve in regulating their activities for their overall interest, and the use primarily of the strategic instruments for projecting policy and objectives. But we must also bear in mind the needs for flexibility and the necessities calling for change in the goals we adopt: some will be redirected or reshaped when conditions or the externalities of the environment of space and the activities taking place there evolve.

Private law - the making of law by and between parties following procedures similar to that of contract - is also a matter of growing importance. Private law is law among private purposes for their own needs, even though it is made enforceable and effective through public law including in the notion of public law the common law or its equivalent in other legal systems. Private law making by private agreement usually supported by public law has long been recognized by most states. In some incidents the private law may be a major factor shaping public law. The politically relevant participants, critical to law shaping, may include individuals and other entities.

These are the active participants in the law-shaping process: what they contribute to custom and practice or what is ultimately to be found in custom and practice as elements in the shared expectations of states is of great importance in how the law of outer space will develop.

Of particular interest is the perception that the activities of private entities and states may be similar, so that the law-shaping element in the conracts or arrangements of private parties or in the private actions of states will support the shaping of our public law. It also goes without saying that much of the law-shaping activity is informal flowing into state practice directly from their actions or indirectly from the activities of the private entities. Because the treaties and agreements are informal dimensions or sources of law, the law-shaping activity is not necessarily law that comes from the treaties, or formal negotiations of states in concluding treaties, or from matters taken before the international courts, tribunals and boards.[7]

Hence to have the impact and effectiveness of law for a community of states, states acting in behalf of that community must reach a consensus on the objectives and constitutive framework of their community, the constitutional framework that allocates power and authority, regardless of where that community exists - i.e., whether in space or elsewhere. States presently exhibiting differing attitudes and perspectives about the legal order are compelled to work through the amalgam created by the interaction of legal orders. Major states seeking to act, for example, to attain their own exclusive interests will run afoul of such standards and of the support of the community. Therefore the community of forces of states acting in outer space supports common attitudes about space and its exploitation. The significance of the community role lies in the fact that state practice - especially as an element of expected state conduct - is the foremost means that states invoke to communicate what they claim to be doing is permissible or impermissible.

The reach and claims for controls by the legally authorized organs among the states themselves opened a major starting point for the public law to be formulated and applied in outer space. As these controls, and the manner of exercising them, have taken root, the necessities of regulation under the controls have turned to the means that will more effectively exercise needed regulation over state conduct in space pertaining to the public order, and almost at the same time, they must also turn to the peaceful activities of states in outer space.

Put briefly, the law of outer space has gradually emerged at the earliest stage of the public law with a stress on the constitutive or constitutional law of space, allocating authority, power and jurisdiction among states and their international institutions with respect to a comprehensive framework for regulation. The public law and public order among states charges them with the responsibilities of community: hence their concern is drawn toward supporting, by force or coercion if necessary, the public global law. This is particularly important for outer space, because transgressions of these standards are likely to lead to major dimensions of instability, and to major incentives to embark upon unacceptable or impermissible power-oriented and power-controlling technologies and apply them for exclusive power.

Law of this nature has drawn upon the public international law - that is, the law largely of treaties and international agreements shared among states. This law unable to fill the gaps of regulatory authority was supplemented from the outset by the municipal, or internal public law, of states. Such law, notably in the United States, included the organic acts establishing and empowering the National Aeronautics and Space Agency or NASA. But in view of the importance of maintaining public order and the security of states in outer space, the municipal law also reached toward municipal law concerned with the security of the state. This was not surprising: The government of the United States, the oldest of democratic governments, was formed with the guidance of political thinking which included the deep concern with the security or "defense" of the nation as the *sine qua non*, appearing in such formative documents as THE FEDERALIST PAPERS. These among other things sought to promote a strong and intended bias and predisposition toward the element of security laying the support for regulating the use of force especially in the operational form of a standard that distinguished permissible and impermissible uses of force, and that promoted the institutions and practice for applying this standard as a balancing principle to accommodate the necessities in using force by putting in that balance the goals of promoting human dignity. Aristotle had early informed us that matters before us for decision necessarily involve choices - hence a balancing between those choices.

From the application of a balancing principle, the outcome was law that was shaped by use - law that became an instrument to achieve policy objectives, law that amounted to a strategic instrument with strategic implications that were refined by all of the activities that we associate with the shaping of law as well as the shaping of the instruments that strengthen or promote our law.

We can therefore anticipate as a shared concern activities involving access to space, and also access to its resources and energy. At the outset, at least, states promoted the notion that outer space was an arena to be shared amongst themselves. But they left open much of the constitutive element: the public order of outer space has not advanced in the conceptual sense beyond that attained terrestrially. Hence we must anticipate that the future of law of outer space depends upon the effectiveness of that law, upon the effectiveness of enforcement and sanction, and upon the effectiveness of law as a vehicle to achieve our goals. This general perspective is applicable to both the conceptual features of this regime of public law, and to the operational including enforcement features. And stress must be given to the high degree of flexibility, even to the notion that a major factor in adopting and strengthening law of space is not unlike the notions and principles of equity and justice, calling for states to behave in accordance with these precepts even when the precision of rule is missing.

This process of law - largely and primarily the continuing

flow, or outcome, of a process of claim and shaping of tolerances among the participants involved in space - was further characterized by the interaction of the legal process. Thus, states and their citizens found that their activities involved problems of the constitutive or global public order even when the space activities and ventures in which they were engaged were those of private as opposed to public matters. This process of law has a further characteristic: the so-called sources of law come from the traditional sources including custom, the treaties and agreements, general principles, and so on. But with our current perceptions of custom as an expanding base, we can also include in the shaping of the public law of outer space the principles inherent in the municipal legal systems throughout the world. This was already apparent in the adoption of Article 38 of the Statute, and earlier in the Permanent Court of International Justice, especially where such principles have developed in parallel, or along the same paths, the principles of those systems, and so on. It is evident that principles are formulated, applied and intended for providing regulatory guidelines, and that they are drawn in such terms as to provide for regulating a general range of activities by invoking the common objectives and purposes of the states involved.

Moreover, we can look for consensus in the resolutions taken by international bodies such as the General Assembly or Security Council of the United Nations. Other "sources" such as the positions, observations or claims of statesmen and public officials may be traced for law-making impacts, but great caution must be exercised in adopting them as "law." Even so, our primary attention is toward expectations, i.e., toward patterns of expectation that occur for repeated situations, and patterns of outcome from the regulatory regime adopted or applied in those situations. Therefore, we are aware of an expanding application of custom. But we were also aware that the various interactions of municipal public orders would have their effect, and this activity is presently occurring among states.

A further element affecting the growth of law occurs in the methodologies invoked: we are aware that law is conveyed, shaped and supported through the decisions of the major participants (i.e., participants in the relevant action under review) and is made effective through those decisions that apply enforcement measures. Full awareness of the relevance of this comprehensive problems calling for choices and decision has led us toward strengthening the decision making processes and toward the now well developed means for assessing such decisions. The decision processes have become the paramount framework for conceiving and applying law. Our concern in law arising from the decision and policy process is naturally occurring: all law is future-oriented policy, so that it is policy that must command our primary and mutual attention. Our perceptions about law, our perspectives in general, the controls adopted and applied, the assurances and consensus arising in overarching community standards of human dignity all have a place, even though the law of outer space always seems, on first glance at least, to demand a law to rule rapidly advancing technologies as the major task. Instead, the tendency thereby to put law into the "abstract" must be avoided and replaced by the recognition that space activities and the law of outer space serve human purpose and human action or they will become the conceptual elements that will serve only a few.

The problems that lead to the application and interpretation of the law of outer space raise a host of fundamental issues: the authority behind the decisions involving space law, the controls imposed, the custom leading to customary international law, and so on. Law of this nature we have learned long ago is not the easily articulated precision of the law of the rule book, or the black-letter rules. It is conveyed as part of a controlling flexible instrument of the social order. It operates and imposes its controls in part to achieve social order controls. If not it would be meaningless - in part because the authority [some use variants of the term "legitimacy"] to which we trace law is the guiding element. Because that authority is the community itself and its consent, we distinguish between such communities and their consensual characteristics, and the authoritarian communities under the control of a few, not responsible to the members of the community at large. For this reason, the tracing effort to a law of human dignity becomes, in parallel with the maintenance of international peace and security, the charging elements of the fundamental community objectives.

In a simple example, it might be argued that using these precepts, even in the absence of agreements spelling out liability in detail, such liability will exist and be imposed upon those states that act in reckless disregard in their use of space, or operate in negligence or acts of carelessness in going about their activities. Formal agreements, in writing, or in meeting the requirements of jurisdictions that impose mandates regarding what is to be in writing, are likely to be by-passed as states become accustomed to certain practices in space. When this occurs the law-shaping activities in space may be through oral statements, policy documents of the entities involved, regulations and directives of the regulating agencies, and so on. The important element here is that the consensual element is promoted and strengthened by these factors, and the law-element is strengthened by the adoption of common perspectives and interests. Such situations are governed by familiar general principles and the regulation applied under such principles innately lead to outcomes consistent with human dignity - not with excusing or justifying naked power. In a real sense our interest is in a standard of community reasonableness: outcomes or activities that fall short of this standard belie our efforts at supporting human dignity, or, in supporting the major treaty among states, to wit, the United Nations Charter.

With the foregoing in view, it is apparent that the interaction of states involving their activities, decisions and policies, is a law shaping process. Without the application of deliberate choice of objectives and deliberate direction toward common objectives, states intent upon tapping their influence or power short of using force will be drawn toward means of maintaining power, at least, most often by balancing power in either the military or "political" sense or both. Hence we can turn to law in outer space as the outcome of the actions aimed at common objectives, to the use of such standards to guide the shape or content of future law, and to the application of such standards when we seek not to apply the law as such but only the probes or tests with regard to law that we might seek to adopt. But problems that are relatively new to the larger social processes arise with these developments: the attempt, for example, to distinguish between "political" and "legal" disputes may become ambiguous and unclear, the use of dispute settling mechanisms may turn from those that rely upon the formal legal orders and their processes to those that depend upon the cooperative enterprise and the means shaped within the enterprise to overcome the dispute problems that they must inevitably face.

We must become aware that our law will be shared as part of the same or common enterprise, and that the underlying power invoked by states is the power to impose sanctions to achieve the goals sought. The overall "checks and balancing" system so often mentioned in connection with the United States constitutional practices and in shaping the structure of its constitutional law and practice is thus expanded in the claims process among states where a fundamental concern raised is the threat that other states must refrain from impermissible conduct or suffer that conduct to be treated as a form of aggression or criminal activity, or in lesser forms, as conduct not supported or supportable by the community at large.

Attempts to strengthen the common interests in the application and interpretation of law may be supported by implemented institutional measures: we can, for example, reach undertstandings to decree among all of the participating states to reach understandings regarding the formulation or shape of the municipal or international law or both prescribed specifically to serve our activities and regulation of activities in space. To ease this process into fruition, we can seek out those interests that can be expressed as common interests, formulate their regulation through treaties and international agreemennts, provide for their enforcement, and so on.

Other devices are available. But in the absence of workable institutions of this kind, there is likely to be the chaos created by overlapping and competing legal regimes, i.e., municipal or state regimes, perhaps regional regimes, and of course a global or international regime. This overlap of regimes necessarily will lead to the often intractable problems of private international law or the conflict of laws that is part of resolving issues of competing claims among legislative entities or among those promoting a particular legislative instrument.

But should they desire, states can draw upon the sophisticated developments in data banks, available through word processors and computers, and to refined procedures for finding the relevant, and authoritative, sources of the newly developing space law. The use of computers may assist us in overcoming the burdens of a detailed development of law, and also enable us to require the registration of the private agreements, contracts and arrangements that are part of the understandings and undertakings of states for activities in space. These agreements, registered, and stored in their entirety, thus become part of the practice of the newly shaped space law.

From this inquiry the necessities of law and regulation of activities in outer space become clear: states must attend to their actions and decisions in the context in which law wells up both as a constitutive leg and a leg of decision relating to the incidents or events in question. States must achieve the effective law they need for social control from close attention to the policies they are adopting partially through their claims. This is especially true in outer space where the objectives are in the form of an effective cooperative and collective effort among states.

1. Cf the Serbian and Brazilian Loans Cases: Any contract which is not a contract between States in their capacity as subjects of international law is based on the municipal law of some country. The question as to which this law is forms the subject of that branch of law which is at the present day usually described as private international law or the doctrine of the conflict of laws. The rules thereof may be common to several States and may even be established by international conventions or customs, and in the latter case may possess the character of true international law governing the relations between States. But apart from this, it has to be considered that these rules form part of municipal law. [P.C.I.J., Ser. A., No. 20/21, at 41 (1929).

2.According to McDougal and Associates in STUDIES IN WORLD PUBLIC ORDER, Hereinafter WPO, Yale Univ Press: New Haven, 1960. p.39: the challenge to scholars of international law is twofold, i.e., (1) to develop a jurisprudence - a comprehensive theory and appropriate methods of inquiry - which will assist the peoples of the world to distnguish public orders based on human dignity and public ordes based either on a law which denies human dignity or a denial of law itself for the simple supremacy of naked force; and (2) to invent and recommend the authority structure and functions (principles and procedures) necessary to a world public order that harmonizes with the growing aspirations of the overwhelming numbers of the peoples of the globe and is in accord with the proclaimed values of human dignity enunciated bvy the moral leaders of mankind.

The thesis of these commentators is that the security of nations is intimately linked to the human dignity protected and preserved in the social orders among nations, and in their commitment to human dignity in global activities and processes.

3. McDougal and Associates, WPO, p. 778 Footnote in citation omitted.

4. According to McDougal et al, WPO, p.948:

> In a world shrinking at an ever accelerating rate because of a relentless expanding, uniformity-imposing technology, both opportunity and need for the comparative study of law are unprecedented. In this contemporary world, people are increasingly demanding common values that transcend the boundaries of nation-states; they are increasingly interdependent in fact, irrespective of nation-state boundaries, for controlling the conditions which affect the securing of their values; and they are becoming ever more realistic in their consciousness of such interdependences, and hence widening their identifications to include in their demands more and more and more of t4.heir fellow men.

5. According to McDougal and Associates, it is fortunately becoming increasingly recognized that what must be compared, if comparisons ae to be relevant and useful, is not doctrine merely but doctrine and practice, not a flow of rules merely, but a flow of decisions. Comparison cannot relevantly and usefully be confined to rules alone, both because rules are not the only variables that affect decision and because, as embodiments of policy crystallizations of the past, they may not offer adequate description of the effects of new decisions. The process of decision making is, indeed, one of continual redefinition of doctrine in the formation and application of policy to ever-changing facts in ever-changing contexts. ftn omitted. At page 954-5 of WPO, op.cit.

6. According to the military courts at Nuremberg:

> The law of war is to be found not only in treaties but in the customs and practices of states which gradually obtained universal recognition, and from the general principles of justice applied by jurists and practised by military courts. This law is not static, but by continual adaptation follows the needs of a changing world. 1 Trials of the Major War Criminals before Int Mil. Trib. at Nurember 221 (1947).

The Department of the Navy of the United States has observed that the principle of operating under law and applicable to military necessity is introduced into a balancing process to gain a kind of equilibrium with regard to what uses of force are permissible and which are not:

> [It will be observed that the principle is not so broad as to permit all uses of force]. The principle of military necessity permits a belligerent to apply only that degree and kind of regulated force, not otherwise prohibited by the laws of war, required for the partial or complete submission of the enemy with the least possible expenditure of time, life, and physical resources. From U.S. Dept of Navy, LAW OF NAVAL WARFARE, Sec. 220()a) 1955, omitted footnotes.

See also, Osgood, LIMITED WAR: THE CHALLENGE TO AMERICAN STRATEGY 4 (1957), linking the military force principle with the general principle of reasonable or rational use of resources.

> It [the principle of economy of force] prescribes that in the use of armed force as an instrument of national policy no greater force should be employed than is necessary to achieve the objectives to which it is directed; or, stated in another way, the dimensions of military force should be proportionate to the value of the objectives at stake.

The principle of military necessity and the principle of humanity may each be seen to express a genuine, inclusive interest of states and peoples. Each territorial community has a most direct and immediate interest in maintaining its security, that is, in protecting the integrity of its fundamental bases of power and the continued functioning of its internal social processes from the obtrusion of unlawful violence....{in balancing the principles and objectives of humanity and necessity] There is no ineluctable necessity for postulating the priority of one of these basic, complementary interests over the other. The point which does bear emphasis is that the whole process of authoritative decision with respet to combat situations is a continuous effort to adjust and accommodate the speciifc requirements of both these interests in a series of concrete contexts.....

Engaged in the decision process in looking into those principles truly complementary in nature, that can express the opposing poles, each with its support in law, policy and human perspective, in finding the operational law in the balancing process in which exercising the reasonable standards of reasonableness the factors for each component is brought into play; in finding among the "givens" those that represent the common interests of mankind.

7. There are those that find that international law is deficient, and not real law, because it lacks sanctions. McDougal and Feliciano in their text, LAW AND MINIMUM WORLD ORDER, Yale Univ. Press, New Haven, 1961, clarifying the link of legal principles to the principles of real world activities observe:

> The opinion has been so often urged that the law of war is not law at all that it may be worth while to observe that the effective sanction which supports the law of war is the same sanction which supports all law: the common interest of the participants in an arena. The common interest which sustains the law of war is the interest of all participants in economy in the use of force - in the minimization of the unnecessary destruction of values. Unnecessary destruction of values constitutes uneconomical use of force not only because it involves, by definition, a dissipation of base values which yields no military advantage; it will also, by operation of the condition of reciprocity, result in the offending belligerent sustaining a positive disadvantage in the shape of at least an equal amount of destruction of its own values. [ft omitted.]

IISL-97-3.10

BILATERAL AGREEMENTS ON SHUTTLE CONTINGENCY LANDING SITES PRACTICAL APPLICATION OF THE BASIC CONCEPTS AND PROVISIONS OF THE OUTER SPACE TREATY AND OTHER AGREEMENTS IN AIR AND SPACE LAW

Kazuhiro NAKATANI
Associate Professor of International Law
University of Tokyo, JAPAN

[Abstract]

Since 1980s, the United States has been concluding a series of bilateral agrements on shuttle contingency landing sites agreements with Senegal, Spain, France, Chile, Morocco, Gambia, Germany and Japan in order to secure and facilitate the necessary assistance in the event of an emergency landing of the space shuttle. These agreements provide for the airport in service, notification in advance, priority in cleaning of the air space, acceptance of personnel, supply of equipment, responsibility for compensation and settlement of disputes. They embody the basic concepts and principles of the Outer Space Treaty as well as other agreements of air and space law. Therefore, naturally, they offer good examples of the practical application of the fundamental space treaties in addition to their practical importance in the event of emergency landing. Nevertheless, space lawyers have paid very little attention to these agreements. In this paper, the above-mentioned application is made clear in detail. Legal characteristics of each agreement are also considered through comparative study.

Contents

1. Major Characteristics and Provisions of the Bilateral Agreements on Shuttle Contingency Landing Sites

The Bilateral Agreements on Shuttle Contingency Landing Sites which the Unites States has concluded so far are as follows : [1] Agreement between the United States of America and Senegal, Effected by Exchange of Notes at Dakar on 15 December 1982 and 31 January 1983, Entered into Force on 31 January 1983 (TIAS 10811, Extended, Effective on 1 January 1988 by Exchange of Notes at Dakar on 2 February and 3 September 1988), [2]Agreement on Space Cooperation between the United States of America and Spain, Effected by Exchange of Memoranda at Madrid on 31 August and 4 September 1984 (TIAS 11067), [3] Agreement between the Government of the United States of America and the French Republic Concerning Emergency Use of the Combined Forces Base at Hao, French Polynesia, by the United States Shuttle, Signed at Paris on 6 September 1984 and Entered into Force on the same day (TIAS 11163),

[4] Agreement between the Government of the United States of America and the Government of the Republic of Chile Concerning the Use of Mataveri Airport, Isla de Pascua, as a Space Shuttle Emergency Landing and Rescue Site, Signed at Santiago on 2 August 1985 and Entered into Force on 6 November 6 1985 (TIAS 11248) , [5] Agreement Between the United States National Aeronautics and Space Administration (NASA) and the Royal Moroccan Air Force (RMAF) Concerning the Use of Ben Guerir Air base as a Space Shuttle Emergency Landing Sites, Signed at Rabat on 21 January and at Washington, DC on 28 January 1987 and Entered into Force on 28 January 1987 (KAV 1428), [6] Agreement between the United States of America and the Gambia Concerning the Use of Banjul International Airport as a Space Shuttle Emergency Landing Site (TIAS 12148), [7] Federal Republic of Germany, Foreign Office, Doc.No.423-455.000 USA of 22 September 1983, Ref.: Emergency Landing of the US Space Transportation System " Space Shuttle " in the Federal Republic of Germany (M.Benkö and K-U Schrogl, International Space Law in the Making, 1993, pp.120-121), [8] Agreement between the United States of America and Japan, Effected by Exchange of Notes at Tokyo on 28 January 1980 and Entered into Force on the same day (TIAS 9915), Amended and Extended by Exchange of Notes at Tokyo on 11 November 1982 and Entered into Force on the same day (TIAS 10607. On 24 January 1985, another agreement was effected and entered into force. by exchange of notes. KAV 1108. However, it has not been published yet.)

First of all, these agreements except [8] provide for specific sites for shuttle contingency landing, namely Dakar-Yoff International Airport ([1]), Rota Naval Base and Zaratoga and Moron Air Bases ([2]), Combined Forces Base at Hao, Tuamotu archipelago, French Polynesia ([3]), the Matayeri International Airport, Isla de Pascua ([4]), Ben Guerir Air Base ([5]), Yundum (Banjul International) Airport ([6]), Cologne/Bonn Airport ([7]). The contracting States shall authorize the landing and the recovery of Space Shuttles in the event of an emergency. An emergency is understood to be any event after the launching of a Space Shuttle which endangers its crew, such that there is no possibility of returning to the United States or reaching outer space ([4], Art. 1).

Secondly, as to the flight programme. Some of the agreements provide that the United States shall notify the Landing States, in advance of the launch, of the flight programme([1] Art.3, [3] Art.2, para.1, [4]Art.2, [5] Art.3, [6] Art.4). In the case of notification, the consent of the landing State is not necessary. [1] provides for the deadline (three months in advance, Art.3), while others do not have. As to [2], the Unite States shall submit the correspondent request for each flight at least 90 days in advance of the launch (Art..2). Spain shall reply 60 days before the launch (Art.4), which means that, quite different from other agreements stipulating for notification, it can refuse the request for the possible emergency landing.

Thirdly, when the emergency landing site is a airport for civil aviation, the cleaning of the air space is necessary. On this point, [1] Art.3 and [6] Art.6 provide as follows : Due to Space Shuttle's restricted maneuverability, the needed priority in cleaning of air space shall be provided by the (landing State's) authorities. [4] has a unique provision to the effect that both parties shall establish acceptable procedures to exchange information on regular flight schedules and to resolve any conflicts between the itineraries of the commercial airlines which use the Airport (Art.15).

Fourthly, as to the dispatch of personnel prior to the launch and/or in the emergency landing. [1](Art.4), [4](Art.3, para.A), [5](Art.4) and[6](Art. 7) provide that the United States shall, prior to each launch, deploy advance support personnel to the other Contracting State in order to perform landing site preparatory activities. The number of the personnel is not no exceed 30 ([1]), 23 ([4], for a period of not more than three weeks) or 40 ([5]). The United States is authorized to send (additional) personnel and equipment to prepare the Space Shuttle, its crew and its payloads for return to the United States, following an emergency landing ([1] Art.5, [3] Art.4, [6] Art.8) or in the event of an emergency landing ([4] Art.3, para. A, [5] Art.6). [4] provided that the number of the personnel will not exceed 450 persons, for not more than 120 days each, with not more than 400 persons at any one time (Art.3 para.A). As to [2], the dispatch of the personnel is included in " the specific measures which might have to be adopted in Spain before, during and after the flight "(Art.3, para. a) and subject to the consent of Spain (Art.4). Personnel whose entry is authorized by Spain shall not exceed 400 persons and 90 days (Art.7). The personnel whose entry is authorize shall be employees of the Government of the United States or its contractors and shall be citizens of the United States or citizens of the landing State ([1]Art.6, [2]Art.7, [5]Art.8, [6]Art.9) or simply nationals of the United States ([4] Art.3, para.B), unless agreed otherwise. Even when the visa is required for their entry, it can be obtained, in an emergency, at the airport ([1] Art.6, [6] Art.9). The personnel shall enjoy no diplomatic immunity ([4] Art.3 para.C).

Fifthly, as to the obligations of the landing State. The landing State shall take appropriate measures to assure the safety of the Space Shuttle and its payloads ([1] Art.7) , within the base unless the space vehicle and its payloads are placed in a support installation ([2] Art.8), or shall assist the United states or shall make best efforts in the search and the rescue operation, and in the protection of the Space Shuttle, its astronauts and all property and personnel under the United States jurisdiction ([3] Art.7, [4] Art.11 para.A, [5] Art.8, [6] Art.10, [7], [8]). The landing State shall authorize the United States to use the radio frequencies ([3] Art.6) or the communication services of the landing State and its instrumentalities shall be used ([5] Art.12, [6] Art.13). Transport to the landing site shall be arranged ([1] Art.5, [4] Art.6 para.A, [6] Art.8) . As to duties and taxes, they shall not be levied on imported property nor removal of Space Shuttle, its payloads and the personal effects of the members of the crew ([1] Arts.11 and 12, [4] Art. 6, [5] Arts. 13 and 14, [6] Art.14).

Sixthly, the United States shall reimburse the landing State for all costs incurred directly related to fulfilment of the terms of the agreement. ([1] Art.16, [3] Art.8, [4] Art.12, [5] Art.19, [6] Art.21). Indirect costs incurred by one side shall not be charged to the other Party ([5] Art.19, [6] Art.21).

Seventhly, on the responsibility for compensation for damage and loss caused. Most of the eight agreements have a provision to the effect that the United States shall be liable for compensation for damage and loss caused by the emergency landing and recovery in accordance with the terms of the Convention on International Liability for Damage Caused by Space Objects ([2] Art.10, [3] Art.9, [4] Art.13 para.B, [5] Art.11, [6] Art.12, [8]). As to [1], Art.9 concerning compensation does not mention the Convention, although Art.21 provides that the claims commission shall apply the rules established by the Convention. The extent to which the landing State shall be liable is not the same. [1] provides that " the Government of Senegal shall not be held responsible unless damage or loss to persons associated with the space shuttle program in Senegalese territory results, wholly or partially, from gross negligence, or an act, or an omission committed with malicious intent by Senegalese people " (Art.9), while [2] provides that " the Government of the United States shall waive any claims against the French Government for any damage that could be caused to its own personnel and equipment and those of its contractors " (Art.9, para.2).

Eighthly, on the settlement of disputes. Many of the agreements have a provision on consultation. Namely, either party may request consultation concerning any matter related to the interpretation or application of the agreement ([1] Art.21, [3] Art.10, [4] Art.16, [5] Art.23, [6] Art.26). [1], [3], [5] and [6] provide that the consultations shall be held ...in 60 days of receipt of such request. [3] and [4] have a provision to the effect that disputes arising from the interpretation or application of the agreement which cannot be settled through the consultations shall be submitted to the mandatory arbitration composed of the three arbitrator at the request of the either Party ([3] Art.9 para.3 and 4, [4] Art.17). As to the matters relating to claims for compensation, the Claims Commission is provided in [1] Art.21 and [3] Art.10 para.2. [2] is unique in providing that " in every case, prior to definitive determination of any claim, the concurrence of the PMAA(Joint Committee for Politico-Military Administrative Affairs) will be obtained ".

These are the major characteristics and provisions of the agreements. Also to be pointed out additionally here are unique provisions. As to [2], the United States reaffirms to Spain its long-standing obligation not to place in orbit around the earth any object carrying nuclear weapons, install them on celestial bodies, or station them in outer space in any other manner (Art. 5). As to [3], two unique provisions are worth referred. One is Art.2 para.4, which provide that the United States agrees to cease, for periods not exceeded 48 hours, all activities conducted on the Base, whenever it is required to do so by the France. The other is Art.3 para.3, stipulating that France reserves the right to refuse access to its territory or to terminate the visit of the persons for reasons of law and order or security. As to [4], the preamble refers to the environment, stating that the interest of the both Parties should allow the maintenance and normal development of the ecological environment of Isla de Pascua. Another remarkable provision is Art.5, which provides : " With the purpose of making possible and enhancing safety in the operations and activities foreseen in this Agrement, the Government of the United States shall extend and improve the present lending strip of the Airport-- . " As to [6], it is agreed that aircraft of the United States Government may be predeployed to perform whether monitoring, provide search and rescue capability, and medical evacuation support in advance of specific Space Shuttle flight missions (Art.18). [7] and [8] are concerning the emergency landing in highly developed States. They lack the detailed provisions, although confidential side agreements can exist.

In total, the agreements with developing States, namely [1] and [6] are more favorable to the United States than other agreements. Reciprocity does not exist in the agreements. However, it is easily conceivable that the favorable agreements are in exchange for the economic aid the United States gives to the developing States.

2. The Relevance to the Outer Space Treaty and Other Agreements of Air and Space Law

Art.5 of the Outer Space Treaty provides that the State Parties shall render to astoronauts all possible assistance in the event of emergency landing on the territory of another State Party. Art.2 of the Rescue Agrement provides that, if owing to emergency or unintended landing, the personnel of a spacecraft land in territory under the jurisdiction of a Contracting Party, it shall immediately take all possible steps to rescue them and render them all possible assistance. The bilateral agreements on shuttle contingency landing sites are in line with these basic provisions and embody detailed rules on emergency landing. Art.5 of the Outer Space Treaty is mentioned in [4]and [8], while the Rescue Agreement is mentioned in [4].

As to the notification, Art.1 of the Rescue Agreement includes even the case when a totally unexpected emergency landing occurs without any notification of the flight programme. However, in reality, the existence of a prior notification of the flight programme will make prepare the territorial State for potential emergency landing and lessen the psychological opposition to the emergency landing that might cause damage.

The cleaning of the air space is and should be in line with Art.9 of the Chicago Convention on prohibition zone.

When an emergency landing occurs, the personnel of space shuttle shall be safely and promptly returned to representatives of the United States in accordance with Art.4 of the Rescue Agreement. The Space Shuttle itself is returned to or held at the disposal of the representatives of the United States in accordance with Art.8 of the Outer Space Treaty and Art.5 para.3 of the Rescue Agreement.

Expenses incurred in fulfilling obligations to recover and return the Space Shuttle shall be borne by the United States in accordance with Art.5 para.5 of the Rescue Agreement. The provisions, contained in some bilateral agreements, concerning the reimbursement to them landing States for costs incurred are in line with the above-mentioned basic rule.

As to the responsibility for compensation for damage and loss caused, Art.2 of the Liability Convention provides that a launching State shall be absolutely liable to pay compensation for damages caused by its space object on the surface of the earth or to aircraft flight. [2],[3],[4],[5],[6] and [8] mention the Liability Convention. Thus, the United States owes the absolute liability. Exception to Art.2 is provided by Art.6 para.1, which states that exoneration from absolute liability shall be granted to the extent that a launching State establishes that the damage has resulted from gross negligence or from an act or omission done with intent to cause damage on the part of a claimant State. [1] Art.9 is in line with the provision.

As to the claims for compensation, the Claims Commission under Art.14 of the Liability Convention is established after a one year diplomatic negotiation. This is also applied in the case of emergency landing. in general.

3. Emergency Landing under General International Law

Landing of a space object without a permission of the territorial State can be wrongful because it might violates the territorial sovereignty, although there are controversies whether a space object can enjoy innocent passage over the foreign air space. However, according to the Draft Articles on State Responsibility prepared by the International Law Commission (UN Doc. A/51/10), the wrongfulness of an act can be precluded if the act comes under one of the following circumstances : consent(Art.29), countermeasures in respect of an internationally wrongful act(Art 30), force majeure and fortuitous event (Art.31), distress (Art. 32), state of necessity (Art. 33) and self-defence (Art. 34). The existence of a [illegible]al agreement on shuttle emergency landing sites gives a clear and ample basis of consent. The Outer Space Treaty and the Liability Convention can also be interpreted as giving consent on emergency landing. Thus, between or among the Contracting States, emergency landing itself is not wrongful.

The problem lies when a landing State is not a Party to the Treaty or the Convention, when general international law governs. As a matter of interpretation, the possible wrongfulness of emergency landing can be precluded because it can come under force majeure and fortuitous event or distress. Art.31 para.1 of the Draft Articles on State Responsibility provides : The wrongfulness of an act of a State not in conformity with an international obligation of that State is precluded if the act was due to an irresistible force or to an unforeseen external event beyond its control which made it materially impossible for the State to act in conformity with that obligation or to know that its conduct was not in conformity with that obligation. Art.32 para.1 provides : The wrongfulness of an act of a State not in conformity with an international obligation of that State is precluded if the author of the conduct which constitutes the act of that State had no other means, in a situation of extreme distress, of saving his life or that of persons entrusted to his care.

Thus, the possible wrongfulness of emergency landing can be precluded irrespective of the bilateral agreements, the Outer Space Treaty or the Liability Convention. However, the preclusion of wrongfulness does not include that of responsibility. Art.35 of the Draft Articles provides : Preclusion of the wrongfulness of an act of a State by virtue of the provisions of articles 29,31,32 or 33 does not prejudge any question that may arise in regard to compensation for damage caused by that act. Therefore, compensation for damage or loss caused is a must.

Even if the wrongfulness of an emergency landing is precluded under general international law, this does not affect the importance of bilateral agreements on shuttle contingency landing sites. Even the Outer Space Treaty and the Rescue Convention are not enough for an emergency landing. To provide details in a bilateral agreement is essential for a smooth and successful emergency landing, which insures the launching of Space Shuttle.

Thus, the contracting States of the bilateral agreements on shuttle emergency landing sites are greatly contributing and cooperating to the Space Shuttle projects. When the contracting States will be involved in the launching of their Space Shuttles in the future, the United States should conclude a similar agreement and offer emergency landing sites for them in accordance with good faith and reciprocity, the very fundamentals of international relations.

IISL-97-IISL-3.11

THE DEVELOPMENT OF FINANCING OF SPACECRAFT

Prof.Dr. I.H.Ph. Diederiks-Verschoor
International Institute of Air & Space Law, University of Leiden, The Netherlands

Abstract

When in 1958 the first Colloquium took place in The Hague, the construction of spacecraft was still in its infancy. The only states, which where involved, the United States and the Soviet Union, were paying for their own expenses. Since then not only the construction of spacecraft but also the financing of it has gone through a big development so that now all kind of financial cooperation taking also the form of leasing has come into being.

Introduction

At this Colloquium we are (re)considering the History and Development of Space Law.

With Dr. Eilene Galloway, Prof. A. Cocca and Prof. S. Gorove I am one of the surviving participants of the first Colloquium on the Law of Outer Space, held in The Hague in 1958 in one of the rooms of the prestigious Knights Hall, where the Queen of the Netherlands has the custom of opening the new year of the Parliament.[1]

Dr.Haley was the President and Dr. Pépin and I were Vice-Presidents.

Then I spoke about the responsibility of States for the damage caused by launched space-bodies but today I would like to say some words about the developments related to financing of spacecraft.

In 1958 only the United States and the Soviet Union were active participants in space constructing satellites, each paying for the expenses of their own activities. In later years the United States stressed the importance of the Shuttles and the Apollo plan, the Soviet Union giving their attention to the construction of the space station Mir, till now the only space station in action.

With the development of satellites and in addition the more and more complicated construction of space engines, while outer space commercial activities such as the development of information technology and the great demand of telephone services are increasing, cooperation in the field of financing also became more and more necessary.

Means of financing

In their book Astrobusiness, Finch and Moore cite a scheme of Raeder mentioning as financial sources corporations, banks, institutions, venture funds, individuals and the public market.[2]

Space cooperation in the field of financing can be realized by governmental agreements, memorandums of understanding and individual contracts. In the USA the Federal Communications Commission (FCC) which was established in 1934 by the Communications Act, serves as licensing authority for communications satellites.[3]

One of the first Agreements between the USA and Europe was already concluded in 1973 between NASA and ESRO. Besides scientific cooperation one of the keynote provisions is 'no exchange of fund' rule according to which each partner is financially responsible for project elements it develops.[4]

Cooperation in Europe

In Europe there is a cooperation between States in the European Space Agency (ESA). ESA has been founded in 1975 and the ESA Convention is in force since 30 October 1980. 'The ESA Convention provides that member-States have to vote for the level of resources for a five-year period, a level which is reviewed after three years, in order to carry out ESA's mandatory activities, namely the ones relating to the general budget and to the scientific program.'[5]

'The Ministers established a Council working group responsible for reviewing the tables establishing a ceiling for common operation costs, allowing Partners to plan in the longterm and with a certain accuracy their financial responsibilities based on fair contribution.'

There are also optional and operational activities. The states are free to participate in optional activities contributing also in the costs, whereas operational activities are effectuated for the requirements of other organizations as f.i. for Arianespace.

Transfers of responsibility have taken place between Eutelsat, Eumetsat and Arianespace.[6]

Before ESA, there was some scientific cooperation between its predecessors, the European Space and Research Organization (ESRO) and the European Organization for the Development and Construction of Space Vehicle Launchers (ELDO) but with the creation of ESA not only scientific research but also financing found a good base. Funding levels are approved by ESA's sovereign legislative council. Individual national space organisations are funding specific ESA spacecraft.

Farand observes that 'ESA's energies also concentrated on negotiations, formally begun in April 1994, to bring Russia into the Association of Partners. These negotiations are complex.'[7]

As 'for example Europeans are demanding the adoption of rules allowing a Partner to pay its portion of the common exploitation costs in goods and services it produces itself thereby reducing to a minimum the exchange of funds between interested parties and rules for the calculator of contributions used to finance ESA's mandatory activities and asked the Director General to elaborate proposals based on the group's conclusions for the implementation of appropriate measures as of 1 January 1997.' After a transition period running from 1997 to 1999, during which contributions in national currencies will continue on a decreasing basis, member States will have to pay their entire contribution in ECU.

Space Station

The Space Station project is a striking example of international cooperation. In 1982 discussions started between NASA's Space Station Task Force, USA, the Canadian Ministry of State and Technology, Europe (ESA) and Japanese experts for a construction and cost sharing of an International Space Station and lead to the Memorandum of Understanding in 1988. As Focke mentions, the legal instruments for this cooperation took about three years to be concluded.[8]

In Article 15 of the Intergovernmental Agreement, concluded on 29 September 1988 (into force on 30 January 1992) the funding has been regulated. The basis is to be found in Paragraph 1 whereas the other four paragraphs elaborate the financial obligations. Paragraph 1 states 'Each Partner shall bear the costs of fulfilling its respective responsibilities under this Agreement, including sharing on an equitable basis the agreed common system operations costs or activities attributed to the operation of the Space Station as a whole, as provided in the MOUs and implementing arrangements'.

In an Annex has been mentioned for which activities of the space station each Partner will provide.

Article 24 entitled: Space Station Cooperation Review states that 'In view of the long-term, complex, and evolving character of their cooperation under this Agreement, the Partners shall keep each other informed of developments which might affect this cooperation. Beginning in 1999, and every three years thereafter, the Partners shall meet to deal with matters involved in their cooperation and to review and promote Space Station cooperation'.[9]

General Cooperation

To stimulate the international cooperation and exploration and use of outer space the United Nations delivered a Declaration[10] in which Item 5 says: 'International cooperation while taking into particular account the needs of developing countries, should aim, inter alia, at the following goals, considering their need for technical assistance and rational and efficient allocation of financial and technical resources: Promoting the development of space science and technology of its applications, fostering the development of relevant and appropriate space capabilities in interested States, facilitating the exchange of expertise and technology among States on a mutually acceptable basis.'

Industrialised countries do provide considerable amounts of money for building satellite utilization capacities in developing countries, especially in the field of telecommunications and remote sensing activities.

Most cooperation is to be found in the field of telecommunication and in international launching activities. Therefore international agreements were necessary to establish institutions for specific operational space functions as Intelsat, Inmarsat, Arabsat and Eutelsat.

Although governments bear the large fixed costs of developing launch systems, some companies may incur significant costs for expension of production capabilities beyond those established by the government, replacement of worn-out tooling equipment and buildings; and improvements to existing vehicles and launch infrastructure. General Dynamic reports that it has invested more than US$ 400 million in upgrading its commercial and military Atlas programs. Governmental needs related to space actitivites could hamper commercial needs.

Financing in Organisations

It is interesting to see how the financial contributions of the States are incorporated in the operating agreement relating to the International Telecommunications Satellite Organisation (Intelsat).[11] In Article 4 it is determined that 'each Signatory shall make contributions in proportion to its investment share as determined pursuant to Article 6 of this Operating Agreement and shall receive capital repayment and compensation for use of capital in accordance with the provisions of Article 8 of this Operating Agreement.' Intelsat is in force since 12 February 1973.

Inmarsat (Convention of the International Maritime Satellite Organisation), in force since 16 July 1976, has stated in article 5 that each Signatory shall have a financial interest in the Organization in proportion to its investment share which shall be determined in accordance with the Operating Agreement.[12]

Arabsat (Arab Corporation for Space Communications) is in force since 16 July 1976. In this agreement has been stated that the contribution by the Member States to the capital of the Corporation shall be in accordance with the proportions set out in the appendix attached to this agreement.[13]

Eutelsat (European Telecommunication Satellite Organization) is an international organization based on an intergovernmental convention to which all European states can accede. It has the same regulation as Intelsat. Article 5C states: 'Each Signatory shall have a financial interest in Eutelsat in proportion to its investment share and this shall correspond to its percentage of all utilization of the Eutelsat Space Segment by all Signatories as determined under the Operating Agreement. However, no Signatory, even if its utilization of the Eutelsat Space Segment is nil, shall have an investment share less than the minimum investment share specified in the Operating Agreement.'[14]

Eumetsat (Exploitation of Meteorological Satellites) is in force since 19 July 1986. Eumetsat is an international organization based on an intergovernmental convention to which all European states can accede. The expenditure of Eumetsat shall be covered by the financial contributions of the Member States and by any other Eumetsat income (art. 9).[15]

The International Telecommunication Union (ITU) is a very important organization. As Lyall observes[16] 'the original purposes of the ITU were first technical specification, standards, compatibility and intercommunication and second, question of rates for communications.' The third purpose was to help development in countries not able to provide for telecommunications themselves. These purposes are represented in three sections.

The membership as big 'M' member of ITU can only be obtained by States. The abovementioned organizations (Inmarsat, Intelsat, etc.) are not allowed to participate. The ITU has a scale of units to regulate their finances.

The members elect the class of contribution they want to make. Although Lyall has no objections against this method of financing itself, he points out that the discrepancy between the conventional highest unit class and the minimum unit class is too big. Moreover he states that more than 80% of the ITU budget is contributed by less than 10% of its members but that every member has one vote.[17]

UNCOPUOS

Of course as a form of international cooperation has to be mentioned also the Uncopuos (United Nations Committee on the Peaceful Uses of Outer Space) established in conformity with Art. 2, section 1 of the UN Charter.[18]

Based on UN resolutions the Treaty of 1967 emerged and was followed and elaborated by four other Treaties and three Resolutions on Principles on practical appliations.[19]

However none of these regulations does contain provisions on financial means.

Launch Services

Prior to the 1980's the US Government was essentially the only provider of space launch services to the western world. In 1979 the Ariane (European satellite) had been launched. Pursuant to the Commercial Space Launch Act the US President established an office within the Department of Transportation dedicated to commercial space transportation.[20]

Competition between the US Government and Arianespace had driven prices down to levels where private US launch companies had difficulties to compete. Commercial launch markets were encouraged to compete with Arianespace.

Competition takes also place with the launching vehicle Long March in China which delivers services rather cheaply.

As Gabler observes 'Terms affecting prices of launch services need to be considered in light of the payment and financing terms attached to it.'[20]

Payments for launch goods and services are almost always made in cash usually in US dollars. Lower European interest rates enable Arianespace to compete against the US Shuttle. Financing for launches is usually arranged through commercial lending channels. Export-import loan guarantees can be arranged through national governments.

Nesgos mentions that also a number of national projects have been supported by export financing such as Brazil's Brazilsat, Mexico's Morelos and Indonesia's Palapa telecommunications satellites.[21] Business of space insurance has been almost exclusively involving communications satellites which operate in geostationary orbit. According to Nesgos the fact that debris in this area are relatively scarce is the cause that 'debris has not had a material adverse effect on the placement of insurance'.[22]

According to Space News France and Japan are with the United States, Russia, China and India important space powers, each investing the equivalent of about $ 2 billion annually in space programs.[23]

The United States

In the United States the space policy is the task of the Government. Regarding communications the Communication Act of 1934 (revised by the Telecommunications Act of 1996 which made no significant change on Telecommunications satellites) and later the Communication Satellite Act of 1962 are applicable whereas also the Federal Communication Commission is supervising the activities of these satellites. This Commission provides also rules for the owning and operating of telecommunications satellites. The NASA Act of 1958 provides for the transportation and manufacture sector.

US Launch services are covered by the Commercial Space Launch Act of 1984. Remote Sensing is regulated by the Land Remote Sensing Commercialisation Act of 1984 and by bilateral agreements whereas on 28 October 1992 US President Bush signed the Land Remote Sensing Policy Act.[24]

As an illustration of the bilateral approach the Landsat agreements may best be taken as an example.[25] These are agreements concluded between the National Aeronautics and Space Administration (NASA) of the USA and a number of countries spread over the world: Argentina, Australia, Brazil, Canada, India, Italy, Japan and Sweden. For those countries, the agreements involve the building of ground stations for the acquisition and processing of SRS data, at their own expense. The authority for NASA to enter into such agreements derives from the NASA Act of 1958, which provides 'that activities in space should be devoted to peaceful purposes for the benefit of all mankind', while Section 205 of that Act empowers NASA to 'engage in a program of international cooperation in work done pursuant to this Act, and in the peaceful application of the results thereof'. In addition to each country paying for its own share in the project, it is also agreed under Landsat arrangements that data obtained from experiments will be made available to the international scientific community. Besides, countries without Landsat facilities but within range of the ground stations will be served with information.

The Land Remote Sensing Policy Act of 1992 is a compromise between several views which have been expressed along the years, since the Land Remote Sensing Commercialization Act of 1984 concerning the operation of the US Landsat Program.[26]

Russia

The Law of the Russian Federation on Space Activities gives in Section III, Article 12 some rules about Financing Space Actitivies and Foreign Investment:

'1. The financing of space activities for scientific and national economy purposes shall be realized within the republican budget of the Russian Federation in acordance with the Federal Space Program of Russia, and figured in the republican budget of the Russian Federation as a separate item. The financing of space activities for the purpose of defense and security of the Russian Federation shall be provided by the republican budget of the Russian Federation as part of defense expenditures.

2. Space activities shall be financed from the republican budget of the Russian Federation in a purpose—oriented manner through state customers for works to create and use space technics, and shall be distributed between contractors in accordance with state contracts. The state customer and the contractor shall have the right to resort to non-budget sources of financing, including their own resources, provided this does not contradict the objectives of the space project.'[27]

The only US contribution is for a special project as e.g. for Mir.

China

China is a developing country with limited funds. Its investment into space technology is low. But the development of China's space technology has its own characteristics and the Long March is now again rather successful.[28]

Asia

In Asia, the Palapa System of Indonesia is well known, whereas India was also one of the first countries which used space activities as telecommunications and remote sensing for the education of its extensive population. As Kosuge mentions 'In 1992, Palapa offered the ASEAN countries satellite coverage of Japan, Korea, Taiwan, Hong Kong, the Pacific Islands and coastal China. The Palapa system itself has plans to expand its network further over the next five years to add new services including VSATs. Palapa will be subject to competition from a consortium comprising an Indonesian company, PT Mediacitra Indostar and International Technologies, who intend to launch a satellite, Indostar 1, which will provide TV and radio services throughout Indonesia.'

A combination of companies also launched in 1990 the Asiasat I satellite which is positioned over Indonesia and cover India, Pakistan, Thailand, Malaysia, all of China, Japan, Mongolia, Korea, the former Russian republics, and even into the Middle East.[29]

Australia

As Johnson-Freese observes 'In the 1960s, Australia was one of the world's leading nations in space research. Not only was it the third country to launch a scientific satellite from its own territory, Australian scientists were also deeply involved with experimental projects. But when the United States and ELDO stopped using the Woomera launch site in the early 1970s, Australian politicians seemed quickly to lose interest[30] in funding further space activities.'

In the mid-1980s, the State of Queensland had some interest to enable several companies under which the Martin Marietta corporation to build a space port and to finance support structures. Till now plans seem not to have been realized but new activities are in the making.

South America

As Monserrat Filho observes a New Brazilian Space Agency has been established in 1994.[31] There is some cooperation in South America for instance between Brazil and China.[32]

Conclusion

Our conclusion is that the financial means are organized and obtained now by national and international measures. Cooperation has become a necessity and will increase more and more, given the fast development and the complexity of space activities and the limitation of economic resources which have to be found in public as well as in private sources.

Notes

1. Proceedings 1st Colloquium on the Law of Outer Space, (The Hague, 1958).

2. E.R. Finch Jr. and A.L. Moore, Astrobusiness, A Guide to the Commerce and Law of Outer Space, 1985, p. 51 et seq. See for more information K.J. Madders and W.H. Thiebaut, 'Two Europes in One Space: the Evolution of Relations Between ESA and the EC in Space Affairs', Journal of Space Law, Vol. 20 (1992), pp. 117-132.

3. P.L. Meredith, 'A Comparative Analysis of US Domestic Licensing Regimes for Private Commercial Space Activities', Proceedings 32nd Colloquium on the Law of Outer Space, (Malaga, 1989), pp. 373-381.

4. E. Sadez, J.P. Lester and W.Z. Sadez, 'Modeling International Cooperation for Space Exploitation', Space Policy, August 1996, pp. 297-223, at p. 216.

5. See for text of the Convention for the Establishment of European Space Agency, N. Jasentuliyana & R.S.K. Lee, Manual on Space Law, Vol. II, 1979, pp. 259-289.

6. G. Lafferranderie, 'The European Space Agency - Present and Future', Proceedings of the 38th Colloquium on the Law of Outer Space, (Oslo, 1995), pp. 189-198, at p. 196-197.

7. A. Farand, 'Recent Developments at the European Space Agency', Annals of Air and Space Law, 1996, Vol. XXI, Part II, pp. 371-381.

8. K. Focke, 'International Cooperation in Space - Legal Aspects for the Future', Proceedings of the 33rd Colloquium on the Law of Outer Space, (Dresden, 1990), pp. 192-194.

9. See for the text of this agreement Proceedings of the Colloquium on Manned Space Stations - Legal Issues, (Paris, 7-8 November 1989), ESA SP-305, ISBN 92-9092-062, January 1990.

10. UN Doc. A/AC105/L211 of 11 June 1996. See also M. Benkö and K.-U. Schrogl, 'The 1996 UN Declaration on "Space Benefits" ending the North-South Debate on Space Cooperation', Proceedings 39th Colloquium on the Law of Outer Space, (Beijing, 1996), pp. 183-186.

11. See for text of this agreement N. Jasentuliana & R.S.K. Lee, Manual on Space Law, Vol. II, 1979, pp. 216-238.

12. See for text of this agreement N. Jasentuliana & R.S.K. Lee, Manual on Space Law, Vol. II, 1979, pp. 291—344.

13. See for text of this agreement N. Jasentuliana & R.S.K. Lee, Manual on Space Law, Vol. II, 1979, pp. 345—362.

14. See S. Courteix, 'EUTELSAT': Europe's Satellite Telecommunications' in: Regulation of Transnational Communications, European Perspectives, Michigan Yearbook of International Legal Studies, 1984, pp. 85-101.

15. See M.G. Bourely, 'EUMETSAT - A New European Space Organization for Cooperation in the Field of Meteorology', Proceedings of the 26th Colloquium on the Law of Outer Space, (Budapest, 1983), pp. 195-196.

16. F. Lyall, 'The International Telecommunication Union Reconstructed', Proceedings of the 36th Colloquium on the Law of Outer Space, (Graz, 1993), pp. 78-88. See also for the History F. Lyall, ibidem, p. 78 a.f.

17. F. Lyall, 'The Role of the International Communication Union', in: Outlook on Space Law over the Next 30 Years, (G. Lafferranderie and D. Crowther, eds.) 1997, pp. 253-267.

18. Resolution No. 1348 (XIII) of 13 December 1958.

19. F. von der Dunk, 'The Undeniably Necessary Cradle - Out of Principle and Ultimately Out of Sense', Outlook on Space Law over the Next 30 Years, (G. Lafferranderie and D. Crowther, ed.) 1997, pp. 401-414.

20. E. Gabler, 'Product and Service pricing: Launch Vehicles in Space Economics', Progress in Astronautics and Aeronautics Space Economics (J.S. Greenberg and H.R. Hertzfeld, eds.), 1992, pp. 263-293.

21. P.D. Nesgos, 'Commercial Space Law - Practical Examples Relating to Contracts, Insurance and Finance', Proceedings 37th Colloquium (Jerusalem, 1994), pp. 305-314.

22. Ibidem.

23. P.B. de Selding, Space News, 2-8 June 1997, p. 5.

24. P. Salin, 'The New Landsat Set of Regulations as per the Land Remote Sensing Policy Act (1992)', Issues in International Air and Space Law, and in Commercial Law (Essays in Commemoration of Prof.Dr. Doo Hwan Kim's Sixtieth Birthday), 1994, pp. 355-364.

25. See E. Galloway, 'Remote Sensing from Outer Space: Legal Implications of Worldwide Utilization and Dissemination of Data', Legal Implications of Remote Sensing from Outer Space, Proceedings of a Symposium (Montreal, 1975), N.M. Matte and H. DeSaussure, eds., pp. 91-104 at p. 91 a.f.

26. See Salin, op.cit., note 24, supra.

27. The text that is used is the English translation contained in the ESALEX database with the European Space Agency in Paris, accessible by Internet or modem; slightly revised as to grammatical aspects and lay-out.

28. Zhu Yilin and Xu Fuxiang, 'Status and Prospects of China's Space Programme', Space Policy, February 1997, pp. 69-76.

29. T. Kosuge, 'Satellite Communications Systems and Legal Issues in the Asia-Pacific Region', Issues in International Air and Space Law, and in Commercial Law (Essays in Commemoration of Prof.Dr. Doo Hwan Kim's Sixtieth Birthday), pp. 305-316.

30. J. Johnson-Freese, Changing Patterns of International Cooperation in Space, 1990, (Orbit series) at p. 71-77.

31. J. Monserrat Filho, 'The New Brazilian Space Agency (AEB): a Political and Legal Analysis', Proceedings 37th Colloquium on the Law of Outer Space, (Jerusalem, 1994), pp. 262-267.

32. J. Monserrat Filho, 'Brazilian—Chinese Space Cooperation: An Analysis of its Legal Performance' Proceedings of the 39th Colloquium, (Beijing, 1996), pp. 164-175.

MARS 96 PLANETARY PROTECTION PROGRAM AND IMPLEMENTATIONS FOR MARS ENVIRONMENT PRESERVATION

A. DEBUS - CNES Toulouse, France
J. RUNAVOT - CNES Toulouse, France
G. ROGOVSKY - NPO LAVOCHKINE Moscow, Russia
V. BOGOMOLOV - NPO LAVOCHKINE Moscow, Russia
N. KHAMIDULLINA - NPO LAVOCHKINE Moscow, Russia
V. TROFIMOV - LEKBIOTECH RESEARCH AND DEVELOPMENT CENTER Moscow, Russia

ABSTRACT

In the frame of extraterrestrial exploration missions and since the beginning of the Solar System exploration, it is required**, according to the article IX of the OUTER SPACE TREATY** (London / Washington January 27., 1967) to preserve planets and Earth from biological contamination. Consequently, COSPAR (Committee of Space Research) has established Planetary Protection recommendations in order to protect other worlds environment from biological contamination by terrestrial microorganisms and to protect Earth environment from back contamination. According to these recommendations, projects involved in planetary exploration programs include in their development plans specifications for biological cleanliness.

After the last planetary protection program applied for the Viking missions two decades ago, the Mars 96 mission has taken in account the last updated Cospar recommendations and a biological decontamination program has been implemented commonly by Russia and France. This program include sterilizations, biocleanings and landers integration in sterile environment in order to ensure Mars environment protection.

HISTORIC OVERVIEW FOR MARS PLANETARY PROTECTION

Cospar recommendations for Mars Exploration

Before the upcoming Mars missions planned at the beginning of the last decade of this century (Mars 96, Pathfinder, Mars Global Surveyor, Planet B), Cospar recommendations were the following:

Orbiters: According to the COSPAR Policy[1], Mars orbiter is a category III mission. The main goals is to limit the impact probability and to control the bioload. The requirements were:

- documentation related to contamination control and organic inventory,
- implementing procedures such as trajectory biaising, cleanroom integration and bioload reduction if necessary.

Landers: According to the COSPAR Policy[1], Mars landers are in category IV. The main goals is to limit the probability of nonnominal impact and to limit the bioload. The requirements were:

- detailed documentation related to microbial reduction plan and organic inventory
- implementing procedures such as cleanroom assembly, bioload reduction, bioshield
- monitoring of bioload via bioassays

Evolution of recommendations and application for Mars missions

Since the first Martian missions, the allowed probability of contamination of Mars by terrestrial microorganisms has been set by COSPAR and specified to 10^{-3} for all probes entering the Mars atmosphere[1]. Each mission had to remain within a determined probability of contamination (10^{-4} for each Viking lander for example) and the sum of all probabilities related to each planned mission was not permitted to exceed 10^{-3}. The permissable number of microorganisms for each vehicle before launch was determined by a probabilistic approach, based on the probability of survival of the terrestrial bioburden during the different events of the mission (launch, travel, UV exposure, planet atmosphere, ...etc). The most important factor in this calculation was Pg, the probability of growth and proliferation of terrestrial microorganisms under environment conditions of Mars. As knowledge of Mars increased, estimated Pg decreased and it appeared, after the Viking mission, that Pg was very low. The evolution is:

- 1963 : $P_g = 1$
- 1964 : $P_g = 10^{-3}$
- 1967 : $P_g = 10^{-4}$
- 1971 : $P_g = 10^{-6}$, probability taken into account for the Viking landers
- 1978 : $P_g < 10^{-10}$ ($< 10^{-7}$ for the polar caps and 10^{-8} for sub-surface deeper than 6 cm), probabilities estimated after the Viking missions.

Consequently, the decontamination level of a martian lander is the result of the biolooad reduction induced by different events of the mission, of the probability of survive on Mars and of the active bioload reduction performed by the project teams. The sterilization specification is consequently dependant of Pg and the decreasing of Pg in function of Mars environment knowledge induces evolution of recommendations and specifications. The consequences for missions were, for the first missions, to perform hard sterilization, for Viking to realize soft sterilization (except for life detection experiments) and for Mars 96 bioload reduction with selective sterilizations.

The Viking Planetary Protection Program

The program was built at the beginning of the 70th and one single method was specified: dry heat sterilization. All components were chosen in function of their compatibility with high temperature and the entire process has been validated and qualified. The principal events were the following:

1- Determination of the initial bioload of each Viking lander: estimated to 10^9 microorganisms

2- Integration at equipments level in cleanroom with biocleaning procedures

Dry heat sterilization (120°C - 54h) to 10^{-6} level of life detection experiments only

Estimated bioload after this step: 3.10^5 microorganisms (300 spores / m^2)

3- Landers integration in cleanroom with biocleaning procedures, integration of the bioshield

Final global sterilization step (111.7 °C - 40h) at lander level

Final estimated bioload: 100 microorganisms for each lander

With a probability of growth and proliferation of 10^{-6} (avalaible value for the Viking mission), the probability of Mars contamination for each Viking lander was about 10^{-4} (one microorganism for 10 000 landers).

COSPAR NEW RECOMMENDATIONS FOR MARS ENVIRONMENT PRESERVATION

Space Studies Board proposal

Because of the Mars exploration missions planned during this decade and taking into account new informations on Mars from the Viking mission, NASA's Planetary Protection Office requested that the Space Studies Board, represented by the concerned Scientific Community, make recommendations for this upcoming Mars missions[6,9]. It appeared, after the last Mars mission (Viking), that the probability of Mars contamination was very low and consequently, this fact was discussed in order to propose for COSPAR approval new recommendations. The conclusion was that the biodecontamination constrains decreased, but active bioload reduction stay necessary.

COSPAR new recommendations

After several meetings, held in 1991 and in 1992, the Space Studies Board proposed the following recommendations[5,8]:

The category IV is divided into two sub-categories:

Category IVa: Spacecrafts (including orbiters) without exobiological experiments as Mars 96 must be subject to at least Viking presterilization procedures, but such spacecrafts don't need sterilization.

Category IVb: Landers and vehicles carrying instrumentation for in-situ investigations of extant Mars life should be subject to at least Viking level sterilization procedures.

MARS 96 MISSION PLANETARY PROTECTION PROGRAM

Mission overview

MARS 96 mission was an international scientific mission to Mars lead by Russia, with significant participation of France and other countries such as

Germany, Finland and USA. It included an orbiter, two small scientific stations and two penetrators deployed on the Martian surface. The launch was planned on november 16., but unfortunatelly, technical problems induced the lost of the spacecraft just after launch. Nevertheless, the planetary protection program has been entirely realized, because as it will be described lower in this paper, all actions were done before the final spacecraft integration. This program, its definition, specifications, plans and implementions is the result of remarquable common work between russian and french teams.

MARS 96 specifications

For category IV missions without life detection experiments as MARS 96, requirements contain:

- cleanroom assembly: the acceptable class is 100 000 (FS 209) or better.
- bioburden reduction: the maximal acceptable level of decontamination is 300 spores per square meter and 3.10^5 spores per vehicle (values delivered from VIKING data). Only the external and internal exposed surfaces are concerned. It must be noticed that MARS 96 landers are smaller than the Viking landers and consequently, the representative maximal level (more stringent case) is 300 spores per square meter and this last value is define as the Mars 96 specification. The specification is given in function of a spore level, because spores are the most resistant of all microorganisms. Consequently, performing a method able to reach a level of 300 spores per square meter (from an initial level of about 10^4 spores/m^2 to 10^6 microorganisms/m^2) will kill all other microorganisms.
- selective bioburden reduction. Large surfaces and subsystems capable of disseminating terrestrial life on the planet (parachutes, balloons, wheels, guiderope external structures, ...) must be sterilized.
- the limit on probability of accidental impact by hardware other than lander, probe or orbiter must be less than or equal to 10^{-5}.
- recontamination prevention.

Remarks:

- the required recontamination level concerns each category IV orbiter, probe, vehicle or landing system. A landing system is defined as all subsystems included in a single landing event.
- for orbiters of category III as MARS 96, trajectory biaising and orbit lifetime computations are required. But Mars 96 orbiter did not need any implementation of sterilization procedures because the probability of spacecraft crash did not exceed 10^{-5} and because it's orbit was in accordance with quarantine requirements (orbit lifetime with 0.9999 confidence for the first 20 years and 0.95 confidence during the next 20 years).
- in case of a lander crash, the affected area must be limited and localized.
- processes and techniques used for establishing burden levels must be the same as those used for Viking or other improved assay methods [7]. In each case, the project teams have to demonstrate equivalence of techniques other than those used for Viking.

Mars 96 Planetary Protection Plan

The general integration philosophy of landers is described on table 1.

Table 1: implementations for Mars 96 landers during different step of integrations

	INTEGRATION PHASES	IMPLEMENTATIONS
1a	Landers equipments and sub-systems	Sterilization or biocleaning + ultraviolet rays exposure Sterile packaging Microbiological control
1b	Experiments	Sterilization or biocleaning + ultraviolet rays exposure Sterile packaging including sterile test interfaces Microbiological control
2	Landers integration in NPO Lavochkine Moscow	Sterile class 100 cleanrooms Sterile procedures for equipement and operators (clothes, tools, ...) Surface biocleaning / UV exposure Microbiological control Sterile packaging
3	Recontamination prevention	Bioshields

MARS 96 PLANETARY PROTECTION IMPLEMENTATIONS

Landers integration

According to the Planetary Protection Plan described on table 1, implementations can be divided into different levels:

Instrument and Subsystem level: Equipment, instruments, and subsystems have been sterilized to 10^{-3} level (1 microorganism for 1000 items) or better, or biocleaned to a level of less than 300 spores per square meter. The applied methods are described on table 2. All biodecontamination or sterilization were subject to the following scheme:

1/ After a first compatibility study, sterilization methods are chosen in function of each equipement

2/ For each method, procedures have been used in order to reach the required decontamination level (verification of microbiological efficiency). For all methods, procedures have been validated.

3/ For each method, the compatibility with the concerned sub-systems or instruments have been validated in order to verify that the effect of sterilization on equipments was acceptable. Validation tests have been performed on critical materials and for the french and some of the russian payload, the hydrogen peroxide gas plasma has been qualified at instrument level[2].

4/ The decontamination level was controlled on every sub-system or instrument using the following procedure:

* for sterilized items:

- initial bioburden assessments
- packaging
- sterilization through the bags (as usual) with microbiological efficiency indicators (sheets of specific materials loaded with a definited number of reference spores (typically a few millions of the most resistant spores)
- analysis of microbiological efficiency indicators and determination of the sterility levels

* for biocleaned / UV exposure equipments

- initial bioburden assessments
- final bioburden assessments
- sterile packaging
- determination of decontamination levels

Packaging: Because of testing on different sites, equipments must be packaged in a minimum of two bags in order to preserve the cleanliness conditions. For experiments a special packaging system was used in order to conduct acceptance tests without impacting the biological cleanliness. A third antidust bag was required for final protection during transportation.

Landers level: The integration site was a cleanroom in class 100 000 in which the small stations and penetrators integration was conducted in NPO Lavochkine (Moscow) under two 100-class laminar flow installations. They were required to control surface recontamination and to ensure a surface bioburden level of less than 300 spores per square meter. The contamination level was controlled by the means of bioburden assessments, periodic surface cleaning, and by use of sterile procedures for operators (clothes, handling), equipment, and the working environment. A bioshield finally covered each probe in order to prevent recontamination during the integration steps.

Table 2: Used decontamination methods for Mars 96 landers sub-systems and experiments

DECONTAMINATION METHODS	CONCERNED ITEMS
Gamma rays sterilization	Landers parachute and airbags systems Some of small station and penetrators sub-systems
Dry heat sterilization	Finnish meterological sensor of the small stations
Hydrogen peroxide gas plasma sterilization	Small stations french payload Penetrators ARS abd BSS experiment units Some of penetrators sub-systems
Biocleaning (alcohol or sporicide cleaning completed with ultraviolet rays exposure) performed in sterile class 100 cleanroom	Small stations finnish experiment Small stations US experiment Small stations russian experiments Some of small stations and penetrators subsystems and housekeepings

* *Class 100,000 Cleanroom:* All integration operations were done in a class 100,000 cleanroom (see Figure 1). This facility has been constructed to have the 100,000 features, particularly: positive pressure, dust filters, access by SAS (small autonomous stations), class 100,000 clothes, and particular cleanliness control (cleaning procedures). In this large cleanroom, two class 100 laminar flow facilities have been installed. Since the class 100 cleanroom has transparent walls, most of the control instruments and operators necessary for control during integration may stay outside the class 100 cleanrooms thus avoiding possible recontamination.

* *Class 100 Laminar Flow Cleanrooms:* The integration of the 2 small stations and the 2 penetrators were done in two class 100 cleanrooms (see Figure 1) as defined by FS 209B.

Figure 1: Integration site layout in NPO Lavochkine

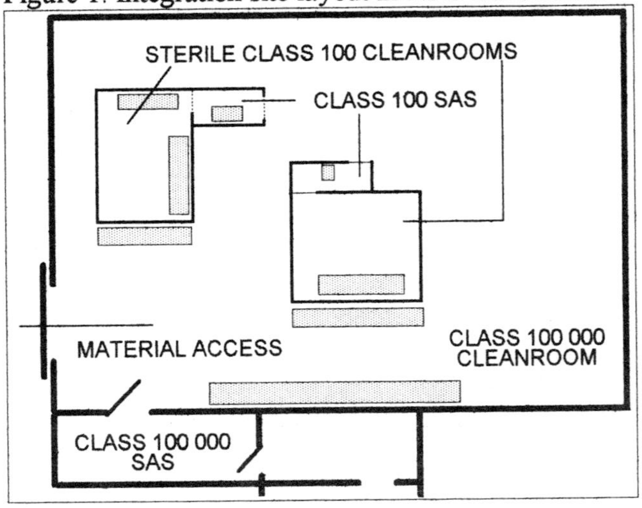

Positive pressure was induced by a laminar flow blowing from top to bottom, which avoids particles (dust and microorganisms) settling. The size of the integration area under the laminar flow was about 10 m^2 for a height equal to 3.5 meters, allowing the integration of two landers at the same time. The access to the class 100 cleanrooms was through a common SAS joining each cleanroom across a corridor. An integration official was in charge of the cleanroom utilization and of the application of procedures. Access to this area, for operators and equipment, was done by means of the SAS where operators change their 100,000 class clothes for the sterile class 100 clothes, according to stringent changing procedures. The transparent walls of these cleanrooms lended a contribution to cleanliness because all ground support equipment remain outside while their screens or numeric information displays were visible from inside.

* *Cleanroom Cleanliness Control and Maintenance.* The integration area was kept clean by means of a daily cleaning with alcohol and sporide of all working surfaces (tables, chairs, etc.) except probe surfaces and other equipment (tools, control instruments, electrical cables, etc.) as well as a weekly cleaning with special disinfecting agents for the cleanroom structures.

* *Material Access Conditions.* Equipments were moved from the storage rooms into the 100,000 class cleanroom according to technological plans. All equipment and probe instruments have been decontaminated, as defined by the specifications. The external bag was removed for acceptance tests and the internal bag was first cleaned in the SAS and then opened in the class 100 cleanroom. All surfaces of other control instruments, tools, etc. have been cleaned with alcohol and/or sporicide wipes.

* *Sterile Clothes.* Class 100,000 clothes were typically suits designed for this level of cleanliness. For the class 100 laminar flow cleanroom some specific sterilized clothes were worn in the class 100 SAS. At each sterile cleanroom access, operators have to wear decontaminated overboots, overalls, hood, mask, and gloves. Except for the gloves which were thrown away, all other clothes were decontaminated after each change. Taking into account that the mission specification did not impose absolute sterility, the clothes were not sterilized, but were washed and completely decontaminated using chlorine water solutions. Typically one wash in ten was followed with actual sterilization. In this case, the clothes were washed and dried before packaging into bags. The bags used depend on the sterilization method. The two methods used were:

- steam sterilization (20 min at 120 °C)
- gamma sterilization (25 kGy).

* *Microbiological Recontamination Control:* Contamination of the lander surfaces were monitored by means of periodic microbiological assessments performed by the Lekbiotech Research and Development Center of Moscow. Bioassays were made according to an assessment plan and the

results of these analyses determined the method of surface biocleaning. Assessments were perfomed usually twice a week on landers surfaces and cleanrooms, and more often when it was required by integration processing. The bioassays method was similar to the Viking one and levels were given as an average of 6 to 8 assessments. In function of the determined levels, corrective actions were defined (biocleanling, time of ultraviolet exposure) in order to keep the following specifications:

- class 100 000 cleanroom: less than 2000 micro-organisms/m^2
- class 100 sterile cleanroom: less than 250 micro-organisms/m^2
- lander surfaces: less than 250 microorganisms/m^2

* *Surface Cleaning:* External surfaces were cleaned by the following methods:

- - Ethyl alcohol was used daily to remove spores. Since alcohol is not a sporicide the alcohol flasks were typically changed every 5 cleanings to guard against contamination of the ethanol by spores.
- - Ultraviolet exposure was used as a sporicide in order to complete biocleaning in case of excessive recontamination. Surface exposures occurred overnight and were stopped two hours before starting work in order to flush from the room the ozone generated by the process. The results of the analyses and estimates of the total contamination produced by all the integration factors (including microorganisms settling on the surfaces of the installation coming from the air, microorganisms produced by the personnel during their work, microorganisms coming with the supplied units and elements, and microorganisms introduced by the technological process itself) at NPO Lavochkine allowed the calculation of the necessary minimal power levels of UV radiation for different cleanliness classes. The results of this study enabled the choice of the characteristics, the number of the UV lamps and the exposure durations. In anycase, protocols were defined in order to reach on probes surfaces a level of 100 spores / m^2 just after exposition.

* *Packaging and Transport of Landers:* To carry out transport operations, the descent modules were packaged into a sterile envelope and placed on the transport devices within a sealed shield with a respiratory valve. These devices were placed in a special container, including cooling and dust / spores filtration subsystem, during transportation to the launch pad in Baikonour.

* *Bioshield Description:* After their reintegration, the landers were placed under a metallic cover. Gas exchange between both sides of the bioshield occured through a valve equipped with biological filters.

* *Results:* Before integration, all subsystems and experiments presented a contamination level of less than 300 spores / m^2. During the integration, the surfaces recontamination was limited and controlled by the above described means. Nevertheless, the level sometimes increased above the specification and correction were brought, as required, by surfaces biocleaning and ultraviolet exposure. The final values were the following:

Small station flight model 1:	270 spores / m^2
Small station flight model 2	210 spores / m^2
Penetrator flight model 1:	240 spores / m^2
Penetrator flight model 2	270 spores / m^2

Documentation

Documents related to planetary protection was included in the project documentation. The principal types of documents were specifications, plans and procedures. Detailed documents are listed in the summary on table 3.

Added to the procedures, it is important to note that inventories of materials, and particularly of organic materials were done through the material lists of landers.

Studies

The principal goals of studies were to validate sterilization methods and procedures and to establish protocols. Detailed studies are described in the summary on table 3.

Manpower

Except the subcontractors the planetary protection team includes 5 peoples. Their tasks were the following:

- NPO Lavochkine planetary protection responsible: it was in charge, commonly with the french responsible, to define specifications, procedures and to follow the work of its subcontractors. During landers integration, its task was to ensure the decontamination specification.

Table 3: MARS 96 Planetary Protection russian / french cooperation scheme

TASKS	RUSSIA	FRANCE
IMPLEMENTATIONS		
- Class 100 000 cleanroom	X	
- Class 100 cleanroom for small station integration		X
- Class 100 cleanroom for penetrators integration	X	
- Cleanrooms sas	X	
- Class 100 sterile clothes		X
- Usual equipment for cleanrooms (carpets, bags, sterility indicators......)	X	X
- Ultraviolet rays sterilization device	X	
- Hydrogen peroxide gas plasma sterilizator		X
- Gamma rays implementations for validation tests	X	X
- Gamma rays implementation for flight models equipments sterilization	X	
- Cleaning agents and sporicides	X	X
- Microbiological assessments	X	X
DOCUMENTATION		
- Mars 96 quarantine program	X	
- Sterilization / biodecontamination specifications	X	X
- Class 100 cleanroom biocleaning procedures		X
- Class 100 clothes biocleaning and sterilization procedures		X
- Cleanroom access procedures for operators and equipment		X
- Class 100 sterile clothes wearing procedures		X
- Payload bioburden assessments procedures	X	X
- Landers bioburden assessments procedures	X	
- Surfaces cleaning procedures	X	X
- Ultraviolet irradiation protocols	X	
- Microbiological assessment plan	X	
- Landers sterile integration plan	X	X
- Decontamination certificates	X	X
- Hydrogen peroxide gas plasma sterilization procedure		X
STUDIES		
- Equipments and materials compatibility with sterilization methods	X	X
- Hydrogen peroxide plasma gas sterilization cycle biological validation		X
- Materials validation testing with hydrogen peroxide gas plasma sterilization process	X	X
- Materials validation testing with gamma rays sterilization	X	X
- Ultraviolet surface dose exposure times	X	
- Specific alcohol / formaldehyde sporicide validation	X	X

- CNES planetary protection responsible: it was in charge, commonly with the russian responsible, to define specifications, procedures and to follow the work of its subcontractors. It was supported by CNES laboratories in order to performed validation tests.

- NPO Lavochkine cleanroom responsible: it was in charge of the particle and biological cleanliness of all integration implementations. Its task was to control if all procedures were correctly applied.

- Two NPO Lavochkine representative in charge of studies

Associated subcontractors

Microbiology and sterilzation are not specific spatial tasks and consequenly, subcontractors participated actively to this programs and it is important to note the remarquable implication and interest that all of them brought in this project. The subcontractors were:

- Lekbiotech Research and Development Center of Moscow: It was in charge of active sterilization and of microbiological assessments done in NPO Lavochkine.

- Paris V Pharmacy University: It was in charge of bioassays performed in France (french and a part of russian payload) and support CNES for procedures writing (sterilization, cleanroom cleanliness control, sterile clothes biocleaning).

- Johnson and Johnson Medical of Chatenay Malabry (France): it performed and followed the sterilizations by hydrogen peroxide plasma gas.

- Caric Mediris of Orsay (France): it performed sterilization by gamma rays for validition tests, with the help of its belgian unit. Most of them were done for the validation of the martian aerostat. This part of the project was unfortunatelly cancelled in 1994.

PARTICULAR REQUIREMENTS AND CONSTRAINS FOR MARS EXPLORATION MISSIONS

The most of scientific programs need classic integration procedures according to requirements for dust cleanliness, compatibility with space environment and with mission specifications. Planetary exploration programs, and particularly martian programs for which planetary protection is required need unusual tasks. They must be taken into account early in the mission program in order to planned all works, personnals and funding. The specific tasks are:

- to include planetary protection in mission development plan
- to build specifications including biological decontamination requirements
- to build specific procedures for biological cleanliness at different levels
- to make studies in order to validate methods and procedures
- to specify all materials choice taking into account the effect of decontamination methods
- to follow the tasks of subcontractors. Subcontract are indispensable to support project teams for specific and unusual tasks as sterilization and bioburden assessments.
- to realize integration in unusual environment. In sterile cleanrooms, operators have to deal with clean and sterile procedures for wearing, handling and working. They must at every time control their work and control themselves, bodies entirely covered except eyes. Working in such conditions during a few hours is difficult and teams rotations must be organized in order to limit the duration of teams work to four consecutive hours maximum.

Taking in charge these tasks need additionnal personnal and funding. The total cost of planetary protection tasks for the entire mission has not be calculated, but on the french side, the CNES participation may be estimated, including personnals, at about 1.5 MF (300 000 USD).

BENEFITS FOR FUTURE MARS EXPLORATION MISSIONS

At this time, american MGS 1 and Pathfinder has been subject to planetary protection requirements. For future missions, all works will be usefull for future Mars exploration missions, because new sterilization methods for space application have been validated. Working in sterile conditions, with stringent procedure will also be usefull, and particularly for the next sample return missions for which hard specifications will probably be built.

REFERENCES

[1] Debus, A., J. Runavot, G. Rogovski, V. Bogomolov, J.C. Darbord, M. Ivanov and A. Victorov, Planetary Protection Program for Mars '94/'96 Mission, *Adv. Space Res.*, **18(N1/2)**, 323 (1995).

[2] Debus, A., J. Runavot, G. Rogovski, V. Bogomolov, and J.C. Darbord, Mars '94/'96 Mission Planetary Protection Program, Techniques for Sterilization and Decontamination Level Control, IAF-95-I.5.10, Oslo, October 02-06, 12p (1995).

[3] Debus, A., J. Runavot, C. Tarrieu, G. Rogovski, V. Bogomolov, Mars '94/'96 Mission Planetary Protection Program, IAF-93-U4.565, Graz, October 16-22, 11p (1993).

[4] Devincenzi, D.L., P.D. Stabekis and J. Barengoltz, A Proposed New Policy for Planetary Protection, *Adv. Space Res.*, **3(8)**, 13 (1983).

[5] Devincenzi, D.L., P.D. Stabekis and J. Barengoltz, Refinement of Planetary Protection Policy for Mars Missions, *Adv. Space Res.*, **18(N1/2)**, 311 (1996).

[6] NASA, *Planetary Protection Implementation on Future Mars Lander Missions*, NASA CP-3216, (1993).

[7] NASA, *Viking '75 Program: Microbiological Assay And Monitoring Plan*, NASA M-75-148-0, (Feb; 1974).

[8] Rummel, J.D., M.A. Meyer, A Consensus Approach To Planetary Protection Requirements: Recommendations For Mars Lander Missions. *Adv. Space Res.*, **18(N1/2)**, 317 (1996).

[9] Space Studies Board, National Research Council (US), *Biological Contamination of Mars: Issues and Recommendations*, Task Group on Planetary Protection, National Academy Press, Washington, D.C. (1992).

[10] Debus, A., J. Runavot, G. Rogovski, V. Bogomolov, J.C. Darbord, M. Ivanov, V Trofimov, N. Khamidullina and B.J. Plombin, Mars 96 small stations biological decontamination, *Adv. Space Res.*, waiting for edition

Legal Analysis of Military Observation from Outer Space

HASHIMOTO, Yasuaki*
The National Institute for Defense Studies
Tokyo, JAPAN

Abstract

While the legality of military related observation from outer space, done by reconnaissance satellites, was not considered in detail, the actual activities have played an important role in the cold war era. In legal sense, these activities remains in the gray area. In the United Nations, however, there can be seen a new tendency for introducing more ideal peace and security in outer space. Through this tendency, the legalized activities might be thought. On the other hand, civil remote sensing technology now can provide the adequate data and information for security purposes. In order to promote its active role in international peace and security, new legal framework for those usage will be needed in near future.

Introduction

In the field of military related observation activities from outer space, while their legal implication has not fixed up yet, *de facto* usage is continuing. This paper looks at the general discussion around this matter and considers one of the fundamental principles which regulate the future activities. And also, this paper points out some problems rising from the civil remote sensing activities from the military viewpoint.

Military Observation Activities from Outer Space

The idea that for the national security, the outer space is the most adequate sphere for observing the military related activities on the earth surface began when the space utilization started in 1950's. Reconnaissance satellites have mainly been used for that purpose.

The space law system which consists of some treaties including the Outer Space Treaty of 1967 set no concrete regulations regarding the existence and operation of reconnaissance satellites. The activities that have been prohibited are to put the weapons of mas-destruction on the earth orbit and military activities on the moon and other celestial bodies (Article 4).

The views expressed in this article are those of author and do not represent the ones of any organization with which the author is affiliated.

* Associate Professor, The National Institute for Defense Studies, Japan. Member, IISL. Member, SOLAPSU(Society for the Study of Law & Policy on Space Utilization), Japan.

The U.S.A. who owns and operates the reconnaissance satellites for the national security expressed that though the Outer Space Treaty forbade some activities clearly provided in the article 4, the other activities should be done according to the international law including the United Nations Charter (Article 3). In this opinion, the U.S.A. can launch and operate the reconnaissance satellites, because the international law did not treat that kind of activities and reconnaissance does not violate sovereignty of the subjacent countries. They insisted that the observation from outer space should be allowed like the observation and information gathering from the high seas because they both are done from the areas to which national jurisdiction does not extend.

In fact, the latest US National Space Policy in 1996 declared that "the United States rejects any limitations on the fundamental right of sovereign nations to acquire data from space."[1]

Under this interpretation and following practices, the absence of explicit claims against the military data gathering from outer space means the implicit acceptance in legal sense. The only one protest by Egypt against The U.S.A. when the Suez area was detected by US reconnaissance satellites in 1970 is thought to be an exception[2].

On the other hand, there can be another idea that the activities which are not prohibited directly should be controled by other basic principles in the space law system. Whether reconnaissance activities violate the common interest principle, as a main pillar of space law system, in the article 1 of Outer Space Treaty as well as the article 3 is considered. In general, it is an interpretation assumed to be a wide concept that this principle contains peaceful use of outer space. The understanding that international cooperation and peaceful use are included in common interest principle was born in the recent United Nations resolutions adopted by General Assembly. For instance, one resolution during the 51st session is "Recognizing the growing scope and significance of international cooperation among States and between States and international organizations in the exploration and use of outer space for peaceful purposes" in the sixth sentence of preamble[3] . And this understanding is also shown more clearly in the following phrase; "Deeply convinced of the common interest of mankind in promoting the exploration and use of outer space for peaceful purposes and in continuing efforts to extend to all States the benefits derived therefrom" in the next resolution of the same session [4]. In addition, the other resolution titled "Prevention of an arms race in outer space" is, at the beginning of the preamble, "Recognizing the common interest of all mankind in the exploration and use of outer space for peaceful purposes." [5] Here, the trend in the United Nations can be read. That is, the common interest principle should be activated by the peaceful use and international cooperation concepts. In this context, for considering whether all the space activities including observation activities by reconnaissance satellite can conform with the common interest of all mankind, two basic concepts, peaceful use and international cooperation will become a judgment standard for accepting its legality. Therefore, a triangle shaped tester which is composed of common interest of mankind on the top, peaceful use and international cooperation on both sides of the base, can be thought for the legal judgement.

Using this tester, there might be the room for questioning legality of the present reconnaissance satellite activities because those are for the national security purpose, not for the international cooperation. If so, the above mentioned protest by Egypt can be treated as an evidence for denying the implicit acceptance of military observation from outer space.

Actually, each agreement of the outer space law system did not restrict observation activities of reconnaissance satellite directly, though a number of reconnaissance satellites exist and are used for mainly national security. The existence of reconnaissance satellites themselves has been ambiguous because the nations have not informed the Secretary General of the United Nations of real functions of reconnaissance satellites according to the registration Treaty. Until now, military observation activities from outer space has not been treated by any space law treaties, and remained in "gray" existence in the outer space law system.

On the other hand, observation activities have been legalized in part. Concretely, it is an allowance in disarmament treaties including the ABM treaty (1972) as verification means. In addition, the International Satellite Monitoring Agency (ISMA) proposal of 1978 where the use of reconnaissance satellites as international treaty verification means and crisis observation measures was shown does not collide with the Outer Space Treaty though it is not established. There also is a similar allowance about various fields for the maintenance of international peace and security like UN Peace Keeping operations. Because these activities are thought to be legalized usage through the above-mentioned judgment tester. It can be evaluated internationally that the allowance as verification means contributes for improving the effect of disarmament, decreases the possibility of a military collision and is related to the crisis evasion. The verification and crisis monitoring by ISMA will contribute to the international peace and security more widely than verification between the contracting parties and be connected with common interest of all mankind. Observation from outer space for Peace Keeping activities is sure to become common interest of mankind for the peaceful purpose, too. Those activities from outer space with such specific purposes can become one of the legal precedent for the future observation activities, because such activities can successfully pass the test (peaceful use and international cooperation) of the common interest principle derived from the article 1 of the Outer Space Treaty.

Civil Observation Activities from Outer Space

The civil remote sensing activity which uses now a technology similar to reconnaissance satellites in part did not become a problem regarding their existence. The problem treated in international space law was the method of distributing the data and information. Remote sensing principles of 1978 adopted by the United Nations[6] shows basic principles about the distribution of the data and information. These principles are thought to follow the common interest direction showed in the Outer Space Treaty of 1967.

Recently, high resolution civil remote sensing satellites appears because of transfering military technology after the collapse of the cold war structure in the end of 1980's. However, even if those satellites achieve the high resolution, fundamental purposes of civil remote sensing is not security. It is an improvement and control of the natural resource, the land use, and the environment (Principle I). It seems clear that the Outer Space Treaty allows that idea as a peaceful use and international cooperation.

When the nations use those civil remote sensing data and information for their national security, there is no legal problem in international law. First of all, the data and information are marketed in general according to the UN remote sensing principles. Principle XII and XIII provide that those are distributed to any nation on an indiscriminately basis. After purchasing those data by some countries, the data usage is legalized by the general international law. The activities done by the Western European Union Satellite Centre which uses marketed remote sensing data for regional security is a good example to be pointed out. This centre should be recognized positively because it plays an important role in European regional security and cooperation [7].

It is time when cause the problem that information about the specific region where the tension rises is opened to the public by the third parties like media. There is a possibility to increase the needless tension and cause the actual collisions and disputes. An arbitrary opening such information to the public might be an unstable factor for the peaceful settlement and not agree with common interest principle of the space law.

Actually, the reasonable restriction in distribution should be considered in the above mentioned case though the data and information are basically accessible by any nations according the UN principles. There also is a problem "who does restrict data distribution?" It is assumed that the countries who are responsible for supervising the private entities' activities and/or the United Nations (the Security Council) who are responsible for international peace and security decide, judging the influence of the related data according to the Outer Space Treaty. In addition, the previous consent of the country which receives the disadvantage by opening information to the public, should be considerable. Oppositely, the data supply which profits one country of the disputed area cannot be admitted by the common interest principle. These problems are not discussed relating the civil remote sensing, when the civil activities have not been connected directly with the security problem in initial stage. The access right secured in the United Nations principles concerning the remote sensing data and information is the one which related to the problem of sovereignty over the development which was the concern from developing countries. Now since the civil remote sensing satellites achieve the high resolution ability, the new influence to the international peace and security will be the problem to be discussed according to the common interest of all mankind principle.

Conclusion

There is room that the new style of lawful observation from outer space is considered because the clarification of common interest principle is progressing in the United Nations while there is no direct regulations which limit and prohibit observation activities by the military satellites as a whole. Though there also is no rules to forbid the civil remote sensing for the military purposes, when the international peace and security seems to be harmed by high resolution data and information, the additional restriction is thinkable. And it will be the time in near future that the positive use of civil remote sensing for the international peace and cooperation should be considered.

Notes

1. National Space Policy, National Science and Technology Council, The White House, 1996, (3).
2. Frankfulter Allgemeine zeitung, Montag, 7 September 1970, Nr.206, Seit 4.
3. A/RES/51/122 on 4 February, 1997.
4. A/RES/51/123 on 10 February, 1997.
5. A/RES/51/44 on 7 January, 1997.
6. A/RES/41/65 on 13 December, 1986.
7. Y. Hashimoto "MULTILATERAL VERIFICATION ORGANIZATIONS -CASE OF WEU SATELLITE CENTRE", The Proceedings of the 38th Colloquium on the Law of Outer Space, pp.262-265, AIAA, Washington, DC.

IISL-97-IISL 3.14

DANGERS FROM ASTEROIDS AND COMETS: RELEVANCE OF INTERNATIONAL LAW AND THE SPACE TREATIES

by

Eugene Brooks
Plainview, New York, USA

ABSTRACT

This paper sets out the scientific basis of the danger to Earth from impacts with asteroids and comets, and whether the hazard requires international action. It then summarizes cooperative searches underway to detect these bodies, and proposed interception techniques.

International laws and the space treaties are examined to determine whether obligations have been created requiring States to detect celestial threats, to notify other States and population of danger and to consult with each other to mitigate any danger.

The paper recommends a stepped up coordinated international network to detect rogue asteroids and comets but suggests caution in developing measures to deflect them.

Treaties banning nuclear explosions in space may have to be reconsidered. Ultimately, international structures may be needed to decide whether, when; and how to test deflection methods, to intercept asteroids if required, and to prevent potential misuse or miscarriage of such activities.

Finally, the desirability of combining asteroid and comet surveys to avoid impacts with the appraisal and use of their resources, is recommended.

SCIENTIFIC BACKGROUND

Barring an unexpected apparition of a large asteroid or comet on collision course with Earth, one about 1 kilometer or more in diameter, the chance of a total catastrophe is one third of a million years away[1] It is estimated that based on the geological record of platinum group layers, especially iridium, and the record of marine extinction (plankton) there have been about 24 peaks - 5 major and 19 minor - extinctions of over 25% -90% of extant genera in the last 540 million years[2]

The best known extinction is the one that took place about 65 million years ago at the boundary between the Cretaceous and Tertiary (KT) periods. The dominant theory is that the impact of a 10 kilometer asteroid or comet traveling at least 30 km per second on the Yucatan Peninsula in Mexico created an immense shock wave, an expanding fireball of vaporized rock, and a further ejecta of rock which heated the region of the air and surface. A blanket of dust settled down through the atmosphere, blocking sunlight, chilling the earth and killing most life. Months of cold persisted, then water vapor and carbon dioxide released from calcium layers of rock, led to a greenhouse effect heating earth, while large quantities of acid rain poisoned air, water, and soil. This extinction is thought to have killed the dinosaurs and 60% of other animals and plants, leaving earth to those smaller mammals that survived[3]

An impact of Yucatan magnitude is thought to occur once every 23 million years. A long interval surely but, smaller objects crash down at greater frequencies - once every century a 50 meter (104 foot) body capable of destroying a city if it landed there, is believed to fall to earth, or explode above it.[4] One asteroid of this size exploded several km over the Tunguska Valley in Russia in 1908, leveling 2000 square kilometers of forest (about 50 kilometers across). There are an estimated .5 million to 1.5 million asteroids larger than 50 meters whose orbits cross the orbit of earth.[5] It is the statistical scope of damage, offset by the infrequency of its occurrence, that presents the problem: how much effort shall be given to the detection and mitigation of damage from celestial bodies?

ORIGIN OF ASTEROIDS AND COMETS

The standard theory of cosmology states that the universe came into being with a sudden explosion of energy from a central point - the Big Bang singularity. The universe, including space and time, rapidly inflated, and cooled. Various forces (gravity, the strong, the weak, and electromagnetic) separated, until atoms were formed. Immense gas clouds, mostly hydrogen and helium, contracted under the force of gravity to become galaxies and stars. The larger stars burned their hydrogen into helium, and then into heavier elements quickly, then exploded, seeding space with gas, dust and heavier elements.[6]

In our part of the universe, a large cloud of gas, dust, and heavier elements condensed, and ignited to become the sun, rotating by virtue of the rotation of the galaxy, the pull of other stars, and turbulence of the original gas cloud. The centrifugal force of this rotation prevented all the matter from becoming part of the sun. Some of the matter formed a disk of small particles, mostly ice and rock, which accumulated over a period of one hundred million years to form the planets. As the earth took shape it was bombarded with extremely large objects, and heated, but over time as the random objects were used up, the collisions became less frequent, and this allowed the earth to cool and, ultimately, life to begin.

Not all the matter gathered into planets, but remains as asteroids and comets, orbiting the sun at varying distances in the large expanse of the solar system[7]

ASTEROIDS

An asteroid is a small planetary body in orbit about the sun. The vast majority, several million, ranging in size from micrometers to several hundred kilometers in diameter circle the sun between the orbits of Mars and Jupiter,[8] a distance between 2 and 4 Astronomical Units (AUs) or between 196 and 392 million miles from the sun.[9] These, the main belt asteroids, are rocks that were kept from forming a planet by the gravity of Jupiter.

In addition to the main belt, swarms of other asteroids orbit the sun closer to earth than the orbit of Mars. These are called Near Earth Asteroids (NEA's); there are about 100,000 over 50 meters across in size, but approximately 1000 to 2000 whose diameters are 1 kilometer or larger[10] and about ten, which are 10 kilometers wide or larger.

A subset of NEA's are Earth Crossing Asteroids (ECA's) whose orbits cross the orbit of earth at some point. As of 1995 those discovered numbered 250,[11] but estimates put the number of asteroids that have the potential to be Earthcrossers as follows: 1500 larger than 1 km[12]; 4-8 thousand larger than 500 meters and .05 to 1.5 million larger than 50 meters.[13]

The source of NEA's and ECA's is probably asteroids in the main belt. Collisions with each other and gravitational forces replenish the numbers of NEA's and ECA's. Thus the population is in steady state, older asteroids have either crashed on a planet or have been ejected from the solar system[14] and have been replaced by newer ones.

Asteroids vary in composition as well as in size - a significant factor in determining the method of asteroid deflection, and of course, prospects for mining of asteroid resources. Generally NEA's appear similar in composition to main belt asteroids, as inferred from radar and spectroscopy, as well as by the actual composition of recovered meteorites. Uncertainties prevail. No spacecraft has landed on an asteroid, but Gaspra, Eros and Mathilde have been photographed. Near Earth asteroids are diverse in composition, ranging from low density carbonaceous, (C type), containing volatile elements (oxygen, carbon, hydrogen, nitrogen, sulfur and chlorine) with a metallic alloy of iron, nickel, cobalt, platinum,[15] to stony and stony iron objects (S-Class), metallic bodies containing nickel-iron alloys, and other rocky bodies.[16] Some asteroids preserve evidence of the early sun history and the original composition of primitive planetesimals[17] Many contain extinct comet cores containing ice useful as propellant for space missions[18] The size, composition and velocity of an asteroid determines whether it will break up and explode in the atmosphere or penetrate it to hit Earth, and these factors bear on the magnitude of damage, selection of asteroids for on-site inspection; selection of methods to deflect them, and use of resources.

Eventually over a period between 10 million to 100 million years, near-earth asteroids will collide with a planet, another asteroid, or will spin out of the solar system by gravitation, particularly Jupiter's. Earth-crossers have a life span according to differing estimates of from 93 to 244 million years.[19] This means that the hazard will continue to the indefinite future with a supply of asteroids furnished by the main belt.

The destructive effect of an asteroid depends upon its size, composition, velocity, and place of impact. A Tunguska like object (about 50 meters in diameter), between 10-20 megatons of energy would destroy a city. Correspondingly, a 350 meter object might destroy a small State; a .7 km object lay waste to a large State like Virginia. It has been estimated that the threshold for a global catastrophe (agricultural collapse and mass starvation, but not human extinction) would be reached by the impact of an asteroid over 1.7 kilometers in diameter striking earth at a speed of 20 km per second. This event might kill about one-quarter of the world's population.[20] An oceanic impact would cause large Tsunamis and correspondingly immense damage to coastal areas.

The frequency of such impacts is said to be from one to several times every million years. Obviously the risk of this happening in the near future is negligible. A Tunguska-like event may occur about once a century. The likelihood of its exploding over a populated area of Earth is once every 3000 years. Every thousand years a celestial object may detonate on Earth with an energy release equivalent to fifty megatons. The possibility of a global or regional catastrophe has led NASA, and others to assert the dominant view, that an asteroid search program should focus on the detection of the estimated 1500 asteroids over one kilometer in diameter, and later to extend the search to diverse smaller bodies.[21] The minority view was that the detection effort be expanded to search out smaller objects (50 to 100) meters as well.[22]

COMETS

Comets are bodies orbiting the sun in highly elliptical orbits. Unlike the asteroids, they were formed in the outer reaches of the solar system. An estimated 100 billion to one trillion comets populate the proposed Oort "cloud" reaching out to one fifth the distance to the nearest star and another theoretical swarm, the Kuiper belt, beyond Neptune's orbit. Their nuclei are composed of organic materials - carbon, ice, dust and other materials, in various degrees of compactness.

Occasionally, a comet is thrown out of its distant orbit by the gravitational force of a passing star, or massive interstellar clouds, or in case of the Kuiper belt, by Neptune, and veers into the solar system. As it approaches the sun, the volatile elements on the outer surface of the comet evaporate to form a tail or tails, visible to the naked eye or telescopes. At length the sun burns away the volatiles, and the comet becomes dead - a virtual asteroid.[23]

Comets may be classified as short period comets if their orbital periods are less than twenty years, intermediate period comets if their periods are between twenty and 200 years and long period comets if their periods are over 200 years.[24] Halley's comet, about nine km wide, returns every 76 years[25] Another large comet, Swift-Tuttle, parent of the annual Perseid meteor showers, is twenty-five km in diameter and has a period of 130 years. Though both comets cross earth's orbit, neither is likely to hit Earth in the next millennium.[26]

There appear to be about twenty-five known short and intermediate earth crossing comets, (ECC's) and an estimated 100 intermediate period comets larger than 1 km in size in earth crossing orbits, constituting about 5% of the 1 kilometer impactor (asteroid and comet) flux to earth[27]

While gravitational effects of the planets on asteroids introduce orbital uncertainties over time spans greater than two centuries, cometary outgassing results in perturbations which result in much greater uncertainties for those bodies.[28] They fit Yeats' description of "disheveled wandering stars."[29] Giant comets are believed to have brought volatiles, including water, to Earth after the early bombardment sterilized it.[30] Recently NASA announced that its Polar spacecraft discovered thousands of "dirty snowballs," diminutive comets the size of houses exploding in the upper atmosphere every day, sending down rain for billions of years.[31]

Though the comet threat appears to be less than that posed by asteroids, their larger velocities make their impacts more catastrophic.

PRESENT SEARCH EFFORTS

The 1908 Tunguska, incident stirred no extraordinary research effort.. The remoteness of the site and political turmoil in

Russia delayed an expedition until 1927.[32] Meteor Crater in northern Arizona, 1200 meter wide, 170 meters deep, was not conclusively recognized as a meteor impact site until 1936.[33] Thereafter, a confluence of factors stimulated interest in the reality of celestial impacts. The late Dr. Eugene M. Shoemaker, a geologist of the United States Geological Survey, among others had studied Meteor Crater. He and his colleagues in the 1960s systematically analyzed lunar cratering rates from Apollo data with a view to estimating the cratering rates on earth. In the 1980s the father-son team of Luis W. Alvarez and Walter Alvarez hypothesized from the disappearance of microfossils, the appearance of iridium at the K/T boundary sediment layers, and the mapping of other impact craters, that the cause of these phenomena was an impact of a large comet or asteroid with earth. Further research pinpointed the site of the now eroded crater on the Yucatan peninsula. The US Geological Survey confirmed that the Chicxulub melt rocks were of K/T age.[34]

The first Earth crossing asteroid was discovered in 1932. Further searches were interrupted by World War II, then resumed. In 1972, Dr. Shoemaker along with Dr. Eleanor Helin initiated the Planet Crossing Asteroid Survey (PCAS) using a Schmidt telescope and photographic exposures; later in 1982 Dr. Shoemaker expanded this into the Palomar Asteroid Crossing Survey (PACS). In 1981 under the auspices of NASA a workshop at Snowmass, Colorado considered the impact flux on Earth.[35]

In 1981 Tom Gehrels of the University of Arizona started a "Spacewatch" team at Kitt Peak using a 0.9 meter Schmidt telescope; a 1.8 meter telescope will soon be added, to be used by Dr.Gehrels and astronomer Dr.Robert McMillan. A new process of light detection uses electronic detectors called charge coupling detectors (CCD's). This process, reading 80% more photons than photographic film was adapted by Dr. Gehrels for Spacewatch[36] and with the help of computers discovers 2000 new mainbelt asteroids and 3 NEO's a month; but only the latter are pursued for further observation.[37] However, Spacewatch can cover less than 20% of the sky per month[38]

Until January 1997 the Southern hemisphere sky was scrutinized by the Anglo American Earth Asteroid Survey (AANEAS) headed by Dr. Duncan Steel at Siding Spring, Australia.[39]

In 1995 the US Air Force, at Congress' urging, upgraded its Ground Based Optical Deep Space Surveillance System (GEODSS) and a one meter telescope at Mt. Haleakali in Hawaii with CCD's. This effort is now in effect under the banner of the Near Earth Asteroid Tracking System (NEAT), a joint effort of the Air Force, NASA and the Jet propulsion Laboratory (JPL) under the leadership of Dr. Eleanor Helin formerly of PCAS. (Both PACS and PCAS have been discontinued as has, as previous indicated, the Australian effort.) [40] Earth crossing asteroids larger than one km in diameter are being discovered at a rate of twenty per year. One new addition is the Lowell Observatory Near Earth Object Survey (LONEOS) at Flagstaff, Arizona led by Dr. Edward Bowell. Another is Lincoln Laboratory, New Mexico, the Lincoln Near Earth Asteroid Research (LINEAR) which uses a telescope similar to NEAT's.

When Spacewatch, NEAT and LONEOS become fully operational, it is anticipated that 90% of Earth-crossers larger than one km could be discovered and catalogued in about twenty to thirty years with current financial support. Dr. Shoemaker recommended a 40 % higher appropriation, reducing the period for 90% one km earth crossing asteroid discovery to ten years.[41]

Outside the United States an NEO search started up in 1995 at Xing Long, China; and France too looked skyward in 1996 at Cote D'Azur.[42]

The above surveys constitute dedicated search programs to which may be added the findings of observatories around the world and amateur astronomer efforts.

THE INTERNATIONAL "CLEARING HOUSE"

International ad hoc non governmental institutions now serve as a clearing house or houses for the reception and coordination of astronomical information gathered by the world wide efforts described above. Newton's laws of gravitation and his laws of motion are so well understood that, within limits, the orbits of asteroids and comets can be predicted and plotted with almost unerring accuracy.

The International Astronomical Union (IAU) has by agreement arranged with the Institute of Theoretical Astronomy located at St. Petersburg, Russia to maintain and publish the ephemeredes (celestial positions) of all numbered minor planets. The worldwide center for receiving observations of newly discovered and established asteroids and comets is the Minor Planet Center, located at the Smithsonian Astrophysical Observatory at Cambridge, Massachusetts under the leadership of Dr. Brian Marsden. The center collates incoming observations, assigns designations, calculates orbits and publishes results on a monthly basis through Minor Planet Circulars.[43] Appropriate institutional networks are already in place. The question arises as to whether they should be expanded and coordinated or consolidated with other institutions.

DEFLECTION OF ASTEROIDS AND COMETS

The developments that led to increased surveillance of asteroids - the Alvarez hypothesis, moon crater scrutiny implicating large meteors, and computer predictions of nuclear winter in case of all out war - stimulated discussion of impact damage. At the same time, Spacewatch discoveries indicated that the number of earth crossing asteroids known and undiscovered, was larger than previously thought.

In 1990 the US House of Representatives commissioned NASA to set up a workshop to define systems to destroy Earth-bound asteroids and comets - in tandem with the detection workshop. NASA and the Department of Energy's Los Alamos National Laboratory collaborated in fashioning a report which was presented to the House of Representatives in 1992. Hearings were held by the subcommittee on Space (now named the Subcommittee on Space and Aeronautics) in March, 1993. The subcommittee's then chairman, Rep. Ralph M. Hall of Texas, was impressed by the information that on March 23, 1989, an asteroid over 100 feet in diameter and traveling 8 miles per second, missed earth by six hours.[44] This study and other studies have explored the subject exhaustively and in depth.

There is at present no national or international project in place to deflect asteroids and comets. Whether or not any classified deflection plan exists is, of course, unknown.

It is beyond the scope of this paper to describe in detail each proposed technical measure to intercept, destroy and /or change the orbits of asteroids and comets. The following conclusions may be drawn from current publications:

- Present technology exists or can be adapted to deflect or destroy an offending body. This includes rockets, launch vehicles, tracking and homing, and finally the energy delivery to the offending body.

- A precondition of successful interception and deflection is early detection and observation. The approaching asteroid's orbit must be known as early as possible.

- The ability to deflect depends on: the size, distance, velocity, and composition of the target body. Obviously, a large body, over 1 km in diameter, would require a large force to deflect it. For the same reason, the velocity of a body would determine the ability of a force to change the body's orbit. The distance from earth of an approaching mass would dictate the ease of interception, and the way a counteracting force would be applied. The optimum case would be one in which the orbits can be predicted years, or decades in advance, particularly if the object was large (greater than one kilometer in diameter). A long lead time would permit a precursor mission to the object to determine more precisely the size, rotation and composition.

- The cost of a deflection system depends on the programmatic effort adapted from: a few million dollars per year for laboratory and theoretical studies; \$10 million a year for telescopes and intensified studies; \$100 million per year for robotic spacecraft missions, and additional funds for preparation of an implementation program which would include launch infrastructure vehicles, explosive devices, etc.[45] One estimate ascended to the range of ten million to 100 million dollars per year worldwide as a "balanced response" to the NEO problem - an amount that is a sizable fraction of funding for air safety and control.[46]

- Workshop opinion was divided as to whether planning for testing of methods to alter asteroid orbits on "representative objects" go forward or whether to defer the testing until an actual impact was threatened.[47] In any case, international consultations and participation would be required.

Among deflection methods proposed were the following:

◊ Nuclear explosives: buried, surface, or stand off.
◊ Kinetic energy, deflection or pulverization.
◊ Attaching a thruster to the asteroid .
◊ Mass drivers (steam rockets, conveyor belts, electromagnetic guns) to thrust up from the asteroid masses of material, the thrust reaction of which would move the asteroid.
◊ Solar sails - huge arrays to "tow away" the asteroid.
◊ Crack - outgassing: drilling holes in the body to induce outgassing to propel the asteroid to a new orbit.
◊ Laser deflection - earth or moon based to "blow off" a portion of the object's surface.

The consensus was reached that kinetic and other non-nuclear measures might be appropriate for small objects only, but were too costly for larger objects. The preferred method suggested was the "stand off" nuclear explosion.[48] The literature on deflection stressed the danger of fracturing an asteroid so that the sum of the pieces falling on earth would do more damage than the original object.

LEGAL AND POLICY IMPLICATIONS

GENERAL OBERVATIONS

The survival of all or a considerable portion of the world's population touches the supreme interest of every nation. One would therefore expect positive action by the international community to mitigate catastrophic dangers.

The subject has attracted extensive media speculation, including television documentaries which emphasize the perils rather than the science, to say nothing of elaborate science fiction motion pictures.[49] One might therefore make an effort to de-sensationalize the topic.

One might also observe that there are many pressing and more immediate international and national problems. Among these are: problems of the global economy, the breakdown of nations, famines, nuclear proliferation, malnutrition and disease. Serious ecological hazards include over-population, global warming and possible climate change.[50] The eventuality of global warming and climate change is estimated to occur within the next century or two, a much more imminent world wide danger than asteroid impact.

The overall asteroid-comet impact has a low rate of probability. Still, the possibility of global or regional impact catastrophe still remains and scientists concede its eventual inevitability. Prudence demands some action. The real question is how much.[51] As indicated earlier, specific action has been taken on a national level (NASA-Air Force) and on an international non-governmental level (Spacewatch and the Minor Planet Center). But what further obligations are imposed on States?

DETECTION, DISCLOSURE AND MITIGATION OF DANGERS

CUSTOMARY LAW

There appears to be no overarching positive rule of international law that binds States to protect the world or any region of the world from a natural disaster including dangers from asteroids and comets.

However, one may say that a fundamental obligation of a State is to protect the lives of its subjects. Whether the Universal Declaration of Human Rights, which asserts that everyone has the right to life, liberty and security of person[52] states customary law too broadly in view of repressive acts of some governments throughout history, it is clear that each State feels obligated to protect its own nationals as distinct from those of other nations, from peril. The preamble of the United States Constitution is expressive of this obvious rule. This would certainly include the obligation to notify its citizens of any possible threat to life. States have, however, cooperated on an ad hoc basis to deal with natural and man made disasters. Absent binding agreements, this assistance is granted or withheld as policy dictates.

INTERNATIONAL AGREEMENTS - GENERALLY

The answer to whether States have a duty to notify and protect other States and the world from catastrophe is somewhat different when international agreements and the activities of the United Nations are considered. These agreements and activities may be taken to be declaratory of an emerging practice of States to detect, share information about, and take measures to avoid natural (and some man made) disasters.

The Antarctic Treaty [53] and the Protocol to the Antarctic Treaty[54] both provide exchange of research information. The Protocol mandates sharing of information on potential environmental risk

(Art.6) and provides for Environmental Impact Assessments (EIA's) (Article 8 and Annex I) with notification to the public and the parties of activities anticipated to have more than minor transitory impacts. The United Nations Convention on the Law of the Sea[55] requires States to preserve the marine environment and to notify other States and competent international organizations of potential or actual damage by pollution.[56] The amended convention has now been ratified by the United States and other major maritime States and, the notification provision will soon be, if it is not now, declaratory of inter-national law.

The United Nations' various programmes and specialized agencies have each in their spheres marked out both preventive and remedial measures for natural and man made catastrophes.

Among the specialized agencies, the Food and Agricultural Organization (FAO) exists to ensure "freedom from hunger;" the World Health Organization's (WHO) mission is, among other things, to prevent the spread of disease; the World Meteorological Organization (WMO) facilitates cooperation of national networks for the rapid exchange of meteorological information. Recently the WHO, UNEP and WMO collaborated on a 1996 report that climate instability poses serious and widespread risks to human health.

One may point, in particular, to treaties to provide for preventive action with respect to environmental perils as being analogous to the asteroid hazard: i.e., the Ozone Convention[57] and the Montreal Protocol on Substances that Deplete the Ozone Layer.[58] The Ozone Convention requires cooperation in research, adoption of agreed measures to control human activities injuring the ozone layer (Art.2), provision for systematic observations of the ozone layer, collection and transmission of research (Art.3). It provides for a conference of parties to meet at regular intervals to review scientific information, and promote harmonization of policies. The Montreal protocol sets targets for phasing out production and consumption of ozone depleting substances.

In 1988, the UN General Assembly declared that global climate change was a common concern of mankind. It urged the United Nations Environmental Programme (UNEP) cooperate with W.M.O. and the International Council of Scientific Unions. An Intergovernmental panel set up by the WMO and UNEP provides coordinated assessments of the magnitude and impact of climate change.

The Framework Convention on Climate Change in force in 1994 created a Conference of Parties to make "decisions necessary to promote the effective implementation of the convention." One subsidiary body of scientific and technological advice would make assessments of climate related knowledge;

while another subsidiary body consisting of government representatives who were experts in climate related knowledge, would help the conference implement recommendations.[59] All major nations are signatories. The commitments of the treaty appear not to bind the parties as yet to specific numerical emission targets. Controversy among rich and poor countries and among special interests, has marked negotiations. A critical meeting in Tokyo in December, 1997 may lead to specific reductions and timetables.[60] The Conference of Parties and its subsidiary organs suggests an institutional structure that may be applicable, in a less elaborate form, to deal with the asteroid problem.

The foregoing review suggests that on the basis of the treaties cited above, a consensus may be emerging that all States are required to take measures to notify all others of natural perils, and to take action to prevent them.

OBLIGATIONS OF THE SPACE TREATIES

The various space treaties contain provisions directly in point on the matter of State obligation to deal with the asteroid/comet problem.

As to space dangers the Outer Space Treaty[61] requires States to immediately notify other States of phenomena constituting danger to the life and health of astronauts (Art.V). This falls short of notification of dangers to the population of the world at large. States are further required to notify the Secretary General of "activities conducted in outer space" (Art.XI). This provision does not directly cover the case of mere surveillance of the sky from earth to detect asteroids but would apply to any activity in space directed toward an asteroid.

The 1979 Moon Treaty governing the activities of States on the moon and other celestial bodies[62] goes much further than the Space Treaty in several respects. First, it makes clear that the treaty applies to all celestial bodies in the solar system except where "other specific norms" apply (Art.1). If this were not obvious before, asteroids and comets are swept up into the treaty. In addition the interest of "future generations" is taken into account.

Article 5 is on target with respect to a State's obligation to notify the world of space dangers. This article provides in part:

> "....3. In carrying out activities under this agreement, States Parties will notify the Secretary General as well as the public and the international scientific community of any phenomena they discover in outer space, including the moon, which would endanger life and health."

An asteroid on trajectory toward Earth would be such phenomenon to endanger life and health. The quoted article requires each State to notify all others of prospective danger. This obligation may seem obvious in real life, but it was wise to assert it clearly to avoid an unforeseeable future situation in which one country might decide, for whatever reason, not to disclose an impending danger to another region of the world.

LIMITATION AS TO MITIGATION ACTIVITIES

As indicated earlier, the mitigation of asteroid danger - barring some future method to manipulate gravity - requires the destruction of an asteroid, or altering its orbit. In the case of all but the smallest asteroids this requires one or more stand-off nuclear explosion, preferably years or decades in advance of probable impact.

Several treaties restrict proposed measures to deflect an asteroid or comet.

1. The Nuclear Test Ban Treaty[63]
2. The Space Treaty[64]
3. The Environmental Modification Treaty[65]
4. The Moon Treaty[66]

The Test Ban Treaty flatly bans any nuclear explosion in outer space. The Space Treaty prohibits the installation of "nuclear weapons or any other kinds of weapons of mass destruction" on celestial bodies, or stationing such weapons in outer space in any manner. The Environmental Modification Treaty bans changing "through the deliberate manipulation of natural processes," the dynamics, composition or structure of earth, or of outer space. The Moon Treaty precludes the placing of objects carrying nuclear weapons or other kinds of weapons of "mass destruction in orbit around or other trajectory to or around" any celestial body or placing them on a celestial body (Art.3.3) It also bars the testing of any type of "weapon" in space (Art 3.4).

With respect to all these treaties, a good argument may be made that the parties did not anticipate the danger to earth from asteroids, and that therefore the treaties have no application to the deflection of asteroids.

The Treaty on treaties declares as a general rule that a treaty shall be interpreted in accordance with the "ordinary meaning" to be given to the terms of the treaty in their context and in the light of its objects and purposes.[67] Among other things to be taken into account were "any subsequent agreement between the parties regarding the interpretation of the treaty or the application of its provisions."[68]

As to the Test Ban Treaty, certainly the "ordinary Meaning" of a "nuclear explosion" in "outer space" is quite clear. Yet it is apparent that in 1963, acute international concern was directed to the arms race on Earth, not to protect the Earth from a celestial marauder.

The Space Treaty prohibitions on the installation of nuclear weapons or weapons of mass destruction on celestial bodies or stationing such weapons in outer space left open several important questions relating to arms control. It is beyond the scope of this paper to revisit these contested questions: the interpretation of "peaceful" in relation to military uses or the definition of "weapons" and "mass destruction" or "stationing." These questions have been exhaustively discussed elsewhere.[69] The weight of authority appears to be that any nuclear or other object not only intended but capable of widespread damage, is objectionable. This interpretation would preclude sending or placing nuclear or any other devices that could cause mass destruction (kinetic deflection, lasers) in or to any part of space, including asteroids. It should be noted that an asteroid deflected toward Earth may itself be a weapon of mass destruction. Another interesting question was whether the Space Treaty ban on "weapons" includes trajectories to or from celestial bodies.[70] The trajectory question, as indicated earlier, is settled by Articles 1 and 3 of the Moon Treaty, by which reference to the moon applies to all solar system celestial bodies, including orbits around and trajectories to or around them.[71]

The combination of the Nuclear Test Ban Treaty and the Space Treaties would seem to prohibit any use of a nuclear explosion to alter an asteroid's orbit. The ban might also apply to any other device capable of causing mass destruction. In order to come to a different conclusion, one would be required to interpret the Test Ban Treaty to refer to outer space itself, not to celestial bodies, to interpret "weapons" in all treaties as referring to intended uses (asteroid deflection) rather than merely capable of use as a weapon, and/or to pronounce asteroid deflection outside the scope of any treaty, on the ground that asteroid deflection was not contemplated in the writing of the treaties.

Perhaps the time is ripe for the UN Space Committee to initiate discussions on whether an exception should not be made for nuclear or other drastic means to alter asteroid orbits. The Space Committee may also review whether nuclear explosions might not be used on celestial bodies for mineral extraction, with provisions for international consultations and agreed safeguards.

If the international community decides to clarify or modify the strict ban on nuclear explosions, the preferred path is one the Space Committee has already taken - forging new treaties, rather than amending or adding to existing treaties.

INTERNATIONAL ORGANIZATION TO DEAL WITH THE ASTEROID/COMET HAZARD

All the questions raised in the prior discussion should begin to be examined, first, on an institutional level (the space and military agencies, and the scientific organizations of the space nations and other interested countries); and then on a national and international level. The following questions should be highlighted:

- Shall States enter into new or more organized arrangements for asteroid detection. If so what arrangements?
- Shall States and the international community enter into preliminary discussions as to asteroid deflection? If so, in what manner - by informal national discussions, by articulation of a set or principles through the Space Committee and the United Nations, or by treaty?
- How shall matter of dissemination of information of detection activities and dissemination of particulars of an asteroid threat be accomplished?.
- How shall decision making as to asteroid deflection be made?
- Who will take action and under what conditions and safeguards?

PRESENT INTERNATIONAL PROGRAM FOR DETECTION

As noted previously, detection activities where and when done, are well done by non-governmental institutions with government financing and support, as in the United States, by its Congress and Air Force. These operations are incomplete because of the magnitude of the task and the cross section of the sky covered (the Southern hemisphere sky is not well represented). On the other hand, coordination of discoveries and prediction of asteroid and comet orbits are well advanced in the capable hands of the Institute For Theoretical Astronomy (ITA) and the Minor Planet Center at the Smithsonian Astrophysical Observatory at Cambridge, Massachusetts.

What remains to be done in the detection area? The scope of the discovery program is presently limited. In 1995 NASA recommended a ten year census of asteroids and short period comets larger than 1 km in diameter at a cost of $24 million for the first five years, and about $17.5 million for the next five years. This program posited the discovery of 70% of such objects the initial five years, and 90% the next five years, expanding the search to smaller objects as well . Several dedicated telescopes along with detectors and software were deemed sufficient.[72] By contrast one author suggests a search program reaching the 200,000 two hundred meter and up objects with 85 computer driven telescopes, each costing $3 million at a total cost of $170 million, or better still 150 telescopes at a cost of $300 million. An additional cost of $600 - $900 million would be required for operations over 20 years to be 90% complete in that time.[73] That author compares the cost to the $1 billion bill for the Voyager outer planet flyby. Other mainstream astronomer collaborators see the need for a 20 year search using 12 telescopes in a worldwide network of three telescopes in each hemisphere. This search proposal envisages flushing out most larger ECA's in a 125 year search that would also identify most smaller Tunguska like asteroids. No cost estimate was given for this suggestion.[74]

The House of Representatives in April, 1997 recommended $3.4 million in 1998 and $3.4 million in 1999 for the Near Earth Object (NEO) survey, somewhat less than recommended by NASA in 1995; however, this does not take into account the Air Force contribution, nor the contribution of NASA's flyby and asteroid/comet probes projected now underway or contemplated.

In addition to Earth-based observations, space probes have already begun. Three perfunctory main belt asteroid flybys took place as part of the Galileo program - of Gaspra, dimensions 6x12 km in October 1991; of Ida 52 km long, in August, 1993; and of Mathilde 53 km in diameter in June, 1997.

Projected encounters include

- A Near Earth Asteroid Rendezvous (NEAR) with mainbelt asteroid Eros 14x40 km in January 1999 through February 2000.
- A flyby combining close-ups of mainbelt asteroid McAuliffe, in January, 1999 and Comet West-Kohoutek-Ikemura, in June 2000 under the Millennium Program.
- The "Muses C" joint cooperative mission between US and Japan which will touch down on Nereus (Earth crosser) in January, 2006, collect surface samples, and return them to Earth.
- A proposed comet tail sampling ("Stardust") mission to be launched in 1999 and to collect and return dust samples to Earth, in 2006.

The European Space Agency has proposed the Rosetta Mission, which includes a Mars flyby, two earth flybys, two asteroid flybys, and a comet rendezvous in 2011.[75]

Germany has in mind high resolution imaging of Near Earth Objects through four exploratory flybys. Different imaging systems, a magnetometer and infrared spectroscopic equipment, would be used, and the dust environment would be investigated.

Intensive radar imaging of a number of near earth asteroids has recently been commenced with the refurbished Aricebo dishes at Aricebo, Puerto Rico. The research includes accurate tracing of asteroid orbits.[76]

In regard to international cooperation, NASA suggested operational coordination among institutional telescope users, - scanning of complementary sky areas, with some stations concentrating on asteroid discovery, and others performing confirmation and follow up. At some point an international center might be required for data analysis and coordination.[77] NASA also recommended NEO computational and data centers.[78]

DEVELOPMENT AND ASSEMBLY OF DEFLECTION MEASURES

Deflection technology in the form of rockets and nuclear warheads is said to be theoretically effective against most asteroids up to a few kilometers across. The cost and effectiveness depends on the size, composition and speed of the asteroid, and the time between discovery and deflection. The longer the time available for mitigation the more chances for success and amortization of the cost.[79]

With these factors in mind, decision makers must consider the cost and effort of preparedness against the infrequency of possible need for action.[80]

Though the theoretical basis for deflection is well developed, there is no evidence that any government has made concrete plans for this activity. In 1996 China insisted on conducting underground nuclear tests to prepare for possible defense against doomsday rocks[81] among other possible uses.

Most writers would defer testing and development for several reasons. One reason is that more must be known of the physical characteristics of asteroids to determine the appropriate diversion method. Thus flight to representative asteroids by robot or geologists should precede hardware development. Another reason is that any present elaborate system of mitigation would be shortly out-dated.[82] Dr. Edward Teller believes that experiments involving stand-off nuclear explosions to obtain necessary data be not delayed except for strong reasons.[83] Carl Sagan was particularly wary of any precipitous project "to push asteroids around." He recommended a combination of accurate orbit estimation, realistic threat assessment and effective public action. Yet in the end he favored robotic flybys and landings, sample returns, human landings, and finally cautious asteroid orbit deflections starting with smaller explosions, then nuclear fusion engines. Finally he somewhat contradictorily approved of "inserting small asteroids made of precious and industrial metals into Earth Orbit" - but in the 22nd century.[84]

The substantial cost of development of deflection technology must be added to the policy consideration mix; the low estimate runs to $100 million per year.[85]

THE DEFLECTION DILEMMA

The "deflection dilemma" (named by Gregory H. Canavan and Carl Sagan), based on statistical analysis, suggests that "the frequency of opportunities to misuse an interception system (by diverting an asteroid toward an enemy nation) is greater than the frequency with which you would anticipate needing to use the system to push an impactor away…" and that the defense system would be more dangerous to mankind than the asteroid itself.[86] More particularly since asteroid near misses are more than one hundred times more frequent than predicted impacts, the occasion or temptation to cause an impact arises one hundred times more frequently as the necessity to avert an impact.[87]

The "dilemma" aspect arises from the fact that if no deflection capability is developed in advance, the Earth is vulnerable to that fraction of NEO bodies or short period comets that may then present an immediate danger, while a complete system built in advance may present opportunity for misuse. This misuse may come from a rogue leader or nation or from an irrational group of people privy to the system,[88] and the opportunity for misuse would be as frequent as the number of new asteroids and the rate of discovery and would increase with the effectiveness of the system.[89] A dark view of mankind perhaps, but, strangely near unanimous.

It is clear therefore that an international consensus would have to be reached for modification of current treaties. David Morrison and Edward Teller recommend that any testing of target objects be taken by an appropriate international body.[90]

INTERNATIONAL ORGANIZATION FOR ASTEROID/COMET DANGER

In the UN Committee on the Peaceful Uses of Outer Space 1996 session some delegations urged that the Committee strengthen the scientific content of its work by promoting international cooperation in other space activities, "especially in the field of disaster warning and mitigation and global search and rescue activities."[91] No doubt disaster warning referred to remote sensing detection of terrestrial disasters, such as hurricanes, volcanoes, aircraft crashes, etc., but it could be expanded to include asteroid detection. The Committee might discuss amplification of asteroid/comet search efforts in its projected Third UN Conference on the Peaceful Uses of Outer Space to be convened in Vienna in 1999 but possibly 2000 AD.[92] One of the topics is likely to be environment and disaster warning, mitigation, and relief.[93]

WITH RESPECT TO DETECTION

What type organization, if any, shall there be for a world wide sky survey of asteroids?

Present discovery efforts are eminently professional but admittedly incomplete.[94] The international mechanisms work very well at the present level of activity. Whether these activities should be more structured may depend on the degree of reliable funding by individual nations by formal agreement.[95]

The asteroid/comet search is primarily astronomical and electronic and there appear to be no precise functional models to turn to. Perhaps the World Meteorological Organization (WMO) which in addition to meteorology, encompasses aviation, shipping, geological and agricultural matters comes closest in terms of observational activity,[96] but because of its myriad activities, its organization is far too complicated to be a role model.

As suggested earlier, the most important precept to follow would be to avoid administrative overload. Therefore, no elaborate organization for asteroid detection and information dissemination is required. National networks can multiply on their own. A more formal coordinated search program which stresses follow up observations like the detection consortium proposed earlier may be desired. If so, it should not be part of any international space agency or under the aegis of any international regime set up by the Moon Treaty. The Minor Planet Center at the Smithsonian Astrophysical Observatory at Cambridge should probably be expanded with a "central nexus.... to coordinate the efforts of different teams...."[97]

An important adjunct to any institute created should be a public information division that would disseminate information frequently and in detail. The integrity of the information should be guaranteed by independent outside data review centers to verify information that might suggest close asteroid approaches.

WITH RESPECT TO TESTING AND HAZARD MITIGATION

As noted earlier, a totally different set of legal and policy problems arise in testing and deflection. The magnitude of the effort and the peril of the process indicates a totally different decision making structure is required.

Flyby missions to asteroids have already taken place and will continue bilaterally or unilaterally. Rendezvous and sampling missions are on the drawing board. Ultimately actual testing of deflection methods will occur. Close international cooperation and coordination are essential because of the cost and the necessity of agreeing on such matters as asteroid selection, technical aspects (orbital determination, propulsion, instrumentation) and financing.

Deflection of an asteroid on earth trajectory may not be needed for hundreds, if not thousands of years. A matter of such importance would imply a decision making organ requiring the highest degree of public confidence.

What would an international organ for activities leading to asteroid deflection, look like?

It should not take the shape of a cooperative enterprise like the Mir Orbiting Station where Russia, the owner and operator of the station, made vital emergency decisions in July, 1997, with the concurrence of the United States and France.[98] Recent Mir misadventures disqualify an overtaxed binational arrangement.

A tiered decision making structure would be necessary. One might suggest a three layered organization:

- The first layer, dealing with exploratory missions to asteroids, including flybys, rendezvous orbiters and sampling - whether robotic or manned - can be composed of a collegial body of scientists and astrodynamical experts, obtaining input from the detection community, and guided by a council composed of scientific advisers and Vice Presidents or other delegates of the space nations and delegates of non-space nations selected by pooling.
- The second layer would deal with the testing of deflection techniques. The present consensus is that any deflection testing should be done well away from Earth, to avoid dangerous errors. Carl Sagan suggested waiting several decades to change the orbits of small (100 meter wide bodies) while developing technology to deflect a large asteroid. Sagan would prefer to have us wait to the 22nd century to move small asteroids around the solar system, with nuclear fission engines.[99] Others think that large objects (1 km or more) should be testing targets because of their destructive capacity.[100] The second tier would require the counsel of senior scientists and political figures, in part drawn from the first tier, with consultations among the top leaders of nations (Presidents, Premiers, etc.). Input might also be encouraged from independent scientific and citizen bodies.
- The apex tier would be a body consisting of an executive council limited to 25 or so top world leaders and scientists chosen by the second tier and assisted by a larger advisory assembly, to make decisions on the deflection of asteroids on collision course with Earth. All decisions on testing of deflection and deflection itself would be carried out by "layers of safeguards."[101]

The maximum dissemination of information should be given to the public with respect to every activity.[102] All proposed activity reports should be accompanied by summaries of the possible effects on the space environment, similar to Environmental Impact Assessments for environmental matter. The United Nations Space Committee should soon begin to discuss the subject of detection and mitigation.

EXPLOITATION OF ASTEROID AND COMET RESOURCES

Most expert commentary on planetary resources focuses on the Moon and Mars. The reason for this is that those bodies may also serve as research bases and ultimately, manned colonies. Asteroids and comets have not, however, escaped attention as possible resource reservoirs.

RESOURCES

Since no asteroid and comet has yet been visited by man, the composition of these bodies has been deduced by other means. In the case of both, asteroids and comets telescopic observations are coupled with spectrographic interpretation to provide preliminary information. In the case of asteroids, the study of meteorites, made mostly of iron and stony irons, helps characterize some asteroids or their interiors. Meteorites, though, do not represent the full range of asteroid content.

Near Earth asteroids are compositionally diverse, and their resources are more varied than lunar materials. Current knowledge of asteroid resources is limited, however, and further research would be required merely to single out some asteroids for further scrutiny, leading to exploration and exploitation.[103]

About four-fifths of asteroids are metal rich, containing iron, magnesium, nickel, cobalt and platinum, while one-fifth contain volatile materials, such as hydrogen, carbon, sulfur and nitrogen compounds. The volatile products of carbonaceous asteroids would be particularly useful for extracting water used for life support, shielding against solar flares and cosmic rays, and for conversion to hydrogen and oxygen. Various mixtures of volatiles may also be used as rocket fuel; in fact, most asteroid material would initially be used in space for space related projects. One advantage of asteroid exploitation is an asteroid's low gravity, which allows the return trip to Earth or elsewhere to be made with less fuel or greater payload.[104]

Short period comets are extraordinarily rich in volatiles, particularly water, carbon monoxide and hydrocarbon dust grains. A typical cometary nucleus contains water ice, hydrocarbons, silicates and other volatile ices[105] A fraction of near Earth asteroids are extinct nuclei of short period comets and their interiors may also contain volatiles.

Asteroid and comet resource

development would be subject to financial and investment limitations, whether conducted by State agencies or private entities. The cost of an asteroid mine would run from one billion to five billion dollars with research, development, testing and delivery of components spread over fifteen years.[106] A long period of testing of processes for resources uses would be required, most effectively by computer simulation.

THE LEGAL REGIME RELEVANT TO ASTEROID AND COMET RESOURCE EXPLOITATION

Prior to the space treaties, States might obtain sovereignty over portions of celestial bodies as on Earth by effective occupation. Effective occupation to uninhabited and desolate regions is determined by some physical activity on the territory coupled with purposeful exercise of political rights purporting to cover the whole of a disputed terrain.[107] Under customary law, resources rather than the territory itself could be used exclusively with or without acclaim of ownership. Thus, prior to the Space Treaties, a State or an entity acting under its authority might take exclusive possession of a portion of an asteroid or an entire small asteroid and might use the iron, nickel or water as well.

The Space Treaty and the Moon Treaty appear to have introduced new concepts that broke away from traditional precepts of "res nullius" and "res communis," although echoes of both theories still vibrate in discussions of property rights on celestial bodies. Distinguished scholars have thoroughly surveyed the extent and limitations of property rights on celestial bodies, most recently in the Proceedings of the Thirty Ninth Colloquium on the Law of Outer Space.[108] These considerations apply to asteroids and comets.

In summary the Moon Treaty, which applies to all celestial bodies, extends the Space Treaty. It says that the exploration and use of celestial bodies shall be carried out for the benefit and in the interests of all countries (Art.4) that States have the right for scientific purposes to collect and remove "..... samples of its mineral and other substances" which shall remain at the disposal of the collecting States and may be used by them for scientific purposes and in support of their missions (Art.6-2). Article 11 (2) repeats the non-appropriation clause of the Space Treaty and continues:

> "Neither the surface nor the subsurface of the moon, nor any part thereof, or natural resources in place, shall become the property of any State, international, intergovernmental, or non-governmental organization, national organization or non governmental entity or any natural person"

Article 11(6) further obliges States "to undertake to establish an international regime ...... to govern the exploitation of the natural resources of the moon as such exploitation is about to become feasible" and with respect to the resources, sets out four purposes: orderly and safe development, the rational management of those resources, the expansion of opportunities, and equitable sharing of benefits, giving special consideration to the interests and needs of the developing countries, and the efforts of countries contributing to the exploration of the moon.

The Moon Treaty opened for signature in December, 1979, came into force in 1984 and as of 1996 has been ratified by nine States but the major space nations have not acceded to the treaty, despite the passage of eighteen years, so that the above provisions are not binding on them.

Intense controversy has swirled about portions of the Moon Treaty, particularly Article 11, generally pitting non-space nations against space nations. Disputes center on whether national or commercial ownership of natural resources has in fact been legally impaired, and to the extent that it is, whether these restrictions stifle incentive and thereby the development of resources, or whether these restrictions are necessary to prevent resource monopoly and profit running to solely private commercial entities. An ancillary legal question spotlights the prohibition of acquiring property to "natural resources in place" - is sanction given to property or ownership once the resources are extracted and removed?

Then, the whole matter of an "international regime" is supported or questioned. The Moon Treaty does not create an international regime, it merely asks States to promise to establish it once the exploitation of resources becomes feasible. Still, shall the international regime take the form of an enterprise or be merely a licensing authority? Shall it be the sole authority and/or eventually be one of the actors, as the Law of the Sea Treaty contemplates for its International Seabed Authority and its organ, the Enterprise, dealing with the seabed.[109]

The mineral resources of the Antarctic area presented the same institutional questions. They were addressed by the draft convention on Regulation of Antarctic Mineral Resources Activities.[110]

The solution there was to fold supervision of all Antarctic mineral resources into a Commission which would among other things entertain State sponsored applications for prospecting and ultimately exploration in designated areas under permits issued by Regulatory Committees for each area. No activities would be permitted outside the convention and exploration would proceed under a "Management Scheme." The Convention provided for commission membership and various decision modalities including a three quarters membership super majority on questions of substance. All activities were circumscribed by various financial, environmental and operational criteria.

Because of persistent environmental concerns, the draft convention has been sidelined for fifty years and may never go into force.

Different views of resource exploration can probably be reconciled. Apprehensions of suffocating incentive are probably overdrawn, particularly since even in the most free market economies, corporations are surrounded by desirable restrictions (wage hour laws, fraud statutes, securities regulations, anti discrimination statutes, price fixing and (some) monopoly barriers). With respect to an international regime, it is likely that the distance between Earth and the planets requires a firm hand, particularly in security and ecological matters which can be exercised by an organ in which the space nations dominate. There must certainly be financial incentives - wealth must be created before it can be distributed; and private enterprises may come to be major participants, dominating through investments and even operation. One the other hand,, one may also take notice of "socialized costs", or collectivized loss, in which tax payers provide the funding while a small group reaps the profits. And developing nations are best served through the World Bank and other agencies directly, rather than by a functional enterprise.

Applying the above remarks to asteroids and comets, the number of worthwhile exploitation objects would be relatively few, the cost immense, and the likelihood of exploitation of resources far in the future. The Moon Treaty provisions against national appropriation and those restricting the creation of property interests should, for the time being, be strictly interpreted, and be implemented through the international regime, separate and distinct from any organ dealing with asteroid deflection testing and mitigation. Security concerns would be paramount as to asteroids and comets so as to prevent misuse discussed in the section on the "Deflection Dilemma."

CONCLUSION

The number of worthwhile human activities is enormous, and scientific advances in all fields including space, have been superlative. Planetary exploration is a significant part of this progress. The asteroid/comet peril, and opportunity, occupies a small corner of worldwide attention or concern but in the long future, should not be wholly disregarded. It may ultimately become a test of human ingenuity.

REFERENCES

[1] Morrison et al: The Impact Hazard, in Hazards Due to Comets & Asteroids, T. Gehrels, ed. The University of Arizona Press, Tucson, Arizona, 1994, P.59-91, the book hereinafter cited as Hazards. This source is to date the most extensive scientific and technological collection of articles on the subject.

[2] Rampino & Haggerty: Extraterrestrial Impacts and Mass Extinctions of Life in Hazards supra, P.827-857; Duncan Steel: Rogue Asteroid and Doomsday Comets, John Wiley & Sons, Inc. NY 1995 - P.100-104; Alvarez: T.Rex and the Crater of Doom, Princeton U. Press, 1997, P.140-142.

[3] T.Rex etc. supra note 2, P.4-11; John S.Lewis Rain of Iron and Ice, Addison-Wesley Publishing Co., NY, 1996 - P.102-3; Steel 155-62; Sheehan & Russell: Faunal Change, in Hazards P-879-891.

[4] T.Gehrels: Collisions with Comets and Asteroids, Scientific American, March, 1996 at p.56.

[5] Report of the Near Earth Object Survey Working Group - National Aeronautical & Space Administration (NASA), Solar System Exploration Division, Office of Space Science, Washington DC hereinafter cited as NASA Survey 1995, P.6

[6] Carl Sagan: Cosmos Random House Press, NY 1980 P.246-248; Steven Weinberg: Life in the Universe, Scientific American, Oct, 1994, P.44-49; Peebles et al, The Evolution of the Universe Id P.53-57; Steven W. Hawking;: A Brief History of Time, Bantam Books, NY 1988 P.117-121; James Trefil: 1000 Things Everyone Should Know about Science, Doubleday, NY P.260-264

[7] Eugene H. Levy: Early Impacts Earth Emergent From its Cosmic Environment in Hazards, P.34; Allegra & Schneider: The Evolution of the Earth in Scientific American, October, 1994, P.166-167; Gehrels, supra, note 4 P.54

[8] Levy, note 7, P.5

[9] An AU is the mean distance from Earth to Sun, about 92 million miles.

[10] Gehrels, note 4, p.56.

[11] NASA Report, note 5, P.6

[12] Statement by Eugene M. Shoemaker for the Subcommittee on Space and Aeronautics, House of Representatives, April 10, 1997, US Govt. Printing Office H.R. Report 105-65.

[13] Supra, note 11.

[14] William A. Bottke et al: Collisional Lifetimes And Impact Statistics of Near Earth Asteroids: Hazards, P.337-357.

[15] Lewis, note 3, P.83

[16] NASA, note 5, P.7.

[17.] Jones et al Human Exploration of Near Earth Asteroids in Hazards P.686.

[18.] Lewis et al Using Resources From Near Earth Space in Lewis (ed) Resources of Near Earth Space, University of Arizona Press, 1983, P.8-10; Nelson et al; Review of Asteroid Composition Id, P.493, 504, 508-20.

[19.] Bottke note 14, P.337-555.

[20] D. Morrison et al; The Impact Hazard in Hazards P.59-91.

[21] US House of Representatives, 130th Congress, First Session, No.8, March, 1993, US Govt. Printing Office; Statement by Eugene Shoemaker, on Near Earth Asteroids, April 10, 1997 supra note 12

[22.] Steel, supra note 2, P.232-233

[23].Gehrels, supra note 4, at P.34-36; NASA, Report, note 5, P6-7

[24].Hazards, supra, note 1, Glossary, P.1246

[25.] NASA Report, note 5, P.7

[26.] Steel, supra, note 2, P8-13

[27.] Id, P.31-36

[28.] Yoemans & Chodas; Predicting Close Approaches of Asteroids and Comets to Earth in Hazards, note 1, P.241-257.

[29.] Yeats, Who Goes With Fergus, Collected Poems of W.B. Yeats, MacMillan Co., 1968, P43.

[30.] Chyba, et al, Impact Delivery of Volatiles and Organic Molecules to Earth in Hazards , note 1, P.9-49.

[31.] New York Times, May 29, 1997, P.A18, Col.1;Id June 3, 1997, P.C1, Col.8.

[32.] Lewis, supra note 3, P.53-54, Steel, supra note 2, P.174-178.

[33].Steel, note 2, P. 81-86.

[34] Alvarez note 2, P.124

[35] Morrison et al The Impact Hazard, in Hazards, note 1.

[36] Steel, note 2, p.200-203

[37] Carusi et al: Near Earth Objects: Present Search Programs in Hazards, note 1, P.127-135.

[38] Steel, note 2, P.203

[39] Id. P.197-198

[40] Report of the Committee on Science, US House of Representative, Report 105-65 April 21, 1997, P.38; JPL Press Release, April 24, 1996.

[41] Statement of Dr. Shoemaker to Subcommittee on Space and Aeronautics, House of Representatives, April 10, 1997.

[42] Communication to author by Dr. Brian Marsden, June 18, 1997; and from Tom Gehrels, April, 1997

[43] NASA Report, Note 5, P.30, Carusi, in Hazards note 1, P.142-5

[44] Supra note 21, P.4, P.7

[45] Id, P.50-53

[46] Ahrens & Harris: Deflection and Fragmentation of Near Earth Asteroids, in Hazards, P.897-927, at P.900.

[47] NASA Report, supra, note 5, P.50.

[48] Id, P.51; P.84-85, See also various articles on Hazards, P.897-1111.

[49] One 1998 cinema project, entitled Armageddon, deals with astronauts placing explosives into the interior of an asteroid to blow it up.

[50] P.R. Weissman: The Comet and Asteroid Impact Hazards in Perspective in Hazards. P.119-1212, at P.1199.

[51] Id 1193-1196

[52] GA Res. 217A (III) 3(1) UNGAOR Res. 71, UN Doc A /810(1948)

[53] December, 1959, in force June 23, 1961, 402 UNTS 71, UST 794, TIAS 4790

[54] Madrid, October 4, 1991, ILM 1455 (1991)

[55] 1983, Official Text 21 ILM 1261 (1982)

[56] Id, Part XII, Section 2, Arts. 197-201

[57] Convention for the Protection of the Ozone Layer, Vienna, March, 1985, 26 ILM 1529 (1987) in force Sept. 22, 1988

[58] Montreal, September 1957, 26 ILM 1550 (1987) in force January 1, 1989

[59] Done at New York, May, 1992, 31 ILM 849, Arts.7-10

[60] New York Times, June 28, 1997, P.37

[61] Treaty on Principles Concerning the Activities of States in the Exploration and Use of Outer Space Including the Moon and Celestial Bodies, 18 USTS 410, TIAS 6347, 610 UNTS 5205 The various space treaties are collected in UN Publications A/AC, 105/572, 1994

[62] Done in New York, December 18, 1979, in force July 11, 1984, ILM 1434, UN Dec. A/Res. 34/68, Dec. 14. 1979 Annex.

[63] Treaty Banning Nuclear Weapons Test in the Atmosphere, in Outer Space or Under Water 14 UST 1313; TIAS 5433, 40 UNTS 43 (1963)

[64] Treaty on Principles Governing the Activities of States in the Exploration and Use of Outer Space, Including the Moon and Other Celestial Bodies. 18 UST 2410, TIAS 6347; 610 UNTS 205 (1966)

[65] Convention on the Prohibition of Military or any other Hostile Use of Environmental Modification Techniques (ENMOD) 1108 UNTS 151,31 UST 333, 161 ILM 188 (1978)

[66] Agreement Governing the Activities of States on the Moon and other Celestial Bodies, 1363 UNTS 3,18 ILM 1434, (1979)

[67] UN Dec A/Conf; 39/27;25 ILM at P.562, Sect.3, Art.31,1

[68] Id, Sect.3, Art.31,3(a)

[69] See especially, The Proceedings of the Twenty-Fifth Colloquium on the Law of Outer Space, 1982, P.69-185 among these articles by E. Galloway, S. Gorove, C.G. Hasselman; also, Gorove: Article IV on the Outer Space Treaty and Some Alternatives For Further Arms Control in Maintaining Outer Space For peaceful Uses, Jasentuliyana ed., United Nations, Tokyo, 1984, P.77-89

[70] Jasentuliyana The Moon Treaty in Maintaining Outer Space For Peaceful Uses, supra note 69, P.126-127

[71] Id, P.130

[72] NASA Report supra note 5, P.1-2

[73] J.S. Lewis, Rain of Iron and Ice, Addison Wesley Pub. Co. 1996, P.209-270, note 3

[74] Bowell and K. Muinonen, Earth Crossing Asteroids and Comets in Hazards, note 1, P.180-185

[75] A.F. Cheng et al: Missions to Near Earth Objects, in Hazards, note 1, P.667-669

[76] NY Daily News, July 19, 1997, P.20

[77] NASA Report note 5, at P.31-32

[78] Id. P.33-34

[79] G.H Canavan: Cost and Benefit of Near Earth Object Detection and Intervention, in Hazards, P.1157-1189

[80] Morrison & Teller: The Impact Hazard: Issues for the Future in Hazards, P. 1135-1142

[81] New York Times, May 14, 1996, P. CI, Col.1. Ibid., April 27, 1996, P4, Col.1

[82] Morrison & Teller, supra, note 80 P.1134-43.

[83] Id. P.1142

[84] Sagan Pale Blue Dot, Random House, NY (1994), P.323-326.

[85] Cost plotted against prospective losses from different sizes of asteroids range from $100 million to $400 million a year - to billions if kilometer size asteroids are considered. An elaborate mathematical analysis is presented by G.H. Canavan in Costs and Benefit of Near Earth Object Detection and Interception in Hazards, P.1157-1188; P.R. Weissman: The Comet Asteroid Hazard in Perspective, in Hazards, Id; P.1191-1212.

[86] Steel, supra note 2, P.236-237

[87] Lewis: Rain of Fire and Ice, supra note 3, at P.221

[88] Harris, Canavan, Sagan, Ostro: The Deflection Dilemma: Use Verses Misuse, etc. in the Hazards note 1, P.1145 et seq.; Sagan, note 84 above P.320-324; Sagan was particularly sensitive to the "madman" possibility.

[89] Id., 1150-1155

[90] Morrison & Teller, supra note 80, P.1144.

[91] Report of the CPUOS, Gen. Assembly Official Records 51st Session, Supplement 2020; A/51/20 at P.5 (1996).

[92] Id., Par.176, P.28

[93] Jasentulyiana, Recent Developments in United Nations Activities Relating to Outer Space. Thirty Eighth Colloquium 1995, at P.208

[94] Hearings before Subcommittee on Space, Committee on Scheince, Space & Technology, U.S. House of Representative, 103rd Congress, First Session, March 24, 1993 (No.8) - The Spaceguard Survey P.158.

[95] Id. P.159

[96] Convention of the World Meteorological Organization, 14 UNTS, P.185-285, Geneva, 1977.

[97] Carusi et al Near Earth Objects: Present Search Programs in Hazards, note 1, P.127-147 at P.146

[98] New York Times, July 22, 1997, P.28 Col. I

[99] Sagan, note 84, P.325-326

[100] Morrison & Teller, note 80, P.318

[101] Sagan, note 84, P.325

[102] Morrison & Teller, note 80, at P.1142

[103] J.S. Lewis: Asteroid Resource Opportunities Suggested By Meteorite Data, in Resources of Near Earth Space, Lewis et al eds.; University of Arizona Press, Tucson, Arizona, hereinafter Resources P.523-542;

[104] J. Lewis at el.: Using Resources From near Earth Space in Resources supra note 103, P.3-10.

[105] P.R. Weissman & H. Campins; Short Period Comets in Resources supra note 103 P.569-617.

[106] C.R. Nichols: Volatile Products From Carbonaceous Asteroids in Resources, note 103, P.543-568 at P.547 et. seq.

[107] East Greenland Case, PCIJ Ser. A/B No 53 (1933); see E. Brooks, Control and Use of Planetary Resources, Eleventh Colloquium on the Law of Outer Space (1968) P.339-350.

[108] Oct. 7-11, 1996, Beijing, China, Amer. Inst. Of Aeronautics and Astronautics, Reston, Virginia, P.1-60

[109] UNCLOS note 55, Section 4, especially articles 156, 157 and 170.

[110] Done at Wellington, New Zealand, June 2, 1988, 27 ILM 859 et. seq.

Launching Spacecraft from the Sea and the Outer Space Treaty: The Sea Launch Project.

Armel Kerrest
Professor of International Law
Faculty of Law, University of Western Brittany.
Centre de Droit et d'Économie de la Mer (CEDEM)
IISL ECSL

Abstract :

During 1998 we may witness an important innovation in the field of launching spacecraft: the Sea Launch Project. Directed by Boeing, this new technique will challenge the existing parameters of the law of the sea and space law. It furthers the trend towards private activities in space. This evolution may be increased by the fact that the launch is scheduled to be take place from the High Sea, i.e. outside the territorial jurisdiction of any state. Until now, launching spacecraft was either a State activity or State controlled activity. On the whole the launch installations were State owned or in any case closely State controlled.

Two main questions are to be highlighted. The first is the definition of "launching state" and such State's potential liability for a launch from the high sea. The second is the question of control by the "appropriate state" of spacecraft launched from the sea.

In the case of the Sea Launch project, the US Commercial Launch Act will be applicable in accordance with the US-Ukraine agreement on commercial launches. The legal status of the operation is obviously more complicated but will be governed on a similar basis to those rules applicable to land launches. In a more prospective way it is possible to notice a trend to deregulation in Spacecraft launching activities if a space port is located in the High Sea.

When defining the launching state, the Outer Space Treaty, confirmed by the Liability convention, gives a very special status to the State whose territory is used for launching spacecraft. Every other criteria, such as "State that launches", "States that procures the launch", or the nationality of installations being used, may easily be chosen by a private Company.

The possible threat is a move to a situation similar to that experienced under the law of the sea regime where an international consortium wishes to avoid the control of capable states and, in doing so, benefit economically. As such, the rationale for existing space law will be hindered, jeopardising the liability of a real and fully responsible launching state, and thus the control of the state of registration which, in accordance with the registration convention, must be one of the launching states.

Introduction

There are some technical innovations which open the way to major juridical changes. The Sea Launch project seems to be one such instance. It is a private commercial programme which intends to launch satellites into orbit from a semi-submersible launch platform positioned on the Equator.

The three stage rocket is composed of the well known Soviet, now Ukrainian, two stage Zenit Rocket, and the Russian Bock DM Upper third stage. Both are very reliable and successful pieces of technology. The Zenit rocket is one of the most automated launch

---

Email: Armel.Kerrest@univ-brest.fr

vehicles. It has an important capacity in that it has the ability to send into orbit a pay load similar to Ariane 5, the very new European launcher.

A self propelled former North Sea oil drilling platform being modified in Stavanger shipyard will serve as a launch pad. It has a empty displacement of 31000 tons. An Assembly and Command Ship is being built at the Kwaerner Govan shipyard in Glasgow Scotland which will displace more than 30000 tons, and provide accommodation for a crew of up to 250. It will be used both to transport the rockets and satellites from the home port to the launch site, to assemble the three stages of the rocket and to serve as the mission control and communication centre.

The home port is located in Long Beach, California. The platform and the Assembly and Command ship will be docked there during the inter-launch period. The home port will also shelter the administrative and commercial services.

Launch operations are to be conducted in a very rationally organised way. The platform is to be positioned at the Equator in the Pacific Ocean about 1000 miles south of Hawaii. The rocket and payload are assembled within and transported to the launch site by the Assembly and Command Ship (ACS). An automated transporter and erector equipment is used to erect the launcher on the launch pad. The rocket is automatically fuelled and made ready for launch. During the launch phase, everything is conducted from the ACS located a few miles away. After the launch, or the launches in the case of multiple launch, the platform and the ACS sail back to the home port. As we can see these procedures seem to be fully integrated and much more economical than the huge launch pad used today.

The Programme aims to take advantage of two main advantages: launching from the Equator by the mobile platform, and the use of the former Soviet Union's space industry whose rockets are reliable and value for money.

The Sea Launch Co is a limited duration company incorporated in the Cayman Islands which is a British Crown colony. It is owned by several international companies including Boeing Commercial Space Company (40%), a subsidiary of the Seattle-based Boeing Company, and Kvaerner a.s. of Oslo, Norway, an important company involved inter alia in shipbuilding and which will provide the platform and the ACS (20%). In addition RSC Energia of Moscow, Russia (25%), supplies the third stage and some parts of the first and second, and conducts the launch, and NPO Yuzhnoye of Ukraine (15%) supplies the Zenit rocket used as first and second stage (this rocket has the facility of being self erecting which is very useful here). Boeing leads the team, furnishes the home port, some parts of the launcher and payload accommodation, and commercialises the launches.

The Sea Launch project already has significant orders from the satellite manufacturing companies Hughes (10 launches with an option for more), and Space Systems/Loral of Palo Alto California (5 launches). The programme is already underway, the first launch being scheduled for the end of 1998.[1]

This new programme poses some interesting juridical problems:

I The legality of launching from a territory outside any State's sovereignty.

A The freedom principle

As far as Space law is concerned,[2] the

1 Information may be found on the Internet Web on the Sea Launch project specially on the sites of the companies member of the consortium.:
Boeing: (http://www.boeing.com/sealaunch.html)
Kvaerner: (http://www.kvaerner/pr/pre-fnd.html)
See also: Monte Enbysk: "*Boeing To Reach New Heights With Commercial Satellite Venture. Sea Launch's international partners will be the first to launch satellites from sea.* at: http://fivash.com/ceo/ceo.network/ceomain.htm

2 Treaty on Principles Governing the Activities of States in the Exploration and Use of Outer

territory of the launching state is in fact one of the criteria determining the launching liable state, but even if the territory of launch is a major point specifically quoted in the liability convention article V (3), there is no obligation that states launch from their territories. There is also no such restriction in Outer Space Treaty (1967) (OST) article 1(2) which deals with the freedom of use of the outer space.

Issue concerning the freedom of the high sea is not so evident but may also be recognised as of relevance. The Montego Bay Convention states that *"Freedom of the high sea (...) comprises inter alia (...) a) freedom of navigation", and "b) freedom of* overflight". As such it does not cover either the freedom to launch spacecraft or a freedom for aircraft to land or take off. However, as stated by the words "inter alia", the list is not exhaustive, and launching spacecraft is not directly prohibited.

B The use of the high sea for launch activities.

The main problem in this respect may arise from the need to install a platform, and to establish limitations on other uses of the high sea.

Article 87 of The Montego Bay convention includes within the freedom of the sea: the *"freedom to construct artificial islands and other installations permitted under international law."*[3] Even with the restriction in the last part of the sentence, this disposition seems clear on the question of the legality of installing a platform. The real question centres on whether any platform may be used for any purpose. At the time of the Third United Nations Conference for the law of the Sea, the question of installing platforms in the high sea was not pertinent because of the creation of the Exclusive Economic Zone within which it was envisaged that such installations would be installed.

What is the meaning of the words: "installations permitted under international law" ? Do such installations have to be expressly permitted or just not prohibited ? Is not the installation of a large platform in the high sea a kind of occupation of such area with possible legal significance under the high sea regime? How is this installation connected with the legal status of the International sea bed (the Zone as qualified in part XI of the Montego Bay Convention) ? [4] Would the rules set in force by Article 147 (1), (2) and (3) "Accommodation of activities in the Area and in the marine environment" be applicable ? [5]

Utilisation of the high sea in this way is unknown at present and, as such, some difficulties may arise if such potentially dangerous activities are to take place in the high sea. Projects involving the installation of non-environmentally friendly industries such as smelters or nuclear plans have already taken place.[6]

Space, Including the Moon and Other Celestial Bodies, opened for signature January 27, 1967, 610 U.N.T.S. 205 18 UST. 2410 (1967) 6 ILM 386 (hereinafter referred to as OST).
Convention on International Liability for Damage Caused by Space Objects oct 9 1973 961 U.N.T.S. 2389 (hereinafter referred to as "Liability Convention")

3 United Nations Convention on the Law of the Sea open for signature on 10 December 1982 in Montego Bay Jamaica (hereinafter referred to as "Montego Bay Convention")

4 Montego Bay Convention: Article 137 (Legal status of the Area and its resources) and Article 139 (Responsibility to ensure compliance and liability for damage)

5 See: Armel Kerrest: "Les aspects juridiques du projet Sea Launch de lancement de satellites depuis la haute mer" revue Droit et défense n°97/1 Paris. And more generally: A. Kerrest "The Launch of Spacecraft from the Sea" in: "An Outlook on Outer Space Law in the Coming Thirty Years". Lafferranderie et Crowther edit. (Kluwer 1997).

6 Even if the launch of such a powerful rocket is always rather dangerous because of the quantity of energy involved, Sea Launch does not seem to be excessively harmful for the marine environment

However the main question should be the limitation on the freedom of navigation. The platform itself, but also more importantly the need to prohibit navigation in a very large area during the launching periods around the launch site impose a significant limitation on navigation. The zone to be used by Sea Launch is far way from the main maritime zones, but, nevertheless, this issue is not strictly dealt with by the Law of the Sea Convention which, within the high sea, deals only with platforms used for mineral resources exploitation and scientific research. The first must be authorised by the Sea Bed Authority, and the second may be established but cannot create a security zone of more than 500 meters. It seems rather difficult to recognise that a private commercial project should be able to limit freedom of navigation to a greater extent than those involved in scientific research, a well respected activity, are allowed to do.

II The consequences in space law of a High Sea launch.

A) Definition of the launching state.

One of the major difficulties regarding the Sea Launch project is in connection with the notion of "launching state" which, as every body knows, gives a prominent place to the launching activity for purposes of identifying liability and responsibility.

First of all, with regard to private launch activities, it is necessary to keep in mind that, in accordance with OST article VII and VI, activities carried out by private entities are deemed to be carried out by the liable state. This State is not only responsible, i.e. obliged to authorise and control any space activity conducted by a non governmental entity having its nationality, but also the "launching State" is, as such, directly liable for any damage cause by the space object as if it were launching it itself. The fact that the launching authority could be a private company is a new situation which has to be taken into consideration. The existence of the Commercial Space Launch Act in the United States legal system is of course of great help. It should certainly not be possible in other countries where the domestic laws are either too general or non-existent.

Such a situation raises the question as to which state or states are the launching states in the case of a launch from the high sea by a private entity. Because of the consequences of liability in case of damage, this point is of major interest and concern. If we use the criteria of OST and Liability Convention, we have: "a) the State which launches". Here, no state launches but rather a private company. From the rule before mentioned, the nationality of the juridical person (in this situation the Sea Launch Co.) renders the state of such nationality the "Launching State". The issue whether the effective control of the company should be considered instead of the legal personality of the such company is unclear in international law.

The Sea Launch Co. being a registered company of limited duration under the law of Grand Cayman Island, a British Crown colony, the UK could be regarded as a launching state[7]. But if we look at the real management of this company, we can see that the project is being carried out by Boeing Commercial Space Co., a subsidiary of the US Boeing Company. Thus the United States should be also regarded as a launching state.

Issues relating to the other parties are not so clear.[8] As far as RSC Energia is concerned,

[7] The introduction of the British "Outer Space Act" in this Crown Colony is unclear. But it is a domestic question not modifying the situation in international law (the possible liability of the UK).

[8] The discussion after the oral presentation of this paper showed that the notion of launching state is not clear especially as far as the amount of participation in the launching activity is concerned. Some lawyers having a wide conception including any participation and others thinking that the participation must be of some special importance in order to qualify the State as a "launching state".

it plays an important role in the Company, furnishes some main parts and is in charge of the technical direction of the launch. As such, Russia should also be considered as a launching state. In this situation both the US and Russia should be considered as launching states either as the "states that launch" or as "the states whose facilities are used".

The Ukrainian company NPO Yuzhnoye supplies the Zenit rocket. As to whether this is enough to consider the Ukraine as a state which launches, this remains a matter for discussion. As this Company takes only a 15% of the shares, the point is not certain. The same situation applies to the Norwegian company Kvaerner, especially if, as seems to be the case, the ship and platform even though provided by Kwaerner, does not fly the Norwegian flag.

The second criterion: "the state that procures the launch" cannot be considered here as it depends on the payload. Taking into account the commercial purpose of Sea Launch, the entity procuring the launch may be either a state or a private company qualifying its national state as a launching state.

The two last criteria which determine the launching states are the territory and the facilities from which the space object is launched. The territory is high sea, i.e. a territory where no state exercises its sovereignty[9]. We therefore have to consider the "state from whose facility a space object is launched", here in fact the nationality of the company whose "facilities" are used. The facilities must first of all include the platform and ACS, both of which fly a flag of convenience, perhaps Liberia's. In this respect, Liberia should be a launching State.

We must then ask the question whether the home port installations are "a facility" in that sense. It would be a good solution to link the space activity to a state's territory. But, is it really a "facility from which the launch is done" ? It is difficult to support this wide interpretation of the text, especially due to the consequences of such an interpretation on the issue of liability.

B) The United States as one of the launching states

The application of the US Commercial Space Launch act (1984) (US CSLA)[10] embraces a lot of people. The application of the various national space acts is always very wide and, in most cases, extraterritorial. States want to avoid being liable as launching state in situations where they are unable to control the space activity through a licensing procedure. It is one of the positive side-effects of the objective liability rule in space law. The US CSLA distinguishes between activities carried out by "citizens of the United States" defined in paragraph USC 2603 (11) A and B as "natural or legal persons of American nationality", on the one hand, and, on the other hand, activities carried on by "citizens of the United States" defined in paragraph C of the same provision as legal persons of foreign nationality controlled by a natural or legal person of the United States.

In the case of Sea Launch, there is a special point which must be highlighted : the project is using the former Soviet Union's technology which is very reliable and rather

9 The theory of the quasi-territoriality of the ships is no more recognised in international law. The nationality of the facility is the only criterion to be taken into consideration here.

10 United States Commercial Space Launch Act 1984, amended 1988 at 49 U.S.C. app. §§ 2601-2623 (hereinafter referred to as CSLA: (USC §2303)

(11) "United States citizen" means-

(A) Any individual who is a citizen of the United States;

(B) any corporation, partnership, joint venture association, or other entity organized or existing under the laws of the United States or any State; and

(C) any corporation, partnership, joint venture, association, or other entity which is organized or exists under the laws of a foreign nation, if the controlling interest (as defined by the Secretary in regulations) in such entity is held by an individual or entity described in subparagraph (A) or(B)

cost effective. As you know, fearing some abuse, the United States have imposed on the states which have not yet an established market economy, agreements concerning the selling of commercial launches. Such states include China, Russia and the Ukraine. [11] These agreements limit the sale of launches from those countries, but special provisions apply in situations involving co-operation between US companies and the foreign rocket manufacturers. In that case, the consortium must be led by the US Company and has to obtain a licence from the US government and to be established within the structure set up by the US Commercial Launch Space Act 1984. The consequence of such a rule is that the US should be a launching state. The launching consortium has no interest in avoiding the application of the US law in this way. The potential victims of an accident have the great advantage of such US liability.

III The possible trend

It seems that the commercialisation of the launching activity, and especially the launch from an international located pad, will change dramatically the application of one of the most important rules of Space law. The territory of launch is, at the moment, a "lock" in the system. As space-crafts are launched from the territory of a technically and economically powerful country,[12] those countries are volens nolens liable launching states. The potential victims have good chances to obtain compensation[13]. If the territory criterion is left aside because of a launch from international domain, this lock is being abandoned, the only criteria left are those connected with nationality which may be chosen by the launching private companies. When they choose their nationality, the entrepreneurs can choose the launching state and thus the registration state[14]-as this state has to be one of the launching states- and the "appropriate state"[15] responsible for the control of the space activity. We are not very far from the situation in the high sea where the companies can -in fact if not in law- choose the flags of their ships and thus the applicable law, the domestic law or even the international one. We all know the very detrimental effect of such "anarchy": the great difficulty to set any efficient rule in those international territories.

Moreover, the situation should be more catastrophic in space than it is on the high sea. A ship is used to link two harbours in two land territories. If they want, the states may intervene at each end. As far as Space activities are concerned, as soon as the spacecraft is

11 Agreement Between the Government of the United States of America and the Government of Ukraine Regarding International Trade in Commercial Space Launch Services signed : 15 Oct. 1995 in force : 21 Feb. 1996. In Journal of Space Law 24 1996 -2 (hereinafter referred to as US-Ukraine agreement) The « integrated space launch services provider » is defined in Article II.3 : "*a joint venture that includes Ukrainian and U.S. companies and provides commercial space launch services or commercial space launch vehicles through implementation of joint projects in the field of rockets and space technology where financing for such projects comes from investments, commercial loans, and other means. In this joint venture : a) the venture receives a commercial launch licence issued by the U.S. Department of Transportation ; b) the US partner maintain a significant equity interest in, and control in fact of the joint venture and the United States is a source of a significant share of the goods and services employed by the joint venture in any space launch ;c) a majority of the goods and services employed by the joint venture in any space launch have their origin in market economy countries....* »

12 This country may be a "developed" or a "developing" country such as India, Brazil or China.

13 Even if the system of settlement of dispute is not efficient enough in the liability convention as in the international society in general.

14 Convention on Registration of Objects Launched into Outer Space:
article 1: *"The term "State of registry" means a* launching State *on whose registry a space object is carried in accordance with article II." (emphasis added)*

15 In the meaning of article VI of the OST 1967

launched, until it falls down to earth, it has no more physical contact with any territory. It remains for ever in an international location, and is linked for ever to a certain or some launching state(s).

Some very talented authors suggest that the present system of the launching state liability should be modified[16] and that the burden on the launching state should be removed. This does not seems to be the case. Of course, the commercialisation of space activities must be taken into consideration, the liability of the private companies acting in space have to be made certain, but the safety net which is the absolute liability of one or some launching states must be maintained. This protection has been set in force as a distinction between the space activity which is by nature highly risky on the one hand and potential victims, not directly involved in space activity who are protected through the absolute liability set by the convention on the other hand. This rule has two advantages: the first is to give a guarantee to victims, and the second, of more interest, is to ensure that a state will control the whole activity. This obligation to control is already established in article VI of the OST but the fulfilment of this obligation is only a matter of international responsibility which is certainly not as efficient as absolute liability. Incurring liability, this state will make sure that the activity is conducted with care. The victim can sue the real actor, but, as we know, it may be very difficulty either to obtain a decision proving fault or even to obtain a proper indemnity from the company. Beyond that possibility it is good to maintain the safety net of the launching state's liability. So much the better if it is not used as the insurance system works well, but, if it is not the case, the victim can sue the launching state, either directly before a domestic judge or through the state he or she is a national of in accordance with application of the liability convention.

Conclusion.

Sea Launch launches from the high sea. To comply with the Ukraine-US agreement, it must act under the US law and be licensed by the US authorities, making the US a liable launching state. Were this particular situation not to exist, another company, more affected by commercial competition, may have chosen to escape from any real control, choosing a flag of convenience for the ship, a nationality of convenience for the company...etc.

If, as it is already the case at sea, competition become fierce in space launch activity or even in any space activity, it will be useful to avoid strong controls. By launching from an international territory, it is much easier to choose the connecting links with the "launching", "appropriate" or "registration" state. We have to keep in mind that choosing a state is also choosing the law, either national or international, applicable to the operation and choosing the authority in charge of the control. Difficulties would arise if states without any space capacity were chosen for this very reason, as unique launching state. If we want to maintain a high level of security and a control over space activities, such a mess is to be avoided. This is quite possible, as the main users of space activities are powerful countries either "developed" or not. Even if the largest freedom of use has to be respected and private enterprise encouraged, in outer space like elsewhere, we have no long term interest in opening the Pandora box of a total deregulation of fast growing activities in Outer Space or connected with it.

---

16 See Henri A. Wassenbergh *A launch and a transportation law separate from outer space law ? It is time to legally unburden the "launching state"* Air and space law 21 1996 1 p. 28-32 . and: Frans von der Dunk: *Commercial space activities: an Inventory of Liability, an inventory of Problems,* Proceedings of the thirty-seventh Colloquium in the law of Outer Space (1992) 161-71. And: *Loopholes in Liability ? Aspects of Liability for Damages Sustained in the Course of Satellite Telecommunications Activities.* 2 Telecommunications & Space Journal (1995) 153-74.

48TH INTERNATIONAL ASTRONAUTICAL CONGRESS , 6-10 OCTOBER, 1997

IISL-97-IISL.3.17

GNSS AUGMENTATION : LEGAL ISSUES

by Paul B. Larsen*)

Georgetown University Law Center

I. Background

GNSS is a satellite positioning and navigation system which provides worldwide three dimensional position, velocity and time data. Users are drawn from a great varriety of professions. The GNSS provided by the United States, known as the Global Positioning System (GPS; in the following discussion the words GPS and GNSS will be used interchangably) provides two kinds of service: Standard Positioning Service (SPS) and Precise Positioning Service (PPS). The less precise of the two services is the Standard Positioning Service. It is freely available to all users worldwide. It became operational for all uses in 1993. By contrast, the Precise Positioning Service is only available for military users and is not here relevant for discussion except to mention that the United States has decided that the Precise Positioning Service will become generally available in the year 2006. In fact it could become available as early as the year 2000 if the U.S. Department of Defense (DOD) cannot justify its exclusive use. However, neither SPS nor PPS may satisfy all users. GPS systems need to be sufficiently precise to satisfy the very demanding requirements of both transportation and non-transportation users. GPS augmentation will be necessary for additional GNSS signal precision. When sufficiently augmented to satisfy the requirements for accuracy, coverage, availability and integrity, the GNSS will become the primary U.S. radionavigation system. 1/ A description of the technical environment within which legal issues arise, follows.

A. U.S. Aviation Augmentation Program

The aviation industry and the Federal Aviation Administration (FAA) are very keen to make augmented GPS the primary radionavigation system. Therefore the FAA has begun to augment GPS. The FAA's Wide Area Augmentation System (WAAS) will provide the accuracy, integrity and availability that are necessary for GPS to become the primary radionavigation system for all kinds of flight from enroute flight to Category I approaches. 2/ First WAAS availability for use is scheduled for 1999 and in that year WAAS will be certified by the FAA as the primary means of navigation for en route flight and for terminal operations and for limited precision approach service. Full WAAS capability is planned for the year 2001. The Local Area Augmentation System (known as LAAS and consisting of ground reference stations) will provide additional accuracy, integrity and availability for Category II and Category III approaches. (See Appendix I for accuracy requirements for air navigation; in particular note the great accuracy required for Category III landings). The FAA has canceled the so-called Microwave Landing System in reliance on availability of a GPS stand-alone navigation system in the future. In the year 2003 the FAA will decide whether to terminate other existing ground based systems, depending on how well the WAAS and the LAAS perform. The FAA also will take into consideration how well users accept the new technology and the extent to which users have the necessary new equipment on board. Needless to say the FAA is motivated by budget constraints to eliminate redundant services. 3/

B. U.S.Maritime Augmentation Program

A maritime augmentation system was declared operational by the U.S. Coast Guard in 1996. It is already functioning successfully. The maritime augmentation system provides service along the U.S. coasts and on navigable rivers such as the Mississippi River. Maritime augmentation operates by use of radionavigation beacons. The maritime augmentation system is based on 4/

"accurate knowledge of the geographic location of one or more reference stations, which is used to compute corrections to GPS parameters, error sources, and/or resultant positions. These differential corrections are then transmitted to GPS users, who apply the corrections to their received GPS signals or computed positions. For civil users of SPS, differential corrections can improve navigational acuracy from 100 meters (2drms) to better than 7 meters (2drms)."

The maritime augmentation system may well be extended inland in the United States. One important reason is that several serious train accidents have caused concern with rail safety and there is interest in using the maritime augmentation system to effect positive train control. Many other potential users also could use this augmentation. Great public benefits would result from such a nationwide augmentation system. For surface traffic the maritime system has significant advantages over the WAAS system. The maritime augmentation system can better reach surface vehicles; because WAAS is specifically designed to meet the needs of aircraft in flight. (See Appendix II for accuracy requirements for land transport positioning and navigation, and see Appendis III for accuracy requirements for surveying, timing and other applications)

C. European Augmentation

The European Commission, Eurocontrol and ESA have agreed to cooperate to implement a European GNSS augmentation system, the European Geostationary Navigation Overlay System, known as EGNOS. Based on a December 1994 European Union Council Resolution, the European Commission, EUROCONTROL and ESA agreed to augment GPS and GLONASS for the purpose of civilian users. The planned augmentation service will use Satellites in Geostationary Orbit (GSO) to provide " supplementary ranging signals, integrity and differential correction data;" 5/ EGNOS is a multimodal satellite service. The Commission provides political and financial support under the Telematics Research Program to demonstrate the feasibility of GNSS augmentation for air, sea and surface uses, in particular for traffic management. ESA provides the ground segment and operates EGNOS during the initial testing phase. EUROCONTROL will develop the aviation certification requirements. EGNOS has obtained access to the navigation payloads of INMARSAT III by virtue of an ESA agreement with INMARSAT though Deutsche Telekom and France Telecom. 6/

The EGNOS action plan contemplates possible extensions of EGNOS to other regions including Africa, India, Eastern Europe and China. 7/ Furthermore, the Commission engages in extensive consultation and coordination with the United States on GPS issues. 8/

D. Japanese Augmentation

Japan is a heavy user of GPS for multiple purposes, including disaster monitoring, intelligent vehicles and for air and maritime navigation. Therefore the Japanese Ministry of Transport will provide supplemental service though its Multifunctional Transport Satellite program. The service is known as the MTSAT Satellite-based augmentation system. The MTSAT-1 satellite is scheduled to be launched in 1999. Specifications generally are based on the U.S. WAAS described above. Thus WAAS and the MTSAT Satellite-based Augmentation System (MSAS) have been coordinated. Japan will study and intends to coordinate its multitude of needs for satellite positioning and navigation. Within that study Japan will consult closely with the

United States and other interested parties regarding future developments. 9/

E. Other Augmentation

Many other countries, especially including India and China 10/ are studying GNSS and its augmentation as they become potential users of satellite positioning and navigation. Clearly this is a rapidly expanding area of technology.

F. GPS and GLONASS Coordination

Improved GNSS availability and thus improved positioning and navigation accuracy may also be obtained by combined use of GPS and the corresponding Russian GNSS system called GLONASS. The United States and Russia have entered into agreement, therefore, on a combined GPS/GLONASS receiver which would take advantage of all the 48 satellites of the two systems. 11/

II Legal Issues

A. Introduction

GNSS augmentation concerns the establishment of additional reference points which are used to improve the accuracy of navigation and positioning. GPS users such as airplanes and ships receive signals directly from the GNSS satellite. Additionally, GPS users (airplanes and ships) receive augmented satellite information indirectly via another object whether that other object be a radio beacon located on earth, such as in the maritime augmentation system described above; or whether it is processed via a satellite as in the case of the WAAS. The new legal consideration with augmented GNSS is that it introduces the additional consideration of satellite information received via other sources. For example in the case of the WAAS there is the additional factor of possible erroneous processing of the GNSS information because either the processing itself, or the manufacture of processing instrumentation, is negligent.

B. Legal Institutions for Augmentation.

1. International Institutions.

GNSS augmentation by an international institution such as EGNOS represents interesting possibilities which were discussed at a conference at Leyden University in 1996. 12/ EGNOS itself, as an international institution, may have international legal status. Otherwise the contracting countries themselves and the international institutions (EUROCONTROL, ESA , European Union) would form the legal basis for the augmentation . During the interim period while EGNOS becomes operational the sponsoring countries and organizations may provide the institutional basis.

The International Civil Aviation Organization (ICAO) and the International Maritime Organization (IMO) are also relevant international organizations because they establish standards for use of navigation satellites by aircraft and by ships. 13/ Furthermore, the International Telecommunication Union (ITU) is relevant because it develops world telecommunication standards and registers GNSS radio frequencies .

International government organizations, that provide augmented GPS, such as EGNOS, as discussed during the University of Leyden Proceedings, would be immune from liability, unless they permit themselves to be held liable by national or international laws.

2. National Organizations

Under the Federal Aviation Act, 14/ the FAA is charged with maintaining air safety; this law provides the FAA's legal basis for operating the WAAS and for the LAAS systems. Under Federal maritime legislation, 14 USC 2 and 81-90, the U.S. Coast Guard is provided the legal basis for the U.S. maritime augmentation system (DGPS). Both systems may be extended to uses other than aviation and maritime if other legal authority is provided either by new legislation or by the existing legal authority of other organizations, for example the Federal Railroad Administration's authority for positive train control. 15/

In the United States the FAA establishes standards for aviation uses of augmented GPS. Likewise the Coast Guard oversees the use of augmented GPS by ships. The Federal Communications Commission (FCC) regulates the use of radio frequencies. 16/ Because there are many uses of GPS, other than navigation, a number of additional Government agencies have oversight functions.

Because the FAA and the Coast Guard, as operators of the augmented GPS, are U.S. Government agencies, they are subject to the U.S. Federal Tort Claims Act (FTCA). 17/ The FTCA provides that the U.S. Government is immune from liability for its negligent acts unless the FTCA specifically waives immunity. The FTCA waives immunity of the FAA for negligent acts except for discretionary acts. Air traffic control is defined by case law as not being discretionary. Thus the FAA can be held liable. Likewise the Coast Guard can be held liable for negligent maritime navigation control. 18/ Consequently, the FAA and the Coast Guard may be held liable for augmented GPS to the extent that the FAA's operation of WAAS and LAAS, and the Coast Guard's operation of its maritime augmentation system are legally analogous to traffic control of airplanes and of ships. 19/

3. Private Organizations.

There are several private augmentation systems. 20/ Their legal bases are the charters of their corporations and national legislation. Private organizations would be subject to government oversight of their augmentation services

Private organizations providing augmentation services would not have immunity for their negligent augmentations. They could be held liable for negligent augmentation under the applicable law of Torts. Sometimes the ascertainment of the applicable torts law would require a conflict of laws analysis. 21/

III. GNSS Augmentation and the 1967 Outer Space Treaty

This is the thirtieth anniversary of the 1967 Treaty on Principles Governing the Activities of States in the Exploration and use of Outer Space, including the Moon and Other Celestial Bodies. 22/ The treaty is the 'Constitution' for outer space activities. This year we are specially encouraged to review space activities from the perspective of the treaty and to consider how space activities may be better assisted and accommodated by, or under, the 1967 Outer Space Treaty.

Augmented GNSS is quickly on its way to replace cumbersome surface-based traffic controls of transportation. Analogy can be made to satellite communication which in a short time relieved communication from dependence on earthbound copper wires. Augmented GNSS likewise appears to constitute a great leap forward. Augmented GNSS should quickly become available to all countries. GNSS is one of the most financially promising outer space activites. U.S. studies estimate future economic activities in the range of 30 billion dollars annually. 23/ The treaty, Article 1, states that use of outer space "shall be carried out for the benefit and in the interest of all countries, irrespective of their degree of economic or scientific development." This may be a suitable subject for implementation of Art. 1 of the 1967 Outer Space Treaty. I have previously proposed 24/ that the great variety of users of GNSS give rise to a great variety of institutions that regulate GNSS and that only one institution has competence to consider all the GNSS uses without giving preference to one over another -- that is the United Nations Committee for the Peaceful Uses of Outer Space (UNCPUOS) Thus I have proposed that UNCPUOS consider preparing overarching principles, as it did for remote sensing satellites in 1987. Such principles could insure the use of GNSS for the benefit of all countries, irrespective of their degree of development. In particular I point to principles on non-discriminatory access to GNSS, technical assistance, protection of the earth's environment, warning of disasters, international responsibility and dispute resolution. 25/ Overall U.N. guidance would leave specialized agencies and governments free to focus on technical issues. 26/

IV. Conclusion

GPS is a valuable tool which promises to be of great benefit to mankind. As uses become more diverse there needs to be refinements of the basic GNSS service. Augmented GNSS is not only a refinement but also an improvement of the basic GNSS service. Legally, augmented GNSS is a GNSS service for which international organizations, such as EGNOS, national governments, or private companies have assumed responsibility. The international and national reponsibilities for augmented GNSS are similar to those for GNSS. In the future further efforts may be required to standardize augmented GNSS. International technical standardization would take place through specialized agencies such as ICAO and IMO. Finally, the United Nations Committee for Peaceful Uses of Outer Space should consider establishing some basic overarching principles including non-discriminatory access to GNSS, technical assistance, protection of the earth's environment, warning of disasters, international responsibility and dispute settlement.

*) The views expressed in this article are the author's and should not be attributed to any organization with which he is associated.

1. 1996 Federal Radionavigation Plan(FRP) at 1-10. The FRP is available through the National Technical Information Services, Springfield,Virginia 22161. The FRP is updated biennially. It describes policies and plans for radionavigation services provided by the U.S. Government. The three appendixes to this paper are from the FRP; Appendix I at 2-5, Appendix II at 2-20, and Appendix III at 2-38.

2. Definition: The WAAS, system consists of WAAS certified aircraft receiving a signal-in-space providing 1: integrity data on GPS and Geostationary Earth Orbit (GEO) satellites; 2. differential corrections of these satellites to improve accuracy; and 3, a ranging capability improving availability and continuity. The GPS satellite data is processed by wide-area master stations "which process the data to determine the integrity, differential corrections, residual errors, and ionospheric information for each monitored satellite and generate GEO satellite parameters. This information is sent to a ground earth station and uplinked along with the GEO navigation message to the GEO satellites. The GEO satellites downlink this data on the GPS L1 frequency with a modulation similar to that used by GPS." The WAAS " will result in area navigation for all phases of flight and a ten-fold increase in runways approved for precision approaches." See Federal Radionavigation Plan at 3-11.

3. Federal Radionavigation Plan, id.

4. Id. at A-9.

5. Barbance, Bergquist, Cheli, Hood and Nordlund, Satellite Navigation Activities, The International Context, 1996 Space Communications, at 158.

6. Id.

7. Id. at 159.

8. Madrid Agreement between the European Union and the United States creates a consultation subcommittee.

9. 1996 Space Communications, supra n. 5, at 160-161. The Japanese augmentation service is known as the MSAT Satellite-based Augmentation System (MSAS).

10. Id.

11. Federal Radionavigation Plan, at 3-24.

12. Proceedings of the 1996 Leyden University Conference.

13. Larsen, Positioning Satellites: Current Institutional Issues, Proceedings of the 1994 Proceedings of the Colloquium on the Law of Outer Space, at 32.

14. 49 USC 40101, et seq..

15. 49.USC 103.

16. 47 USC 151 et seq.

17. Federal Tort Claims Act (FTCA), 28 USC 2671, et seq.. Larsen, Liability for Global Positioning Navigation Satellite Systems, Proceedings of the 36th Colloquim on the Law of Outer Space, at 70-71.

18. Dalehite v. U.S., 346 U.S. 15 (1952); Eastern Airlines v. Union Trust, 221 F. 2d 62 (affirmed per curiam sub nom. U.S. v. Union Trust, 350 U.S. 907 (1955)); Indian Towing v. U.S., 350 U.S. 61 (1955).

19. Id. supra n. 17. Note that the FTCA does not apply outside of the United States.

20. Federal Radionavigation Plan supra n. 1.

21. Restatement of the Law of Conflicts.

22. Treaty on Principles Governing the Activities of States in the Exploration and Use of Outer Space, including the Moon and Other Celestial Bodies, 18 UST 141; hereinafter the 1967 Outer Space Treaty.

23. National Academy of Public Administration Report.

24. Larsen, supra n. 13,

25. 1967 Outer Space Treaty, Art. VI.

26. Larsen, supra n. 13.

APPENDIX I

Table 2-1. Controlled Airspace Navigation Accuracy Requirements

PHASE	SUB-PHASE		ALTITUDE FL/FT	TRAFFIC DENSITY	ROUTE WIDTH (nm)	SOURCE ACCURACY CROSS-TRACK (95%, nm)	SYSTEM USE ACCURACY CROSS-TRACK (95%, nm)
EN ROUTE/ TERMINAL	Oceanic		FL 275 to 400	Normal	60*	12.4*	12.6*
	Domestic		FL 180 TO 600	Low	16	2.8	3.0
				Normal	8	2.8	3.0
			500 FT to FL 180	High	8	2.8	3.0
	Terminal		500 FT to FL 180	High	4	1.7	2.0
APPROACH AND LANDING	Nonprecision		250 to 3,000 FT	Normal	N/A	0.3	0.6
	Precision	CAT I	N/A	Normal	N/A	+/-18.2 ** / +/-7.7 to +/-4.4 *** CAT I Decision Height Point ****	N/A
		CAT II	N/A	Normal	N/A	+/-6.5 ** / +/-1.7 *** CAT II Decision Height Point ****	N/A
		CAT III	N/A	Normal	N/A	+/-4.1 ** / +/-0.6 *** At Runway Threshold ****	N/A

* *North Atlantic Track System requirements.*
** *Lateral position accuracy in meters.*
*** *Vertical position accuracy in meters.*
**** *Assumes a 3° glideslope and 8,000 ft. distance between runway threshold and localizer antenna. It may be possible to meet CAT III touchdown performance requirements while applying the CAT II requirement down to the runway.*

APPENDIX II

Table 2-3. Current Maritime User Requirements/Benefits for Purposes of System Planning and Development - Harbor Entrance and Approach Phase

REQUIREMENTS	MEASURES OF MINIMUM PERFORMANCE CRITERIA TO MEET REQUIREMENTS								
	ACCURACY (meters, 2drms) PREDICTABLE	ACCURACY (meters, 2drms) REPEATABLE	COVERAGE	AVAILABILITY	RELIABILITY	FIX INTERVAL	FIX DIMENSIONS	SYSTEM CAPACITY	AMBIGUITY
SAFETY OF NAVIGATION - LARGE SHIPS & TOWS	8-20***	-	US harbor entrance and approach	99.7%	**	6-10 seconds	Two	Unlimited	Resolvable with 99.9% confidence
SAFETY OF NAVIGATION - SMALLER SHIPS	8-20	8-20	US harbor entrance and approach	99.9%	**	***	Two	Unlimited	Resolvable with 99.9% confidence
RESOURCE EXPLORATION	1-5*	1-5*	US harbor entrance and approach	99%	**	1 second	Two	Unlimited	Resolvable with 99.9% confidence
ENGINEERING & CONSTRUCTION VESSELS HARBOR PHASE	1****-5	1****-5	Entrance channel & jetties, etc.	99%	**	1-2 seconds	Two and Three	Unlimited	Resolvable with 99.9% confidence
BENEFITS									
FISHING, RECREATIONAL & OTHER SMALL VESSELS	8-20	4-10	US harbor entrance and approach	99.7%	**	***	Two	Unlimited	Resolvable with 99.9% confidence

* Based on stated user need.
** Dependent upon mission time.
*** Varies from one harbor to another. Specific requirements are being reviewed by the Coast Guard.
**** Vertical dimension.

Table 2-8. Requirements for Surveying, Timing and Other Applications

Surveying

TASK	MINIMUM PERFORMANCE CRITERIA								Remarks
	Accuracy - 1 Sigma				Coverage %	Availability %	Interval		
	Position						Measurement Recording (seconds)	Solution Fix	
	Absolute (m)		Relative (cm)						
	Horizontal	Vertical	Horizontal	Vertical					
STATIC SURVEY	0.3	0.5	1.0	2.0	99	99	5	30 min.	0 - 25 km
GEODETIC SURVEY	0.1	0.2	1.0	2.0	99	99	5	4 hrs.	0 - 6000 km
RAPID SURVEY	0.3	0.5	2.0	5.0	99	99	1	5 min.	0 - 20 km
"ON THE FLY" KINEMATIC SURVEY	0.3	0.5	2.0	5.0	99	99	0.1 - 1.0	0.1 - 1.0 sec.	0 - 20 km Real Time
HYDROGRAPHIC SURVEY			300	15	99	99	1	1 sec.	

Timing and Other Applications

REQUIREMENTS	MEASURES OF MINIMUM PERFORMANCE CRITERIA TO MEET REQUIREMENTS								
	ACCURACY (2 drms)			COVERAGE	AVAILABILITY	FIX INTERVAL	FIX DIMENSION	SYSTEM CAPACITY	AMBIGUITY
	PREDICTABLE	REPEATABLE	RELATIVE						
COMMUNICATIONS NETWORK SYNCHRONIZATION	-	1 part in 10^{-10} (freq)*	-	Nationwide	99.7%	Continuous	N/A	Unlimited	N/A
SCIENTIFIC COMMUNITY	-	1 part in 10^{-10} (freq)	-	Worldwide	99.7%	Continuous	N/A	Unlimited	N/A
METEOROLOGY	Velocity 1m/sec	-	-	-	TBD	TBD	TBD	-	TBD
POWER NETWORK SYNCHRONIZATION	-	1ms**	-	North America	99.7%	1 second	Two	Unlimited	Resolvable with 99.9% confidence

* Proposed ITU Standard based on American Telephone and Telegraph "Stratum 1 Requirement".

** At any substation. 8ms (1/2 cycle) systemwide.

IISL-97-3.18

OUTER SPACE TREATY AS A FRAMEWORK FOR THE REGULATION OF SPACE DEBRIS

Dr. Mahulena Hošková[*]

The problem of reducing space debris with the help of legal instruments belongs to the most discussed issues since the end of the 80ties. These discussions led to the elaboration of the Draft of the International Law Association for a Convention on Space Debris submitted to the 66 Conference of the ILA in Buenos Aires in August 1994, the provisions of which seem to constitute a useful starting point for further regulatory activities. The basis of this document are the pertinent provisions of the Outer Space Treaty together with the principles of the 1972 Liability Convention.

This paper deals with the question whether and under which conditions the international instruments already in force, primarily the Outer Space Treaty itself, have the capacity to influence the behavior of the space providers in respect to the space debris or whether a supplementary set of rules is necessary. The analysis of this question leads to the conclusion that, with regard to the specific character of space debris, the already valid rules seem to be too general to be able to cope effectively with this phenomenon.

I. Introduction

Celebrating the Thirtieth Anniversary of the Outer Space Treaty of 1967 (OST) the UN Committee on Peaceful Uses of Outer Space (COPUOS) reviewed the scientifical and technical as well as the legal aspects of the exploration and the peaceful uses of outer space, space debris being among the highlighted topics.

---

* JUDr. (Prague), CSc.; Member IISL; Alternate Member, Space Law Committee of the International Law Association.
Max Planck Institute for Comparative Public Law and International Law, Heidelberg, Germany.

One of the questions connected with this phenomenon was, to what extent the OST can cover very specialised items not envisaged in the years of its elaboration and how intensive is the necessity of its amendment, in order to cope with the new conditions: the growing quantity of debris on one side and the improving technical possibilities of tracking and managing space debris on the other side.

The Czech delegation to the Legal Sub-Committee of the COPUOS put on its 1996 session as an unofficial background note a set of questions, the answer to which could help to catalogue the main groups of legal problems connected with space debris.[1] The following problems have been defined:

- Does the definition of „space object" as contained in instruments of space law cover space debris?
- Do provisions of the 1967 OST concerning the avoidance of harmful contamination of outer space and adverse changes in the environment of the Earth apply to the problem of space debris?
- Should the protection of ownership of space objects, and their component parts, also be extended to space debris?
- Should liability for damage caused to a space object and/or its crew by space debris depend on the proof of fault as in the case of a collision of two space objects?[2]

Nothwithstanding the changing political and technological approaches and constellations, the respective conclusions could help to measure the extent of flexibility of the OST for this particular area of environmental protection and to signalise the urgency of a legislative action. As a model of an already formulated solution of the problem of space debris, the 1984 International Law Association Draft „Buenos Aires International Instrument on the Protection of the Environment from Damage Caused by Space Debris" may be mentioned[3].

II. The term „space debris"

De lege lata, the question whether the definition of „space object", as contained in the valid instruments of space law, covers space debris, does not depend merely on the fact what should be - in the scientific aspects - understood under the terms „debris". It depends in the same way on the interpretation of the category of „space object" under the international law in force. Because of the fact that the OST was a „pragmatic" instrument corresponding to the political and technological conditions of the period in which it has been drafted, there was no necessity to elaborate such definition. In order to determine precisely the scope of liability for damage to another State Party or to its natural or juridical persons, the term „object" has been included into Art. VII OST, differentiating between „objects launched into outer space" and „component parts" of these object.

The provision of Art. 1 of the 1972 Convention on International Liability for Damage Caused by Space Objects describes a „space object" as including „component parts of these object as well as its launch vehicle and parts thereof." *Mutatis mutandis*, the 1975 Convention on Registration of Objects Launched into Outer Space stipulates in Art. Ib that the term „space object" includes component parts of a space object as well as its launch vehicle and parts thereof.

The partial „in circle" definition of space objects together with many unsolved problems concerning the regime of aero-space objects did not bring any precise line between the category of space objects and another objects which exist in outer space. It is is *e.g.* unclear whether a permanent installation affixed to a celestial body constitutes a space object. Another open question is whether an object must be launched into outer space to be considered a space object within the terms of the OST or whether it is sufficient that it can be assembled in outer space? Or: Is the space shuttle a space object?[4] However, the repeatedly underlined fact that the term „space objects" include in the valid international space instruments its component parts, should permit to include the „space debris" in the category of the „space objects" and their - however often unclear - regime under present international law.

De lege ferenda, in the course of many years in which the legal aspects of space debris have been under discussion, many scientific definitions have been elaborated.

The 1984 Buenos Aires Document defined „space debris" as man made objects in outer space, other than active or otherwise useful satellites, when no change can reasonably be expected in these conditions in the foreseable future (Art. 1 c). Based upon the same approach that space debris fall under the notion of „space objects" and nothwithstanding that the 1984 definition contains terms which could theoretically lead to certain misinterpretations, the intent of the drafters of the document is clear.

The same can be said about the definition prepared by the Scientific and Technical Sub-Committee of the COPUOS in 1996, which stipulates that space debris are all man made objects, including their fragments and parts, whether their owners can be identified or not, in Earth orbit or after re-entering the dense layers of the atmosphere that are non-functional with no reasonable expectation of their being able to assume their intended functions or any other functions for which they are or can be authorised. However, the uncertainity concerning the „reasonable expectation" and the institution which should define such a constellation persists.

It is well known, however, that the success of a legal definition does not depend only on the scientific accuracy of the described phenomenon; it should primarily serve as an expression of the consensus of the future parties to the legal document as to the scope of the regulation in question. If there exists a strong controversy among the authors of the document in preparation about the width or depth of the proposed regulation, even the best scientific definition cannot bring any realistic results.

III. Harmful Contamination

De lege lata, one of the approaches to solve pending legal problems connected with space debris was to introduce their regime under the provisions of OST concerning the harmful contamination and adverse changes in the environment of the Earth[5]. The particular provision of the Art. IX OST reads as follows: „States Parties of the Treaty shall pursue studies of outer space, including the Moon and other celestial bodies, and conduct exploration of them so as to avoid their harmful contamination and also adverse changes in the environment of the Earth resulting from the introduction of the extra-terrestrial matter and, when necessary, shall adopt appropriate measures for this purpose."

It is obvious that this provision of Art. IX OST connects the legal consequences in case of harmful contamination with their source such as „studies of outer space" or its „exploration", excluding its exploitation *e.g.* in the form of telecommunication satellites. However, the sphere of space telecommunications is one of the most important fields in which the non-functioning space objects can pose serious problems. Because of the pending re-structuralisation of the ITU - notification procedures concerning the use of the slots on the geostationary orbit, the question even arises at present whether there is a possibility of a co-ordinated approach of both regulatory systems, thus avoiding overlapping activities and the misspenditure of scarce resources.

As far as „adverse changes in the environment" are concerned, the application of Art. IX OST depends on the introduction of the „extra-terrestrial matter" in the Earth environment. The problem can be, therefore, reduced to the question whether space debris can be included into the category of „extraterrestrial matter".

Theoretically, both possibilities exist. In practice, however, the real effect of inserting space debris into the category of „extra-terrestrial matter" should be analysed carefully. In the affirmative case, Art. IX stipulates the following consequences: The State pursuing studies and exploration of outer space is obliged to do so in a manner avoiding such negative effects; if necessary, it shall take „appropriate measures" for this purpose. Thus, it seems so that the positive effect of including space debris into the category of „extraterrestrial matter" is not very considerable.

The second provision of Art. IX OST concerning harmful effects of space activites deals with „harmful interference with activities of other States Parties in the peaceful exploration and use of outer space". Without any doubt, space debris are in a position to cause such a „harmful interference". The consequence of this fact would be the duty of a State Party „which has a reason to believe that an activity or experiment planned by it or by its nationals in outer space would cause potentially harmful interference with activities of other State Parties in the peaceful exploration and use of outer space" to undertake consultations before proceeding with any such activity or experiment. The problem of the applicability of this provision is, however, that the existence of space debris can hardly be described as a part of a „planned" activity or experiment. The practical impact of this provision on the legal regime of space debris seems to be, therefore, quite limited indeed.

The effect of the third „harmful contamination" provision of the OST has already been diminished in the period of elaboration of the OST: Under this part of Art. IX OST, „a State Party to the Treaty which has a reason to believe that an activity or experiment planned by another State Party in Outer Space" ..."would cause potentially harmful interference with activities in peaceful exploration and use of outer space"..."may request consultation concerning the activity or experiment." The hypothesis of this provision seems to be applicable to the problem of space debris, because it includes not only exploration, but also exploitation of Outer Space. On the other side, as in the previous case, it concerns only the activities which are only „planned" by the States Parties and, therefore, does not solve the question of already existing forms of exploitation. The effect of this rule is, however, reduced since it does not provide for an obligation of the other State Party to enter into the consultation.

The most important rule of Art. IX which is applicable to various aspects of the space debris problems, is the general provision under which „in the exploration and use of Outer Space...States Parties...shall conduct all their activities in outer space...with due regard to the corresponding interests of all other States Parties to the Treaty". This rule fully justifies the demand of other States parties to the OST to keep outer space free of space debris for further „exploration and use of outer space".

De lege ferenda, the proposed concepts offer to fill the existing gaps of Art. IX by formulating more precise rules concerning especially the duty to cooperate to protect the environment and the obligation to enter into consultation in case of danger caused by space debris: Art. 3 of the 1994 Buenos Aires document stipulates a general obligation of states to cooperate to protect the environment and to take appropriate measures to prevent, reduce, and control any damage or significant risk arising from activities under their jurisdiction which are likely to produce space debris. In Art. 4 of this draft, the obligation of States and international organisations to hold consultations in case of risk to the environment from space debris is further expressed, together with the right of these parties to request consultations in case of such a risk. The refusal to hold consultations, or their breaking up without justification, shall be interpreted as bad faith. If Art. IX should be taken as a basis for the regulation of the consequences of space debris, this methodological approach is surely appropriate.

IV. <u>Should the protection of ownership of space objects, and their component parts, also be extended to space debris</u>?

De lege lata, in te Art. VIII OST the principle was laid down according to which „a State Party to the Treaty on whose registry an object launched into outer space is carried shall retain jurisdiction and control over such an object...while in outer space or on a celestial body. Ownership of objects launched into outer space, including objects landed or constructed on a celestial body, and of their component parts, is not affected by their presence in outer space or on a celestial body or by their return to the earth."

The interpretation of „space debris" as an sub-category of space objects or as their component parts does not seem to allow any other conclusion than that the principle of permanent ownership of space objects extends also to space debris. This approach, empfasizing the link between the launching State or State of registry and the not-functioning space object, seems to be strenghtened by the following part of article VIII OST[6]: Even „such (space, MH.) objects or their component parts found beyond the limits of the State Party to the Treaty on whose registry they are carried shall be returned to that State Party, which shall, upon request, furnish identifying data prior to their return."

The practical consequence of this conclusion is that even the space object that has not been functioning for a long time cannot be removed from orbit without the consent of the state of registry.

This approach of the OST matches, however, to the relatively large objects, the state of registry of which can be easily identyfied[7]. In despite of the scientific progress, it seems to be unsuitable for the regime of relatively small particles the trajectory of which can be observed, but the „owner" of which is hardly to be identified.

De lege ferenda, a new regulation must draw the line between the objects the regime of which respects the provisions of Art. VIII OST and those, which should be - by agreement - be excluded from the protection of ownership and can be removed from the orbits without consent of the „owner". The Buenos Aires International Instrument does not approach this dilemma; its „centre of gravity" lays in its catalogue of co-operation provisions and its solution of internatioonal liability issues.

V. <u>Should liability for damage caused to a space object and/or its crew by space debris depend on the proof of fault as in the case of a collision of two space objects?</u>

Art. VII OST reads as follows: „Each State Party to the Treaty that launches or procures the launching of an object into outer space,...and each State from whose territory or facility an object is launched, is internationally liable for damage to another State Party to the Treaty or to its natural or juridical persons by such object or its component parts on the earth, in air or in outer space...". This provision has been extended by the relevant provisions of the Convention on International Liability for Damage Caused by Space Objects of 1972 (Liability Convention)[8].

The *de lege lata* liability regime established by the Liability Convention may be divided into four distinct classifications: absolute liability, fault liability, joint and several liability, and exoneration from liability. In case space debris causes damage „...on the surface of the earth or to aircraft in flight", an absolute liability of the launching State occurs, without necessity to prove fault or negligence (art. II of the Liability Convention).

This regime is limited, however, to land, water and air space. In the event of damage being caused elsewhere than on the surface of the earth to a space object of another launching State or to persons or property on board such a space object of another launching State, the latter shall be liable only if the damage is due to its fault or the fault of the persons for whom it is responsible (Art. III of the Liability Convention). The central term of this provision is the notion of „fault", the attributes of which being not defined in the Convention. Traditionally, fault liability comes into effect when there is „a failure to exercise the degree of prudence considered reasonable under the circumstances.[9]" One of the problems connected with the application of this rule is to determine when a negligence occurs. The decision how to determine the fault will have to be settled either through negotiations or by a decision of a claims commission[10].

This regime applies, however, only to the cases in which the liable State is known. Concerning space debris, this situation does not occur so often: more frequently the damage will be caused by particles the owner of which remains unidentifiable. For such a case, the construction of art. III of the Liability Convention does not suit.

For both reasons, the lack of precise criteria enabling to determine the fault of the launching State, as well as the specifica of many space debris, the concept of liability based on „fault" does not seem appropriate for regulating space debris.

De lege ferenda, the Buenos Aires International Instrument provides in its article 8 that „each State or international organisation ...that launches or procures the launching of a space object is internationally liable for damage arising therefrom to another State, persons or objects, or international organisation party to this instrument as a consequence of space debris produced by such any object".

This approach enables to compensate damages caused by space debris of known launching States. In order not to deprive a damaged State from any compensation in case the launching State is unknown, the idea of a special fund arose[11] which would compensate victims of such damage, each launching State contributing proportionally according to the number of launches it provides. This idea seems to be an appropriate measure to cope with this problem and its future formultion in precise legal terms would be a constructive approach how to regulate the legal consequences of the damages occured.

VI. Conclusion

The crucial issue of the foregoing analysis was the question to which extent the OST is able to regulate the specific issue of the legal regime of space debris.

The response is twofold: It is, but only to some extent. No doubt concerning the applicability of its provisions arises in the case of damage caused by relatively large objects the launching State of which is identifiable, although the principle of fault liability could be discussed. The rule of Art. IX OST under which the States Parties shall conduct their activities in outer space with due regard to the corresponding interests of all other States Parties is a very important element of this legal framework.

The phenomenon of the damage caused by particles the provider of which remains unknown does not seem to be specifically regulated. In order not to deprive a damaged State from any compensation, the idea of a special fund compensating victims of such damage appears as an appropriate measure to cope with this problem; its formulation in precise legal terms would represent a constructive approach how to contribute to the progress in this field.

---

[1] A/AC.105/639, 11 April 1996, Annex E.

[2] See Lubos Perek, Space debris: Discussion in the United Nations in 1996, IISL-96-IISL-4.08.

[3] See: Karl-Heinz-Böckstiegel, The Draft of the International Law Association for a Convention on Space Debris, IISL-95-IISL.2.0.3.; Id. ILA Draft Convention on Space Debris, ZLW 1 (1995), pp. 29-34.

[4] See: Dimitri Maniatis, The Law governing Liability for Damage Caused by Sopace Objects: From State Responsibility to Privat Liability. In: Annals of Air and Space Law (1997), p.380.

[5] Cf. Manfred Hintz, Weltraumrechtlicher Umweltschutz im völkerrechtlichen Regelungs-zusammenhang (1995), p. 33 *et seq.*

[6] This approach was precised in the Agreement on the Rescue of Astronauts, the Return of Astronauts and the Return of Objects Launched into Outer Space of 1968.

[7] To the question of the identification of space debris cf. Carl Q. Christol, Space Law: Past, Present and Future (1991), p.250 *et seq.*

[8] See e.g. George T. Hacket, Space Debris and the Corpus Iuris Spatialis (1994), p.153 *et seq.*

[9] S. e.g. P.B. Dembling, „International Liability for Damages Caused by the Launching of Objects into Outer Space, in: Proceedings of the XI. Colloquium on the Law of Outer Space, (1968), p.236 et seq.

[10] See Dimitri Maniatis, *supra note* 4, p. 33 et seq.

[11] H.A. Baker, Liability for Damage Caused in Outer Space by Space Refuse, annals of Air and Space Law (1988), pp. 183-192.

Session 4:
THE FUTURE APPLICATIONS OF THE OUTER SPACE TREATY

Chairmen: K. H. Bockstiegel, Germany
A. D. Terekhov, USA

Rapporteur: O. Ribblelink, Netherlands

IISL-97-4.01

OUTER SPACE TREATY IN PERSPECTIVE

Luboš Perek[1]
Astronomical Institute
Academy of Sciences, Prague, Czech Republic

Abstract

In the course of the forty years which elapsed since the beginning of the space era, new facts, not known in the 1950's and 1960's became important. Space debris are playing an important role not foreseen by the Treaty. Space law as a whole has to reflect the present state of affairs and of knowledge. An important addition to the exisitng instruments of space law, the Agreement on Space Debris, seems to be necessary to preserve the main function of space law: to provide a framework for an orderly conduct of space activities for the benefit mankind.

1. Introduction

The Outer Space Treaty[2] has a very high ethical value. Many problems which could arise from space activities were foreseen and solutions based on international cooperation were proposed. It is by no means certain that an equally valuable document would reach consensus at present. Therefore the Treaty should be preserved as a necessary basis for an orderly conduct of space activities. Possible changes or additions should be done in a form not requiring the opening of the Treaty.

The treaty entered into force in October 1967. It was based on the Declaration of Legal Principles Governing the Activities of States in the Exploration and Use of Outer Space[3] adopted in 1963. At that time, space activities consisted mainly in launching scientific and experimental satellites. In the first three years, 1957-1959, the number of payloads exceeded that of space debris. It was only the breakup of the satellite 1961-Omicron into some 270 trackable debris on 29 June 1961, which started the overwhelming preponderance of debris over active payloads. This breakup occurred presumably too late to have any influence on the drafting of the Principles. It was tacitly assumed that the few debris are lost in the immensity of outer space and that the problem in hand concerned

[2]UN GA Resolution 2222(XXI) of 19 December 1996.

[3]UN GA Resolution 1962(XVIII) of 13 December 1963.

active satellites. These were considered so valuable that even after their fall to the ground, they had, according to Article VIII of the Treaty, to be returned to the owner in case they were found outside the territory of the launching state. It was Article IX of the Treaty, second sentence, which dealt with the environment. It says:

State Parties to the Treaty shall pursue studies of outer space, including the moon and other celestial bodies, and conduct exploration of them so as to avoid their harmful contamination and also adverse changes in the environment of the Earth resulting from the introduction of extraterrestrial matter and, where necessary, shall adopt appropriate measures for this purpose.

The sentence reflects the concern about contamination of the environment by extraterrestrial matter. It can hardly be interpreted as applying to space debris which are of terrestrial origin. This follows from the definition of orbital debris adopted in the IAA Position Paper on Orbital Debris[4]:

Orbital debris is herein defined as any man-made Earth-orbiting object which is non-functional with no reasonable expectation of assuming or resuming its intended function or any other function for which it is or can be expected to be authorized, including fragments and parts thereof. Orbital debris includes non-operational spacecraft, spent rocket bodies, material released during space operations, and fragments generated by satellite and upper stage breakup due to explosions and collisions.

"Orbital debris" was the subject of the IAA Position Paper, while another term, "space debris", appears in the title of the item on the agenda of the UN Scientific and Technical Subcommittee. The term "orbital debris" refers to the objects while they are in orbit. The IAA Position Paper thus does not deal with debris which are no longer in orbit, i.e., which are falling through the dense layers of the atmosphere and with those which impacted on the ground or into the ocean. The term, "space debris", can be understood as slightly more general then "orbital debris" and including the decaying phase and the phase after the impact.

A definition of space debris, agreeing in substance with the definition above, has been adopted in the Buenos Aires International Instrument on the Protection of the Environment from Damage Caused by Space Debris[5]. Its text appears in Section 3 a.

2. Facts to be recognized by space law

The purpose of this paper is to point out which new facts, not taken into account in the Treaty, should be recognized by space law. Most of the new facts were not known or not appreciated in the early years of space activities but gained significance and recognition in the course of later developments.

In the first place, the **existence of space debris** should be reflected in space law because they constitute some 95% of all objects in space and because they pose a hazard to active spacecraft as well as to manned missions.

In the second place, those who come into contact with debris should be able to **determine whether the object in question is or is not space debris.**

[4]Acta Astronautica, vol. 31, October 1993, p. 169.

[5]Annex to statement by Prof. Dr. K.H. Böckstiegel, as the representative of the ILA at the 39th session of the COPUOS, Vienna, June 1966.

There should be no problem with **fragments or detached parts**. Their size and shape, as far as it can be recognized in their radar signature, and their orbital elements attributing their origin to a breakup event, are sufficient in most cases to reveal their true character. It may be different with satellites which became **inactive but are still orbiting more or less intact**. By definition they are space debris but they cannot be distinguished from dormant objects. Another example of seemingly inactive objects are scientific satellites used as indicators of the gravitational field, their sole activity consisting in revealing their location by reflecting incoming light. Only the owner or operator is familiar with the true status of his satellites. And only he can state that he has no more interest in the satellite, that he is abandoning it, that he has no means of de-orbiting it, and that it should be considered a space debris.

In the third place, the **status of space debris** has to be different from the status of active spacecraft just as the status of functional industrial products is different from that of garbage or junk. If the ownership rights of garbage were preserved, it would be impossible to collect garbage. On the other hand, the **original owner should remain liable** for potential damage caused by space debris which originated from a space object he used to be liable for.

In the fourth place, the difference of status has to be expressed in terminology. The same term should not apply to functional products and to garbage or junk as well. In our case, if confusion is to be avoided, the term **space object** should not apply to **space debris**.

Last, but quite important, the **legal climate has to be friendly** not only to those who prevent debris from being generated but also **to those who dispose of orbiting non-manoeuvrable debris**. This seems to be a necessity because preventive measures may not be enough to restrict sufficiently the hazard of space debris. There are too many inactive objects in orbit and at least some of them might have to be removed. Present law forbids the removal of an object by someone who is not its owner. Moreover, non-functional objects have very little value, if any, because their recovery is extremely difficult, if possible at all, and expensive. Removing debris from space should not be considered as disposing of foreign property but rather as collecting discarded garbage. And the devices used for removal of debris from orbit should by no means be confused with anti-satellite weapons.

3. Agreement on Space Debris

Let us assume that the international community decides one day (let us hope soon) that a new **Agreement on Space Debris** will be added to the existing instruments of space law. Its title already recognizes the existence of space debris satisfying our first requirement. The agreement should deal with the following questions:

a) Definition of space debris

The purpose would be served by a technical definition of space debris as it appeared in the IAA Position Paper (see Section 1 and footnote 4) or by the definition introduced in the ILA document (footnote 5). Here we quote the definition and the taxative enumeration of sources of space debris from the ILA document:

"Space debris" means man-made objects in outer space, other than active or otherwise useful satellites, when no change can reasonably be expected in these conditions in the foreseeable future.

Space debris may result, inter alia, from:

- Routine space operations including spent stages of rockets and space vehicles, and hardware released during normal manoeuvres.
- Orbital explosions and satellite breakups, whether intentional or accidental.
- Collision-generated debris.
- Particles and other forms of pollution ejected, for example, by solid rocket exhaust.
- Abandoned satellites.

b) How space objects become space debris

Fragments or detached parts of space objects become space debris at the time of breakup, explosion, fragmentation collision, or when they leave the parent body.

More or less intact satellites become space debris when the relevant launching agency announces, in accordance with Article IV, section 2 and 3, of the Registration Convention[6] that the activity of the relevant space object has been terminated, will not be resumed and that the object has no technical means for landing or de-orbiting. In particular, the inactive **component parts of a space object** and **parts of its launch vehicle**, appearing in Article I (c), would presumably be announced as space debris soon after launch, unless they decay within a specified period of time.

Possibly, the term "abandoned space object" should be introduced. It would be consistent with the last item on the ILA list of sources of space debris and it might simplify the wording.

c) Liability

Provisions for liability, in particular Article VII of the Outer Space Treaty and relevant articles of the Convention on International Liability of Objects Launched into Outer Space[7] should apply to space debris, in agreement with the ILA document. No change of the Liability Convention itself would be required.

d) Return of Objects

States should not be under the obligation to return worthless debris to the owner. Article VIII of the Outer Space Treaty and Article 5, section 1 to 4 of the Agreement on Rescue and Return[8] should apply to space debris only in exceptional cases, e.g., if that particular piece is valuable. If the object has to be removed from the environment because of its hazardous or deleterious nature, it should be either returned to the original owner or disposed of in an agreed way.

A suitable wording would have to be found which would obviate the necessity of making any change in the Agreement on Rescue and Return.

[6]UN GA Resolution 3235(XXIX) of 12 November 1974.
[7]UN GA Resolution 2777(XXVI) of 29 November 1971.
[8]UN GA Resolution 2345 (XXII) of 19 December 1967.

e) Preventive measures

General principles for restricting the generation of debris should be stated. Technical details and parameters which may be subject to relatively frequent changes should be left to a special body. A convenient — and already existing — body may be the Inter-Agency Space Debris Coordinating Committee. Space agencies participating in the IADC are interested in preventing the generation of space debris while, on the other hand, they would adopt only measures which are compatible with the development and successful operation of space missions. The UN COPUOS could provide a political roof for such technical measures. In a way, it would be a certain analogy to the Radio Regulations and the ITU in the field of radio-communications.

f) Disposal of space debris

The disposal of orbiting space debris by third parties should be permitted under safety conditions to be agreed upon.

g) Opening of instruments of space law

As was explained in the preceding sections, neither the Outer Space Treaty, nor the Agreement on Rescue and Return, nor the Liability Convention would have to be opened for renegotiation. As regards the Registration Convention, the additional announcements, proposed in Section 3 b, could be made within the context of the present wording of the Registration Convention. And the Moon Treaty[9] is concerned with activities on the Moon and on other celestial bodies within the solar system, other than the Earth, including orbits around these bodies and trajectories to or around them. It does not apply to activities and to objects in orbits around the Earth.

4. Conclusion

This proposal to elaborate an **Agreement on Space Debris** is just an outline. Its details and legal provisions would have to be elaborated by the same process which other instruments of space law have undergone in the COPUOS and its subcommittees. The Agreement or some other elaboration of the ideas contained in this proposal could be a useful addition to existing instruments of space law. The Agreement:

- Would recognize 95% of objects in space which at present have no legal status,
- Would legalize the disposal of non-manoeuvrable debris from the orbit, a method of cleaning outer space which might prove to be indispensable,
- Might introduce the concept of "abandoning a space object",
- Would state general principles for preventing the generation of new debris while leaving the technical details to a competent technical body,
- Might revitalize the Registration Convention, and
- Would not require the opening of the Outer Space Treaty or of other instruments of space law.

[9] UN GA Resolution 34/68 of 18 December 1979.

IISL-97-4.02

PROTECTION OF THE SPACE ENVIRONMENT UNDER THE OUTER SPACE TREATY

Mark Williamson
Space Technology Consultant
The Glebe House, Kirkby Thore, CA10 1UR, UK
Tel/fax: +44 1768 361040

ABSTRACT

Space development requires the institution of many safeguards, not least the protection of the space environment for the use of future generations. This paper analyses the relevant articles of the Outer Space Treaty and concludes that it should be amended, supplemented or otherwise reinforced to offer a degree of environmental protection that is currently lacking. It is argued that, as the expected development of Moon and Mars bases is pursued in the 21st century, this aspect of space law will become equally as important as the current formative measures regarding orbital debris.

Mark Williamson is an independent Space Technology Consultant, advising the space industry and the space insurance community. He is the author of two student texts ('The Communications Satellite' and 'Dictionary of Space Technology') and the editor of 'SPACE & Communications', an international space industry magazine.

INTRODUCTION

Mankind has spent some 40 years exploring the solar system, beginning with near-Earth space and the Moon, then extending its reach to the rest of the planets. Although much has been achieved by remote sensing, for example using the Hubble Space Telescope, by far the most detailed information has been gained by sending spacecraft to the planetary bodies in question and using what, in medical terminology, would be described as invasive techniques.

The first stage of lunar 'exploration', for example, involved launching hardware to impact on the Moon, an occupation in which success was measured by how close the spacecraft got to the target. It was only later that the techniques for orbiting and landing were perfected.

This led to the surface of the Moon being littered with spacecraft debris and spent rocket stages - ostensibly in the name of space science. Even the Apollo lunar seismic experiments involved purposely crashing Saturn V third stages onto the surface, so that previously sited seismometers could detect the resulting lunar reverberations. The practice of utilising what could be termed 'disposable spacecraft' was subsequently extended to some of the planets and their moons...and continues today.

In an era of growing global consciousness, in which protection of the Earth's environment is a key concern, should we not extend that concern to outer space? After all, on a universal scale the solar system is our own back yard. As a new phase of scientific and commercial development begins, with the expected Moon and Mars bases of the 21st century, should we not begin to develop guidelines for its protection, or at least an awareness of the importance of keeping it tidy? We have already begun the discussion as far as orbital debris is concerned, but what of the debris that already litters the surfaces of the planetary bodies?

ROLE OF THE OUTER SPACE TREATY

It has often been the case that resource development here on Earth has led to the degradation of the natural environment, which in some cases has encouraged the formulation of legislation to guard against it. The Outer Space Treaty of 1967 encapsulates mankind's main attempt to guard against such degradation in space through its basic principles of "common interest to all mankind" and "benefit of all peoples", as stated in the preamble to the Treaty.

Space exploration and development has evolved considerably since the introduction of the Outer Space Treaty and will continue to evolve in the early years of the 21st century. The Treaty must therefore be continually reviewed and updated to take this evolution into account, since where the articles of law may be content to remain unchanged for centuries, the 'articles of space technology' are not!

The following section analyses the relevant Articles of the Outer Space Treaty to identify the parts of the Treaty which are currently lacking and suggests potential areas for improvement.

ANALYSING THE OUTER SPACE TREATY

Title

It is evident from the official title of the Outer Space Treaty -

> **"Treaty on Principles Governing the Activities of States in the Exploration and Use of Outer Space, including the Moon and other Celestial Bodies"**

- that it is intended to cover much more than is currently of direct commercial or scientific interest. The use of the term "other celestial bodies", which includes the other planets of the Solar System and their moons as well as comets and asteroids, indicates the intended all-inclusive nature of the Treaty.

Since the Treaty was written, commercial space endeavours have been concentrated in various orbits around the Earth, largely in geostationary orbit (GEO) where the majority of communications satellite reside. Scientific endeavours have reached further into the Solar System, but since these have been pursued in a piecemeal fashion (compared to the commercial explosion of space-based communications systems) they have not attracted the same attention in terms of potential harm.

We are now familiar with the potential for orbital debris to restrict access to some of the low Earth orbits and cause damage to comsats in GEO, but there has been very little publicity attached to the environmental damage caused by scientific spacecraft. This is probably because these spacecraft are seen as benevolent and because they are not part of a coordinated effort in exploration, exploitation or colonisation. The spacecraft debris which already litters the lunar surface provides evidence against this rather rose-tinted view.

Furthermore, if mankind decides to mine the Moon or colonise Mars, the environmental impact will increase by at least an order of magnitude. As terrestrial experience has shown, when exploration becomes exploitation the environment tends to suffer.

The title of the Treaty is all-inclusive and, as such, cannot be improved. There is, however, ample scope (allowed by the title) to develop the content of the Treaty itself.

Article I

Article I states that:

> **"The exploration and use of outer space, including the Moon and other celestial bodies, shall be carried out for the benefit and in the interests of all countries...and shall be the province of all mankind."**

In view of the developments that have taken place, the question arises: can the littering of planetary surfaces with discarded rocket stages and disused spaceprobes be defined as "for the benefit...of all mankind"? The question is akin to those concerning the exploration of the further reaches of the Earth's surface, for example whether the littering of the slopes of Mount Everest with discarded oxygen cylinders is an acceptable sacrifice to the altar of man's continual battle with nature. The discussion is a philosophical one, but no less important for that!

Article I goes on to state that:

> **"...there shall be free access to all areas of celestial bodies."**

Unfortunately, the deposition of manmade debris on various celestial bodies (mainly the Moon) has made access to certain areas difficult, if not dangerous. Although no-one has personally surveyed the sites of the crash-landings, one can safely assume that there will be shards of metal and other materials surrounding the impact craters. A future 'Users' Guide to the Outer Space Treaty' might therefore include the warning that this free access is granted 'at one's own risk'!

The problem with the Article rests with its (admittedly inescapable and laudable) democratic nature, which demands the "freedom of scientific investigation in outer space". Perhaps, in future revisions, a proviso should be attached, stating that this freedom is allowed only if it does not, thereafter, limit the freedom of later investigators. In other words, you may visit if you tidy up afterwards.

Article IV

A similar proviso would be a useful addition to Article IV, which states:

> **"The use of any equipment or facility necessary for peaceful exploration of the Moon and other celestial bodies shall...not be prohibited."**

The problem with the Article as it stands is the potentially broad interpretation of the phrase "peaceful exploration". It was certainly not considered at all 'warlike', during the Apollo programme, to cause Saturn V third stage boosters to impact the Moon as part of a seismology experiment; it was simply an element of the peaceful investigation of the internal structure of our nearest celestial body.

Many aspects of life which were considered acceptable and harmless during the 1960s and 1970s have since become unacceptable and harmful - and surely impacting large rocket stages on the Moon is a part of that evolution - but Article IV's broad acceptance of "any equipment or facility necessary" is rather too broad a definition. Not only does this Article allow a broad interpretation of the word "peaceful", it also begs the definition of "necessary". Do we allow, for example, the operation of mining equipment which permanently disfigures the lunar surface? And, if so, to what extent: as viewed with the average naked eye; through the average amateur telescope; or from a 100km lunar orbit?

Article VI

The matter of which activities are allowed is closely connected to who is responsible for sanctioning those activities. The matter of responsibility is addressed, to some extent, in Article VI:

> **"States Parties to the Treaty shall bear international responsibility for national activities in outer space...whether such activities are carried out by governmental agencies or non-governmental entities, and for assuring that national activities are carried out in conformity with the...present Treaty."**

This is good in theory, but only to the same extent that the International Telecommunications Union (ITU) is *responsible* for the administration of radio frequencies; it has no powers to stop 'rogue users' causing harmful interference, for example. Moreover, if particular States have had *responsibility* for their national activities since 1967, when the Treaty was enacted, does this mean that the relevant States are responsible for the 'space debris' deposited on the Moon and other planetary bodies - and are they therefore responsible for clearing it up?

Of course this begs several questions, among them: is this 'space debris' actually 'unwanted' in the eyes of the international space community and is there actually a requirement to clear it up? As with all environmental legislation, a sufficiently common understanding of 'undesirability' must be arrived at before a given type of pollution can be outlawed. Obvious terrestrial analogies are the undesirability of smoking and of the lead content of vehicle exhaust emissions, both of which have taken many years to be recognised as deserving of legislation.

To give this Article any real meaning, the implications of "responsibility" need to be discussed, agreed and spelled out in an appendix to the Treaty.

Article VII

The matter of responsibility, indeed liability, is further addressed in Article VII:

> **"Each State Party...that launches or procures the launching of an object into outer space..., and...from whose territory or facility an object is launched, is internationally liable for damage to another State Party to the Treaty or to its natural or juridical persons by such object or its component parts on the Earth, in air space or in outer space, including the Moon and other celestial bodies."**

Although this covers liability for damage to other States and their citizens, no responsibility for damage to the natural environment is included. Perhaps it is as difficult to legislate for damage to the space environment as it is for damage to an undeveloped part of the *terrestrial* environment. For example, is the agency responsible for the impact of a spacecraft on a planetary body (intentionally or otherwise) any more liable for the resultant environmental damage than a pilot who crashes his aircraft in a barren desert?

Further consideration of liability is covered in the Liability Convention of 1971, but this is beyond the remit of the current paper.

Article IX

With Article IX, we are at the nub of the environmental issue:

> **"States Parties to the Treaty shall pursue studies of outer space, including the Moon and other celestial bodies, and conduct exploration of them so as to avoid their harmful contamination..."**

Theoretically, this Article should provide the protection of the space environment called for in this paper. Unfortunately, the phrase "harmful contamination" is open to interpretation and argument: it is likely, in the context of the Treaty, that it means harmful to humans rather than harmful to the environment, but this is not clear.

Apart from that, what sort of contamination is classed as harmful and what is not? When space probes, such as Viking or Pathfinder, are sent to Mars to analyse the surface, it is assumed that they are analysing chemical constituents native to Mars, not contamination from Earth. If a future probe should crash, or break up as it de-orbits, and spread its debris over a wide area, it will contaminate the 'pristine' Martian environment. This will not immediately be harmful to humans, but it could prove extremely harmful to science.

Moreover, it is well known that whereas the Earth's environment has a natural ability to repair itself, the lunar environment has no such capability. Since the Moon has no atmosphere and no weather, it will bear the scars of man's intervention for the foreseeable future, a future which may be measured in hundreds of thousands of years. It was concern about the lunar environment, and the realisation that it should be protected, that produced the Moon Agreement of 1979.

Interestingly, the Moon Agreement contains what amounts to a rewrite of Article IX of the Outer Space Treaty (in Article 7, paragraph 1):

> **"In exploring and using the Moon, States Parties shall take measures to prevent the disruption of the existing balance of its environment, whether by introducing adverse changes in that environment, by its harmful contamination through the introduction of extra-environmental matter or otherwise."**

As with Article IX of the Outer Space Treaty, this is extremely well-intentioned, but avoids the important definition of "harmful".

Further, in Article 7 paragraph 3 of the Moon Agreement, it is suggested that consideration be given to the designation of areas of special scientific interest, or "international scientific preserves", which, it states, would require agreement "in consultation with the competent bodies of the United Nations". This is not an end in itself because of the need for further consultation, but, suitably extended to include other celestial bodies, would make a useful addition to the Outer Space Treaty.

CONCLUSIONS

Even a cursory analysis of the Outer Space Treaty shows that, while in general it is well-meaning and democratic, it provides insufficient protection for the space environment. In particular, it makes insufficient provision for the protection of the planetary bodies from the creation of further debris, which will undoubtedly accompany future development and exploitation. As the expected development of Moon and Mars bases is pursued in the 21st century, this aspect of space law will become equally as important as the current formative measures regarding orbital debris. The length of time it takes for legislation to become enacted suggests that we would be well advised not to wait until this development has begun.

This paper has indicated why the Outer Space Treaty - which is thirty years old this year - should be amended or supplemented. It will be up to legal practitioners, assuming they agree, to decide upon the form of words of the supplement; however, they would be well advised to consider the testimonies of science and technology practitioners in their deliberations. In this way they might be better assured of creating a Treaty that will be applicable thirty years hence.

The practical problem in the application of the Outer Space Treaty, of course, is one that is common to the body of Space Law - the relative lack of power - and it is this, as much as any other matter, which deserves the attention of the space community. There is no practical use in a law, however well written, if it cannot be applied in a given situation, and unless all space-faring nations have signed up to the Treaty, it cannot be universally applied.

As far as protection of the space environment is concerned, two possible activities may help to lay the groundwork for improvement:

- a continual review of space missions to the planetary bodies to compile a catalogue of the debris from spacecraft impacts that may still be found on their surfaces

- consideration of the future of scientific and commercial exploration and exploitation of the planetary bodies, in both ethical and pragmatic terms, with a view to a sustainable balance between the productive activities of mankind and the desire to retain the purity of the space environment.

Eventually we shall have to decide whether Outer Space should be declared an 'International Park', in which all commercial and industrial development is prohibited, or whether the Planetary Bodies constitute a body of resources to plunder? The most likely outcome will be some combination of the two.

For those who prefer the concept of the International Park, a slogan used here on Earth seems apt:

> **Take only photographs, leave only footprints.**

OUTER SPACE TREATY IN 21ST CENTURY
A CHANGE OF CONCEPT

Ali A. Golrounia, Managing Director
Atlas Aviation Group

Mohsen Bahrami, Associate Professor
Aerospace Engineering Department
Amir Kabir University of Technology
Tehran, 15 I. R. Iran

ABSTRACT

Fantasy of space exploration in 1960's has evolved to an ever increasing scientific strife and real life commercial arena for private venture.[·]

Political competition and gaining national prestige in the beginning, which grow into cooperation and collaboration of governments in development of engineering and scientific knowledge has changed to multinational investment and deep private interest in a very promising new industry with almost no limits for growth, compare what is known in human history.

The pioneers in space explorations and those who put together the treaty have brought us something of value, concept that for duration of written history mass has been longing for. Now those concepts are being challenged and many aspect of life, and space technology will bring a devastating blow in next century, if we choose to take things as they come.

In this paper we review the fundamental concepts of Outer Space Treaty and discuss how they are being or going to be challenged. We also introduce some measures that can e taken to prepare for the 21st century in space.

INTRODUCTION

On October 4, 1957 a long awaited dream came true. What Isaac Newton had stated about four centuries ago of how to put an object into earth's orbit, had become reality. Sputnik soon after its lunch started beeping from space. That was all it did. This made true what Arthur C. Clark, in an article had anticipated in 1945 about space repeaters or as he called them "extra terrestrial relays". Of course U.S. Naval Research Laboratory had successfully transmitted the first voice message through space, using moon to reflect signals[1].

An era of intense and exciting competition started. Soviets had shown their scientific and technological superiority by successful launch of a satellite which was about nine time heavier than U.S. proposed one, and was injected into an orbit twice the altitude of what U.S. had planned[2].

In less than four months Explorer 1 was launched, on January 31, 1958. In the same year NASA was created as a civil agency to undertake space science and exploration programs.

The race has just started. The first human, Soviets cosmonaut Yuri Gagarin was launched into orbit on April 12, 1961. Astronaut Neil Armstrong became the first human to set foot on the surface of the moon on July 20, 1969. Space program was a sign of national prestige and honor.

In 1971 Soviet Union was the first to soft land a probe on Mars, which was able to communicate for only 20 seconds. It should be mentioned that after 26 years Pathfinder successfully landed on Mars in 1997 and has been able to communicate and execute commands from earth rather flawlessly.

But soon it was realized that for going along this road successfully, cooperation was needed. Of course for real cooperation to start some time was needed, but it was obvious cooperation is part of the development of this technology.

In 1973 European Space Agency(ESA) was created. Even though this was a replacement for ELDO(European Launcher Development Organization, created in 1961,) but it was a turning point regarding cooperation in development of technology.

This fever of cooperation took many different faces in particular in application of space technology. INTELSAT, INMARSAT and EUTELSAT are the ones created by space fairing nations. The developing countries also tried to get into the game. Indonesia was the fist to have its own satellite in 1979. ARABSAT was launched in 1985, providing service to twenty two Arab countries. This list is of course is very long and growing.

In the midst of all dreams and activities the need for cooperation grew to international levels. Several treaties and principals resulted as the consequence. Outer Space Treaty, as the major achievement has prove to be of historic value. This treaty has been the subject of many discussions[3].

MAIN CONCEPTS OF OUTER SPACE TREATY

Many interesting and powerful points were considered and included in the treaty, some of which are as follows:

- The exploration and use of outer space including the moon and other terrestrial bodies shall be carried out for the benefit and in the interest of all countries.
- Peaceful use of outer space
- Outer space shall be the province of all mankind.
- There shall be the freedom of scientific investigation in outer space.
- There shall be free access to all areas of celestial bodies.
- outer space including the moon and other celestial bodies is not subject to national appropriation.
- States shall pursue studies of outer space including the moon and other celestial bodies, and conduct exploration of them, so as to avoid their harmful contamination, and also adverse changes in the environment of the earth, resulting from the introduction of extra terrestrial matter.

The history of human being is a long story of struggle to own and possess, which for the most part has meant to fight, kill and destroy. Many different doctrine and theories were developed to solve or at least ease the situation, which all meant more of the same.

For the first time a phrase has been used, which is unique and brings the hope that perhaps we have found the solution, i.e. "the province of mankind," which is to be "used peacefully." This phrase is supported by another one to make sure it can be materialized, i.e. "not subject to national appropriation." The picture is perfected by indicating clearly that "the exploration and use of outer space... shall be carried out for the benefit of all countries," regardless of their state of development. Of course the protection of environment is also taken into account. This by itself is of some value, because at the time protection of environment was not a popular and well established idea.

This kind of phrasing certainly shows a consistency in picturing a future different to what the human race is used to. Even if this has come out of political negotiation, it has come out the way it should have. We owe this to ourselves to try to use the opportunity to make a different future for us and coming generations, a future which is an environment for growth, free of wars for possessing of what is the province of mankind. The Outer Space Treaty seems to be the appropriate starting point.

RESPECTED CONCEPTS

Of all provisions considered in the treaty, some have worked well. Through the course of the development process of space technology they have shown to be meaningful and applicable. These are as follows:

- Freedom of scientific investigation in outer space.
- Free access to all areas of outer space and celestial bodies.
- Outer space not being the subject of national appropriation (at least for the time being)

It is quite clear away from technical obstacles that there are no barriers for scientific investigation and free access to all areas of outer space and celestial bodies.

Even the movement of Equatorial States with regard to some specific rights for them in geostationary orbit, failed[4].

It is obvious that states which enjoy this freedom that outer space treaty has provided for them, are those which have more active roll and more significant technical abilities.

One could argue that freedom of scientific investigation and free access to space are so far exercised, mainly due to the fact no conflict of interest has come up in any situation. But whatever the reason the free nature of the current status is something desirable and of value.

So far non of the participant in the space activities either in exploration and/or application area, has made any claim of appropriation of moon or any other celestial bodies either in whole or in part. This on one hand is a success for the treaty, and on other hand this could be due to the fact that no state has enough establishment any where in the space to justify confronting the national and international public opinion.

Certainly this freedom is essential for exploration and exploitation. Thus if we value this notion, and do not wish to have a change of situation in the course of future development, we have to regulate internationally the outer space activities, in a meaningful manner.

NOT RESPECTED CONCEPTS

There are some provisions in outer space treaty which are not necessarily respected in practice, such as:

- Peaceful use of outer space
- Outer space as a province of all mankind
- Environmental issues in outer space

Even though we have not witnessed a war or a forced conflict in outer space, so far, but military investment in outer space in terms of satellite, and equipment is not only enormous, it is by all means much more capable than nonmilitary ones. These abilities are obviously not accessible to civilian use. The military activities not only outperform the civilian ones in terms of accuracy and coverage, but also in terms of number.

Even star war technology development in 1980's was part of US international and long range policies.

The gap between space fairing nations and the rest of the world with regard to outer space activities and benefit sharing has an ever increasing pace.

Distribution of data from outer space has become more and more selective, restrictive and expensive, especially for developing countries. States can be kept in darkness with regard to data collected by military and civilian satellites from their strategic and much needed information about their natural resources. This information can be misused very easily without any noticeable international objections. In other word mankind is loosing his privacy to satellites and space technology which are out his control and jurisdiction. In fact the notion of jurisdiction is no longer of any significance when it comes to space technology.

CONCEPTS FOR 21ST CENTURY

All the provisions of the outer Space Treaty, either those that are being respected, or those that are ignored, do not cover what is needed for the development trend of space technology. There are some new concepts in the space arena that need to be considered, for the sake of saving the underlying principles of Outer Space Treaty. Among the more important ones one can easily notice the following:

- Commercialization and privatization of outer space.
- Liability and insurance.
- Intellectual property rights

Space activities like many other technology development processes started as a scientific curiosity but very soon entered a serious application period both civil and military. Now a days space activities are very much an everyday routine happening. MIR space station renting out its facilities. Many launches are taking place each month, from different launching facilities in different countries. The course of development is such that, ideas like Motorola's IRIDIUM which once was considered very aggressive, soon was a very ordinary project when TELEDISC was proposed to place 1000 satellite in orbit. If reusable launch vehicle was a far reaching idea, using helicopter autorotation concept for landing such vehicle certainly made it look a normal reality. Even though there is still some ways to go before real applications.

This fast pace course of development in space technology has prepared the ground for many investors and entrepreneurs to look at the commercialization of space activities as an exciting and rewarding one. The volume of investment is certainly noticeable both in government and private sector compare to any other investment sector, one underlying point is that space technology is covering many fields, such as communication, earth observation, micro gravity, remote sensing, telemedicine, navigation... This already immense volume of investment is growing everyday, bearing in mind that exploring space is just taking off, not all its aspect is known yet, and commercialization is in its infancy.

One important issue is the legal frame work necessary for the current and future development. Some indication can be seen that even though the space law has had a fast course of development, there is a tendency to apply national law of one or the other space fairing nations. Of course this tendency has been seen in terrestrial activities, too. The current international law covering space activities are in the form of conventions, which are of a public law character, and to which states are parties.

These conventions do not make directly or indirectly any references to commercialization of outer space, and private sector involvement; neither make any direct or indirect references to prohibit such activities. In some provisions there are references to freedom of exploration and exploitation of outer space, which can be interpreted as a legal frame work for private sector activities in outer space. However, this is certainly insufficient, and need a considerable amount of work.

It should be noted that commercialization and private sector activities can be very beneficial for mankind, in many different ways, provided that there will be sufficient meaningful control and monitoring over these activities, to make sure all states regardless of their degree of development will benefit from space technology advancement.

The interest of private sector in space activities, is certainly accompanied by many concern about some points. One issue of importance is the risk involved, and as a direct consequence the insurance is of main concern.

Clearly the main objective of private sector activities in outer space is to make and maximize profit. This policy is not necessarily in line with article 1 of outer space treaty, i.e. "the exploitation of outer space, including moon and other celestial bodies, shall be carried out for the benefit and interest of all countries, irrespective of their degree of economic and scientific development and shall be the province of mankind." Since the activities of private sector is of international character, to avoid misuse of the current and future opportunities, proper mechanism of control, coordination and monitoring is necessary.

Many work has been done on the subject of organizational structure for activities in outer space[5]. Reference 6 has a good and comparative review of different proposals. The main problem of all the proposed organization is that they are not thought out to be fully compatible with current treaties, particularly the Outer Space Treaty.

Here we suggest to create an agency to manage the outer space properties and resources including the moon and other celestial bodies.

OUTER SPACE AGENCY

It is proposed that this agency is created under the auspices of United Nations, for the main purpose of management of outer space properties.

The creation of this agency is based on the following generally accepted facts:

- Outer space is the property of mankind.
- Outer space is to be explored for the benefit of all states regardless of their degree of development.

The Notion that space technology has opened a new era for people of earth, which is going to change all aspect of human life, and it necessary to have future developments of all countries based on space technology justifies the creation of this agency with the concept introduced here.

All properties of outer space including the moon and any other celestial bodies, accessible to mankind, at the present or the future, are under the management of this agency.

Any state or international organization planning to explore exploit or establish base anywhere in outer space inform this agency of its intention and plans.

Using outer space is possible only through renting or leasing properties.

The funds created from this operation is used to make the space technology the basis of development of all countries. The agency can invest the money earned in space industry sector.

United Nation Outer Space Office will set up the agency and prepare the necessary regulation for general assembly approval.

CONCLUSION

An outer space agency, which manages the outer space properties including the moon and other celestial bodies will put the concepts of Outer Space Treaty in an applied frame work. While countries have the freedom to explore the outer space, a practical order is created, and some revenue is generated for development.

REFERENCES

1. H. E. Hudson, Communication Sattelites, Their Developement and Impact, The Free Press, London, 1990, pp2-4

2. ibid, pp15-17.

3. M. Nakamura, "Review of Article I of The Outer Space Treaty," Proceedings of 39th International Colloquium on the Law of Outer Space, IISL-96-IISL.3.11, Beijing, China, 1996.

4. N. Jasentuliyana, "Recent Development in the United Nations Activities Relating to Outer Space," Presentation, 38th International Colloquium on the Law of Outer Space, IISL-96-IISL.3.05, Beijing, China, 1996.

5. A. A. Golrounia, M. Bahrami, "The Draft of the International Law Association for a Convention on Space Debris (Buenos Aires) Can It Meet the Needs of 21st Century," Proceedings of 39th International Colloquium on the Law of Outer Space, IISL-96-IISL.4.09, Beijing, China, 1996.

6. D. J. O'Donnell, " Commercialization by Evolution in the Jurisdiction of Outer Space," Proceedings, 47th International Astronautical Congress, IAA-96-IAA.1.4.04, Oct., 1996, Beijing, China.

IISL-97-4.04

THE TREATY OF 67 IN FRONT OF 21st CENTURY

CARLOS HERNANDO REBELLON BETANCOURT

LAWYER

CALI - COLOMBIA

After 30 years of Treaty of 67 is neccesary to amend it and supplement it too. The conditions of the world in 1967 are not the conditions today. The space knowledge, the advance of space technology, new states, International Intergovernmental Organizations, natural and juridical persons in space activities require a new space law to 21st century.

I think that is neccesary to amend the Treaty of 67, in differents aspects: To abolish the expression "States Parties" in order to change the Treaty of 67 of states parties by one treaty more general to all states, International Intergovernmental Organizations any nations, associations of states, union of states. As a consequence for instance, the article XIV will be change in order to open for signature by International Intergovernmental Organizations. The space is not only for states parties of Treaty of 67 or any other treaty of outer space activities but is for all mankind. The article II, is neccesary to amend it, in order to International Intergovernmental Organizations, natural or juridical persons can not claim of property, by means of use or occupation or by other means.

A new article: The sovereignty and jurisdiction over the Moon, other celestial bodies and any natural space resource will take exercise by mankind.

I think that is neccesary supplement the Treaty of 67 to develop the concepts "Astronauts as envoys of mankind and common heritage of mankind". The first one, today only develop the aspects of security activities but not exist a law of people into outer space: Principles, rights, warrants, etc. The second one is other rich and important concept that is neccesary to develop with special treaty.

The next century requiere a space law to new conditions of the world.

Introduction

The history of relationship of the human people teach to us that the in the first time was make it between a short group: The family, the tribe, after arise the polis in Grace, after the empire, the feudal time, the kingdoms, the states and now in end of 20st century arise states union, states association and we see in front of 21st century that the activity of the mankind principaly will be do it with the last system. The concept of the states go to dissapear and emerge new concept of people association. In the other hand I think that in middle of 21st century, human people will be live and will be born into outer space, maybe on the Moon or Mars, for this reason is important to develop the outer space law about the concept of astronauts as a envoys of mankind, and common heritage of mankind.

There is the future situation that the Treaty of 67 will applicate in 21st century. It is necessary adapt it to new time.

To amend the Treaty of 67

When the space activities was start, only states was do it and the space law only was make it by states for states, but now after 30 years of Treaty of 67 the situation are different, not only the states individual consider have interest in outer space, but States Union Association, Intergovernmental Organizations, private companies have too. We must to see the space not for states, but by mankind, for mankind and from mankind.

The space law time by time must to adapt to new time in from of future, the space law is actual, dinamics and base of future.

Pacta Sunt Servanda. The Treaties oblige to the parties and must to fulfill them of good faith. Express that all parties in Treaty is oblige by it. The international doctrine and international jurisprudence recognize this norm as a juridical base of the peace, security and union life between staes or any subject who signature a Treaty. This law as in the preamble of charter of united nations, American States Organizations and other International Organization. This norm is a rule independence of Treaty recognize it as a general norm of international law, is universally recognize by all states of the world.

With base of Pacta Sunt Servanda and the new and future situation of the world, I think that is important to amend the Treaty of 67 in order to abolish the expression "States Parties and States" in all articles of the Treaty and change for expression "Parties" in some articles and in other for the expression "mankind". Is time to see the outer space from other perspective different of states, but of union states, association states, intergovernmental and other specialist international organism.

The Treaty of 67 must to open to signature to other international subject moreover the States, even the autonomy territories recognize it by the United Nations, but without total independence with base of 1514 resolution of General Assembly. If the Treaty of 67 is open to signature to other international subject moreover of states, the expression States Parties is not appropriate, the correct is only Parties. The Convention of Viene, article 26, express that the Treaty in force oblige to the parties. In the other hand, any subject that signature any Treaty that is in force, is oblige by it independence of that Treaty have to expression Parties, because for all Treaties is recognize this important rule of Pacta Sunt Servanda.

The relationship of human groups change time by time. Until middle of 20st Century was do it by States, by now and the future wil be do it by States Union, States Association, International Organism. Then as a consequence of this all international activities of mankind, shall be regulate with a new conception of the international law. The outer space law, is not a exception of the reality and must to be adapt it now. To opening to signature the Treaty of 67 to other international subjects moreover states.

Sovereignty and Jurisdiction

With this paper I present a new different concept of sovereignty and jurisdiction, different of the concept express in the article II and VIII of Treaty of 67, and the articles XI and XII of Agreement governing the activities of States on the Moon and other celestial bodies.

I present the sovereignty and jurisdiction no from states, no by states, no for states, but from mankind, by mankind and for mankind. I think that any State can not made a law in order to apply into outer space, and special on the Moon, Mars or any celestial bodies, include their natural resources without consider the sovereignty and jurisdiction of mankind over them. I do not believe that we go to see the next time in space facilities, stations, installations, settlements as a extraterritorial position of states. One United States facilities the law of the United States; one installation of Rusia, the law of Rusia; one station of Germany, the law of Germany, etc.

The jurisdiction of states is only for maintenance the control of property of objects and control over the personnel into space for that the other states can not appropriate them, but neve in order to conform a new extension of their legal order into the space.

I think that we must to consider one outer space law over the states law. In the other hand, if states law will be applicate into spae, in the same form that exist actually in the earth, we go to see the sovereignty and jurisdiction of states into outer space include the Moon, Mars and other celestial bodies.

The sovereignty of mankind over outer space include Moon, Mars and other celestial bodies and natural resources, arise as a natural consequence of the man is until now the unique intelligent that exist of the universe. The space is for mankind consider as a one thing and not as a one state or some states or several states. The sovereignty of states into outer space is not exist, is one right that never the states can obtain. The sovereignty into outer space is only one mankind right. As a consequence of sovereignty exercise of mankind, the jurisdiction consider as exercise of authority and expression of make a law is other right of mankind and the states law must to sunmited to mankind outer space law. For natural and juridical person, the real rights as a property mortgage, inheritance, etc. over surface or subsurface of the moon or other celestial bodies and any natural space resources, never will be arise into any place of outer space. Any kind of action about this rights are forbidden, any document is null.

Is important that we see the outer space as a place where the all activities there will be do it no as a states but by mankind. Only in this form we can think when we draw or develope or comment the outer space law as a envoys of mankind into space. Our condition of nationality of one state is out in front of outer space. The space is a beautiful opportunity to us in order to think, act and to live as a one, as a mankind.

The general norm about third state. One Treaty do not create duties nor rights for one third state without its consent. I think that in front of outer space law is possible to create duties and rights for any state or people even without its consent. The reason is that the all celestial bodies and any natural space resources are common heritage of mankind, then as a result of this, the sovereignty and jurisdiction of mankind over them.

New Treaty

TREATY ON PRINCIPLES GOVERNING THE ACTIVITIES OF MANKIND IN THE EXPLORATION AND USE OF OUTER SPACE, INCLUDING THE MOON AND OTHER CELESTIAL BODIES.

The parties to this Treaty:

Inspired by the great prospects opening up before mankind as a result of man's entry into outer space,

Recognizing the common interest of all mankind in the progress of exploration and use of outer space for peaceful purposes,

Believing that the exploration and use of outer space should be carried on for the benefit of all peoples irrespective of the degree of their economics or scientific development.

Desiring to contribute to broad international co-operation in the scientific as well as the legal aspects of the exploration and use of outer space peaceful purposes,

Believing that such co-operation will contribute to the development of mutual understanding and to the strengthening of friendly relations between States and peoples.

Recalling resolution 1962 (XVIII), entitled "Declaration of Legal Principles Governing the Activities of States in the Exploration and Use of Outer Space", wich was adopted unanimously by the United Nations General Assembly on 13 December 1963,

Recalling resolution 1884 (XVIII), calling upon States to refrain from placing in orbit around the Earth any objects carrying nuclear weapons or any other kinds of weapons of mass destruction or from installing such weapons on celestial bodies, which was adopted unanimously by the United Nations General Assembly on 27 October 1963,

Taking account of United Nations General Assembly resolution 110 (II) of 3 November 1947, which condemned propaganda designed or likely to provoke or encourage any threat to the peace, breach of the peace or act of agression, and considering that the aforementioned resolution is applicable to outer space,

Convinced that a Treaty on Principles Governing the Activities of Mankind in the Exploration and Use of Outer Space, including the Moon and Other Celestial Bodies, will further the purposes and principles of the Charter of the United Nations,

Have agreed on the following:

Article I

The exploration and use of outer space, including the Moon and other celestial bodies, shall be carried out for the benefit and in the interests of all parties irrespective of their degree of economic or scientific development, and shall be the province of all mankind.

Outer space, including the Moon and other celestial bodies, shall be free for exploration and use by all mankind without discrimination of any kind, on a basis of equality and in accordance with international law, and there shall be free access to all areas of celestial bodies.

There shall be freedom of scientific investigation in outer space, including the Moon and other celestial bodies, and parties shall facilities and encourage international co-operation in such investigation.

Article II

Outer space, including the Moon and other celestial bodies is not subject to national appropriation by claim of sovereignty, by means of use or occupation or by any other means

- Any other international subject, natural or juridical person can not claim of property, by means of use or occupation or by other means.

New Article III

The sovereignty and jurisdiction over the Moon other celestial bodies and any natural space resources will be take exercise by mankind.

Article

Parties to the Treaty shall carry on activities in the exploration and use of outer space, including the Moon and other celestial bodies, in accordance with international law, including the Charter of the United Nations, in the interest of maintaining international peace and security and promoting international co-operation and understanding.

Article

Parties to the Treaty undertake not to place in orbit around the Earth any objects carrying nuclear weapons or any other kinds of weapons of mass destruction, install such weapons on celestial bodies, or station such weapons in outer space in any other manner.

The Moon and other celestial bodies shall be used by all mankind exclusively for peaceful purposes. The establishment of military bases, instalations and fortifications, the testing of any type of weapons and the conduct of military manoeuvres on celestial bodies shall be forbidden. The use of military personnel for scientific research or for any other peaceful purposes shall not be prohibited. The use of any equipment or facility necessary for peaceful exploration of the Moon and other celestial bodies shall also not be prohibited.

Article

Any States of the World and - or Parties to the Treaty shall regard astronauts as envoys of mankind in outer space and shall render to them all possible assistance in the event of accident, distress, or emergency landing on any territory of the world or on the high seas. When astronauts make such a landing, they shall be safely and promptly returned to the State or Party of registry of their space vehicle.

In carrying on activities in outer space and on celestial bodies, any astronaut shall render all possible assistance to the other astronauts or any States of the World.

Any person or parties to the Treaty shall inmediately inform the other Parties to the Treaty of the Secretary-General of the United Nations of any phenomena they discover in outer space, including the Moon and other celestial bodies, which could constitute a danger to the life or health of astronauts.

Article

Any State of the World or any Parties to the Treaty shall bear international responsability for national activities in outer space, including the Moon and other celestial bodies, whether such activities are carried on by governmental agencies or by non-governmental entities, and for assuring that national activities area carried out in conformity with the provisions set forth in the present Treaty. The activities of non-governmental entities in outer space, including the Moon and other celestial bodies shall require authorization and continuing supervision by the appropriate State of the World or State Party. When activities are carried on in outer space including the Moon and other celestial bodies, by an international organization, responsability for compliance with this Treaty shall be born both by the international organization and by the State Parties to the Treaty participating in such organization.

Article

Any State of the World or any Party to the Treaty that launches or procures the launching of an object into outer space, including the Moon and other celestial bodies, and any State of the World or any Party from whose territory or facility and object is launched, is international liable for damage to another State of the World or any Party to the Treaty or to its natural or juridical persons by such object or its component parts on the Earth, in air or in outer space, including the Moon and other celestial bodies.

Article

Any State of the World or any Party to the Treaty on whose registry an object launched into outer space is carried shall retain jurisdiction and control over such object, and over any personnel thereof, while in outer space or on a celestial body. Ownership of object launched into outer space, including objects landed or constructed on a celestial body, and of their component parts, is not affected by their presence in outer space or on a celestial body or by their return to the Earth. Such object or component partes found beyond the limits of the State of the World or any Party to the Treaty on whose registry they area carried shall be return to that State of the World or any Party, which shall, upon request, furnish identifying data prior to their return.

Article

In the exploration and use of outer space, including the Moon and other celestial bodies, all States of the World and the Parties to the Treaty shall be guide by the principle of co-operation and mutual assistance and shall conduct all their activities in outer space, including the Moon and other celestial bodies, with due regard to the corresponding interest of all States of the World and Parties to the Treaty. States of the World and Parties to the Treaty shall pursue studies of outer space, including the Moon and other celestial bodies, and conduct exploration of them so as to avoid their harmful contamition and also adverse changes in the environment of the Earth resulting from the introduction of extraterrestrial matter and, where necessary, shall adopt appropriate measure for this purpose. If a any State of the World or any Party to the Treaty has reason to believe that an activity or experiment planned by it or its nationals in outer space, including the Moon and other celestial bodies, would cause potencially harmful interference with activities of any State or Parties in the peaceful exploration and use of outer space, including the Moon and other celestial bodies, it shall undertake appropriate international consultations before proceeding with any such activity or experiment. Any State of the World or any Party to the Treaty which has reason to believe that an activity or experiment planned by any State of the World or Party to the Treaty in outer space, including the Moon and other celestial bodies, would cause potencially harmful interference with activities in the peaceful exploration and use of outer space, including the Moon and other celestial bodies, may request consultation concerning the activity or experiment.

Article

In order to promote international co-operation in the exploration and use of outer space, including the Moon and other celestial bodies, in conformity with the purposes of this Treaty, the Parties to the Treaty shall consider on a bases of equality any request by any other Party to the Treaty to be afforded and opportunity to observe the flight of space objects launched by those Parties.

The nature of such an opportunity for observation and the conditions under which it could be afforded shall be determined by agreement between the Parties concerned.

Article

In order to promote international co-operation in the peaceful exploration and use of outer space, any State of the World and any Party to the Treaty conducting activities in outer space, including the Moon and other celestial bodies, shall to inform the Secretary-General of the United Nations as well as the public and the international scientific community to the greatest extent feasible and practicable, of the nature, conduct, location, and result of such activities. On receiving the said information, the Secretary-General of the United Nations should be prepared to disseminate it inmediately and effectively.

Article

All stations, installations, equipment and space vehicles on the Moon and other celestial bodies, shall be open to representatives of any State of the World or any Party to the Treaty on a basis of reciprocity. Such representatives shall give reasonable advance notice of a projected visit, in order that appropriate consultations may be heald and that maximun precautions may be taken to assure safety and to avoid interference with normal operations in the facility to bo visited.

Article

The provisions of this Treaty shall apply to the activities of any State of the World and any Party to the Treaty in the exploration and use of outer space, including the Moon and other celestial bodies, whether such activities are carried on by any State of the World a single State or any Party to the Treaty or jointly with other States, including cases where they are carried on within the framework of international intergovernmental organizations.

Any practical questions arising in connection with activities carried on by international intergovernmental organizations in the exploration and use of outer space, including the Moon and other celestial bodies, shall be resolved by the Parties to the Treaty either with the appropriate international organization or with one or more States members of that international organizations which area Parties to this Treaty.

Article

1. This Treaty shall be open to all States, States Association, States Union, international intergovernmental organizations autonomy territories recognized it by the United Nations, but without total independence with base of 1514 resolution of General Assembly, for signature. Any State which does not sign this Treaty before its entry into force in accordance with paragraph 3of this article may accede to it at any time.

2. This Treaty shall be subject to ratification by signatory States and the appropriate proceding of other subjects that signature this Treaty. Instruments or ratification, instruments of accession and other instruments about this Treaty shall be deposited with the Governments of the Rusia, the United Kingdom of Great Britain and Worthern Ireland and the United States of America, which are hereby designated the Depositary Governments.

3. This Treaty shall enter into force upon the deposit of instruments of ratification by five Governments including the Governments designated as Depositary Governments under this Treaty.

4. For States or other subject accordance with paragraph 1 whose instruments of ratification or accesion are deposited subsequent to the entry into force of this Treaty, it shall enter into force on the date of the deposit of their instruments of ratification or accession.

5. The Depositary Governments shall promptly inform all signatory and acceding States of the date of each signature, the date of deposit of each instrument of ratification of an accession to this Treaty, the date of its entry into force and other notices.

6. This Treaty shall be registered by the Depositary Government pursuant to Article 102 of the Charter of the United Nations.

Article

Any - Party to the Treaty may propose amendments to this Treaty. Amendments shall enter into force for each - Party to the Treaty accepting the amendments upon their acceptance by a majority of the - Parties to the Treaty and thereafter for each remaining - Party to the Treaty on the date of acceptance by it.

Article

Any - Party to the Treaty may give notice of its withdrawal from the Treaty one year after its entry into force by written notification to the Depositary Governments. Such withdrawal shall take effect one year from the date of receipt of this notification.

Article

This Treaty, of which the Chinese, English, French, Russian and Spanish texts are equally authentic, shall be deposited i the archives of the Depositary Governments. Duly certified copies of this Treaty shall be transmitted by the Depositary Governments to the Governments of the signatory and acceding Parties.

IN WITNESS WHEREOF the undersigned, duly authorized, have signed this Treaty.

Astronauts as envoys of mankind
Common heritage of mankind

The States Parties to the Treaty shall regard astronauts as envoys as mankind. The Treaty of 67 in the article V, consider respect the astronaut, the aspects of assistance in the event of accidents distress or emergency landing on the territory of another state Party or on the high seas, in carring on activities in outer space and on celestial bodies, the astronauts of one State Party shall render all possible assistance to the astronauts of other States Parties. Other aspect, is the State Party to the Treaty shall inmediatelly inform the other States Parties to the Treaty or the Secretary-General of United Nations of any phenomena they discover in outer space, including the Moon and other celestial bodies, which could constitute a danger to the life or health of astronaut. This article only consider security aspect of activities astronauts.

Astronaut is a person who stay in outer space, who go into outer space or who come back to Earth from outer space.

I think that is important and neccesary to draw a new agreement that consider other aspects of astronauts as envoys of mankind. For instance rights, duties, warrants, etc.

The Parties to this Treaty.

- Recognizing the actual and future interest of all mankind in outer space activities for peaceful purposes.

- Believing that the exploration and use of outer space should be carried on for the benefit of all mankind.

- Considering that the all people of the Earth shall regard astronauts as envoys mankind.

Have agreed on the following:

<u>Article</u>

The astronauts are regard as envoys of mankind in outer space.

<u>Article</u>

Astronauts is any person who go into space, who stay in outer space or who come back to Earth from outer space.

<u>Article</u>

The man and woman as astronaut have in outer space equal rights, duties, warrants and opportunities, irrespective of nationality religious creed, political ideas, sex, color, etc.

<u>Article</u>

States of the World and Parties to the Treaty shall recognize to astronauts the following rights:

The right to life; the right to be born in outer space, the Moon and other celestial bodies; the right to the freedom; the right to the honor; the right to the integrity personnel; the right to the security; the right to live in outer space, the Moon and other celestial bodies; the right to the exploration, investigation, use and explotation of outer space the Moon and other celestial bodies. Death penalty is forbidden. The right to work in outer space.

The astronauts will have freedom of political ideas and religious creed.

<u>Article</u>

The Parties to the Treaty stablish: Astronauts have the following duties:

To live, to act and to think as envoys of mankind in outer space the Moon and other celestial bodies; to fulfill the outer space law; the exploration, investigation, use and explotation of outer space, the Moon and other celestial bodies, will be do it a good faith and for peaceful purposes; to inform to any party of the Treaty or General Secretary of United Nations, about any activities that has reason to believe would cause potencially harmful to life of astronauts, Earth environment, and space environment; to obey the veredict of court of justice of space.

<u>Article</u>

Parties to the Treaty, according that the astronauts conserve their rights, duties, warrants and opportunities of their nationality, even during the time of that person is in outer space, respect of its life on Earth.

<u>Article</u>

The astronauts will fulfill first of all the space law and after the law of their State.

<u>Article</u>

The astronauts have responsability in front of mankind and after in fron of their state, that has international responsability for national activities, wheter such activities are carry out by governmental agency or by non governmental entities.

<u>Article</u>

The person who born in outer space, the Moon and other celestial bodies, would choose its nationality between the fathers nationalities, the state of registry or any state of the world that he like it.

<u>Article</u>

This agreement shall be open to all states, States Association, States Union, Autonomy Territories, for signature. Any State, State Association, State Union, Autonomy Territories which does not sign this agreement before its entry into force, may accede to it at any time.

<u>Article</u>

Any party to the agreement may propose amendments to this agreement. Amendments shall enter into force each party to the agreement accepting the amendments upon their acceptance by a majority of the parties to the agreement and thereafter for each ramaining party to the agreement on the day of acceptance by it.

<u>Article</u>

The original of this agreement, of which the Arabic, Chinese, English, French, Russian and Spanish texts are equaly authentic, shall be deposited with the Secretary-General of the United Nations, who shall send certify copies thereof to all signatory and acceding States, States Associations, States Union

Autonomy territories.

The Astronauts as envoys of mankind tell to us a new kind to think about the future space law, over all with respect a rights, duties, warrants, opportunities, etc. of man and woman into space and future generations of mankind. Is important, is neccesary and urgent to draw the law of regulate all about the persons who will born, will live and will died into space.

The common heritage of mankind

The common heritage of mankind þrincipal express in the Moon agreement of 1979 in force. Have the following means:

The exploration, investigation, use and explotation of outer space, including the Moon and other celestial bodies, will be do it with imprescindible co-operation and participation of all States, that shall to act together as a common activity. The effort of some one is not for benefit for others, but the participation of all for benefit of all. All States and peoples not only have rights but duties too in front of space activities. If space activities require the co-operation and participation of all States in order to obtain benefits for all States. Is not possible to think that until all states of the world make an action in any space program, the States that have participation in space activities have not duties to share benefits from space activities to other States. In this moment, no exist a law that regulate this important aspects. Arise the following questions:

- The develop States that have space program have rights to use and explot the space with out duty of give benefit to other states.

- Until one State do not participate in one space program have not right to obtain benefits from space activities.

- Is neccesary to create an international world organism that have the control and administration of natural resources of outer space, including the Moon and other celestial bodies.

- Any natural or juridical person (private companies) can explore, use and explot free the outer space.

- Is possible that the natural and juridical person have right of property of any natural resources of outer space, the Moon and other celestial bodies.

This questions and others that we can think about the common heritage of mankind principle indicate to us that we must to draw one agreement that the develop this important principle.

Conclusion

As a conclusion, the next 21st century require a new space law that correspondance of new situation of the world, in the other hand, with States Associations, States Union, natural and juridical person develop space activities, a new perspective to work and live into outer space. Time by time is neccesary to adapt the law to new real situation of mankind. The Treaty of 67 is not an exception of this reality. The space in not for States, but for mankind.

With this paper I present the idea of supplement the Treaty of 67 developing the principles of Astronauts as envoys of mankind and common heritage of mankind, with new agreements.

The Treaty of 67 is the more important space law that govern the space activities and is neccesary develop it and adapt it to the new and future time. When the lawyer think about to amend and to supplement any space law, must to do it as a envoys of mankind.

References

- Treaty on Principles Governing the Activities of States in the Exploration and Use Outer Space, including the Moon and other Celestial Bodies, Jan. 27, 1967.

- Agreement on the Rescue of Astronauts, the Return of Astronauts, and the Return of Objects Launched into Outer Space, April 22, 1968.

- Agreement Governing the Activities of States on the Moon and other Celestial Bodies.

- Convention of Viene, 1969.

IISL-97-4.05

REVISITING THE OUTER SPACE TREATY: A RE-EXAMINATION OF THE SOVEREIGNTY - JURISDICTION COMPROMISE[*]

N.C. Goldman
D.J. O'Donnell
United Societies in Space, Inc.[**]

Abstract

International space law is dominated by the Outer Space Treaty, 1967. It neglected the concept of jurisdiction while detailing principles of lesser magnitude, such as ownership, appropriation, usage, and sovereignty. it left a void in governance in outer space that may be revisited as we bridge into the Twenty-first Century. With technology advancing quickly the day when we live and work in outer space is nearby. Perhaps we should re-look at the fundamental first principle of governance -- jurisdiction. There are several alternative models examined herein, all of which may be compatible with the 1967 treaty.

Introduction

The 1967 Outer Space Treaty resolved the important issue of ownership of outer space, including the celestial bodies. The compromise prohibited "national appropriation" of space, but required nations to retain "control and jurisdiction: over their space objects and spacefarers. That solution was widely hailed as a solomonic decision that played a role in preventing greater conflict in space (and on earth). That solution, consistent with the politics and indeed the technologies of the 1960s, may not be so clearly appropriate for the technology and experience of a new millennium. This paper will look at the compromise in light of the changing realities in space and will investigate alternatives that may prove more consistent with the needs of spacefarers in the next century.

The Compromise

Before the 1967 treaty was signed and ratified by more than one hundred nations, many scholars speculated about the prerequisites needed to claim and annex the moon and planets. The Russians had crash landed an unmanned probe on the Moon; soon, the United States would land men on the Moon. Clearly, actual

---

[**]Author Nathan C. Goldman, PhD, JD, specializes in space law and is a professor in Houston, Texas. Goldman is Executive Editor of the journal, *Space Governance*, and is a USIS Regent. Author Declan J. O'Donnell, JD, is a member of the IISL, President of United Societies in Space, President of World-Space Bar Association, and principal attorney in Declan Joseph O'Donnell PC in Denver, Colorado.

settlement -- a most telling determinant of sovereignty -- was not technologically feasible into the foreseeable future. Based somewhat on the earlier Antarctica analogy (in which claims of sovereignty over territory were held in abeyance), the UN Committee on the Peaceful Uses of Outer Space (UNCOPUOS) agreed that space was the province of mankind, for the benefit of mankind, and finally that the moon and the other celestial bodies could not be annexed or appropriated by any means.[1]

This solution prevented squabbling over territory, but seemed to create a lacuna regarding responsibility for space activities. That problem, in turn, was solved by Articles 6, 7, and 8 of the Outer Space Treaty, which placed responsibility with the "launching" or "registering" nation. Article 6 placed international responsibility on the State Parties for the actions of public and private space activities by their nations. Article 7 made nations (states parties) launching or procuring the launch responsible for any damages resulting from the launch. And finally, Article 8 placed control and jurisdiction over space objects and spacefarers with the nation of registry. These three articles, thus, bestowed many of the indicia of sovereignty upon the launching or registering nation.

In the past thirty years, the implications of such "quasi-sovereignty" has not been very fully explored. Yet, in this context, the Bogata Resolution and the lonely French position that a long-termed orbital slot was a form of national appropriation find new philosophical meaning. Indeed, the structural and policy struggles in the ITU and other governmental and nongovernmental organizations (such as INTELSAT) at century's end can be interpreted in the wider context of finding new mechanisms to balance anew the sovereignty equation, and to manage and otherwise supervise the increasingly sophisticated activities in space.

More as an intellectual exercise than an actual agenda, we will begin by looking at new mechanisms to manage space activities deeper into the future. As personnel remain in space for longer time and in larger numbers, as the financial commitment and benefit from space becomes more profound, the very nature of supervision will begin to change. On the one hand, the interests of earth will dictate a greater interest in space activities; on the other hand, the interests of the spacefarers may begin to diverge from those of earth.

Of course, the introduction of the interests of the spacefarers emphasizes a new element into the legal and political equation. It is in this context that we should reopen the debate over the 1967 compromise. And it is in this spirit, that we want to discuss examples of quasi-sovereignty (control and jurisdiction with less than full independence). First, we will look briefly at American Indian Law, the jurisprudential treatment of the semi-autonomous tribes vis a via the sovereignty of the federal government and the limited sovereignty of the states. Then, we will review the legal treatment of "territories" under the Jurisdiction of the UN Trusteeship Council. Finally, we will return to the more near termed needs for a stable legal regime in space. We would like to discuss the EDA, Economic Development Authority (about which O'Donnell has written extensively), in the context of the control - sovereignty continuum as preparatory for the other more long termed solutions.

Analogy -- American Indian Law

The patch quilt of legal relationships between the United States Government and the Indian tribes in America represents a jurisdiction - sovereignty compromise similar to that expressed in the 1967 treaty. In this analogy, the province of mankind at least as represented by the UN or by the world as a whole is the "federal government" and the space settlers are the "tribe." The treaty or contract between the ultimate sovereign (the federal) and the dependent (tribe) creates a "constitutional" and often a "corporate" entity.[2]

The tribes retain those indicia of sovereignty that the treaty permits; moreover, the tribe retains inherent elements of sovereignty that are not negated by the dominant relationship.[3] There is also some authority that the tribe, as "dependent nations," have a higher status than the fifty states of the union have![4] Tribes maintain control over their members and their territory.[5]

Although the comparison is admittedly not complete, this analogy for subordinate authority suggests several issues that are likely to arise in the future between the semi-autonomous space settlements and the sovereignty on earth.

The tribes, subject to treaty provisions, have the inherent right to issue regulations to administer to its needs and to protect its members.[6] The tribal councils can determine membership in the tribe, and can regulate internal matters such as domestic relations and inheritance.[7] Nonetheless, representing the dominant sovereign, the Congress still can intervene and assume control over all Indian affairs as deemed necessary.[8]

This model suggests insights into ways that a "quasi-sovereignty" in space may deal with its citizens as well as with its earth-based visitors. In general, Indian tribes have the inherent sovereign powers to control the activities of non-Indians on the reservation; this authority includes the power to license and tax these activities, and to regulate and prohibit actions that endanger the safety or the welfare of the tribe.[9] In this regard, it should be noted that the federal congress can and has interceded to require tribal councils and courts to meet constitutional requirements of due process and civil rights.[10]

The space settlement in orbit or on another world will similarly need to regulate its own members as well as its visitors. Indeed, the existing agreement among the nations on the International Space Station already provides rules about whose laws apply and who has the authority to enforce them. The large space settlements of the future will need such quasi-sovereign powers to provide for the safety and welfare of the installation and its inhabitants.

Another related (if esoteric) situation that might arise is the case where a criminal or terrorist on earth escaped to the quasi-sovereignty in space and asked for asylum or "citizenship" in the community. Or less dramatically, if a recent space immigrant is involved in a legal conflict, is he or she subject to the laws of earth or of the space settlement?

An analogous situation arises when someone claims membership in a tribe and claims its law applies. Generally the Congress by treaty or statute provides a mechanism to determine who is included on the tribal role (often a commission).[11] The treaty with the Seneca nation, however, permits that tribe to retain authority to determine its own membership roles.[12] Moreover, tribes have been found to have the authority to admit (adopt) "aliens" into the tribe.[13]

These types of issues arise in the event that space settlements will be governed by a distant sovereign. But what if the future determines that at some point the space settlement should become independent or at least has a right to prepare for eventual independence. What earthly analogy offers itself in this eventuality?

Analogy -- Trust Territories

The United Nations Trusteeship Council represents another theory and vehicle with which to address the problem of managing, administering, and otherwise governing spacefarers and space activities in the Twenty-first Century. This theory requires further interpretation as to the province of mankind and even the common heritage of mankind provisions. Since the UN represents almost the entire world, it would not be much of a stretch to vest that world body with the duty of overseeing the "province of mankind." Moreover, that body could determine, at some future date, that it is to the "benefit of mankind" that fully self-sufficient space colonies require their independence and sovereignty.

Created after World War II to head off struggles over conquered territories, the Trusteeship Council administered eleven trust territories. By 1975, ten had become independent. The status of US Micronesia was finally settled in 1990, and the last territory, Palau, became independent in 1994.

The Trusteeship Council originally had 14 member nations divided between administering (colonial) powers and non-administering nations. After 1980, the five permanent members of the Security Council assumed the duty and role of the Trusteeship Council.

As the human and physical infrastructure in space matures, the need to oversee these assets will increase. Indeed, the struggles within and around the ITU are symptomatic of attempts to deal with the maturation of satellite telecommunications. This process will continually repeat itself as other areas of space commerce and space applications mature. The Trusteeship is a tested mechanism that might find applications in the future evolution of space settlements.

Economic Development Authority (EDA)

Space economic development authorities (EDAs), about which Mr. O'Donnell has written extensively, can fit into either scenario - the quasi-sovereign dependency of the tribe, or the preparation for sovereignty. EDAs are important in the near term as a means to organize, finance, and otherwise rationalize the commercial development of space. In the context of the sovereignty dilemma, EDA also offers a solution that practically oversees and encourages space development and can serve either as the terminus or a stopgap until a satisfactory new compromise can be reached between mother earth and its space children.

Courts in the US have found that some Indian tribes have a corporate as well as a political/constitutional existence.[14] Similarly, EDAs could exist within either a corporate or a political or a combined structure. Indeed, INTELSAT and other space international organizations have such dual natures. Somewhat ironically, INTELSAT and INMARSAT are busy trying to divest themselves of much of the political component of their existence.

EDAs can exist as free standing authorities that would help finance and supervise space development. Additionally, it could have a political existence either as a UN-affiliated organization or as a semi-autonomous proto-governmental agency.

Conclusions

In the final analysis, the sovereignty - jurisdiction compromise of 1967 contained two related components: (1) the limited sovereignty for humans in space (non-appropriation), because of (2) the ultimate sovereignty of humankind over human activities in space (province of mankind/common heritage of mankind). As settlements in space become more common and develop the need and desire to exercise more autonomy, one or both components of the 1967 compromise will have to be addressed, amended, or abolished. If the focus will be on the first component (limited, shared sovereignty), the Native American analogy might offer some guidance or analogies for the new regime in space. If the second component (eventual sovereignty) is the focus, the Trust Territory may better address the ultimate determination of sovereignty in space. EDA mechanisms can play an important role in either scenario.

References and Notes

1. Treaty on Principles Governing the Activities of States in the Exploration and Use of Outer Space, Including the Moon and other Celestial Bodies, January 27, 1967.

2. Parker Drilling Co. v. Metlakatla Indian Community, 451 F. Supp. 1127 (D.C. Alaska, 1978).

3. Bottomly v. Passamaquoddy Tribe, 599 F.2d 1061 (1st Cir., 1979); Merrion v. Jicarilla Apache Tribe, 455 U.S. 130 (1982).

4. Colliflower v. Garland, 342 F.2d 369 (9th Cir., 1965).

5. White Mountain Apache Tribe v. Bracker, 448 U.S. 136 (1980).

6. Iowa Mutual Insurance Co. v. LaPlante, 480 U.S. 9 (1987).

7. Montana v. U.S., 450 U.S. 544 (1981).

8. U.S. v. Choctaw Nation, 193 U.S. 115 (1904).

9. Montana v. U.S., 450 U.S. 544 (1981).

10. Indian Civil Rights Act, 25 U.S.C.S. Secs. 1301-1303.

11. U.S. v. LaMottee, 67 F.2d 788 (10th Cir., 1933).

12. Woodin v. Seeley, 261 N.Y.S. 1042 (1931).

13. U.S. v. Hitchcock, 205 U.S. 80 (1906).

14. U.S. v. Rose, 20 F. Supp. 350 (W.D.N.C. 1937).

IISL-97-4.06

ASTRO LAW AS LEX COMMUNIS SPATIALIS*

D. O'Donnell
N. Goldman
United Societies in Space, Inc.**

Abstract

International space law, often called *Corpus Juris Spatialis*, refers to space treaties and selected international law in order to regulate the behavior of nations in their dealings with each other. It does not feature a common law. Astro law is asserted as the extension of common law into space as a venue, thereby supplementing the treaty system with a *Lex Communis Spatialis*, a common law in space. This would focus on the venue itself in order to regulate the behavior of all people of space. Although the essential features of this new body of law are well defined historically, immediate modification would be required in order to use it in space. The astro law result therefore is a common law modified by the rule of A-E-I-O-U and sometimes Y, which is described herein. A proposal on how to initiate astro law is described as a future UN activity that would follow up after a citizen convention as planned on August 4, 2000 AD, in Denver, Colorado, under sponsorship of United Societies in Space Inc. and others.

Introduction

International space law consists of five space treaties,[1] certain UN General Assembly resolutions,[2] dicta from the UN Committee on Peaceful Uses of Outer Space (UNCOPUOS),[3] as well as select multinational agreements contemplated by treaty.[4] These provide for the exploration of space and the use of space resources by nations as well as establish universal principles of behavior in the venue of space as to all nations.

However, there is a void in space law as to a common law of property rules, contract administration, tort law, and public policy beyond the conception of space policy. The every day behavior of people in space is not addressed by the treaties (except inside of the space vessel). Private property as a fundamental starting point in common law is denied under the non-appropriation clause of the 1967 treaty.[5] This codifies a basic tenant of space law that was part of the

---

**Author Nathan C. Goldman, PhD, JD, specializes in space law and is a professor in Houston, Texas. Goldman is Executive Editor of the journal, *Space Governance*, and is a USIS Regent. Author Declan J. O'Donnell, JD, is a member of the IISL, President of United Societies in Space, President of World-Space Bar Association, and principal attorney in Declan Joseph O'Donnell PC in Denver, Colorado.

earliest UN provisions about space.[6]

Denial of the right of private ownership of real and personal property (that classifies as space resources rather than space objects carried into space by astronauts), seriously modifies the entire idea of common law in space. The subject is skewed because of the impressed public interest in space resources. More perplexing, there are at least three differing versions of how that public principle of non-appropriation works.[7] The 1979 Moon Treaty extended this to a larger meaning under the phrase "common heritage of mankind." It also clarified that non-appropriation applied to natural persons and commercial corporations.[8] A UN proposal in 1982 may have catapulted part of these treaty burdens, particularly benefit sharing as a principle of space law, far beyond the earlier conception of it: all nations should actively manage the resource and actually receive dividends from its use.[9] Gorove has observed that the expansion of this simple principle has led to animosity and dissention by the spacefaring and developed nations.[10]

Despite the lack of detail and the disputes that exist among states as to implementation, the space treaty principles are well established as universal principles. Moreover, they permeate all of space resource planning and management so as to affect, if not to determine, the character of such use.[11]

Therefore, our starting point is that treaty burdens known as international cooperation, non-appropriation, and benefit sharing are statutorily overriding principles under any appellation and description. Astro law by definition will have to include these in the formula for common law that it obtains.

Extended Common Law

Perhaps the first rule of common law is that it only applies where statutory law is silent, or where it is so ambiguous as to need reference to the common law to interpret it. The very confusing, conflicting, and changing interpretations of the space treaty burdens suggest a long term ambiguity, particularly in reference to space properties. The need for clarity has been asserted because space policy has come to represent a barrier to space travel:

> "Nevertheless, rules must be established regarding the manner in which property may be acquired and maintained (in space), and to apply those rules to all entities, whether they be individuals, corporations, states, or international organizations."[12]

Historically the common law has not only been a useful tool in solving difficult legal problems with workable rules, it has also been able to grow and extend itself to new venues. In Merrie Olde England, the king's law applied in the outlying villages often caused problems that reflected rather fundamental unfairness. The church court provided relief where the king's court could not respond. This became the Court of Equity and the common law of England became incorporated in due course into British law.

After America became established it too needed a common

law to bridge the gap between statutory law and day to day human experience. It borrowed the British common law and extended it to all of the states. The cut-off date of 1690 was used, conveniently the year that England finally abdicated its claim of sovereignty over the high seas. The federal and state governments both used this *Corpus Juris* to their advantage because it reduced uncertainty.

Congress extended the common law to all of its extraterritorial courts in 1850 AD. Again, this flexible and nevertheless traditional set of rules assisted in the management of people and properties in far flung places. Colonial England and territorial America managed most of the free world for centuries under this legal paradigm.

It is proposed that common law as developed at 2000 AD be extended to space. The newly constructed *Lex Communis Spatialis* would be modified to accommodate treaty burdens and venue specific requirements and, as modified, should be known as "astro law." These changes are described below.

The process of extension is a problem in itself and that is work in process. Since there is no governance structure in space the task is not like previous extensions. It has been proposed that space be managed by a Metanation, a governance structure on earth with jurisdiction in space, in order to fill the void, manage the property, and regulate activity of all for the benefit of all.[13] Extension would be part of that process.

Astro Law Described

The concept of astro law as a necessary future component of international space law has been recognized for over two decades: it would incorporate common law and metalaw principles.[14,15] The third part of this formula is proposed herein: the incorporation of international space law treaty principles. This can be preplanned at the Convention 2000 AD as part of the space paradigm curative activity.[16]

The common law transformation from England to the United States of America in 1690 and its extension extraterritorially in 1850 occasioned some changes in that body of law because the venue required changes.[17] Interestingly, this is well known to be the strength of common law, not its weakness.[18]

Substantial changes may be expected as we transform it into astro law on the high frontier. Not only do the treaty principles need to be incorporated, but, also, we may anticipate future changes as the idea of jurisdiction evolves and matures in that venue.[19] For example, there is no effective municipal governance in place at any space venue currently. When that occurs, astro law may be expected to respond with supporting rules and offer new applications for older principles.

This point is not difficult. If a Lunar Economic Development Authority were to adopt the Moon Agreement of 1979 and require all habitats to contain extra quarters for strangers found in distress on the moon, a new regulation regarding space would become enforceable. A common law contract between two or more developers on the moon to avoid that regulation, thereby suggesting an illegal contract, should be voidable under astro law, rendering that contract unenforceable. The new venue, therefore, presents a new application of an old common law and metalaw principle: illegal contracts are not enforceable among parties to that illegality (*in pari delecto*). The Outer Space Treaty of 1967 probably permits this kind of extension of older principles into space venues.[20] Clearly it does not resolve the riddle as to a developer's duty to over build habitats on the moon to accommodate strangers, however.

One of the major changes to the future of astro law may be anticipated when a Metanation or other direct governance paradigm becomes established in space. Statutes enacted by that entity for the entire jurisdiction would preempt astro law, much like the space treaties do at this time. But for this kind of change, the Metanation plans are to adopt astro law as the common law of outer space at Convention 2000 AD in Denver, Colorado, on August 4, 2000 AD.[21] The work product of this convention and a proposed treaty on jurisdiction in outer space would be tendered to UNCOPUOS for consensus adoption. Eventually the UNGA will be asked to ratify it. The extension of astro law as the common law of space will be part of this proposal, as currently planned.

The intention is to preplan a legal paradigm that can work for developers of the moon and settlers well before the need arises. The fear is that a lack of such planning will delay our going to space agenda substantially, if not lead to chaos and conflict. Work on the anticipated rules that will constitute astro law should begin sooner rather than later because a great deal of debate will be required.

Limitations

Astro law should be limited to situations that occur outside of the space vessel and away from the space habitat of the sponsoring nation.[22] Clearly, international space law has legislated in this area so the sponsoring nation retains jurisdiction over the vessel, its interior, the crew members therein or thereabout, and space objects carried into space.[23] A contractual type of document called "Mission Rules" traditionally covers the conduct of all mission participants. One of the peculiarities of space travel is that governance consists of very lofty treaties and very detailed mission rules and absolutely nothing in between. As we enter the third millennium and technology permits space travel more often by more nations, a wider spectrum of rules is necessary.

The common law would permit equitable estates in space. This is a consequence of the prohibition on legal ownership of space resources. Equitable estates include leaseholds, easements, trusts, and mortgages. Astro law has the juridical capacity to define and protect these traditional estates.

Whatever astro law evolves as rules of property law, it will remain inchoate as to the final paradigm. Someday a king or sovereign or supreme court or conquering nation may establish itself in space. Perhaps it will be the UN as a representative of 185 nations or the Metanation as trustee for 269 nations (all in the world at 1995 per *Town & Country* magazine -- but eight did not order Coca-Cola so there may be only 261) and all of humankind per its charter. Regardless, the void political paradigm in space is subject to change, from within or from without, and that will affect the laws we preplan. The new king may require an easement over all property to support schools, churches, and its own government. Common law estates would shift accordingly.

Because of the absolute nature of this political and governance void, however, all estates would change equally and everywhere. Any space government actually founded would have universal jurisdiction in space until a border was found. Even a station could be a nation because there are no limits.[24] Therefore, inchoate estates would all change equally and, as to each other, their relative bundle of rights and duties would remain equal.

A similar limitation may be anticipated due to universal applicability in outer space of the treaty burdens, non-appropriation, benefit sharing, international cooperation, and disclosure of research results. These permeate everything and must be incorporated into astro law from the beginning.[11]

One of the benefits of the municipal model and a primary reason for designing a Metanation in advance of lunar settlement is treaty compliance by regulation. No development could occur at a station unless and until equitable leases were taken instead of claims to ownership, benefits for future generations were zoned into the project, and international involvement was satisfactory. Without this kind of traditional municipal planning our space policy problems would become exacerbated, rather than cured, by space development activity.[25]

Astro law therefore can be subject to the rule of A-E-I-O-U:

A is in the astro venue only.
E is equitable estates only.
I is inchoate character of equal rights.
O is outside of space vessels only.
U is universal easement quality of treaty burdens -- past, present, and future.

A Caveat

"And sometimes Y," meaning Spanish for "and. . ." must be considered. We must remember that nations can and will assert various novel and important positions in space from time to time. The equatorial nations have asserted their own jurisdiction over geosynchronous orbits, for example.[26]

Other examples of not yet ready for prime time legal positioning for space includes abusive waivers, reciprocal waivers, space constitutions for space stations, and a long list of soft law problems. No solution is proposed for any of these but a recommendation is made clearly: astro law needs to be made available as *Lex Communis Spatialis* in order to help deal with these and future unexpected and untested legal phenomenon.

This kind of extension has occurred before. The communication satellite industry regularly utilizes American governance resources. It has matured into a subculture that has legal significance.[27] This new submission of space law may have no better legal basis than the fact that it exists by consensus and it works for those concerned. Astro law, incorporating the five space treaties, metalaw, and one-thousand years of common law should have its day in court too.

Conclusion

For these reasons it is asserted that astro law become the new common law of space.

REFERENCES

1. The five space treaties are: (1) The Treaty on Principles Governing the Activities of States in the Exploration and Uses of Outer Space, including the Moon and Other Celestial Bodies, also known as "The Outer Space Treaty of 1967"; (2) Agreement on the Rescue of Astronauts, the Return of Astronauts, and the Return of Objects Launched into Outer Space, also known as "The Rescue and Return Treaty of 1968"; (3) The Convention on International Liability for Damage Caused by Space Objects, also known as "The Liability Treaty of 1972"; (4) The Convention on Registration of Objects Launched into Outer Space, also known as "The Registration Treaty of 1975"; and (5) The Treaty Governing the Activities of States on the Moon and Other Celestial Bodies, also known as the "Moon Treaty of 1979."

2. United Nations General Assembly (UNGA) resolutions of importance may include one non-resolution on December 9, 1994, when UNGA decided to not take action to amend the Moon Treaty of 1979, despite a clause therein anticipating amendments, and despite the unpopularity of the treaty among spacefaring nations.

3. The UN Committee on Peaceful Uses of Outer Space acts in an official manner when all 51 represented States reach unanimous consensus. However, published dicta is useful in interpreting its five treaties as adopted previously.

4. The European Space Agency (ESA) and various regional communication satellite organizations are noteworthy products of such authority.

5. The Outer Space Treaty of 1967, Article II, ". . . Outer Space, including the Moon and Other Celestial Bodies, are not subject to national appropriation by claim of sovereignty, by means of use or occupation, or by any other means."

6. UNGA Resolution 1721-16 (1961).

7. E. Fasan, "Dominium Lunae, Properietas Lunas," IISL 96.1.01, Beijing, 1996, surveying the literature and finding only varying degrees of non-appropriation opined upon and concluding: *"It would be detrimental for all mankind if due to unclear legal situation the hiatus in expeditions to the Moon and other celestial bodies would be extended too long. And it would be illogical to protect natural resources on the Moon more strictly than those on Earth."*

8. "The Common Heritage of Mankind" is introduced in The Moon Treaty. Also see, Article II, p. 2, which recites, *". . . Neither the surface nor the subsurface of the Moon, nor any part thereof or natural resources in place, shall become property of any State, international governmental or non-governmental organization, national organization or non-governmental entity, or of any natural person."*

9. UN Agreement on the Law of the Sea, 1982.

10. S. Gorove, "Space Law in the Global Community," *Journal of Space Law,* 1995, Vol. 44, p. 201.

11. See, for example, Gorove, *Journal of Space Law*, 1995, the UNCOPUOS general counsel, Dembling, former counsel for NASA, is reported as clarifying that benefits derived from space resource manufacture follow into the finished product. Therefore, if a lunar habitat consist half of regolith cement and half of earth imported materials, then benefit sharing will attach to half of the structure, its uses, and the rents derived.

12. G.H. Stine and P.M. Stems and L.I. Tennen, "Preliminary Jurisprudential Observations Concerning Property Rights on the Moon and Other Celestial Bodies in the Commercial Space Age," IAF, IISL-96-IISL.1.08, Beijing, 1996, p. 5.

13. Aristotle's principle that society is an image of the individual writ larger will not apply to outer space unless and until that venue is permitted to grow its own head, a govemance structure rather than a void. United Societies in Space, Inc., and others seek such a solution and the *Space Govemance Joumal* records that mission.

14. Nathan Goldman, *American Space Law: International and Domestic*, (2nd ed.) Univelt, San Diego, CA, 1996, Chap. 10.

15. G.S. Robinson and H. White, *Envoys of Mankind,* Smithsonian Institution Press, Washington, DC, 1985, Appendix I.

16. D.J. O'Donnell and P.R. Harris, "Metanation and Other Words," IAF, IAA-96-IAA.5.1.05, Beijing, 1996, p. 8, "Lex Communis Spatialis."

17. 48 U.S. C. 1463 extending common law from America to its territories.

18. J. Holmes, *The Common Law*

19. D.J. O'Donnell, "Commercialization by Evolution in the Jurisdiction of Outer Space," IAF, IAA-96-IAA.1.4.04, Beijing, 1996.

20. Outer Space Treaty, 1967, Article I.

21. D.J. O'Donnell, "Metaspace: A Design for Govemance in Outer Space," *Space Govemance Joumal,* June 1994, p. 5.

22. Registration Treaty, 1975.

23. Outer Space Treaty, 1967, Article VIII.

24. E. Fasan, "Human Settlements on Planets: New Stations or New Nations," *Joumal of Space Law,* Vol. 22, 1994, p. 47.

25. D.J. O'Donnell, "Benefit Sharing: The Municipal Model," IAF, IISL-96-IISL.3.09, Beijing, 1996, p. 4.

26. Bogota Convention, 1972.

27. Martine Rothblatt, "Lex Americana," *Joumal of Space Law*, Vol. 23:2, 1995, p. 123.

IISL-97-4.10

COMERCIALIZATION OF SPACE ACTIVITIES AND APPLICATIONS OF THE SPACE TREATY.......GEOSTATIONARY ORBIT AND FREQUENCY SPECTRUM

Toshio KOSUGE
University of Electro-Communications
Tokyo, Japan

1.Introduction

More than 5,000 satellites have been launched into orbit since 1957. 496 satellites of those were on the geostationary earth orbit (GEO). 113 of those satellites were communication and broadcasting satellites on the orbit. It is clear that most of those communication and broadcasting satellites have been utilized for global, regional and domestic commercial communication network services in order to improve terrestrial network systems revolutionary and very rapidly since 1960's.

In the Asia-Pacific region more than 40 communication satellites were launched on the GEO by countries, like Australia, China (including Hong Kong), India, Indonesia, Japan, Korea, Malaysia, Thailand, Tonga. Within a few years several non-GEO communication satellites will be in service by using low earth orbit(LEO) and medium earth orbit(MEO). Demand for those commercial communication satellites for fixed and mobile services has been very strongly increased by not only developed but also developing countries.

In such circumstances it would be very important for critical issues of international cooperation and coordination to solve conflicts among nations for more efficient and equitable use of those limited natural resources, orbit and frequency spectrum on commercial bases under the Space Treaty and other international laws.

2. Commercial use of outer space and the Space Treaty

Since 1980, the commercial use of communication and broadcasting satellites has been rapidly increasing in both developed and developing countries. From the end of 1990's, many new projects for commercial communication and broadcasting services using GEO and non-GEO orbit will be introduced.

---

*Director, IISL; Professor, University of Electro-Communications, Chofu, Tokyo, Japan; Chairman, SOLAPSU.

The number of communication and broadcasting satellites to be launched in GEO from 1996 to 2006 is expected to reach approximately 300. The market size will be around 28 billion US$. In addition, many projects for mobile satellite communication services using LEO and MEO are planned; ICO, Iridium, Odyssey, Global Star, etc. There will be more than 242 such kind of satellites during ten years till 2006.

Commercial satellite markets in the world will grow largely in 1990's, because of the exploding demands of satellite communication and broadcasting services in the Asia-Pacific and other developing areas in the process of economic growth and also during the replacement period of previous generation satellites. Furthermore, many small non-GEO satellite projects will contribute quick development of the commercial satellite market.

Under the Space Treaty of 1967 there has been development of commercial satellite communication and broadcasting services using GEO. The question has not yet been raised on Article 1 of the Space Treaty for equitable access to the limited natural resources, GEO and radio spectrum. Before 1980, the demand for GEO and spectrum was not so high, and as a result such space resources were used by a few countries on the "first come, first served" basis. In the 1980's many countries, especially developing countries, joined International Telecommunication Union (ITU) and such allocation systems. As a result of both WARC-85 and WARC-88 on GSO, reflecting those developing countries' positions, it was decided that at least one orbit and one frequency band shall be allocated to all member countries for their domestic satellite communication.

Realizing the importance and the value of the space resources, each country tries to insure as many orbits and spectrum as possible in advance and there are overflowing applications for the orbit and spectrum as paper satellites. This situation resulted from the exploding demand for telecommunications services fueled by private sector enterprises; growing congestion in the use of those orbit and spectrum resources; the current ITU procedures that do not discourage hoarding of the orbit and spectrum resources and increased awareness of the economic value of orbit and spectrum resources.

Even though there is the flexibility in the ITU Plan for satellite communications, its technological obsolescence is clear. The standardized parameters used to develop the Plan were based on 1980's technology. The technological advances of the last 10 years since the Conference (WARC) already shows us the potential advances of the next 10 or even 5 years. Technology obsolescence, however, is an inherent danger of any planning approach. It would be quite appropriate for ITU to have WARC-97 Conference for many agenda, including satellite communication.

The following issue could be a potential key element in the WARC-97. It will address international satellite network coordination. Substantive preparatory work has been done in this area and for good reason. It implicitly touches on key elements in frequency coordination, which in general have been carefully elaborated in an evolutionary manner, trying to cope with the ever-accelerating telecommunication developments, without neglecting the principles of economic use of, and equitable access to the spectrum.

Whilst it can be claimed that this intent has been reflected faithfully and in great detail in the current relevant ITU Radio Regulations), a certain uneasiness is building up. The Regulations are considered to be increasingly complex, complicated and non-transparent, except for some respectable specialists. The dispersion of related provisions in the various Regulations and their overlapping different timing schemes with regard to advance publication coordination and notification of satellite systems, leads to confusion and conflicting interpretations.

The classic approach to reliable frequency assignments is careful planning, provided that the necessary parameters are agreed and supported by users, technology and timing requirements. With a streamlined approach such as this, an economic use of the spectrum can be achieved and equitable access guaranteed, at best, with a priori plans.

In the case of implementation of satellite systems over an unspecified period of time, some of these parameters may change in the meantime. The more rigid a plan is, the more critical its value, benefit and justification. Medium and especially long term planning therefore call for appropriate attention and may be confined only to the respective allocation provisions of Article 8 of the Radio Regulations. This would avoid the inherent drawbacks of planned, but not used, parts of the spectrum.

The alternative approach would be to rely on regulations, rules and procedures providing timely, qualified case-by-case coordination with guaranteed access. This brings us to the coordination procedures in general, be they in the core of the Radio Regulations or in annexes. Every effort should be made to consolidate these procedures into basic, transparent, and truly relevant stable instruments, favoring a better general understanding, this would save time and resources. Specific additions for different services should be kept to the minimum necessary. Special attention should also be given to the associated overlapping time-schemes mentioned earlier often a source of controversy.

The specific orbital aspects with respect, for example, to the GEO and the non-GEO also call for careful examination. The physical properties of these orbits and the operational requirements for satellite systems must be reconciled while avoiding inappropriate preferences. Orbital and spectrum economy are closely related issues and conflicts could be rendered less harsh by making suitable separate allocations for GEO and non-GEO---this would require some adjustments. A timely review of the allocation table and service definitions is recommended.

The technical fundamentals assembled and

condensed in the ITU Radio Communication Sector, form the basis and starting point for a realistic and orderly use of the spectrum. Their proper development and application would not only make it possible to accommodate the ever-increasing requirements but would also open the door to fair burden-sharing in cases of bottlenecks when assigning frequencies.

Numerous are the factors determining fair and reliable spectrum usage. The misuse of only a few of them leads to inefficient, one-sided applications to the detriment of others. “First come---first served” should not be followed by “next come---maybe served”.

In the changing environments of the more commercial and private sectors participation into the satellite communications and broadcasting services under the 1967 Space Treaty, the now more frequent possibilities of revising the Radio Regulations of ITU bring with them the most welcome advantage of aligning Regulations with evolving requirements, provided no complications, duplications and implicit hesitations are perpetuated based on over caution, mistrust prevailing fear. Continuous cosmetics and fine-tuning of existing Regulations in their old framework may no longer suffice to overcome present problems and should be complemented by courageous new decisive steps, in the area of satellite networks in particular. Already, the acceptance and implementation of a new concept with guaranteed access when required, replacing, where appropriate, the abstract value of an immediate plan entry may ease the situation.

3. Optimizing the spectrum for satellite communications

In exploding demands for commercial satellite communications and broadcasting services, it would be more important and critical for each country and also ITU to elaborate the optimizing the orbit and spectrum within the framework of limited natural resources for mankind.

It could be useful for development of future implementation of Article 1 of the Space Treaty in the changing environment of more commercial utilization of space resources.

In the case of national level, price-based allocations of the spectrum might be interesting. Allocating spectrum to the user who values it most is a particularly acute problem where demand exceeds supply (probably most evident in the allocation of spectrum for new telecommunications services). Conventional spectrum allocation methods cannot deal adequately with this problem. Over the counter and lottery allocations provide windfall gains to those lucky enough to be allocated scarce licenses, at the cost of the community as a whole. There is, moreover, no guarantee that the most valued and efficient uses will be accommodated. A comparative selection process needs considerable time and resources to work properly, and there is no guarantee that the most economically efficient use will be rewarded.

In these circumstances, some countries like Australia, have used a system of price-based allocations to distribute licenses among competing users. While we have three means of price-based allocations available to us: auction, tender or negotiated price; in practice Australia has only used the auction mechanism. Australia's Spectrum Management Agency (SMA). The SMA conducted during 1994-1996 three auction styles, and they have been extremely successful, with a total of over 550 licenses allocated. The process worked smoothly, was widely regarded as successful in allocating the spectrum to those who valued it most, and contrasted to previous much less successful attempts in Australia to allocate communications licenses by tender.

The simultaneous auction has significant advantages over sequential auctions where a number of related licenses are offered for sale. Because all licenses are on sale at the same time, buyers can adjust their strategies as the auction develops. In a traditional auction, when bidding for a particular lot, buyers have to guess what prices will be paid for later lots. In a simultaneous auction a buyer can much more easily switch to other lots during the course of an auction if outbid on their preferred lots. All lots are on sale until bidding activity ceases on all lots.

This kind of philosophy is that, wherever possible, planning and management decisions are best left to the marketplace. Maximum freedom for licensees to use spectrum in the way they wish is likely to produce the greatest overall efficiency, because the least obstacles will be placed in the way of the highest value use.

Spectrum pricing and market allocations are not the only way to introduce market force into spectrum management. A crucial factor is what licensees can do with spectrum once they have access to it. Traditionally, particular bands are licensed for specific services, with tight operating requirements constraining its use. This has long been thought necessary to minimize interference and maximize spectrum efficiency.

However, there are growing difficulties with this approach. For a start, different technologies increasingly compete to use the same bands, with differing standards for the same basic service developed in Europe, the US and Japan competing for a share in the marketplace. Secondly, technology and use are changing rapidly. And, finally, there is recognition that in a deregulated world, as much freedom as possible should be given to users to make basic planning and management decisions by themselves.

Recognizing this, Australia has provided for a new, more market-oriented form of radio communications license, called a spectrum license. It introduces an access right, akin to a property right, over the spectrum.

Instead of focusing on particular pieces of equipment and its uses, spectrum licensing authorizes the use of spectrum. It entrusts the spectrum to the care of licensees. Licensees will have considerable flexibility to use the spectrum as they wish, provided they comply with the core technical conditions of the license. In other words, the SMA will not determine the use to which the spectrum is put.

Spectrum licenses are fully tradable in the market place, further increasing flexibility for users. They will be allocated using a price-based process, and the SMA intends that this would normally be through an auction. There are of course some major coordination and interference issues to be addressed. Because of the considerable freedom available to licensees to choose their own services and technologies, core conditions have to be developed which minimize the risk of interference regardless of the technology used.

Domestic and international challenges

In order to respond to the great demand to the orbit and spectrum, space resources, at domestic level, a lighter touch, market-oriented approach is the only effective way to deal with challenges facing spectrum managers in a time of rapid technological change, deregulation and competition, and internationalization of communications.

But considerable challenges remain. Traditional planning and licensing techniques have difficulty keeping up with the pace of technological and commercial change, and this is compounded where international planning is required. Domestically we believe that providing maximum flexibility to licensees is the best way of allowing the market to adapt to change. Internationally this is more difficult. Inevitably, the licensing of global mobile satellite services in the countries will be heavily influenced by licensing and spectrum allocation decisions taken by the FCC of the US. While it may be argued, for example, that countries should be able to auction spectrum for such services, such an approach adopted by every country would almost certainly make these systems impossible to implement. This is an area where increased international cooperation may be desirable. One example of such cooperation came early this year when prospective mobile satellite competitors (Iridium, Odyssey and Globalstar) agreed to abide by a spectrum-sharing plan.

Another challenge is that of providing quick and fair means of clearing spectrum which is required for new services. In the case of Australia, the government is providing a mechanism by which spectrum can be sold while still encumbered, with clearance proceeding after sale according to a pre-defined timetable. A similar type of system has operated in the US in the Personal Communication Service bands, but with new purchasers being required to meet the costs of moving incumbents elsewhere. Both systems have been the subject of controversy, but finding the best way to meet both the expectations of new purchasers and incumbents will be a growing challenge for spectrum mangers throughout the world.

4. Conclusion

Since the orbital space and frequency spectrum have been used and also secured by both developed and developing countries, it is quite important for us to optimize the spectrum for further development of commercial satellite communications and broadcasting services. As mentioned in the case of some countries, domestic management of the spectrum could be feasible for implementing more efficient, effective and economical use of limited natural resources like orbit and spectrum at the international level. It would be important for the future development of commercial use of satellite communications and broadcasting services for all countries under the scheme of Articles of the Space Treaty., to improve rules and regulations for the procedures of ITU and other international regimes in order to optimize those resources.

Selected Bibliography

*Arnopoulos, *The International Politics of the Orbit Spectrum Issues,*7 Annals Air & Space Law,215(1982)

*Butler, *The International Telecommunication Regulatory Framework For Satellite Communications.* Proc. 28th Colloq. On the Law of Outer Space 295(1985)

*Cheng, *The Legal Regime of Air and Outer Space: The Boundary Problem, Functionalism versus Spatialism: the Major premises.* Annals Air & Space L.(1980)

*Christol, The Modern International Law of Outer Space (1982)

*Christol, Space Law: Past, Present and Future (1991)

*Codding & Rutkowski, The International Telecommunication Union In a Changing World (1982)

Constitution of International Telecommunication Union,(1994) Kyoto

*Doyle, *Regulating the Geostationary Orbit:ITU's WARC-ORB-85-88,*15 J. Space L.1(1987)

**Final Acts adopted by the First Session of the WARC-ORB-85* (Geneva,1985)

**Final Acts adopted by the Second Session of the WARC-ORB-88* (Geneva, 1988)

*Gobriel, International Space Law Miscellanea, 1995

*Gorove, *Geostationary Orbit: Issues of Law and Policy,* 73 Am. J. Int'l L. 144(1979)

*Jassentuliyanna, *The Developing Countries and the Geostationary Orbit* (paper presented at the 13the Annual Friedman conference on the Global Telecommunications Revolution, Columbia Univ., March 29,1985

*Kosuge, *Legal Problems of Geostationary Orbit,* Law and Outer Space (1990)

*Leive, International Telecommunications and

International Law: The Regulation of The Radio Spectrum (1970)

*Matte, Aerospace Law, Telecommunications Satellites (1982)

*Papp & McIntyre, International Space Policy (1985)

*Rothblatt, *Satellite Communication and Spectrum Allocation,* Am. J. Int'l. L. (1982)

*Smith, D., International Telecommunication Control (1969)

*Smith, D., Communication via Satellite (1976)

*Smith, M., International Regulation of Satellite Communication (1990)

*Zwaan, ed., Space Law: Views of the Future (1988)

*Weiss, *Maximizing Access to the Geostationary-Satellite Orbit,* 53 Telecommunication J. 469 (1986)

*White, R. & White, H., The Law and the Regulation of International Space Communication (1988)

*Williams, *The Exploitation and Use of Natural Resources in the New Law of the Sea and the Law of Outer Space,* 29th Collq. On the L. of Outer Space, 198 (1986)

IISL-97-4.11

IMPLEMENTATION OF ARTICLE IV OF THE OUTER SPACE TREATY OF 1967 DURING THE 21st CENTURY

"And he shall judge among the nations, and shall rebuke many people: and they shall beat their swords into plowshares ,and their spears into pruninghooks: nation shall not lift up sword against nation, neither shall they learn war any more"[1]

*Maurice N. Andem**
Faculty of Law, University of Lapland, Rovaniemi, Finland

Introduction

The second half of the 1990's has not only been a period of anniversaries but also of taking stock world events: In 1995, the 50th anniversary of the end of the Second World War and the birth of the United Nations (UN); the Halifax Summit of the Seven Industrialized Nations was held from 15 to 17 June. This year - 1997, we are participating in the 40th International Colloquium on the Law of Outer Space, in which we are commemorating the 30th Anniversary of the entering into force of the Outer Space Treaty, that is, **The Treaty on Principles Governing the Activities of States in the Exploration and Use of Outer Space, including the Moon and Other Celestial Bodies.**

In other words we are gathered here not only to celebrate but to re-evaluate the effectiveness and the efficacy of the provisions of the 1967 Outer Space Treaty (OST), since it's entering into force three decades ago. It is not my intention to make any detailed review of the law-making process of law of outer space or space law; but to reflect on some of the issues of the Outer Space Treaty of 1967 relevant to the topic of my paper.

It should be borne in mind that the space age began with the successful launching into orbit of **Sputnik-1** on 4 October 1957 by the former Soviet Union, and the launching of the **Explorer-1** satellite on 31 January 1958 by the United States. At the time no state in the world protested the launching of these artificial earth satellites into the orbit around the earth had violated its territorial air sovereignty. Thus the freedom of all states to explore, use and carry out scientific investigations in outer space, including the moon and other celestial bodies, was instantly established as an international law custom, within short period of time.

Moreover, it should be borne in mind that the beginning of the space age coincided with the period of intensification of the Cold War, which was not only characterised by suspicion and mistrust, but also by the arms race and military build-up of the super powers - the former Soviet Union and the United States, including their military bloc members. Thus, it was imperative for the international community of States as represented in the United Nations (UN) to set into motion the process of formulating new principles and norms to regulate the activities of States in the new domain - outer space. This has been a continuous process, as experience has shown during the past 40 years. It will continue to be so as new advances and break-throughs are being made in space science and technology.

United Nations and the Codification of the Law of Outer Space

The United Nations did not fold its hands to allow the super powers' military rivalries and arms race to be extended into the new domain. It had to carry out its obligations imposed upon it under the provisions of The Preamble and Article 1 of the Charter. Even the two super powers themselves, as permanent Members of the UN Security Council, had to acknowledge and recognize the danger of using outer space to foster their national military and strategic policies. In this context, it was objectively mandatory for them to co-operate in the law-making process in the UN General Assembly and other organs within the UN system. This was also demonstrated in the letter written by the US President, Dwight D. Eisenhower, on 13 January 1958 to the Chairman of the Council of Ministers of the former Soviet Union, Mr. Nikolai Bulganin. In the letter, the US President proposed the use of outer space for peaceful purposes.

From 1958 to 1966, before the conclusion of the 1967 Outer Space Treaty, resolutions and declarations adopted by the UN General Assembly had been the only legal instruments available in the process of codification and progressive development of this new branch of contemporary international law - the law of outer space or space law. Pursuant to the provisions of Article 13 of the UN Charter, the General Assembly, through the indefatigable efforts of its Committee on the Peaceful Uses of Outer Space (COPUOS) and its two Sub-Committees, has adopted five treaties and five other legal instruments on principles on outer space.[2]

Outer Space Treaty of 1967: Treaty on Principles Governing the Activities of States in the Exploration and Use of Outer Space, including the Moon and Other Celestial Bodies (OST)

The Outer Space Treaty of 1967 is the Basic Space Treaty or the **Magna Carta** of space law. It stipulates in its provisions nine (9) basic principles. For example, the principles of freedom of exploration and peaceful uses of outer space, non-appropriation of outer space, including the moon and other celestial bodies, application of international law, including the UN Charter, and non-military use of outer space, are the most important principles and the cornerstone of the law of outer space or space law.

These principles are embodied in Articles I, II, III and IV of the 1967 Treaty on Principles Governing the Activities of States in the Exploration and Use of Outer Space, including the Moon and Other Celestial Bodies.

According to Article I:
"The exploration and use of outer space, including the Moon and other celestial bodies, shall be carried out for the benefit and in the interests of all countries, irrespective of their degree of economic or scientific development, and shall be the province of all mankind.

Outer space, including the Moon and other celestial bodies, shall be free for exploration and use by all States without discrimination of any kind, on a basis of equality and in accordance with international law, and there shall be free access to all areas of celestial bodies.

There shall be freedom of scientific investigation in outer space, including the Moon and other celestial bodies, and States shall facilitate and encourage international co-operation in such investigation."

Article II further declares that:

"Outer space, including the Moon and other celestial bodies, is not subject to national appropriation by claim of sovereignty, by means of use or occupation, or by any other means."

The two Articles of the 1967 Space Treaty quoted above are further elaborated in the provisions of Articles 4-6 and 11 of the 1979 Agreement Governing the Activities of States on the Moon and Other Celestial Bodies.[3]
Article 4(1) of the 1979 Moon Treaty, for example, stipulates:
"The exploration and use of the moon shall be the province of all mankind and shall be carried out for the benefit and in the interest of all countries, irrespective of their degree of economic or scientific development. Due regard shall be paid to the interests of present and future generations as well as to the need to promote higher standards of living and conditions of economic and social progress and development in accordance with the Charter of the United Nations."

Paragraph 2 of the Article stipulates that States are to be guided by the principle of co-operation and mutual assistance in all their activities concerning the explorations and use of the moon.

The legal status of the moon is further stipulated elaborately in the provisions of Article 11 of the Agreement. Paragraphs 1 and 2, for example, stipulate as follows:

"1. The moon and its natural resources are common heritage of mankind
2. The moon is not subject to national appropriation by any claim of sovereignty, by means of use or occupation, or by any other means."

Furthermore, pursuant to the provisions of paragraph 3, neither the surface nor the subsurface of the moon, nor any part thereof or natural resources in place, shall become the property of any state, international intergovernmental or non-governmental entity or of any natural person. The placement of personnel, space vehicles, equipment, facilities, stations and installations on or below the surface of the moon, including structures connected with its surface or subsurface, shall not create a right of ownership over the surface or the subsurface of the moon or any areas thereof.

In my candid opinion, the formulations of the legal status of the moon and its natural resources in Article 11, particularly in its three paragraphs quoted above, is a great revolution in the classical theory of international law, which recognized all uninhabited territories on earth as _res nullius_, and, therefore, subject to appropriation by any state. However, it should be noted that this theory of _terra nullius_ has not always been upheld in our present time (e.g. the Antarctica).

The signing of the Agreement on the moon and its entering into legal force,[4]will assist in creating more uniformity and consistency in the interpretation of the legal status of the moon and other celestial bodies in particular as embodied in the provisions of the 1967 Space Treaty.[5]

All States Parties are granted the freedom to carry out scientific investigations on the moon without discrimination of any kind, on the basis of equality and in accordance with international law (Article 6, para 1).

As already mentioned above, the 1967 Space Treaty and the 1979 Moon Agreement provide and guarantee for all states parties equal opportunities and freedom to carry out scientific investigations and experiments in outer space, on the moon, and celestial bodies, without any interference whatsoever from any state or group of states. Moreover, it should be noted that this freedom of exploration and use of outer space, including the moon and other celestial bodies, as provided in these legal instruments, first and foremost, includes the right of every state to launch into the earth's orbit rockets, artificial earth satellites, space vehicles (manned and unmanned), and space objects over the air space of other states; and secondly, the right to carry out scientific investigations and experiments in those areas. This freedom can be analogously compared with the provisions of Article 87 of the 1982 UN Convention on the Law of the Sea, which deal with the legal status of the High Seas and Article II of the 1959 Antarctic Treaty.

It is necessary to point out here that the principle of freedom of exploration and the use of outer space, including the moon and other celestial bodies, however, differs from the principle of territorial air sovereignty as embodied in the national legislation of each state and in bilateral and multilateral agreements, conventions, etc., on international air navigation.

The provisions of Articles 8 and 12 of the Chicago Convention, it should be borne in mind, cannot be analogously applied in regards to the flight of space vehicles and other objects launched into outer space, because the air sovereignty of any state, as we have already observed, does not extend ad infinitum into the universe.[6] It is necessary, at this juncture, to emphasise that the freedom of innocent passage by space vehicles launched from territory of one state through the air space of another state is an international legal right belonging to all states and recognized by states themselves.[7] Moreover, it should be remembered that the absence of a fixed frontier between air space and outer space has not caused any international conflict since the beginning of the space era forty years ago. However, this does not mean that there is no need to establish the border between the two spheres.

Limitation of the Freedom of Exploration

This now brings us to the following question: Does the principle of freedom of exploration and use of outer space, including the moon and other celestial bodies, mean any state is absolutely free to carry out any kind of space activity or experiment which is a threat to international peace and security, or aimed at the intensification of the arms race in outer space? The answer is certainly "No".

Before I proceed further in examining the provisions of Article IV of the Outer Space Treaty, it should, however, be emphasised that the freedom of exploration and use of outer space, the moon and other celestial bodies, coupled with the right to transit of space vehicles, does not mean that, that freedom is absolute. The Outer Space Treaty of 1967, besides granting rights, imposes specific obligations, which are imperative and mandatory to all the states actively engaged in the exploration and use of outer space, including the moon and other celestial bodies.[8]

For example, pursuant to the provisions of Article III of the Treaty, "States Parties to the Treaty shall carry on activities in the explorations and use of outer space, including the Moon and other celestial bodies, in accordance with international law, including the Charter of the United Nations, in the interest of maintaining international peace and security and in promoting international co-operations and understanding."

Article 2 of the 1979 Moon Agreement restates the provisions of the above Article of the Space Treaty with more additional provisions. It stipulates:

"All activities on the moon, including its exploration and use, shall be carried out in accordance with international law, in particular the Charter of the United Nations, and taking into account the Declaration on Principles of International Law concerning Friendly Relations and Co-operation among States in accordance with the Charter of the United Nations, adopted by the General Assembly on 24 October, 1970, in the interest of maintaining international peace and security and promoting international co-operations and mutual understanding, and with the due regard to the corresponding interests of all other States."

This means that any activities in the exploration and use of outer space, including the moon and other celestial bodies, whose objectives are to further international confrontation - for example, the introduction of the arms race or other military activities in the new domain - is outlawed under the provisions of the Outer Space Treaties, including the UN Charter.

For example, in the provisions of the 1970 UN Declaration on Principles of International Law concerning Friendly Relations and Co-operation among States in accordance with the Charter of the United Nations, Member States undertake to pursue, in good faith, negotiations for the early conclusion of a universal treaty on general and complete disarmament under effective international control and strive to adopt appropriate measures to reduce international tensions and strengthen confidence among States.[9]

How can the above lofty goal be achieved by the world community of peace-loving states in view of the intensification of the arms race in outer space by the super powers and their allies? It is, however, encouraging to note that this aspect had been foreseen during the earlier stages of the development of the international law of outer space.

For example, in the preambular part of the Outer Space Treaty, the States Parties express their desire to contribute to broad international co-operation in the scientific as well as legal aspect of the exploration and use of outer space for peaceful purposes. In the Preamble of the Agreement on the moon and other celestial bodies, the States Parties further express their "desire to prevent the moon from becoming an area of international conflict". The moon and other celestial bodies are declared to be used by all States Parties exclusively for peaceful purposes.[10]

This now brings us to the controversial concept or provision of "peaceful purposes" as it concerns the legal status of outer space, the moon and other celestial bodies. It would be recalled that since the launching into orbit of the first man-made object "SPUTNIK" forty years ago, the attitudes and positions of states, international organizations including individual authorities (legal, political, scientific, etc.) have never been harmonious as regards to the meaning of "peaceful purposes".

It should, however, be borne in mind that the provisions of all the international legal instruments on the law of outer space which we have already reviewed above, do not give any comprehensive list of those activities, which are authorized to be carried out by states, international intergovernmental organizations, etc., in the course of the implementation and realization of the principle of freedom of exploration and use of outer space, including the moon and other celestial bodies.[11]

In the final analysis, I will like to point out that any military use of outer space is unlawful and contrary to the spirit of the UN Charter and the Outer Space Treaty of 1967, because peace and the military are contrary to one another. A civilian organization, however, can implement those measures, for example, the WMO, UNEP or ESA.[12]

Article IV of the 1967 Outer Space Treaty (OST)

This now brings me to the examination of the topic of my paper -"Implementation of Article IV of the Outer Space Treaty of 1967 during the 21st Century."

The strengthening and maintenance of international peace and security after the Second World War had been, and still is one of the most important purposes of the United Nations as stipulated in Article 1 (1) of the Charter. Thus, in the first resolution 1348 (XIII) of 13 December 1958, the General Assembly expressed the wish to avoid the extension of national rivalries into the new domain - outer space. Therefore, in the law-making process it had to employ the institution of neutralization and demilitarization in creating a special legal regime for outer space.

According to Professor Vladimir Kopal, "Neutralization and demilitarization of outer space would prohibit the use of any preparation whose immediate or final goal would serve for waging war. These prohibitions would include (1) launching rockets with warheads through outer space; (2) orbital arms and military apparatus in outer space; (3) establishing military bases on the celestial bodies."[13]Professor G. P. Zhukov writes that "Demilitarization of outer space means the prohibition of activities pursuing military aims in peace time."[14]

It would be recalled that in resolution 1884 (XVIII) of 17 October 1963, the General Assembly welcomed the expressions by the Soviet Union and the United States of their intention not to station in outer space any object carrying nuclear weapons or other kinds of weapons of mass destruction. The provisions are embodied in Article I of the Moscow Treaty Banning Nuclear Weapons Tests in the Atmosphere, in Outer Space and under Water of October 1963 (NTBT).[15]

The principle of partial demilitarization of outer space and total demilitarization of the moon and other celestial bodies are embodied in Articles IV of the Space Treaty of 1967 and 3 of the Moon Agreement of 1979. Article IV provides that: "States Parties to the Treaty undertake not to place in orbit around the earth any objects carrying nuclear weapons or any other kinds of weapons of mass destruction, install such weapons on celestial bodies, or station such weapons in outer space in any other manner. The moon and other celestial bodies shall be used by all States Parties to the Treaty exclusively for peaceful purposes. The establishment of military bases, installations and fortifications, the testing of any type of weapons and the conduct of military manoeuvres on celestial bodies shall be forbidden. The use of military personnel for scientific research or for any other peaceful purposes shall not be prohibited. The use of any equipment or facility necessary for peaceful exploration of the moon and other celestial bodies shall also not be prohibited."

It is interesting to note that the interpretation of Article IV has been subject-matter of extensive discussions in various international forums by eminent experts and scholars in space law, during the past 30 years.[16] It is impossible for me to discuss in this paper all the views of the experts as expressed in those forums and books. Nevertheless, it would be useful to make some comments on the provisions of Article IV.

As could be seen in paragraph 1 of the Article, the States Parties <u>undertake</u> not "to place in orbit around the earth any objects carrying nuclear weapons or any kinds of weapons of mass destruction, install such weapons on celestial bodies, or station such weapons in outer space in any other manner." According to the <u>BLACK'S LAW DICTIONARY</u>, Fifth Edition (1979), the word "<u>undertake</u>" means: "to lay oneself under obligation or to enter into stipulation; to perform or to execute; to covenant to contract." "<u>Any</u>", on the other hand means: "Some; one out of many; an indefinite number. One indiscriminately of whatever kind or quantity."

Thus, taking into consideration these definitions, it is submitted that "any other kinds of weapons of mass destruction" should be construed, in the context of total disarmament, to mean all existing weapons and their delivery systems (e.g. ASAT, etc.), including future ones which may be developed for military purposes. The States Parties to the Space Treaty of 1967, had an international duty to ensure that all military activities are not carried out in outer space, on the moon and other celestial bodies. However, it should be noted that the use of military personnel, any equipment and facilities for scientific research and for purposes is permitted (para. 2).

Therefore, the testing of any weapons and their delivery systems by the super powers is a gross violation of the provisions of the Space Treaty and constitutes a threat to international peace and security (Art. III and IV). The 1982 Second UN Conference on the Exploration and peaceful Uses of Outer Space expressed grave concern on the extension of arms race into outer space. All nations, particularly those with major space capabilities, were urged to contribute actively to the prevention of the arms race into outer space and to refrain from any action contrary to that aim.[17]

For example, the weapon systems developed under the U.S. SDI Programme are contrary to the spirit of the Space Treaty and the Charter of the United Nations. In this respect it will be useful to refer to the second stanza of the Preamble of the UN Charter, which reads as follows: **AND FOR THESE ENDS to practice tolerance and live together in peace with one another as good neighbours, and to unite our strength to maintain international peace and security, and to ensure, by the acceptance of principles and institutions of methods, that armed force shall not be used, save in the common interest, and to employ international machinery for the promottion of the economic and social advancement of all peoples.**

It seems to me that we tend to forget or ignore the importance of the above stanza of the Preamble in all problems relating to disarmament, including those relating to the demilitarisation neutralisation of outer space. It should be pointed out that good neighbours, through peaceful means, will always strive to establish friendly and peaceful relations among themselves. The interdependence of states and peoples of the world today demands that maintenace of international peace and security, as **the common interest of humanity** as a whole. From time immemorial, armaments have always created fear, suspicion, mistrust and hatred among nations. Is this what humanity should continue to pursue with space science and technology during the 21st Century?

Therefore the latest laser satellite tests carried by the U.S. Army this year[18] were contrary to the provisions of that Treaty and various arms control agreements between the former Soviet Union (Russia) and the USA. The pictures below show the various types of space weapons developed under the SDI Programme, which in my candid opinion, should be classified as weapons of mass destruction.

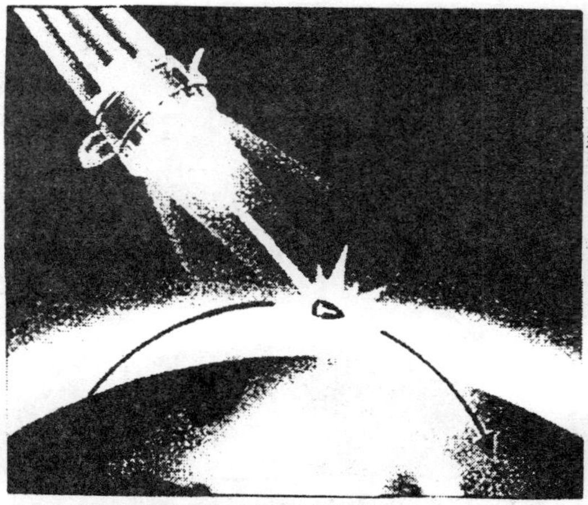

SPACE-BASED LASERS: In theory, these would combine chemicals in the manner of rocket engines to fire beams of concentrated laser energy through space. Lately, they have lost luster because of fears about the vulnerability of space-based objects.

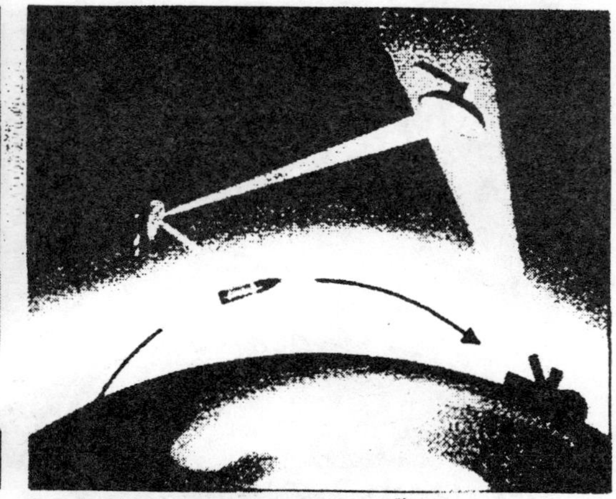

GROUND-BASED LASERS: These devices, especially free-electron lasers, would bounce beams off orbiting "battle mirrors" toward enemy missiles. Such systems, which are cheaper because heavy lasers need not be lifted into space, are viewed as less vulnerable to attack.

RAILGUNS: These devices, based in space, would use electromagnetic fields to accelerate and launch "smart" projectiles to home in on enemy boosters. The small projectiles are envisioned as something similar to those recently tested in American antisatellite weapons.

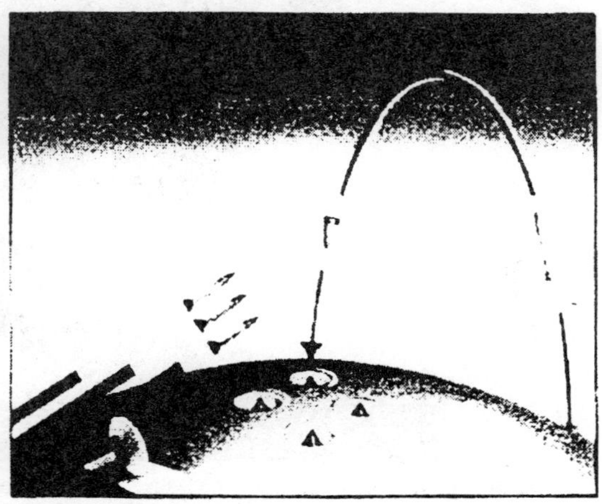

GROUND-BASED INTERCEPTORS: Using conventional rocket technology, these would be used to destroy enemy warheads during final phase of their flight, just before they hit targets. Such interceptors are often viewed as ideal for defending fields of American missiles.

Source: Boffey, Philip M. and co, Claiming the Heavens, The New York Times Complete guide to the Star Wars Debate, p. 98.

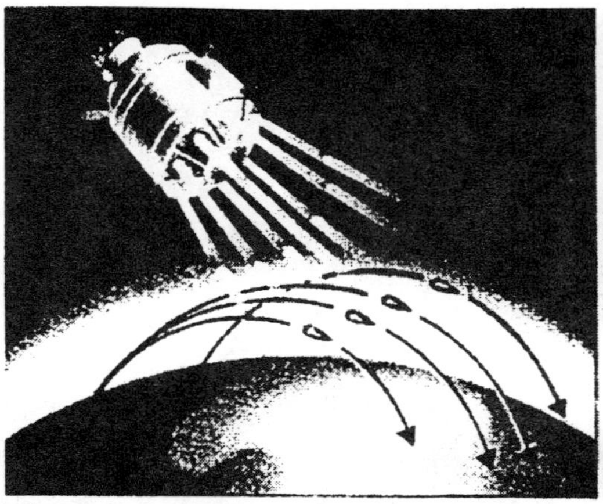

X-RAY LASERS: Powered by nuclear bombs, these would fire beams of X-rays at targets before consuming themselves in fireballs. Small and light, they could be "popped up" into space as needed. But problems in their testing have cooled enthusiasm.

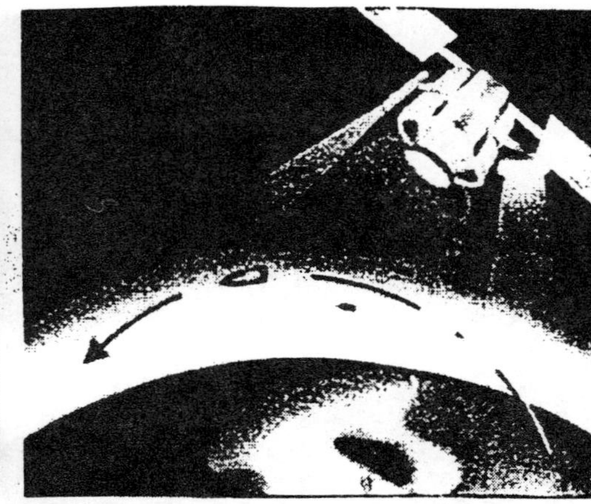

SPACE SENSORS: These "eyes" would be critical for coordinating battles and knowing which missiles and warheads had slipped through the defensive shield. An emerging hurdle is seen as quick digestion of sensor data and its relay to military commanders.

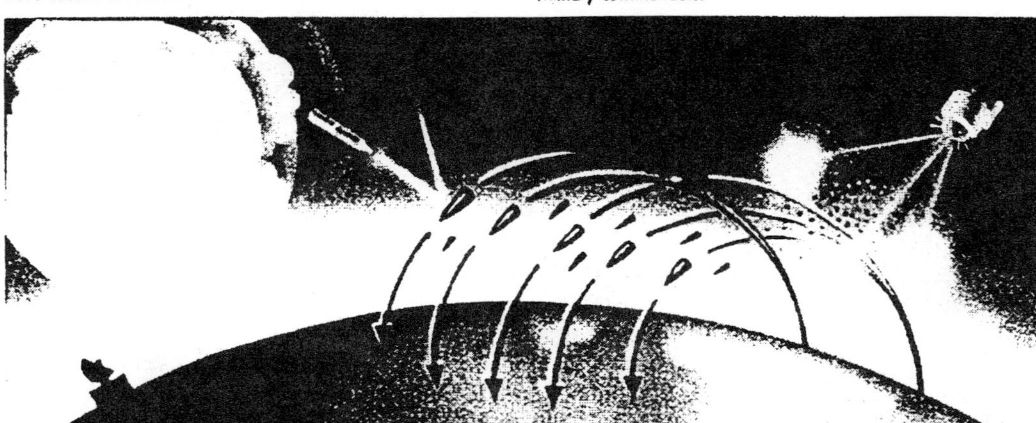

COUNTERMEASURES: An enemy could try to outwit a shield by attacking it or by complicating its job. Chaff dropped from missiles could confuse space-based radars and sensors. Decoys could complicate targeting. Missiles and warheads with mirror-like coatings could reflect laser beams.

Most challenging of all, an exploding nuclear warhead, set off accidentally by defender or intentionally by enemy, would send out electromagnetic pulses that would wreak havoc in electrical systems in space and on earth.

Source: Boffey, Philip M. and co, Claiming the Heavens, The New York Times Complete guide to the Star Wars Debate, p. 99.

It is my conviction that the maintenance of international peace and security will never be achieved by using space science and technology in building strategic weapons systems in the new domain; but by using the wonderful achievement of human genius for the progress and development of the human race.

It is interesting, moreover, to point out that the Treaty on the Prohibition of the Emplacement of Nuclear Weapons and Other Weapons of Mass Destruction on the Ocean Floor and in the Subsoil Thereof (adopted by the General Assembly in resolution 2660 (XXV) of 7 December 1970) embodies similar provisions of Article IV of the Space Treaty of 1967.

This further leads me to reflect on the meaning of the statue of the horseman standing in front of the United Nations Headquarters in New York, which symbolises the wordings of **ISAIAH 2:4** as follows:
"......: and they shall beat their swords into plowshares, and their spears into pruninghooks: nation shall not lift up sword against nation, neither shall they learn war any more".

Furthermore, the above quotation is very important in the interpretation of the provisions of the outer space instruments relating to the peaceful exploration and uses of outer space, the moon and other celestial bodies. I am very convinced that the founding fathers of the United Nations had the wordings of that verse at the bottom of their hearts in the formulation of the Preamble, Principles and the Purposes of the UN Charter. This has been reiterated in UNGA resolution 50/6, entitled **Declaration on the Occasion of the Fiftieth Anniversary of the United Nations**, of 24 October 1996. In the Declaration, the Member States and observers of the United Nations, representing the world:

"Solemnly reaffirm the Purposes and Principles of the Charter of the United Nations and their commitments to them;

Express their gratitude to all men and women who have made the United Nations possible, done its work and served its ideals, particularly those who have given their lives during service to the United Nations;

Are determined that the United Nations of the future will work with renewed vogour and effectiveness in promoting, peace, development, equality and justice and understanding among peoples of the world;

Will give the twenty-first century a United Nations equipped, financed and structured to serve effectively the peoples in whose name it was established."

It is interesting to note that this vision as depicted in the above quoted verse of ISAIAH 2:4 is gradually being realised in the Russian Federation by the industrial conversion of military rockets to launch vehicles for space business operations (**Spaceflight, Vol. 38 No. 8 (August 1996), page 256**). It is submitted that this should be a permanent and mandatory practice in the 21st Century, if genuine world peace and security are to be achieved.

Conclusion

The development of the international law of outer space, it should be borne in mind, is a progressive and dynamic process, which is also directly linked with further development and advancement in space science and technology and their practical application in solving various complex political and socio-economic problems facing mankind on earth. Consequently, the development of the operational rules and principles to regulate the various activities of states in outer space should be part and parcel of this process, as it was in the development of rules in fields of aerial and maritime navigation.

Moreover, it should be reiterated here that contemporary international law, which evolved after the Second World War, is a law of peace, disarmament, mutual and broad international co-operation, as could be seen in the provisions of UN Charter. Thus, space law as a branch of the contemporary law of nations or international law, should be based on the principles embodied in the UN Charter and not on those classical principles of international law, which only promoted individual national interests of powerful states. In fact, outer space has turned the modern world into a global village.

It is, therefore, submitted that the interpretation of the fundamental principles of international law of outer space, as embodied in the provisions of the 1967 Space Treaty, must and should be given in the light and spirit of the provisions of the Preamble and Articles 1 and 2 of the UN Charter. Furthermore, it should be emphasised that the lack of juridical precision, or different omissions in that document (i.e. the 1967 Space Treaty), as Professor Markov correctly pointed out earlier, cannot afford to neglect the fundamental changes and the new patterns it introduces into the law of nations.[19] The existing Post-Cold War Political World Order provides the international community with a better opportunity to strengthen the implementation of the provisions of the Outer Space Treaty of 1967. In this regards, it should be pointed out that the Report of the former UN Secretary-General, Dr. Boutros Boutros-Ghali, on **International Cooperation in Space Activities For Enhancing Security in the Post-Cold War Era**, has provided the framework or guidelines on international cooperation for the 21st Century.

Thus, in the present stage of international development in the nuclear-space era, taking into consideration what has already been discussed above, it should be borne in mind that, the question concerning the definition and determination of the frontiers between airspace and outer space is of great significance in safeguarding the sovereignty and security of all states and at the same time strengthening mutual respect, understanding, mutual trust and international co-operation in the exploration and peaceful use of outer space, including the moon and other celestial bodies. The use of outer space in some forms, as we have already noticed earlier, in one way or the other, infringes on the interests of all states in general.

Thus, in order to carry on with various activities in outer space for the benefit of mankind in general, I will like to stress that international co-operation, through the conclusion of international treaties and agreements concerning the legal regulation of such activities, is indispensable and it is the only method of finding solutions to most of the outstanding legal problems in space law.[20] It should, moreover, be borne in mind that international co-operation is one of the fundamental and mandatory principles of international law of outer space and it is embodied in Article IX of the 1967 Space Treaty and Article 9 of the 1979 Moon Treaty. The importance of international co-operation in the exploration and use of outer space was further emphasised in the 1982 UN Conference on Outer Space.[21]

Furthermore, this principle has been elaborated in the provisions of the **Declaration on the legal aspects related to the application of the principle that the exploration and utilization of outer space should be carried out for the benefit and in the interest of all States, taking into particular account the needs of developing countries**, adopted by the UN General Assembly under Resolution 51/122 of 13 December 1996.

Thus, in view of the existing divergences in the interpretation of the provisions of the basic principles of international law of outer space, particularly the one embodied in Article IV, it is submitted that the International Court of Justice (ICJ) should be involved by the UN General Assembly, pursuant to the provisions of Article 36, paragraphs 1 and 2(a) and (b) of the Statute of the ICJ, in the interpretation of the controversial terms of the space treaties. This, it is believed, would create uniformity in the application and realisation of these principles by states in the course of their individual and co-operative efforts in the exploration and use of outer space, including the moon and other celestial bodies, for peaceful purposes only.

Furthermore, in order to promote the peaceful exploration and use of outer space, it is proposed that the major space powers should encourage the participation of other states, including their nationals, in their future space programmes and flight missions. It is, moreover, encouraging to note that the on-going international co-operation between the Russian Federation and the United States, including other member states of the European Space Agency (ESA), Canada, China, Japan, India, etc., is gathering more momentum each year. The United Nations, as the centre for the co-ordination of international co-operation in the exploration and peaceful uses of outer space, has done a lot in creating the legal principles and norms applicable to outer space activities.

Therefore, it is necessary for all states, including their nationals, to unite and work together in finding the right solution to this problem, in order to ensure that this new area of humankind's activities will not be a domain of conflicting political/legal theories and ideologies, but a zone of genuine peace, security, mutual understanding, friendship and effective international co-operation.

However, this will further require the preparedness of all states to exercise their political will. As we are approaching the end of the 20th Century and the beginning of the 21st Century, the political will, it should be emphasised, will demand that the world policy-makers and their advisers should always search their conscience first, before making decisions concerning international peace and security, bearing in mind the following quotation of the 34th President of the US, General Dwight D. Eisenhower: *"Every gun that is made, every warship launched, every rocket fired signifies, in the final sense, a theft from those who hunger and are not fed, those who are cold and are not clothed. This world in arms is not spending money alone. It is spending the sweat of its labourers, the genius of its scientists, the hopes of its children (Speech in Washington 16 April, 1953)".*

References

[1]**Isaiah 2:4.** This same idea is being expressed in **Psalm 46:9**. Psalm 46:9 reads "He maketh wars to cease unto the end of the earth; he breaketh the bow, and cutteth the spear in sunder; he burneth the chariot in fire".

[2]For details, see, **UNITED NATIONS TREATIES AND PRINCIPLES ON OUTER SPACE**, UNITED NATIONS, New York (1996).

[3] For more details, see United Nations Treaties and Principles on Outer Space, (1994), op. cit., pp. 28-36.

[4] It entered into force on 11 July 1984.

[5] See, e.g., Wassenbergh, H.A., Reality and Value in Air and Space Law, op. cit., pp. 33-36; Zhukov, G.P., "The Problem of Legal Status of Scientific Research Stations on the Moon", in Proceedings, 10th Colloquium on the Law of Outer Space, Belgrade, Yugoslavia (1967), pp. 60-62. Ibid., pp. 21-25, for the "Summary of Discussions on the Moon"; Gorove, S., "Interpreting Article II of the Outer Space Treaty", in Proceedings, 11th Colloquium on the Law of Outer Space, New York, U.S.A. (1968), pp. 42-44; by same author Studies in Space Law: Its Challenges and Prospects, Sijthoff, Leyden (1977), pp. 79-84; Goedhuis, D. "Some Legal Problems Arising from the utilization of Outer Space", in ILA: Report of the 54th Conference, The Hague (1970), pp. 422-434 esp., pp. 405-421, for the view of other participants of the conference.

[6]According to the statement of the Chairman of COPUOS, there is now an average of nearly 1,000 objects actively engaged in orbiting the earth. No state has yet raised any objection against the orbiting of those objects in outer space above its territorial air space. See for details GAOR: 32nd Sess. Supp. No. 20 (A/32/20), p. 22. See also Goedhuis, D., "General Questions on the Legal Regime of Space", in ILA: Report of the 50th Conference, Brussels (1962), op. cit., pp. 76-77.

[7] For detail, see, e.g., Lachs, M. (1972), pp. 59-61.

[8] See Articles IV-XIII of the Treaty. The provisions of these articles are also embodied in the 1979 Agreement on the Moon and other celestial bodies.

[9]For more details, see UN Office of Public Information OPI/424-00253 - Jan. 1971 - 33M, p. 7. See also Art. 1 and (4) of the UN Charter.

[10]See Paragraph 2, Article IV of the Space Treaty; Article 3 of the Moon Agreement of 1979.

[11]Article IV of the 1967 Space Treaty and III of the 1979 Moon Treaty contain specific prohibitions which pertain to demilitarization and neutralization of outer space, including the moon and other celestial bodies. For the opinions of other experts, for example, see Zhukov, G.P., "Freedom of Space and its Limits" in Space Law Perspective (1976), p. 118, Seara Vázquez, M., Cosmic International Law, translated by Malley Elaine, Wayne State University Press, Detroit (1965), p. 174, Cheng, B., "The United Nations and Outer Space" in Current Legal Problems, Vol. 14 (1961), p. 262 et seq.; See also his comments in the ILA: Report of the 50th Conference, Brussels (1962), p. 52 et seq... Goedhuis, D., ILA:Report of the 50th Conference (1962), op. cit., p. 81 et seq.. Meyer, A., "Comment", Ibid., p. 42, Meyer, A., "Interpreting the term 'Peaceful' in the Light of Space Treaty", Proceedings, 11th Colloquium on the Law of Outer Space, pp. 24-29, Fawcett, J.E.S., International Law and the Uses of Outer Space (1968), pp. 32-33, Kish, J., op. cit., p. 188, and Current Legal Problems (1961), op. cit., pp. 262-272, Goedhuis, D., ILA: Report of the 50th Conference (1962), op. cit., pp. 81-82. For instance the Cosmos type of satellites, e.g., Cosmos 390, 396, 597, 603, 674, 1010, 1012, 1537 etc. ILA:Report of the 50th Conference (1962), op. cit., p. 52. Fawcett (1968), op. cit., p. 33. See also for comparison by the same author Outer Space, New Challenges to Law and Policy, Clarendon Press, Oxford (1984), pp. 106-114. See for more details Table of Artificial Satellites Launched between 1957-1976; Table of Artificial Satellites Launched between 1977-1984, published by ITU. See for details International Co-operation and Organization for Outer Space: Staff Report for the Committee on Aeronautical and Space Sciences, United States Senate, U.S. Government Printing Office, Washington D.C. (1965), p. 459. It should be noted that the establishment of a permanent space platform - SALYUT and MIR systems - have given the Soviet Union (Russia) the capacity to carry out various experiments and activities - including reconnaissance and surveillance of any part of the earth, without necessarily launching reconnaissance satellites. The US space shuttles are carrying out remote sensing activities, which are the same with these military activities. For more details, see, Tim Brown, "International Peace through the free market. The effect of commercial remote sensing satellites on international peace", in **Proceedings, IISL 37th Colloquium on the Law of Outer Space, October 9-14, 1994, pp. 201-212.**

[12] For more details, see, for example, Outer Space -Battlefield of the Future?, published by Taylor and Francis Ltd., London (1978), p. 12 et seq.; SIPRI Yearbook (1983), pp. 30-31.

[13]See Kopal, Vladimir, "The Problem of Neutralization and Demilitarization of Outer Space", in Proceedings, 4th Colloquim on the Law of Outer Space (1961), pp. 337 and 339.

[14] See Zhukov, G. P., "Fundamental Principles of Space Law", in Contemporary International Law, Tunkin, G., ed. (1969), p. 276 and Zhukov, G. P. and Kolosov, Yuri, International Space Law, Translated from Russian by Boris Belitzky, Praeger Publishers, New York (1984), pp. 53-54.

[15]Article I stipulates as follows: " 1. Each of the Party to this Treaty undertakes to prohibit, to prevent, and not to carry out nuclear weapon test explosion, or any other nuclear explosion, at any place under its jurisdiction or control:

(a) in the atmosphere; beyond its limits, including outer space; or under water, including territorial waters or high seas; or

(b) in any other environment if such explosion causes radioactive debris to be present outside the territorial limits of the State under whose jurisdiction or control such explosion is conducted. It is understood in this connection that the provisions of this subparagraph are without prejudice to the conclusion of a treaty resulting in the permanent banning of all nuclear test explosions, including all such explosions underground, the conclusion of which, as the Parties have stated in the preamble to this treaty, they seek to achieve.

2.Each of the Parties to this treaty undertakes furthermore to refrain from causing, encouraging, or in any way participating in, the carrying out of any nuclear weapon test explosion, or any other nuclear explosion, anywhere which would take place in any of the environments described, or have the effect referred to, in paragraph 1 of this Article."

[16] For more details, see, for example, Maintaining Outer Space for Peaceful Purposes, Proceedings of a Symposium Held in the Hague, March 1984, edited by Jasentuliyana, Nandasiri, United Nations University (1984); ILA: Report of the 62nd Conference, Seoul (1986), pp. 385-408; Report of the 63rd Conference, Warsaw (1988), pp. 282-380.

[17]Doc. A/CONF.101/10, op. cit., p.paras 13-14. the prevention of an arms race in outer space is an agenda item of the UN General Assembly. In

resolution 48/74 of december 1993, for example, the General Assembly reaffirmed the importance and urgency of preventing an arms race in outer space and readiness of all States to contribute to that common objective, in conformity with the provisions of the treaty on principles Governing the Activities of States in the Exploration and Use of Outer Space, including the Moon and Other Celestial bodies (para. 1) and Boffey, Philip M. and co, Claiming the Heavens. The New York Times Complete guide to the Star Wars Debate.

[18] **International Herald Tribune**, Frankfurt, Friday, October 3, 1997, pp. 1 and 8 and **USA Today**, October 3-5, 1997, p. 1.

[19]For more details, see Markov, M.G., "Against the so-called 'Broader' Interpretation of the term 'Peaceful' in International Space Law", in Proceedings. 11th Colloquium on the Law of Outer Space (1968), p. 81. It should also be borne in mind that any interpretation that conflicts with the purposes and principles of the UN Charter is ultra vires (see Art. 103, UN Charter).

[20] See also Lachs, M. (1972), op. cit., p. 61, for more comments on this points. See also Boutros Boutros-Ghali, International Co-operation in Space Activities for Enhancing Security in the Post-Cold War Era: Report of the Secretary-General. UN Doc.A/48/221.

[21]For details, see UN Doc. A/CONF. 101/10, op cit., pp. 5, 78-106. The various UN General Assembly resolutions adopted on the peaceful uses of outer space do stress the importance of international co-operation.

IISL-97-4.13

THE LUNAR USERS UNION - AN ORGANIZATION TO GRANT LAND USE RIGHTS ON THE MOON IN ACCORDANCE WITH THE OUTER SPACE TREATY

Karen Cramer
Space Policy Institute
George Washington University
Washington DC

Abstract

The Outer Space Treaty leaves a sovereignty vacuum. No one has sovereignty over outer space therefore no one can grant property rights. Without property rights there is nothing to prevent anarchy among potential users of lunar territory. Anarchy is not conducive to large investments of capital. Mining, astronomy, geology, solar power, manufacturing and landing facilities are not all compatible. Some form of land use regulation is needed to prevent interference. One possible solution, within the framework of the Outer Space Treaty, is the Lunar Users Union (LUU). It would deal with lunar territory the way the International Telecommunication Union (ITU) deals with radio spectrum and orbital slots. LUU would regulate use of the moon to insure non-interference between users. Rights would be granted for specified activities in specified areas. Other users who wished to use the area for an activity which would not interfere with the first user could be granted the right to use the same area. If rights to an area are not exercised than they would be lost. Rights would be granted on a first come first served basis. LUU would be governed by a board of representatives of States which were engaged in lunar activities.

Introduction

The Outer Space Treaty states that celestial bodies are not subject to national appropriation by claim of sovereignty, but it does not give sovereignty to any other entity, this leaves a sovereignty vacuum. An unintended consequence is that since no one has sovereignty no one can grant property rights This can pose problems for those wishing to use Lunar territory, since there is no authority to regulate use of territory therefore adds greatly to the risk of attempting commercial lunar mining. There is no way a mining company can legally keep some other company from coming in and mining an area that they spent a large amount of money doing exploratory surveys on. However the moon is a big place and the most of the resources are evenly distributed across the surface, so there is no reason why legitimate organizations need to conflict. The lack of mining rights does make it difficult to justify the expense of exploratory surveys. Also it makes difficult to assure potential investors of legitimacy, seriousness and safety of mining venture.

The lack of sovereignty was not intended to block exploitation but was intended to prevent wars, so there is a way to build a legal framework which facilitates profitable exploitation while staying within the treaty This paper will explore the principles relevant to use of lunar territory stated in the Outer Space Treaty and propose a solution to the uncertainty created by the sovereignty vacuum which is in harmony with those principles. It will also consider the types of lunar resources, the types of interference which will need to be regulated and the infrastructure necessary to facilitate safety and Lunar development.

Principles in the Outer Space Treaty

The Outer Space Treaty allows use, requires free access but implies that access does not mean interference and encourages coordination to eliminate inference. The right of use is stated in Article 1 paragraph 2 "Outer space, including the Moon and other celestial bodies shall be free for exploration and use by all states without discrimination of any kind on a basis of equality and in accordance with international Law and there shall be free access to all areas of celestial bodies" The right of use is restated in Article 4 paragraph 2 " The Use of any equipment or facility necessary for the peaceful exploration of the Moon shall not be prohibited".

The principle that free access does not mean interference is stated in Article 9 Paragraph 1 "shall conduct all their activities in outer space including the Moon and other celestial bodies, with due regard to the corresponding interests of all other state parties." It is also contained in Article 12 "All stations, installations, equipment and space activities on the Moon and other celestial bodies shall be open to representatives of other states on the basis of reciprocity. Such representatives shall give reasonable advanced notice of a projected visit, in order that appropriate consultations may be held and the maximum precautions may be taken to assure safety and avoid interference with normal operation of the facility to be visited."

Coordination to prevent interference is encouraged in Article 9 "If a State Party to the Outer Space Treaty has reason to believe that an activity or experiment planned by it or its nationals ... would cause potentially harmful interference with activities of other States Parties in peaceful exploration and use of ... , it shall undertake appropriate international consultations before proceeding with any activity or experiment."

The Proposed Solution

The Outer Space Treaty clearly encourages coordination to prevent harmful interference between parties using the moon. The question then becomes what form should this coordination take. Bilateral coordination would be highly inefficient of there were three or more States which had a presence on the moon. So it must be multilateral organization to be effective. The organization must include all those countries which will engage in Lunar activity. Countries not engaged in Lunar activity have no reason to be involved in this organization. The organization will regulate interference between activities between states.

There is an organization regulates interference between users of a scarce resource which acts voluntarily and is internationally respected. That organization is the International Telecommunications Union (ITU). The ITU is a good model since it has operated successfully for decades in an area which requires a great deal of cooperation and has high economic stakes for all countries. While it is noted that the ITU is presently having difficulty filling all the requests for bandwidth and has many slots reserved by organizations who are not able to fill them these are minor administrative problems which are solvable by minor administrative changes. These problems are also caused by the spectrum shortage which will not be a problem for lunar territory for a long time.

The Outer Space Treaty allows use, encourages coordination to eliminate inference, requires free access but implies that access does not mean interference. The International Telecommunications Union was set up to insure non-interference between radio signals. The ITU insures that other members of the union will not cause interference with activities. The ITU is joined by all parties since the cost of telecommunications satellites is so high no one wants to go it alone and risk loosing the investment. The same would be true of the Moon. Permission for use from the ITU is not considered a property right. It is legally recognized right of use which can allow potential users to raise capital.

A Lunar Users Union (LUU) could be set up to insure non-interference between groups exploring and using of the Moon. Membership would be open to Sovereign States which have soft landed equipment on the Moon. Countries planning such activities may become affiliate members. The purpose of the Lunar Users Union would be to maintain and extend international cooperation between all members for the rational use of the Moon, and to promote development of technical facilities and their most efficient operation to make lunar habitation and development as safe and profitable as possible, and to harmonize the actions of nations of those ends. To this end the Lunar Users Union (LUU) will effect allocation of uses of Lunar territory and registration of those assignments in order to avoid harmful interference between activities of countries.

The LUU will coordinate efforts to eliminate harmful interference between parties and to improve the use made of Lunar territory both in terms of financial profit and scientific value. The LUU will promote adoption of measures which ensure maximum safety of humans on the Moon, in space and on earth. Each member will have one vote. All members will pay equal shares of the expense of the Union. A Member which is in arrears of its payment shall lose its right to vote.

<u>Territory Allocation</u> Historical and Scientific reserves should be set aside before any other territorial assignments are made. These areas should include historical landing sights, geologically important sights and an area reserved for astronomical observatories. Geographical locations with special potential uses should also be reserved for activities that do not interfere which those potential uses. Areas like equatorial longitude 33.1° E, which is the ideal site for a mass launcher and other areas which are ideal for power beaming to the Earth. Placing 20% of Lunar territory in these reserve should be sufficient to protect all the important historical, scientific and geographic sights.

The possible deposits of water ice on the south lunar pole also should be considered. If they are found to be small, they too may need to receive special protection, but if they are large and over a wide area it may be best to treat them like the majority of the Moon. No entity should have the sole right to use the water ice on the Lunar South pole. A monopoly on lunar water could be very detrimental to lunar development.

Once areas to be preserved are set aside member states should make requests for right of use of territory. Requests should be evaluated to consider possible interference with previous assignments or other requests. If a member has not used the assigned territory in the manner stated in the request for non-interference within 5 years then the right is forfeit and the requests of others can be granted. If a member ceases to use an area and does not resume use within a period of 5 years than the right of use is forfeited and other request for use can be considered.

Some may feel that this is not a suitable legal structure since it depends on cooperation and does not grant private ownership of land. The price of lunar activity will be high, so parties will cooperate. As for the lack of private ownership of land, it is very difficult to achieve under the present treaty.

The main effect of the LUU would facilitate maximum benefits to be gained from the Moon by allowing mining and other uses while protecting of the scientific treasures which could easily be squandered unintentionally.

<u>Resources</u> There are some attributes of the moon which must be considered when setting up a legal regime to allow for its efficient and profitable use. The moon has several potentially valuable resources, a delicate environment which is ideal for science but is a hazardous place for humans. These attributes must be considered so we can use the moon as financially and scientifically profitable and safe as possible The potentially valuable resources must be consider when devising a system of non-interference. The Moon has 3 main potentially valuable resources Helium-3, Oxygen and possible water ice on the South Pole. The availability and value must be considered before they are legally distributed.

Helium-3 is concentrated to parts per billion over all of the lunar surface. It is more concentrated in Titanium rich regolith. Since Helium-3 is implanted by the solar wind it is found in the greatest concentrations in the first 3 meters of soil and is a renewable resource. It is potentially worth a billion dollars per ton if Helium-3 fusion becomes a commercial power source. If Helium-3 is commercially exported to Earth then large tracks on the Moon would need to be strip mined in order to meet the demand. This would suggest granting large assignments for mining.

Oxygen accounts for 60% of the atoms which make the of the Moon[i]. All of the Oxygen is chemically bound in the rocks and soil of the Moon. It will be vital for life support and rocket fuel. Mining oxygen for use as return fuel will greatly reduce the cost of a lunar transportation system. There are many possible techniques for mining the Oxygen using various of the oxide minerals. The value of a given area will depend on the technology available and the minerals present.

Both Helium-3 and Oxygen could be mined simultaneously with properly designed mining equipment. One of the easiest oxides to release oxygen from is Ilmenite which is Titanium Iron Oxide, Titanium happens to be the element which allows for the greatest concentrations of Helium-3. Helium-3 and other solar wind implanted elements such as Hydrogen, Nitrogen, Carbon, Helium-4 (regular Helium), and other noble gases are trapped in the rough surface features of the regolith[ii] and will be released if the regolith is disturbed. These elements are valuable for life support and scientific purposes. All that needs to be done to mine them while mining for oxygen is to seal the equipment, collect and separate the gases. If they are not collected they will pollute the pristine lunar vacuum for kilometers around the mining site. These gases will be extremely valuable for life support and scientific uses.

The pristine Lunar vacuum will need to be carefully protected from "air pollution" the biggest problems will be hydrogen-oxygen rocket use and mining. An electromagnetic launcher will help as will use of alternate fuels such as Sulfur-Oxygen or Sodium-oxygen rockets.

Water ice on the Moon may be an extremely valuable and scare resource. It is vital for life support and can be used as rocket fuel. At present we have some less than optimum data that there is some ice on the South Pole of the Moon. We also have evidence that it can only be in an extremely small area of the Moon. The Lunar Prospector which will soon go to the Moon will give us better data. We must keep its value and scarcity in mind when we build a legal structure for the Moon. It would be a terrible if one company managed to get the right of non-interfered use of all the lunar ice.

There is another possible place water ice might be hiding on the Moon, inside intact lunar lava tubes. These are permanently shaded and temperature stable. If water vapor from meteor impacts seeped in through the cracks it may still be there frozen to the walls. Intact Lunar Lava Tubes are also the best place for human habitation since they have thick stone roofs which block radiation and meteors. They are also believed to be temperature stable at -20° C [iii] unlike the lunar surface which is between -150 ° C in the lunar night to 250° C during the lunar day. These Lava tubes meander for kilometers below the lunar surface. When assigning rights the possible presence of undetected lava tubes must be considered. Lunar lava tubes are a valuable resource but their value would be greatly diminished if they were irrationally divided because of purely surface phenomena.

<u>Interference Considerations</u> There are also environmental concerns. The Moon is a treasure trove of scientific data. Astronomy, geology, solar science and physics could benefit greatly by having access to the Moon. We must be sure that areas are surveyed by geologist before they are disturbed by mining and to keep a reservation undisturbed for the astronomers whose telescopes need to be far away from mining activities and radio waves.

There are many important scientific resources on the Moon. The Far side of the Moon is ideal for radio astronomy since it is in a permanent shadow from man made radio sources on the Earth. This radio shadow needs to be preserved if at all possible. Care must be taken when designing a lunar communication system to preserve as much of the radio silence as possible without greatly compromising safety.

The Moon is covered with interesting geological sites. The youngest rocks on the Moon are older than the oldest Earth rocks. The surface of the Moon is littered with meteors which could provide geological data about the formation of the solar system and other planets.

Scientific reservations should be set aside. Scientific societies from all nations should have input into which areas shall be preserved. The requests for preservation should be ranked in importance so if there are requests for over 20% of the lunar surface area, the most important sites can be protected. All areas of the Moon will be accessible to scientists even once they are allocated for use as long as scientist do not interfere with that use.

Infrastructure Safety is a major concern on the Moon. The lunar vacuum, temperature extremes, radiation exposure, rough surface and total dependence on line of sight radio transmission for communication make human survival difficult. Lunar infrastructure such as emergency radiation shelters and communications systems will help. As will requiring assistance to persons in distress.

Communications on the Moon will be a challenge since radio signals only travel line-of-sight since there is no atmosphere to bend them. Also the horizon is closer since the Moon is smaller than the earth. On the near side of the Moon this is not a significant challenge since the Earth is always in sight and can be used as a communications relay. On the other hand an individual who is on the Far side of the Moon has no way to communicate with anyone who is not within a few kilometers. This problem could be solved with surface relay towers every 10 km or with communications satellites. Any system should be designed with care so as not to cause radio pollution of lunar radio observatories. These observatories will be build on the Far Side because of the radio isolation. Solutions which could keep every one safe and happy include only using scientifically uninteresting frequencies for communication or using only towers to relay radio signals and not having any towers within 10 km of the observatories.

Navigation on the Moon is presently a challenge since there is no magnetic field to reference. The Apollo astronauts had great difficulty navigating on the moon. A GPS like system may need to be developed and deployed to assure safety and unintentional interference between users of lunar territory.

Lunar infrastructure will need to be constructed if there is to be a thriving lunar settlement. Things like communications and landing pads should be open for all to use. Also there is the matter of control over infrastructure. The main Lunar port must be open to all. It must have a solid landing pad to prevent landing ships from sand blasting everything for 2 km around. The main port also ought to have a constantly manned communications center. It should also have a bioregenerative life support system so as to minimize pollution of the pristine lunar environment due to human habitation and maximize use of scarce lunar resources. The port should be a base for rescue personnel and have excellent medical facilities.

Garbage disposal requirements will be necessary since nothing will decompose on the lunar surface. Bioregenerative life support systems can turn any organic wastes into food, fresh air and clean water. Recycling techniques for other types of garbage will need to be used as well. Nuclear waste will need to be carefully disposed of.

The right of non-inference does not mean the right of exclusive access and certainly not the right to block the access of others to the territory which an entity has the right to use. Provisions must be made to construct surface roads and to provide access to all parts of the Moon which are assigned for use. Through ways between claims or easements across claims should be considered when granting right of non-interfered use.

Conclusion

The non-appropriation clause complicates lunar development because it leaves a sovereignty vacuum. There is an existing organizations which has successfully dealt with allocating scarce resources without granting ownership for decades. That organization is the International Telecommunications Union. It should be used as a model for dealing with the need for exclusive right to mine an area while staying within the bounds of the Outer Space Treaty.

The Lunar Users Union modeled on the ITU facilitate lunar development, respect the treaties and protect the scientifically interesting sensitive areas so maximum knowledge can be gained.

---

[i] Lunar Source Book, Edited By G. H. Heiken, D. T. Vaniman and B. M. French, Cambridge University Press 1991, P.363
[ii] Lunar dirt
[iii] Wendell Mendell personal communication cited in Lava Tubes: Potential Shelters for Habitats by F. Hörz Lunar Bases and Space Activities of the 21st Century edited by W. W. Mendel 1985 p.405

TOTAL MILITARIZATION OF SPACE AND SPACE LAW: THE FUTURE OF THE ARTICLE IV OF THE 67'OUTER SPACE TREATY *

José Monserrat Filho**
Brazilian Society of Aerospace Law (SBDA)
Brazilian Society for the Advancement of Science (SBPC)

"Ensuring that outer space is devoted exclusively to peaceful purposes has remained a most important goal, but until now that goal has not been politically possible to bar arms from the new dimension. It is only now, with the cessation of the cold war and with new relations developing between all powers, that it may be possible to give full effect to the principle of the use of outer space for exclusively peaceful purposes." **Manfred Lachs** [1]

This paper attempts to examine the perspectives of total militarization of outer space in the light of what International Space Law provides today on the military use of space and celestial bodies. In this sense, special attention is paid to the theory, very much in vogue, especially in the USA, that the "non-aggressive" military use of space can and must be considered as a "peaceful" use. In short, the article offers some ideas and reflections on how to avoid the installation of weapons in space and avoid its becoming an arena of warring rivals.

Perspectives of total militarization in space

The plans to militarize space to the ultimate consequences — including the qualitative leap to install weapons there and thus create a new battlefield — have survived the end of the Cold War. They may eventually just be a farce, given the current economic hardships against their realization. Yet they continue on the agenda of the powerful leaders and general staff, and of the gigantic aerospace industry complex, all busily defending their technical feasibility, strategic requirement and ethical-legal validity.

A doctrinal veneer covers the issue. It must be examined.

In the 1980s, the clash between the USA and former USSR was on the verge of causing total militarization of space. The two largest powers in the world were about to test anti-satellite weapons. In 1983, the USA announced the construction of an ambitious antimissile defense system, officially called "Strategic Defense Initiative" (SDI), which was soon to be known as the "Star War". The design of a formidable shield against intercontinental ballistic missiles bearing nuclear warheads, consisting of state-of-the-art weapons also distributed in space, breached the principle of strategic balance and deterrence. It adopted the premise that one of the parties in the dispute could become invincible and, therefore, stronger than the other. This party, then, could win a nuclear war. The proposal was not successfully proven feasible or effective, like that of the anti-satellite weapons. Yet the arms race accelerated, building up the tension in the world. Even so, space was not invaded by weapons.

In 1985, a new Soviet leader emerged who proved to be increasingly determined put an end to the situation of rivalry on the edge of an abyss and to open the way to broad negotiations with the USA. The Cold War was becoming a mere shadow of itself.

In less than three years, between 1989 and 1991, the world watched open-mouthed what had perhaps been the greatest all-time political collapse in times of peace. The so-called "Community of Brother Socialist States" in Eastern Europe, guided by the USSR, with its state economy systems and single party, and the "Union of Socialist Republics" itself crumbled like a house of cards. One of the centers of dispute and arms race which for so many decades had threatened the planet with nothing less than a nuclear disaster was disappearing.

The new geopolitical world map seemed to say that there were no longer reasons for militarizing space until new weapons emerge which could turn it into a battlefield.

The impression — or would it be hope? — was to fade all too soon. The Gulf War in 1991 left this quite clear. The "Desert Storm" operation — started by the USA and some allies on behalf of UNO, to force Iraq to leave the illegally occupied Kuwait — grew to dimensions clearly out of proportion with the objective, especially in view of the modern means used.

The vast military mobilization was oriented by images from seven satellites. Never had so many of these devices been used at the same time for the same war-waging purpose. They made an average of 12 passes over the theater of operations each day, producing hundreds of images daily. In addition, he USA armed forces had in operation 15 to 20 signals intelligence satellites, intercepting radio communications of the Iraqis. Furthermore, the USA had at its disposal the services of three weather satellites, at least four military communications satellites and up to 16

**Member of the Board of Directors of the International Institute of Space Law, Master of International Law, Professor of Space Law at the Brazilian Society of Aerospace Law (SBDA) and editor of the newsletter *Jornal da Ciência* (Journal of Science), published by the Brazilian Society for the Advancement of Science (SPBC). Address: Av. Oswaldo Cruz 73/701, Flamengo, Rio de Janeiro, RJ, CEP 22250-060, Brazil. Phones: 55-21-552-9036. Phone/fax: 55-21-295-6198. E-mail: <monserrat@ax.apc.org>.

"Navstar" Global Positioning System (GPS). This armada of military space assets received additional assistance from the images acquired by the French SPOT and US Landsat civilian remote sensing satellites, which were used to update maps for the operational forces. [2]

The US Air Force Chief of Staff, Merrill McPeak, overestimate the "Desert Storm" defining it as the "first 'space war', since it was for the first occasion on which the full range of modern military space assets was applied to a terrestrial conflict". He, of course, lost no time in praising the increase of the military expenditures on space "even during this time of decline in many other dimensions of our activities..."[2A]

In fact, there was no "space war" whatsoever. The "Desert Storm" operation did not occur in outer space but rather on land and in the air space over Iraq, which is very different. The latest space technologies only acted as support to increase the degree of accuracy of the air and land operations. Seeing this as the "first space war" leads to concealing the transcendental steps which still have to be taken to bring about a real space war. We may, however, arrive at it if and when the international community permits putting arms in orbit and converting space into a combat zone. And this possibility will continue to exist as long as there are plots and close-knit plans in high military and political circles, accompanied by intense lobbying, involving the return to anti-satellite weapons, already developed by the former USSR and USA, and to antimissile systems with space segments.

In 1997, we are commemorating the 40 year anniversary of the Space Age which began with the launching of *Sputnik I* on October 4, 1957, and the 30 years of the 1967 Treaty on Principles Governing the Activities of States in the Exploration and Use of Outer Space, including the Moon and other Celestial Bodies [3], known as the "Space Treaty". We can also commemorate the auspicious fact of having kept space free of any kind of weapon.

The space militarization process, however, continued, although having been forced to slow down in the new world context and internal budget setbacks. The so-called "passive" and "non-aggressive" military uses of space have not at all ceased to be enhanced and qualified. These are uses of reconnaissance and observation, communication, navigation, meteorology, geodesics, control and command, which were put to an excellent test in the Gulf and Bosnia wars. With regard to Bosnia, the USA successfully experimented the space observation Jstars system, with which they accompanied the movement of each vehicle in an area of 200 sq.km, in any climate, on a simple computer screen.

There are euphemisms to be examined here. The "passive" military and "non-aggressive" uses can support both equally "non-aggressive" and "aggressive" activities. Taken on their own, of course, they do not configure a war operation. Yet, why separate them from their real function? After all, they are integral and indispensable parts of the central intelligence which prepares and activates all essential elements of a war operation. The result of this is that such uses can only be considered "passive" if taken from the situation in which they operate and from the objectives which they serve — an artifice which minimizes and even hides their effective destination. Active military uses depend entirely on them. They cannot be intrinsically and permanently "aggressive" but potentially never cease to be such. Their raison d'être is always military. They are at the service of the use of force.

The move from partial and "passive" to totally active militarization of space continues to be supported in the USA by influential and tenacious advocates.

"We are going fight **in** space", declared general Joseph W. Ashy, in mid-1996, when he was at the head of the USA Space Command and was asked if outer space would become a battlefield one day. He was emphatic: "Some people don't want to hear this, and it isn't in vogue... but – absolutely — we are going to fight **in** space". [4]

USA Defense Department authorities consider that military control of space must be given a high priority in USA national security, since the country increasingly depends on an enormous group of vital tasks, ranging from fixing targets for missiles to undertaking economic transactions. They maintain that the space control will probably require the development of space-based weapons.

In September 1995, the USA undersecretary of Defense for Acquisition and Technology, Paul Kaminski, created the Space Architect Office, entrusting it to Air Force Major General Robert Dickman. The Office was immediately in charge of preparing a plan for space control to be presented in the first half of 1997 to the Joint Space Management Board of the Defense Department, made up of the military and top experts from the intelligence services. [5]

For some time now, the USA Air Force has struggled to see space control become a national defense priority. In September 1994, the then Air Force secretary Sheila Widnall, stated: "Certainly, part of the Air Force mission is control of space, our ability to deny the use of space if necessary."

Intervening in the 2nd Annual Space Policy and Architecture Symposium on February 12, 1997, general Howell Estes, commander-in-chief of the US Space Command, was careful to note that any decision to renew development of space-based weaponry would not be made by the military, but all too clearly he stated that "we [...] support whatever decisions our elected leadership may arrive at regard to space control and the weapon systems required".

In general Estes' view, the USA must be able to "control space", an action which he defines as "ensuring friendly use of space while denying hostile forces' use of space against us". As, in his opinion, "in purely military terms, the national dependence on space-based systems equates to a vulnerability and history shows that that vulnerabilities are eventually exploited by adversaries, so the US military must be prepared to defend these systems". He mentions that space-based weapons are also "a viable alternative to 'terrestrially-based' systems". [6]

Control of space, according to the USA military sector, can be achieved by a variety of means, ranging from diplomacy – which was used during the Gulf War to prevent the Iraqis from obtaining commercially available satellite imagery — to antisatellite weapons, including bombing enemy satellite ground stations and electronic jamming. [7]

"Facing an Air, Space Challenge", is the title of the article by Col. James Smith, vice-director of the North American Aerospace Defense Command Operations, written in tribute to the 50 years of the USA Air Force.

He says: "As a service, the Air Force can no longer be air or space; it must be both. Traditionally, the Air Force has drawn a distinction between air and space missions; that distinction has been eliminated with the six core competencies for the Air Force of the future: air and space superiority, global attack, rapid global mobility, precise engagement, information superiority and agile combat support."

He also stresses: "As this nation [USA] becomes increasingly dependent on its space resources, the need to defend those resources will become increasingly compelling. To this end, the space side must develop a war fighting mentality."

And adds: "We are at a point in our service history where space is becoming the dominant medium over air. The Air Force must look at the future battlefield from the perspective gained at 30,000 feet and from that gained in geosynchronous orbit... The service must completely operationalize space, so airmen completely understand and control the air and space mediums. As we celebrate the first 50 years as a separate service, we honor our success in dominating the air medium. The next 50 years will require a dominance in both air and space. Such is a job for a US space and air force." [8]

"Pentagon considers space as a new area of responsibility", discloses the magazine *Aviation Week & Space Technology* in March 1997, spokesman for the USA industrial-military complex. The article discusses the view defended by the military in the debate underway at top North American government level, on whether outer space must be promoted or not to "a new area of military responsibility of the USA". [9]

The publication mentions that today the USA has three "areas of military responsibility" — air, land and sea — and announces that, if president Bill Clinton agrees, outer space will become the fourth area. The suggestion is from the USA Space Command and is based on the Unified Command Plan (UCP), renewed every two years and always submitted to the Pentagon. This UCP defines the missions of the five Unified Commands with which the USA covers every important region on the planet: Central, European, Atlantic, Southern and Pacific. This, therefore, deals with creating the Unified Space Command, the sixth in the series.

The Plan, under study at the Joint Chiefs of Staff level, must be submitted to JCS Chairman Gen. John M. Shalikashvili, who will submit it for the White House's final decision.

For those who suggested the idea, the active development over the last ten years of commercial space systems, particularly in the areas of communication, navigation and imaging, is analogous to "historical US economic expansions over land, on the seas and in the air". "As the freedom to operate in each of these mediums became essential to the nation's economic well-being, it was necessary to protect the associated lanes of commerce. That led to development of the US Army, Navy and Air Force. A natural extension of land-, sea- and air-based economic activity is into 'the fourth medium – space', in the words of the publication. The following words of general Howell M. Estes, commander-in-chief of the US or unified Space Command (CINCSPACE) and commander of the Air Force Space Command, seem to reinforce this: "All the commercialization [there indicates] that space will soon be of vital national interest."

In fact, space and its business are already of crucial interest for the USA, as they are also for all other countries. What country today can live normally without telecommunication, meteorology satellites and reconnaissance of onshore natural resources?

Here it is worth stressing: outer space is not destined for national expansions. The vast majority of countries, including the USA, have already decided to exclude from the new environment the ways of conquering new lands, seas and continents by which the most powerful countries divided and colonized our planet. It was also decided that no country is apt to extend its sovereignty beyond its air space. Since the early 1960s the fundamental principle was defined by universal consensus that all mankind is interested in the progress of the exploration and use of outer space for peaceful purposes and that the exploration and use of outer space should be carried on for the betterment of mankind and for the benefit of States irrespective of their degree of economic and scientific development. [10] That is, everything that occurs in space is of general interest to every country and nation.

This is evident from the *Aviation Week & Space Technology*'s own information, despite its parochial view: "Today there are more than 200 US satellites in orbit, worth in excess of US$ 100 billion, and many more scheduled for launch during the next few years. When international and allied spacecraft are included, more than 500 platforms are in Earth orbit..."

In this situation, general Estes' conclusion could not be more one-sided: As these assets become more crucial to national welfare and economic strength, "I, as a military commander, have to say that somebody is going to threat them; and when they [do], we [should] have armed forces to protect them." He believes that, "if there was ever a threat to our national security [in space], the best — and only — way to solve the problem is to take weapons into space".

General Estes, however, acknowledges that, at the moment, so far there is no visible or foreseeable foe. So much so that, he explains, the Space Command has not articulated a requirement for space control or force application systems, "because there isn't threat that [demands] these types of systems in space".

Yet the lack of a real enemy does not seem to discourage the strategists. The aforementioned general Joseph W. Ashy is much more concerned in how to structure the actions to be performed directly in space. He contributes to the special issue of *Aviation Week & Space Technology* which commemorates the 50 years of the USA Air Force, published in April 1997, with an article in which he emphasizes the need to think about the organization of space operations, since "future military operations will be supported not only **from** space (as in the first stages of airplane use, but also within and to space". [11]

The Economist, in a recent cover article on "The future of warfare", states that "the world is in the early stages of a new military revolution", but that this "embryonic revolution" (...) "has not emerged in response to any particular threat to the USA and its allies", having only arisen because "the generals want to play with new technologies in case a future threat emerges", among which are the space threats. [12]

The "toy" of those USA generals, although the text does not say so, is also the plaything of the billionaire arms industry which has done nothing but lose since the end of the Cold War.

The joint happiness of certain generals and industries leads us to suspect that the arms race is not necessarily the result of effective danger, as was claimed, for instance, in relation to the "empire of evil", the former USSR — even when it was already giving unmistakable signs of serious internal distress, carefully disregarded in order not to prejudice the greater interest of furthering the dispute.

Along the same lines, French jurist Monique Chemillier-Gendreau warns in a recent paper that "the arms industries are over-sized in all industrialized countries in relation to their defense requirements" and that "the military machine looks like a crazy machine which escapes from political and social control". [13]

Any arms race always adopts its own dynamics, independent of the real life facts. It does not need real enemies for it to exist and flourish. Ghosts are enough.

In fact, the absence of adversaries, on Earth as in space, causes no embarrassment to the new USA defense secretary, William S. Cohen. His Quadrennial Defense Review, sent to Congress in the second fortnight of May 1997, lists the conditions to be met to accredit the USA to use its military power anywhere in the world.

The fourth condition is as follows: "The United States must retain superiority in space. Global intelligence collection, navigation support, meteorological forecasting, and communications rely on space-based assets. To maintain our current advantage in space even as more users develop capabilities and access, we must focus sufficient intelligence efforts on monitoring foreign use of space-based assets as well as developing the capabilities required to protect our systems and prevent hostile use of space by an adversary." [14]

This stance adopts, essentially, the strategic ideas prevailing in the country. Nevertheless, it has been criticized as "insufficient" by the National Defense Panel, formed by top military officers, which appraised the Quadrennial Defense Review report.

Specifically on the space segment, the Panel stated as follows: "Space is clearly of great importance to national security and we must maximize the effectiveness of functions carried out in space. Moreover, its value and range of uses will almost certainly increase exponentially over the next two decades. Access to space-based information allows us to better apply the military and civil systems we currently have as those in the acquisition stream. Threats to space access and our space-based systems include computer 'hacking', electronic jamming, and future laser and kinetic energy systems. One can expect threats in space to further increase as the technology grows. It is the Panel's view that use of space and vulnerability to space threats received insufficient attention in the Quadrennial Defense Review. The Department needs to develop a strategy for maintaining access to space. Military strategy and doctrine in the 21st Century will be effective and viable only if space is addressed as a frontier vital to the warfight."

It is worth mentioning that the Defense secretary preferred to include in the report that the USA must "retain superiority in space" rather than "control space", as many of the military had suggested. The change is symptomatic. It reveals that the USA government cannot ignore or be indifferent to the universally recognized norms of the 1967 Space Treaty, in whose preparation the USA played an active role and to which it is committed since the beginning. A country's official objective to "control space" does not combine with the terms of the first two articles of this Treaty, considered as the space code.

Article I, § 2, stipulates that "outer space, including the moon and other celestial bodies, shall be free for exploration and use by all States without discrimination of any kind, on a basis of equality and in accordance with international law, and there should be free access to all areas of celestial bodies".

Article II, in its turn, states that "outer space, including the moon and other celestial bodies, is not be subject to national appropriation by claim of sovereignty, by means of use or occupation or by any other means".

Now, controlling space cannot mean anything else but the act or power of exercising supervision, the domain and certain government over space, overseeing and judging what is or is not done there. This is incompatible with the principles of freedom of access, use and exploration of space for every country, without discrimination, and the non-appropriation of space in any form or pretext whatsoever. No country has any right to claim control over space, whatever the projected manner and alleged reason. Space, not belonging to any country in particular, belongs to everyone. The prerogative of adopting any kind of responsibility over it, especially military, either will be the joint competence of the United Nations or will be illegal. This is the text and spirit of the entirely in force international space law.

The USA Defense secretary's caution, therefore, is a positive sign of recognition of the prevailing international legality. He changed the legally unsustainable term "space control" for "superiority in space", an effective situation resulting from the technological advance of a country which, in principle, cannot be questioned from the legal point of view, unless this superiority means actions detrimental to the space activities of the other countries.

It so happens that this substitution does not change the planned military goals at all. *Aviation Week & Space Technology* is very clear in this respect: "The 'space control' theme is consistent with the Defense Department objectives set out in the Joint Chiefs of Staff's 'Joint Vision 2010' and a National Security Space Master Plan, developed by the deputy under-secretary for space, Robert V. Davis". [15]

Yet, there is a more sophisticated line of thought on "space control" in the USA Air Force Space Command itself. Major Cynthia A. S. McKinley, the Command's political and strategy analyst, acknowledges that: "Today many nations and businesses are dependent on space systems, creating a much more complex environment than existing during the Cold War. Space is no longer the 'high ground'; it's more 'common ground'. As a result, in her opinion, "the space warfighters need to start thinking of achieving military and political objectives through 'campaigns' that rely on influencing, deterring, compelling and defeating adversaries". This is what she calls a "coercion spectrum", considering it compatible with Defense Department's emerging space control architecture and doctrine.

Major McKinley's focus seems sensitive to the international nature of space. She understands that "the ongoing rapid commercialization of space is changing the potential combat environment from military domain to an internationally intertwined commercial domain". And she argues that "space systems may not have a sole owner. Some satellites serve as many as 135 consortium members."

This is why, in her mind, "the (US) military's responsibility to control space is not to control the *medium*, but rather to control the adversary's ability to exploit and derive benefit from the medium". After all, the highest military goal is to deny an adversary the use of space systems. [16]

McKinley's words are correct. Not so with the inference. As a self-defense measure — or to be more precise, as a preventive measure for the eventual use of the right to legitimate defense — any country, including the USA, has the right to be perfectly aware of what the other countries are doing in outer space and even what they can or aspire to do there. Yet no country has the right to act unilaterally and preventively against another country — denying it, in the case, access or use of space.

Every and any potential conflict or disputes on Earth or in outer space is under the absolute competence of the UN Security Council. This is the system established by the UN Charter, in its Chapter VII (Action with respect to threats to the peace, breach of the peace, and acts of aggression), whose prevalence in outer space and on celestial bodies is among the peremptory norms, from which no derogation is permitted and which can be modified only by a subsequent of general international law having the same character.

In contemporary international law and, especially, space law, the fact that a country considers another an enemy is not a merely domestic matter. It is international. It has to do with peace and security of the entire community of countries and peoples and, therefore, humanity as a whole.

In space, whose assimilation is defined as "the province of the all mankind" in Article I of the 1967 Space Treaty, the presence of disputes and, above all, potential conflicts is too serious a threat to the group of space activities with their increasingly essential implications to the development of life on Earth. It must be subdued as early as possible, with the explicit support and active participation of the whole community of countries, through their main intergovernmental organizations. The alleged hostile use of space is a problem to be ascertained, examined and face up by every and not by one country alone.

Despite all this, the doctrine and strategy which treat space as "a vital frontier for war" seem to prevail in the USA, especially in the circles of power. Influential sectors desire that the country, which obstinately attributes to itself the right and role of judge in international questions and the world's police, is prepared to act in the same way in space too. They see the new environment, first and foremost, as an area USA's own area of strategic interest. They make no mention of the alternative or possibility at all of resorting to the joint effort of the international community, cooperation between countries, all equally interested in restraining abuse and aggression, as well as in the use and exploration of space for peaceful objectives. The USA's one-sidedness, already so often revealed in arbitrary actions, tends to continue unchangeable at the time of galloping scientific and technological progress which they lead. In these times of such glorified globalization, its implacable nationalism shows no propensity to yield whatsoever.

"No one can be certain where this [military] revolution will end", states *The Economist*. Great mistake. History has a lot to testify. Militarization plans, in general, end by artificially creating, fostering and even aggravating tensions. In particular, in today's world conditions, they are perfectly capable of blocking and postponing the achievement of a more just global legal, democratic and cooperative order, claimed by the absolute majority of nations.

These trends and perspectives in no way whatsoever favor the best performance and consolidation of modern International Law with its focus on the superior mission of maintaining international peace and security. In history's most violent century, when man did his utmost to destroy his own species and the planet where he lives, this law succeeded — as written in the UN Charter — in prohibiting the threat or use of force against the territorial integrity or political independence of any state, demanding an exclusively peaceful solution to the international disputes and admitting the use of arms only and exclusively in the common interest of world community.

The mere attempt to extend to outer space the logic of force and destruction, with such tragic consequences here on Earth, is much more than a dangerous step backwards in relation to these historical advances. It is an insult.

Yet there have been attempts to put the mind at rest with seemingly humanist reasons: "War in outer space would affect almost exclusively material objects and not human lives. It would be crewless satellites which would be destroyed or damaged. In outer space there would not be the slaughter which decorates the sorrowful war history archives. Nothing of Verdun, Stalingrad or Hiroshima. So why condemn it, then?" [17]

There is nothing but to condemn a cunning irresponsible bait on the subject of the utmost relevance for the evolution of the human race, for at least four reasons:

1) It is absolutely impossible to guarantee that the space war would be limited only to space objects and would not cause the death of human lives, as if the interests at stake in space were any different and remote from the interests at stake here on Earth.

2) There is not the least doubt that destruction, albeit partial, of the vast and varied group of satellites which today serve the inhabitants of our planet can bring incalculable damage to the economic, industrial, farming, financial, cultural, educational, technological and scientific activities of countless countries.

3) To categorically assure that there would be no massacres such as Verdun, Stalingrad and Hiroshima is to underestimate, without any real basis, the scope of the disastrous effects caused by the total lack of control of the blind and deaf space systems to be targeted.

4) Even assuming the absurd hypothesis that life would not be affected, nothing would justify the indiscriminate elimination of space objects created at such great cost by mankind over so many years, to enhance its existence and permit it an increasingly in-depth knowledge of the universe.

Space Law on space militarization

The 1967 Space Treaty, cornerstone of the Space Law, adopts, right at the beginning, three standpoints which denote its strong tendency and readiness against the military use of space:

1) It acknowledges "the common interest of all mankind in the progress of the exploration and use of outer space for peaceful purposes";

2) It expresses the desire "to contribute to broad international co-operation in the scientific as well as the legal aspects of the exploration and use of outer space for peaceful purposes";

3) It considers "applicable to outer space" the resolution 110 (II) of the UN General Assembly of November 3, 1947, which "condemned propaganda designed or likely to provoke or encourage any threat to the peace, breach of the peace or act of aggression";

In the same sense, in its Article III, the Space Treaty establishes that space activities shall be carried on "in accordance with international law, including the Charter of the United Nations, in the interest of maintaining international peace and security and promoting international co-operation and understanding".

In fact, the Space Treaty, drawn up between 1963 and 1966, seems to reflect, as far as possible, to its planners the desire for peace and understanding that captured international public opinion and community of nations in the 1960s in the middle of the Cold War. The text and spirit of this basic document are openly pacifist.

In 1984, in one of the worst phases of the Cold War, the judge of the International Court of Justice, Manfred Lachs, who played an active role in preparing the Space Treaty, emphatically reiterated, in the seminar on "the maintenance of outer space for peaceful uses", held in The Hague, Netherlands by the United Nations University: "However one may attempt to interpret the text [Space Treaty], the law is clear and is confirmed by the preparatory work: the goal is the full demilitarization of this new dimension". [18]

Nevertheless, at no moment does the Space Treaty determine that space must be used "exclusively for peaceful purposes". Such a clause would be unrealistic and innocuous since it would clash with the strategic interests of the two conflicting space powers, the USA and USSR. That was when they had proven the excellent performance of the "spy satellites". At the same time and with identical concerns, they would never admit to closing outer space off to the passage of their intercontinental ballistic missiles armed with nuclear warheads, the ultimate in powerful weapons with which they fenced against each other.

Thus, however great were the peaceful desires and effort of selfless jurists from many countries, such as Manfred Lachs himself, against space militarization, it was not possible to fully prohibit its military use. Gaps continued. They were not, in fact, fully utilized during the Cold War. Today, they serve those who consider it inevitable and indispensable to store weapons up there.

These gaps are in Article IV of the Space Treaty, which deals specifically with the military issue. Let us analyze it.

Article 4 says:

'The States Parties to the Treaty undertake not to place in orbit around the earth any objects carrying nuclear weapons or any other kinds of weapons of mass destruction, install such weapons on celestial bodies, or station such weapons in outer space in any other manner.

The moon and other celestial bodies shall be used by all States Parties to the Treaty exclusively for peaceful purposes. The establishment of military bases, installations and fortifications, the testing of any type of weapons and the conduct of military manoeuvres on celestial bodies shall be forbidden. The use of military personnel for scientific research or for any other peaceful purposes shall not be prohibited. The use of any equipment or facility necessary for peaceful exploration of the moon and other celestial bodies shall also not be prohibited."

In the first paragraph there are two very serious gaps:

1. Only mass destruction weapons placed in orbit around the Earth are prohibited. There is no prohibition — and therefore permitted — on semi-orbital flights of ballistic missiles carrying mass destruction weapons, which only leap into outer space and, therefore, do not enter the Earth's orbit;

2. Only mass destruction weapons (nuclear, chemical and bacteriological) are prohibited in space. There is no prohibition on any other kind of weapons, including the latest, laser or particle beans — therefore, permitted.

The only partial demilitarization of space — veto on testing nuclear weapons and putting into orbit mass destruction weapons — contrasts with the complete demilitarization of the moon and other celestial bodies, stipulated in the second paragraph.

The celestial bodies must be "exclusively for peaceful purposes". The moon, Mars and all the rest are free of military bases, installations, fortifications, manoeuvres and tests with any kind of weapons.

The Agreement Governing the Activities of States on the Moon and other Celestial Bodies [19], known as the "Moon Agreement", approved unanimously by the UN General Assembly in 1979 and in force since 1984, gives more detail to the principle of demilitarization.

Its thorough Article 3 states:

"1 - The moon shall be used by all States Parties exclusively for peaceful purposes".

2 - Any threat or use of force or any other hostile act or threat of hostile act on the moon is prohibited. It is likewise prohibited to use the moon in order to commit any such act or to engage in any such threat in relation to the earth, the moon, spacecraft, the personnel of spacecraft or man-made space objects.

3 - States Parties shall not place in orbit around or other trajectory to or around the moon objects carrying nuclear weapons or any other kinds of weapons of mass destruction or place or use such weapons on or in the moon.

4 - The establishment of military bases, installations or fortifications, the testing of any type of weapons and the conduct of military manoeuvres on the moon shall be forbidden. The use of military personnel for scientific research or for any other peaceful purposes shall not be prohibited. The use of any equipment or facility necessary for peaceful exploration and use of the moon shall also not be prohibited."

The Moon Agreement, despite the unanimous approval of the UN General Assembly, has been ratified today by only nine countries, among which there is no space power. This, of course, prejudices its influence and authority. Yet in no way does it weaken the principle of total demilitarization of the Moon and other celestial bodies, set out, first and foremost, in the Space Treaty, of undeniable validity and only given in greater detail in this Moon Agreement.

It so heppens that both Article IV of the Space Treaty and Article 3 of the Moon Agreement have, in the end, an identical text permitting the use of military personnel, as well as equipment and installations of a military origin, in scientific and other activities, as long as they are for "peaceful purposes" and consist of "peaceful exploration" of the celestial body.

This is justified in the dual nature of space technologies, which can attend peaceful or military purposes.

The most important item in this provision is, however, the clearly stipulated contrast between "military" and "peaceful". The military, together with their equipment and installations, can only be admitted on the moon and other celestial bodies if they are not exercising their main, military, tasks, and if they are at the service of "peaceful" projects. It is of little importance if the military activities are "aggressive" or "non-aggressive". They just need to be "military" to be unacceptable.

The same focus has already been confirmed in the Antarctic, the first continent on Earth completely demilitarized through an international agreement. The 1959 Antarctic Treaty stipulates, in Article I, that this region "shall be used for peaceful purposes only", and that in it "there shall be prohibited, *inter alia*, any measures of military nature, such as the establishment of military bases and fortifications, the carrying out of military maneuvers, as well as the testing of any type of weapons".

The whole practical experience of 40 years of a demilitarized Antarctic — without banning the admission and work of military specialists — clearly shows that any effectively military activity is incompatible with its special status as territory devoted exclusively to peaceful projects.

The ocean bed, soil and subsoil areas, outside the jurisdiction of the States, also "is open exclusively to peaceful purposes", according to Article 141 of the 1982 UN Convention on the Law of the Sea, in which arms testing nor their installation is permitted.

In the same way, the multilateral agreement signed in 1988 between the USA, European Space Agency, Japan and Canada to build an international permanently manned civil space station linked the project to "peaceful purposes" in its Article 1.1. These words have been interpreted by Ivan Vlasic, professor of the Aviation and Space Law Institute of McGill University in Montreal, Canada, as follows: "Considering that the major problem in negotiating this agreement was the strong opposition of the majority of the participating States against the military having anything to do with the design or use of the station, the inclusion of the term "peaceful uses" can have only one logical meaning — 'non-military'."

Understanding "peaceful" as "non-military" appears to be obvious. The first resolution of the UN General Assembly on a space issue on November 14, 1957, after Sputnik I was launched on October 4 that same year, introduced the term "exclusively for peaceful puposes' with the unmistakable meaning of "non-military". This view had full support from the USA government. In January 1958, President Dwight D. Eisenhower proposed to the USSR that their two countries should agree to use outer space "only for peaceful purposes" and not for "testing missiles designed for military purposes". [20]

It so happens that, exactly at this time, in the late 1950s, military reconnaissance satellites had become technically feasible. Shortly afterwards, they inaugurated the first military use of space. The merit probably goes to the North American military watch program called Samos (Satellite and Missile Observation System) which launched around 20 satellites between October 1960 and January 1963.[21]

It was at this time, then, that the theory arose in the USA that the military satellites not involved in aggressive operations could and should be considered "peaceful", since they would be harmless. Richard N. Gardner, university professor and vice-secretary of State for the USA, sustained that "the test of the legitimacy of a particular use of outer space is not whether it is military or non-military, but whether it is peaceful or aggressive". [22] With this, the "passive" military satellites, such as those of reconnaissance (espionage), communication, navigation and others, would be part of the respectable "peaceful" list.

This position was adopted by many jurists, especially in the USA. Stephen Gorove, for instance, states that it "is supported by the UN Charter, which does not consider general military activities illegal, and only prohibits the threats to the peace, breaches of the peace and acts of aggression". [23]

Of course, the Charter does not ban all military activities. It permits defensive military activities to qualify the countries to exercise their "inherent right of individual or collective self-defense if an armed attack occurs (...) until the Security Council has taken the measures necessary to maintain international peace and security". as stipulates

Article 51 in the Charter. Nothing fairer than that. Yet this does not turn defensive military actions into "peaceful" ones. By their very nature, they continue being "military", albeit "defensive" and, therefore, "non-aggressive". Military activities can be "aggressive" and "non-aggressive". This does not mean they cease to be military. Nor does it mean that "non-aggressive military" becomes "peaceful".

Hence the ironic reaction of Vladlen Verschchetin, today judge of the International Court of Justice: "Semantic methods can not transform a military activity into peaceful activity and vice-versa; in any language, peaceful activity remains peaceful, and military, military."[24]

Along the same lines, E. Kamenétskaia went even farther and warned that "certain efforts to distort the generally accepted meaning of the word 'military' (making it synonymous, for example, with 'aggressive') are of doubtful legitimacy; and do not help elucidate the truth". [25]

In fact, the effort to restrict the meaning of "military actions" solely to "aggressive" actions and placing non-aggressive military activities on the same plane as peaceful activities, as if following the same dynamics, being stimulated by the same interests and pursuing the same objectives, can be seen as a mere artifice to disguise or hide the true nature of each kind of activity in question.

If in space military activities are not entirely prohibited, as they are on the moon, it does not make the least sense to give certain military operations the title of peaceful. Very much to the contrary, it is necessary to clearly define the military uses permitted today or tolerated in practice. This has the double advantage of granting such uses the necessary guarantee of legitimacy and, at the same time, fixing in a clear cut fashion a reasonable and controllable limit on the space militarization process.

The greatest danger, today, lies in introducing, installing and using weapons in space. If this boundary is crossed, as many would like, it is most probable that the insecurity and uncertainties of the space activities in general and the nations themselves here on Earth increase.

A serious practical problem must not be minimized: if we permit the admission of weapons to space, we will be worsening the situation in such a way that, as Kamenetskaia very well warned, there will be no going back. As historical experience shows, it is much more difficult to remove already installed weapons from somewhere than prevent their installation.

Important legal instruments help us in this task, providing that they are kept enforced, strengthened and even developed.

According to Article V of the Treaty Between the USA and the USSR on the Limitation of the Antiballistic Missile Systems, signed in 1972, known as the "ABM Treaty", "each Party undertakes not to develop, test, or deploy ABM systems or components which are sea-based, air based, space-based, or mobile land-based. From the start, it permitted each country to build two earth-based antimissile systems. Later, under the 1974 Protocol, only one was permitted. This was to prevent the installation of a nationwide antimissile defense system to protect the whole country, since such a structure would be a breach of the principle of strategic parity considered essential for the coexistence between the two powers.

To "legalize" its aforementioned project, "Strategic Defense Initiative" (SDI), an enormous anti-missile system with a vast space segment, destined to make USA territory inaccessible to Soviet missiles, the Ronald Reagan government curiously did not wish to run the political risk of withdrawing unilaterally from the ABM Treaty and preferred to interpret its rules *sui generis*; the anti-missile systems with weapons based on new technologies, such as directed energy, foreseen in the SDI, non-existent at the time of signing the ABM Treaty, would be free of prohibition.

The manoeuvre failed. The new interpretation was rejected by important members of the USA Congress, North American diplomats and specialists who negotiated the ABM Treaty, the country's scientific and academic community and most of the allied countries. Thanks to this, to the end of the Cold War and the need for cuts in government spending, the SDI lost its raison d'être and its character as a megaproject. But it did not disappear altogether.

This year, 1997, the Clinton government announced its intention to double the program's budget over the next six years, from US$ 2.3 to US$ 4.6 billion. It is a solid basis for the qualitative advance of space militarization. However, it will first have to substantially change the ABM Treaty.

As Scott Pace tells in "Economic Interests and Military Space Systems: an American Perspective", "for many supporters of ballistic missiles defences, the ABM Treaty is seen as an anachronism that should be dropped as soon as possible."[26] This is why defending the original integrity of this treaty helps block space from arms installations.

The 1977 Convention on the Prohibition of Use of Environmental Modifying Techniques with Military or Any Other Hostile Uses, considers illegal all "deliberate manipulation of the natural processes", with military uses, and which aim at "modifying the dynamics, composition or structure of the Earth, including its biota, lithosphere, hydrosphere and atmosphere, or outer space". Article 1 of the Convention vetoes the military use of environment modifying techniques with widespread, lasting or serious effects, to cause destruction, losses and damage to other countries. It is not an easy nor direct instrument to apply. Yet it may be assumed, without much difficulty, that the launching of modern weapons against space from Earth, against Earth from space and in space itself, and the consequent conversion of this environment in a battlefield, will surely be extremely detrimental today to the natural processes both of space and Earth.

The 1979 Strategic Arms Limitation Treaty, known as "Salt II" prohibits in its Article 9 the development, testing and deployment of nuclear or mass destruction arms systems in Earth's orbit, including fractional orbit missiles.

The historic first Strategic Nuclear Arms Reduction Treaty (Start I) in 1991 reduces the arsenals of ballistic missiles and, although does not restrict the military use of space, reinforces the ABM and Salt II Talks, as well as all arms control measures adopted by the USA and USSR (today the Russian Federation).

Ironically, on the same day as presidents George Bush and Mikhail Gorbachev signed Start I, the USA Senate approved the Missile Defense Act, planning on the construction shortly in Grand Forks, in the State of North Dakota, of a new anti-missile defense system, described as the first step towards deploying a nationwide system. The Act itself recognizes that, in order to be effectively implemented, the ABM Treaty would have to be amended.

In 1995, the USA Congress approved a law linking the country to the development and installation of a multiple antimissile defense system covering all American territory. [27] They are further proof of how important it is to keep the ABM Treaty intact.

Outer space not only continues free of any kind of weapon but has never been the stage for a single hostile act. To keep it like that, however, the current legal bench mark regulating its military use is, without a doubt, not enough and needs to be substantially enhanced before it is too late.

Ways to avoid total space militarization

"What could and may prevent outer space being transformed into an arena of military contest and arms-protected territory?", asks Kamenétskaia, and she herself replies; "The answer lies not on the technical, but on the politico-legal plane. It is a matter of political will, of political decisions, of legal obligations. In this connection it should be emphasized that the essential factor in restraining the militarization of outer space is international law."[28]

This means that the political will, negotiations and decisions are indispensable elements in the process in question, but not enough. The political vector must be committed to an effort of cooperation and convergence which, by necessity, leads to the joint preparation of clear-cut effective and powerful legal obligations, approved mutually and in complete harmony with the basic principles of international law, consolidated in the UN Charter and, in this case, also in the Space Treaty.

The use and exploration of space — indispensable, extremely risk and increasingly complex — are, more than ever, strategic activities for peace, security and the development of all countries and humanity. The States — and their public or private companies — involved in space programs have the responsibility and obligation to immediately anticipate the facts that may cause damage and loss to the other countries, as stipulated in Article IX of the Space Treaty.

This is why the space law, like other vanguard branches of modern Law, has the special duty of advancing possible or probable contexts of major and recognized danger, adopting clearly preventive measures.

On the other hand, space, as an medium of an already indispensable universal utility, albeit still contaminated with uncertainties and threats, requires strict legal security, that is, the prevalence of objective clear norms which guarantee the domain of the law over the individual will.

This is especially important inasmuch as predictability and objectiveness of the legal norms are not in many times of interest to the stronger countries, accustomed to getting their own way: they prefer individual negotiation, where they can more easily make their strength and interests prevail.

It is worth mentioning that, in the current international scenario, the absence of partners to negotiate with the USA on equal conditions may reduce even further its reluctance to make concessions to the concerns of most of the other countries, which are in no conditions to make corresponding concessions. [29]

Thus, it is not hard to predict: the space law provision which impedes total space militarization will be the crowning achievement of outstanding politico-legal work, able to overcome apparently invincible obstacles in today's world, which is so unequal and still so powerless before frequent arbitrariness.

This crusade, however, has already begun. Let us look at some of its landmarks:

In 1978, the UN General Assembly, in the final report of the Special Session on Disarmament, succeeded for the first time in recommending that "in order to prevent an arms race in outer space, further measures should be taken and appropriate international negotiations held". [30]

In 1979, Italy submitted to the Disarmament Committee the project of a Protocol to the 1967 Space Treaty, proclaming the intention to wholly prohibit military activities in outer space and stipulating the prohibition against "the development and use of earth-based or space-based systems designed to damage, destroy and interfere with the operations of other State satellites", as well as the tests with any kind of weapon.

In 1981, the UN General Assembly asked the Disarmament Committee, consisting of representatives from 40 countries, to give priority and study forms to prevent the extention of the arms race into outer space, with the aim of drafting an effective and verifiable agreement.

In 1981, the USSR proposed the project of "Treaty on Prohibition of Placing Weapons of Any Kind in Outer Space", which, however, vetoes only weapons placed in orbit around Earth, and therefore does not include ground and air-based anti-satellite devices. But the greatest flaw in this project is its Article 3, under which "each Party-State of the Treaty is committed to not destroying, damaging nor interfering in the operation of other countries' space objects, if these objects are placed in space strictly according to Article 1, paragraph 1, of this Treaty [Do not deploy in orbit objects around Earth which carry any kind of weapon aboard]". Such terms lead to the absurd idea that it is admissible for each country to have the right to attack a satellite of another country if it thinks that it carries weapons. This would mean legalizing the use of force in space in the form of a unilateral and preventive action, in breach of the UNO Charter. Another serious flaw: the project permits that the member countries of the Treaty withdraw from it when "extraordinary facts ... endanger their supreme interests". Yet the most valuable point in treaties of this nature is that they have to work exactly at extraordinary moments.

In 1983, the USSR proposed, in a letter from its minister of Foreign Affairs to the UNO secretary-general, the project of "Treaty on Prohibition of the Use of Force in Outer Space and from Outer Space with respect to the Earth". The proposal, broader and more convincing than that of

1981, had the great merit of banning the anti-satellite systems completely, including those based on the ground and in the air. Moreover, it creates a Consulting Committee to solve the problems that may arise while implementing the treaty, and simply does not foresee the possibility of the countries resigning from it.

In 1985, the USSR included in the agenda of the UN General Assembly agenda the item "On International Cooperation in Peaceful Exploration of Outer Space Uder Conditions of the Nonmilitarization Thereof".

In 1986, the USSR government, in a letter to the UN Secretary-General, further developed the document of the previous year and proposed to create a World Space Organization as part of an ambitious global space cooperation program, based on the premise of "nonmilitarization" of outer space. The Soviets defined "nonmilitarization" of outer space as "the rejection by the States of the creation (including the corresponding scientific research work) of tests and development of space arms". But this restricted definition of the expression did not move the USA government, which has always preferred to interpret it as embracing every and any military use of outer space. This, of course, greatly contributed to putting an end to the proposal.

In 1986, the USSR proposed at the Disarmament Talks the signing of an international agreement guaranteeing satellite immunity, prohibiting the development, testing and deployment of anti-satellite weapons, and eliminating those already in existence.

In 1987, the USSR recommended to the *Ad Hoc* Committee that it consider the possibility of forming an international checking system for satellites and other space objects before they are launched, to intercept the dispatch of arms to outer space.

In 1988, the USSR submitted to the Disarmament Committee the memorandum "On the Creation of an International Control System for Not Allowing the Stationing of Weapons of Any Kind in Outer Space", with the aim of on site inspection of all objects ready to be launched into space, against their carrying weapons. The memo omits the phrase "non-militarization". It goes straight to the point. (The proposition is similar to the France one made in 1978 on the creation of the "International Satellite Monitoring Agency".)

In 1989, Venezuela proposed at the Disarmament Talks the increase in scope of Article IV of the Space Treaty, extending its prohibition to all kinds of weapons and attributing to all member-countries the obligation of "not developing, producing, storing or using such weapons'.

In 1992, the *Ad Hoc* Committee report discusses the two basic clashing views on this subject. For the so-called Western powers, among them the USA, "the existing legal regime provided an equitable and balanced responce to the need to promote peaceful uses and arms control in outer space". Most of the Committee's member-countries, however, think that the exinting legal instruments were "not satisfactory"; "Limited in scope, they were utterly inadequate in forstalling an arms race in outer space in that they contained no clear-cut provision on the prohibition of all types of space weapony".

In 1993, France and other countries proposed at the Disarmament Talks an amendment to Article IV of the 1975 Convention on Registration of Objects Launched into Outer Space, seeking to ensure that more specific information on the real mission of each satellite, especially with regard to its civil or military nature, be more rapidly disseminated before the respective launching. At the same time, they suggested the creation, under the auspices of UN, of an international center in charge of collecting and distributing information supplied by the countries launching space objects. The absolute transparency of the essential data of each space launching, if effectively complied with, can encourage an atmosphere of confidence, which may get rid of suspicions between the countries, weakening the argument that defends outer space arms deployment against "possible future enemies".

It is worth recalling that Brazil did not sign the Convention on Registration of Objects Launched into Outer Space, precisely because it considered the data required by it for space launchings insufficient. [31]

Everything indicates that for some years now the *Ad Hoc* Committee has been spending much more time in discussing confidence-building measures than in preparing an agreement or agreements to block outer space against the arms race. It is a deviation from its principal task. There are by no means few countries which criticize this unjustifiable change in direction. One of them is Russia, which maintains the USSR position on this subject. However beneficial it may be, confidence-building does not substitute the fully assumed and written legal commitment, in terms of effectiveness and security.

The new post-Cold War reality permitted the USA and Russia to adapt at lower levels of its military parities, especially in the nuclear field. [32] The cause of global disarmament could advance as it had never done before. The following are the documents achieved in this short period:

1) Treaty on the Elimination of Intermediate-Range and Short-Range Missiles, 1987;

2) First Strategic Nuclear Arms Reduction Treaty, 1991 (Start I);

3) Second Strategic Nuclear Arms Reduction Treaty, 1993 (Start II);

4) Open Skies Treaty, 1992, signed between USA, Canada and 23 European countries, among them Russia;

5) Convention on the Prohibition of the Development, Production, Stockpiling and Use of Chemical Arms and Their Destruction, 1993, in force since April 1997, and which took 20 years to be ready;

6) Comprehensive Test Ban Treaty (CTBT), 1996, also an old claim by most countries.

Thanks to Start I and II, if everything goes according to plan, in 2003 there will only be 30% of the nuclear arsenals calculated in the USA and Russia in 1990, that is, 3,500 warheads in each country, instead of 12,718 and 10,779 respectively. There are also expectations around the negotiation of Start III planned for the USA-Russia summit meeting in Helsinki in May 1997, soon after the ratification of Start II by the Russian parliament. Start III may diminish the nuclear arsenals to around 2,000-2,500 warheads within ten years.

In 1995, it is fair to add that the review mechanism of the Treaty on The Non-Proliferation of Nuclear Weapons (NPT) was reinforced to enable the non-nuclear member-countries to put more pressure in favor of nuclear disarmament.

By the end of 1997, it is expected to conclude an international agreement prohibiting antipersonnel mines, if the USA back this project.

All this, of course, would be unthinkable just over ten years ago.

Curiously enough, so far there has been no legal advance in the scope of the multiple attempts to close outer space to the arms race.

It so happens that the vast majority of UN member-countries wish to prohibit space weapons and their influence only tends to increase, in the coming years, with the increasingly active development of outer space activities in the field of scientific research, industry and services, and with the growing interest and participation of more and more countries and private corporations in such prosperous enterprises.

Thus, from an optimistic viewpoint, perhaps the day is not too far away when beneficial blessed perseverance of the world community will have achieved the mutually agreed political desire and decision to exclude from history the possibility of total outer space militarization.

When that day comes, one of the first multilateral legal tasks will appear logical and urgent: to update the Outer Space Treaty, starting at Article IV, which admits putting arms in outer space, except for nuclear and mass destruction weapons.

Article IV perhaps was useful in the time of deterrence and balance of terror. Today we live in the age of widening the commercial and scientific uses of outer space. The historic mission is to close the sky for the arms race forever. It means – *inter alia* – the elimination of the present incoherence and the discrepancy between article IV, on one hand, and, on the other, articles I (outer space only for benefits and in the interests of all countries) and IX (co-operation and due regard to the corresponding interests of all countries).

It would be quite important for the UN General Assembly to request an advisory opinion from the International Court of Justice on this topic in order to open a better legal way to prevent the deployment of weapons of any kind in outer space, as the one recently requested about the legality of the use of nuclear weapons. After all, these questions are related and their subject is the same: to warrant peace on Earth as in space.

The present generation has the chance to stop the gaps, through which today space actions can infiltrate which could put an end to the outstanding Article I of Outer Space, in which, as a pioneer, the common good of all mankind is established as a law. Will we make good use of this?

Notes

1. **Space Law: Development and Scope**, edited by Nandasiri Jasentuliyana: foreword by Manfred Lachs, Praeger Published. 1992, p. x.
2. **Vlasic, Ivan**, *Space Law and the Military Applications of Space Technology*, in *Perspectives on International Law*. Edited by Nandasiri Jasentuliyana. London: Kluwer Law International. 1995. p. 388.

2A. Id. ibid.

3. **Treaty on Principles Governing the Activities of States in the Exploration and Use of Outer Space, Including the Moon and Other Celestial Bodies**, approved by UNO General Assembly, as Resolution 2222 (XXI) on December 19, 1966, was open for signature by the States on January 27, 1967 and in force since October 10, 1967.
4. *Aviation Week & Space Technology*, August 5, 1996.
5. *Space News*, February 17-23, 1997.
6. *Space News*, February 17-23, 1997.
7. *Space News*, February 17-23, 1997.
8. *Space News*, March 24-30, 1997.
9. *Aviation Week & Space Technology*, March 24, 1997.
10. **Declaration of Legal Principles Governing the Activities of States in the Exploration and Use of Outer Space**, Resolution 1962 (XVIII) of UN General Assembly on December 13, 1963.
11. *Aviation Week & Space Technology*, March 24, 1997.
12. *The Economist*, March 8, 1997.
13. **Chemillier-Gendreau, Monique**, *Humanité et souverainetés — essai sur la fonction du droit international*, Paris: Editions La Découverte, 1995, pp. 293/4.
14. *Defense News*, USA, via Internet.
15. *Aviation Week & Space Technology*, February 26 and March 17, 1997.
16. *Aviation Week & Space Technology*, March 10, 1997.
17. **Grouard, Serge**, *La Guerre en Orbite — Essai de politique et de stratégie spatiales*, Paris: Ed. Economica, 1994, p. 221.
18. **Maintaining Outer Space for Peaceful Uses**, edited by Nandasiri Jasentuliyana. The United Nations University, 1984. p. 7. Manfred Lachs, judge of the International Court of Justice, headed the seminar. As chairman of the Legal Sub-Committee of the UN Committee on Peaceful Uses of Outer Space (Copuos) in the 1960s, he played a fundamental role in the preparation process of the 1967 Space Treaty.
19. **Agreement Governing the Activities of States on the Moon and Other Celestial Bodies**, approved by the UN General Assembly as Resolution 34/68, on December 4, 1979. Open to the signature of the countries on December 18, 1979. In force since July 11, 1984. Today it has nine ratifications (Australia, Austria, Chile, Philippines, Morocco, Mexico, The Netherlands, Pakistan and Uruguay) and five signatures (France, Guatemala, India, Peru and Romania).
20. **Vlasic, Ivan**, ib, p. 390.
21. **Cotardière, Philippe de la, & Penot, Jean-Pierre**, *Dictionnaire de L'Espace*, Paris: Larousse, 1993, p. 204.
22. **Vlasic, Ivan**, ib., p. 391.
23. **Gorove, Stephen**, *Developments in Space Law*, The Netherlands: Martinus Nijhoff Publishers, 1991, p. 257.
24. **Kamenétskaia, E. P.**, *International Legal Problems of Preventing an Arms Race in Outer Space*, in *Perestroika and International Law*, edited by W. E. Butler, The Netherlands: Martinus Nijhoff Publishers, 1990, p. 149.
25. Id. ib.
26. **Pace, Scott**, *Economic Interests and Military Space Systems: An American Perspective*, in Gasparini Alves, Péricles (editor), *Evolving Trends in the Dual Use of Satellites*, Unidir, United Nations Institute for Disarmament Research, Geneva, 1996, p. 146.
27. **Jankowitsch, Peter**, *Legal Aspects of Military Space Activities*, in *Space Law: Development and Scope*, edited by Nandasiri Jasentuliyana: foreword by Manfred Lachs, Praeger Published. 1992, p. 153. Also see SIPRI Yearbook 1996 Armaments, Disarmament and International Security, Stockholm International Peace Research Institute, New York: Oxford University Press, 1996, p. 650.
28. **Kamenétskaia, E. P.**, ib., p. 147.
29. **Jankowitsch, Peter**, ib. p. 155.
30. **Vlasic, Ivan**, ib. p. 401.
31. **Convention on Registration of Objects Launched to Outer Space**, approved by UN General Assembly, as Resolution 3235 (XXIX), on November 12, 1975. Open for signature of countries on January 14, 1976, in force on September 15 that same year. It has four signatures and 39 ratifications. On Brazil's decision not

to sign this convention, see Monserrat Filho, José, *O Brasil e o Direito Espacial (I)*, in Revista Brasileira de Direito Aeroespacial, no. 64, July-December 1993.

32. **Lafer, Celso**, *"Dividendos da paz" no mundo pós-guerra fria*, O Estado de São Paulo newspaper, July 5, 1997, p. 2.

Sources

Andem, Maurice N., *International Legal Problems in the Peaceful Exploration and Use of Outer Space*, Finland: University of Lapland Pubications, 1992.

Cosmic Weapon: Security Dilemma, several authors, Moscow: Mir, 1986 (in Russian).

Chemillier-Gendreau, Monique, *Humanité et souverainetés - essai sur la fonction du droit international*, Paris: Editions La Découverte, 1995.

Christol, Carl Q., *Space Law: Past, Present and Future*, The Netherlands: Kluwer Law and Taxation Publishers, 1991.

Cirincione, Joseph, *Why the Right Lost the Missile Defense Debate*, Foreign Policy, Number 106, Spring 1997, pp. 39-55.

Doyle, Stephen E., *Civil Space Systems — Implications for International Security*, Unidir, United Nations Institute for Disarmament Research, Geneva, 1994.

Felden, Marceau, *La Guerre dans L'Espace — Armes et Technologies Nouvelles*, Paris: Berger-Levrault, 1984.

Ferrer (H), Manuel Augusto, *Derecho Espacial*, Buenos Aires: Editorial Plus Ultra, 1976, pp. 158-161.

Friedman, George and Meredith, *The future of war: power, technology, and American world dominance in the 21st century*, New York: Crown Publishers, 1996.

Gasparini Alves, Péricles (editor), *Evolving Trends in the Dual Use of Satellites*, Unidir, United Nations Institute for Disarmament Research, Geneva, 1996.

Gorove, Stephen, *Developments in Space Law*, The Netherlands: Martinus Nijhoff Publishers, 1991.

Grouard, Serge, *La Guerre en Orbite — Essai de politique et de stratégie spatiales*, Paris: Ed. Economica, 1994.

Humble, Ronald D., *The Soviet Space Programme*, USA: Routledge, 1988.

Kamenétskaia, E. P., *International Legal Problems of Preventing an Arms Race in Outer Space*, in *Perestroika and International Law*, edited by W. E. Butler, The Netherlands: Martinus Nijhoff Publishers, 1990, pp. 147-164.

Lachs, Manfred, *El derecho del espacio ultraterrestre*. Mexico: Fondo de Cultura Económica, 1977.

Maintaining Outer Space for Peaceful Uses, edited by Nandasiri Jasentuliyana, The United Nations University, 1984.

Mantenimiento de la Utilización del Espacio Ultraterrestre con Fines Pacíficos, con contribuciones de Aldo Armando Cocca, Isabella H. Ph. Diederiks-Verschoor, Eilene Galloway, Peter Jankowitsch, Manfred Lachs y Edward W. Ploman, Estudios Internacionales Avanzados, Córdoba, Argentina, 1986.

Martin, Pierre-Marie, *Droit des activités spatiales*, Paris: Masson, 1992, pp. 43-58.

Outer Space, A Source of Conflict or Co-operation?, edited by Bhupendra Jasani, United Nations University Press, 1991.

Pereira, Antônio Celso Alves, *Os Impérios nucleares e seus reféns: relações internacionais contemporâneas*, Rio de Janeiro: Edições Graal, 1984.

Reijnen, Bess C. M., *The United Nations Space Treaties Analysed*, Editions Frontieres, France, 1992, pp. 103-106.

Reynolds, Glenn H. and Merges, Robert P., *Outer Space - Problems of Law and Policy*, USA: Westview Press, 2nd edition, 1997.

Space Law: Development and Scope, edited by Nandasiri Jasentuliyana; foreword by Manfred Lachs, Praeger Published, 1992.

SIPRI Yearbook 1996, Armaments, Disarmament and International Security, Stockholm International Peace Research Institute, New York: Oxford University Press, 1996.

Stares, Paul B., *The Militarization of Space - U.S. Policy, 1945-1984*, USA: Cornell University Press, 1985.

The Non-Proliferation Treaty and Nuclear Disarmament, Pugwash Meeting Rio de Janeiro, edited by Luiz Pinguelli Rosa, Suzana Kahn Ribeiro, Brazil: Coppe/UFRJ, 1997.

Vereschetin V., Vasileveskaia E., Kamenétskaia, *Outer Space - Politics and Law*, Moscow: Progress Publishers, 1987.

Vlasic, Ivan, *Space Law and the Military Applications of Space Technology*, in *Perspectives on International Law*, edited by Nandasiri Jasentuliyana, London: Kluwer Law International, 1995, pp. 385-410.

Weinberger, Caspar & Schweizer, Peter, *The Next War*, Washington: Regnery Publishing, 1996.

IISL-97-4.15

REAL PROPERTY RIGHTS IN OUTER SPACE

Wayne N. White, Jr.
Attorney at Law
4465 Kipling Street, Suite 200
Wheat Ridge, Colorado 80033

Abstract

This paper begins with a discussion of the need for a regime of real property rights in outer space. The author then analyzes the non-appropriation and sovereignty language in Article II of the 1967 Outer Space Treaty. In the following sections, the author analyzes the resource appropriation regimes in the Moon Treaty, the Law of the Sea Treaty and in deep sea mining legislation. The paper then discusses the legality of real property rights within and beyond the confines of a space facility.

Finally, the paper proposes a regime for real property rights in outer space. This regime would not require nations to establish territorial sovereignty, and is consistent with the provisions and principles of the Outer Space Treaty.

Introduction

At some point in the future, private entities will begin to appropriate resources and inhabit outer space. Initially, such activities will be risky and expensive. Existing international law provides limited legal protection and little incentive for investment in outer space. This article proposes a regime of real property rights which would provide an element of legal certainty and incentive for private ventures.

The concept of real property rights is intimately tied to the sovereignty which nation states exercise over territory. The 1967 Outer Space Treaty prohibits states from establishing territorial sovereignty, but authorizes and, in some cases even requires, that states exercise jurisdiction over space objects and personnel. This author therefore proposes a form of property rights which would not require states to establish territorial sovereignty, while remaining within the jurisdictional limitations set forth in the Outer Space Treaty.

Why Real Property Rights are Necessary

The 1967 Outer Space Treaty[1] does not provide a positive regime for the governance of space development. The 1979 Moon Treaty[2] provides a regime for development, but that regime prohibits real property rights. For that and other reasons, most nations have not signed or ratified the Moon Treaty.

A development regime which provides some form of property rights will become increasingly necessary as space develops. Professionals foresee an integrated system of solar power generation, lunar and asteroidal mining, orbital industrialization, and habitation in outer space. In the midst of this complexity, the right to maintain a facility in a given location relative to another space object may create conflict. Such conflicts may arise sooner than we expect, if private companies begin building subsidiary facilities around space stations. Eventually large public facilities will become the hub of private space development, and owners will want to protect the proximity value of their facility location.

It also seems likely that at some point national governments and/or private companies will clash over the right to exploit a given mineral deposit. Finally, the geosynchronous orbit is already crowded with satellites, and other orbits with unique characteristics may become scarce in the future.

The institution of real property is the most efficient method of allocating the scarce resource of location value. Space habitats, for example, will be very expensive and will probably require financing from private as well as public sources. Selling property rights for living or business space on the habitat would be one way of obtaining private financing. Private law condominiums would seem to be a particularly apt financing model--inhabitants could hold title to their living space and pay a monthly fee for life-support services and maintenance of common areas.

Even those countries which do not have launch capability would benefit from a property regime. Private entities from the developing nations could obtain property rights by purchasing obsolete facilities from foreign entities that are more technologically advanced.

A regime of real property rights would provide legal and political certainty. Investors and settlers could predict the outcome of a conflict with greater certainty by analogizing to terrestrial property law. Settlers and developers would also be reassured, knowing that other nations would respect their right to remain at a given location.

The Outer Space Treaty

National Appropriation

Article II of the Outer Space Treaty governs the appropriation of space resources. Article II provides that "Outer Space, including the moon and other celestial bodies, is not subject to national appropriation by claim of sovereignty, by means of use or occupation, or by any other means."

International lawyers differ in their interpretation of the term "national appropriation." Some interpret Article II narrowly to prohibit only national appropriation.[3] Many others interpret the clause broadly to prohibit all forms of appropriation, including private and international appropriation.[4] When Article II is compared to similar provisions in other documents, however, it becomes clear that the narrow interpretation is correct.

Before the Space Treaty was drafted by the U.N. Committee on the Peaceful Uses of Outer Space (COPUOS), four other international legal organizations prepared draft resolutions. All of these documents recommended non-appropriation clauses which are broader than Article II.[5] The terminology in these clauses suggests that at the time the Space Treaty was drafted, international lawyers did not consider "national appropriation" to be an all-inclusive phrase. For example, a resolution of the International Institute of Space Law specifically distinguished between national and private appropriation: "Celestial bodies or regions on them shall not be subject to national or private appropriation, by claim of sovereignty, by means of use or occupation or by any other means."[6]

On the basis of a similar analysis, Professor Gorove has concluded that Article II only prohibits *national* and not private appropriation.[7]

Sovereignty

Article II also refers to claims of sovereignty. Sovereignty is a nation's right to exert exclusive authority over people, resources and institutions. It is exercised to its fullest extent within the boundaries of a nation's territory. Countries also express their sovereignty outside national boundaries, but that authority is limited to certain specific functions, such as jurisdiction over ships, aircraft, and citizens abroad. Thus, international lawyers distinguish between the absolute territorial sovereignty which is exercised within national boundaries, and the functional aspects of sovereignty, which are exercised beyond national boundaries.

The Outer Space Treaty contains provisions other than Article II which actually *require* parties to exercise functional sovereignty. The most significant example is Article VIII, which requires parties to "retain jurisdiction and control over... space objects on their registry... and over any personnel thereof, while in outer space or on a celestial body."

It follows that all aspects of sovereignty cannot be prohibited by Article II.

Possible Territorial Claims

Prior to 1960, many authors addressed the issue of sovereignty in outer space. In the popular literature, authors had fewer reservations about the legality of territorial claims. For instance, the following quote appeared in a 1957 (U.S.) article entitled *Let's Claim the Moon-- Now!*: "Columbus stuck the Spanish Flag into the sands of a West Indies beach--and we or the Russians would be perfectly within the concept of international law to claim possession of the Moon by shooting our national flag there by rocket."[8]

There are signs that the space powers of that era considered the possibility of territorial claims. In September of 1959, the Soviet Union impacted the nose cone of Lunik II on the surface of the moon. The impact scattered medallions inscribed with the Soviet coat of arms.[9] In 1969 Apollo 11 left a plaque on the moon inscribed with the words "we came in peace for all mankind." However, the United States government rejected suggestions that the Apollo 11 crew leave a United Nations flag.[10] Instead Neil Armstrong saluted an American flag.

International law does not require a fixed degree of state activity to establish a valid claim of territorial sovereignty. Traditionally occupation has been the principal method of perfecting territorial claims, but the degree of occupation necessary has varied.

In the past, symbolic occupation, or "discovery" was sometimes sufficient. European countries established claims by planting their national flag, and Russians buried medallions bearing their coat of arms. Later, some nations questioned the sufficiency of symbolic occupation. Eventually, it came to be regarded as an inchoate title which could only mature if reasonably prompt, "effective" occupation followed.[11]

During the twentieth century the concept of effective occupation has broadened and changed in emphasis--from colonization and settlement, to a more political character-- the continuous and peaceful display of state authority. Two prerequisites are necessary to establish a continuous display of authority-- (1) the intention and will to act as sovereign, and (2) some actual exercise or display of such authority.[12]

The degree of control which is necessary to establish a valid claim varies with the circumstances of each claim. International case law provides us with the following guidelines: (1) the smaller, the more inaccessible and uninhabited an area is, the less control a state must display to establish a claim;[13] (2) the area claimed must be a geographical unit-- "a naturally rounded-off region"; and (3) competing claims may either defeat an inchoate title or geographically restrict other claims based on effective occupation.[14]

On the basis of these rules, the symbolic acts of the Soviet Union (scattering medallions and naming features on the far side) would not be sufficient to establish a valid claim on the moon. Nevertheless, on the day when Lunik II landed, Premier Kruschev stated that his country had established "priority" over the Moon, and it appeared that the U.S.S.R. might eventually make a claim. But the Soviets subsequently renounced any territorial claims.[15]

There are four principal reasons why the U.S.S.R. (and later other countries) chose to reject territorial sovereignty: (1) to prevent conflict; (2) to ensure free access to all areas of outer space; (3) because it would be difficult for states to delineate boundaries in outer space; and (4) to enhance national pride, prestige and influence.

The major powers were vying for the allegiance of the many new African and Asian nations. These recently independent former colonies were extremely wary of "superpower imperialism." Consequently, both the Soviet Union and the United States could expect to gain political influence and prestige should they reject territorial sovereignty and its overtones of colonialism.

However, treaty representatives could not expect states to accept responsibility for actions which they could not control. Consequently, parties to the treaty *had* to retain jurisdiction and control if they also wanted to provide for international liability. Thus, COPUOS delegates elected to prohibit only territorial and not functional sovereignty.

Ultimately, Article II *must* be interpreted narrowly. For under international law states may do whatever is not expressly forbidden. "Restrictions upon the independence of states cannot... be presumed".[16] The language in Article II "by claim of sovereignty, by means of use or occupation, or by any other means," refers to the traditional (occupation) and the broader modern (display of authority) standards for establishing territorial claims. The clause does not prohibit other exercises of sovereignty or jurisdiction.

The Bogota Declaration

There is one instance in which nations have asserted territorial claims in outer space. In 1976 eight equatorial states claimed sovereignty over the geosynchronous orbit, which is located 22,300 miles above the equator.

In their document, *The Bogota Declaration*, the countries claimed that the orbit is a physical fact arising from the nature of our planet, that its existence depends upon gravity, and that it therefore should not be considered a part of outer space. On the basis of this rationale, they argued that the orbit formed an integral part of their territory which was subject to their sovereignty.[17]

This rationale was soundly rejected by other nations. Commentators agreed that the geosynchronous orbit is a part of outer space, and that territorial claims violated Articles I and II of the Outer Space Treaty.[18] Ironically, this incident only seems to have strengthened the prohibition against national appropriation.

Jurisdiction

Members of the international community sometimes complain about *de facto* territorial appropriation. When a nation exercises jurisdiction and control over a facility for an extended period of time, they argue, the end result is indistinguishable from territorial sovereignty.[19]

Article VIII confers "quasi-territorial" jurisdiction. It applies to the space facility, to a reasonable area around the facility (for safety purposes), and to all personnel in or near the facility, irrespective of nationality. Space objects occupy locations on a first-come, first-served basis, and personnel have the right to conduct their activities without the harmful interference of other states. In addition, although entities may not claim ownership of mineral resources "in place," once they have been removed (i.e. mined) then they *are* subject to ownership.[20] Former Attorney General Nicholas Katzenbach aptly describes this new hybrid as "primary rights... in a localized facility."[21]

This jurisdiction permits the state of registry to subject its space objects and personnel to any national laws which are not in conflict with international law.

So states may legislate with respect to a broad range of both public and private activities; and, in most circumstances, they exercise as much authority within the vicinity of their space facilities as they would within their territory on Earth.

Under Article VIII, jurisdiction and control is only valid insofar as it is necessary to accomplish the exploration and exploitation of outer space and celestial bodies. Jurisdiction and control is also limited in time. It ceases to exist when activity is halted-- as, for example, when a space object is abandoned or returned to Earth.

Because states only control as much territory as is actually used, the Outer Space Treaty does permit free access to outer space.

The Moon Treaty

The 1979 Moon Treaty contains a non-appropriation clause which is more inclusive than Article II. Although Article 11, paragraph 2 of the Moon Treaty reiterates the language of Article II of the Outer Space Treaty, Article 11, paragraph 3 further provides that "neither the surface nor the subsurface of the moon... shall become property of any state, international intergovernmental or non-governmental organization, national organization *or non-governmental entity or of any natural person*"(references to "the moon" in the Moon Treaty refer to all celestial bodies and areas of outer space other than Earth and Earth orbits).

The treaty also says, in Article 11, paragraph 1, that "the moon and its natural resources are the "common heritage of mankind." Opponents of the treaty note that the developing nations often interpret "common heritage" to mean "common *property*" of mankind. As a result, the Moon Treaty has encountered resistance from countries with free market economies.

The Moon Treaty entered into force (with respect to ratifying and acceding states) when Austria became the fifth state to ratify on July 11, 1984. However, the United States and many other spacefaring nations have decided not to sign the treaty.

The Law of the Sea Treaty

The Law of the Sea Treaty was drafted at the same time as the Moon Treaty, and many governmental representatives participated in drafting both treaties. The Sea Treaty contains a regime to govern appropriation of ocean resources that is very similar to the Moon Treaty's regime governing space resources. Perhaps it is not surprising that the Sea Treaty encountered resistance in many nations.

In the United States, those who argued against the Sea Treaty resource regime prevailed. In 1980 the United States enacted the "Deep Seabed Hard Minerals Resources Act" -- "interim" legislation intended to govern mining until an acceptable international agreement is ratified and enters into force. This act provides for renewable permits to ensure tenure at mining sites, with respect to both U.S. and foreign nationals, a reciprocity provision for similar foreign legislation, and a specific denial of extraterritorial sovereignty.[22] Five other nations have enacted similar legislation: West Germany, the United Kingdom, France, Japan, and the Soviet Union.[23]

Designated Zones of Functional Jurisdiction

Imre Csabafi's proposed "designated zones" of functional jurisdiction would permit unilateral action in outer space, just as the aforementioned statutes allow unilateral action with respect to the seabed. In his book THE CONCEPT OF STATE JURISDICTION IN INTERNATIONAL SPACE LAW, Csabafi suggests that an international agreement is necessary "which would define certain specific cases when a state, being able to show a 'particular and distinctive interest,' may claim the right to exercise functional jurisdiction in a designated zone of outer space or on a celestial body." States would then create "designated areas" of functional sovereignty through unilateral legislation. Csabafi analogizes to the regime on the continental shelf, and to the functional sovereignty which some nations exercise over pearl and sedentary fisheries on the seabed.[24]

Unfortunately, the zones which Csabafi describes are ill-suited to the complex interactions which will occur when industry and habitation become routinized.

Real Property Rights

In general, real property law would seem to provide more appropriate analogies when addressing the problems associated with permanently located space facilities and mining sites. Maritime analogies should only be applied in connection with space *vehicles* and satellites in unstable orbits.

There has been little discussion of property rights in the literature of space law. C. Wilfred Jenks provides one of the few treatments of the subject in his book, SPACE LAW:

> If property transactions should take place in space it would seem appropriate to regard them as governed by the law with which it has the most substantial connection. If anything in the nature of real property rights at a space station

> on a celestial body were to be recognized, the law applicable there would presumably govern them.... Any recognition of real property rights beyond the limits of such a station would... raise a major question of policy concerning the basis of authority to confer or recognize such rights.[25]

Jenks does not explain the distinction between property rights within a facility and property rights outside a facility. Why would recognition of property rights outside a facility "raise a major question of policy" while property rights within a facility would not?

Real Property Rights Beyond a Facility

The relationship between property and sovereignty differs under common law and civil law systems. The common law theory of title has its roots in feudal law. Under this theory the Crown holds the ultimate title to all lands, and the proprietary rights of the subject are explained in terms of vassalage. Civil law, on the other hand, is derived from Roman law, which distinguishes between property and sovereignty. Under this theory it is possible for property to exist in the *absence* of sovereignty.

Article II of the Outer Space Treaty prohibits territorial sovereignty but does not prohibit private appropriation. Hence, private entities may appropriate area in outer space or on a celestial body, although states may not. Under the common law theory of property rights, however, states (lacking sovereignty), would not have any rights to confer on private entities. Conversely, under the civil law view, property rights would exist independent of sovereignty, and therefore could be recognized.

This is why "[i]n the discussions leading to the conclusion of the [Outer Space] treaty, France [a civil law country] indicated more than once that she was not altogether satisfied with the wording of Article II...." France's representative was "thinking in particular of the risks of ambiguity between the principle of non-sovereignty--which falls under public law--and that of non-appropriation, flowing from private law."[26]

It follows that any recognition of real property rights beyond the confines of a facility would, as Jenks observed, "raise a major question of policy." Because a private entity could conceivably establish control over an area over an area of the same magnitude that a country might control, recognition of real property rights beyond facilities would raise issues similar to those raised by territorial sovereignty.

Although proponents of space development would undoubtedly welcome the economic incentive of unlimited appropriation, such claims should not be recognized. This form of property rights could potentially preclude free access to outer space in the same manner as territorial sovereignty would preclude free access. Finally, as a point of law, recognition of real property rights beyond the confines of space facilities would be inconsistent with the common law theory of property.

Real Property Rights Within a Facility

Jenks stated that property rights within a facility would be permissible under international law. Nevertheless, in light of the maxim that entities cannot transfer a greater right than they have, these property rights would be, in common law jurisdictions, necessarily more limited than traditional property rights. The common law sovereign could only confer title to the extent of its own sovereignty; thus, under the functional sovereignty conferred by Article VIII of the Outer Space Treaty, property rights would be functionally defined and limited in time.

Spitzbergen

Could nations, however, confer property rights which *are* limited, to the extent of the sovereignty conferred under Article VIII of the Outer Space Treaty? Terrestrial governments have never actually conferred or recognized property rights predicated on functional rather than territorial sovereignty. Nevertheless, the idea is not without precedent. Functional property rights have been considered in connection with the Spitzbergen Islands, which are located off the Norwegian coast.

At the turn of the century, coal deposits were discovered on these islands. Simultaneously companies of several different nationalities began mining operations. Because the arctic climate discouraged permanent occupation, the nations which had previously used the islands never considered it worthwhile to claim them. So these islands were generally recognized as *terra nullius*-- a "no-man's land." Suddenly, when coal was discovered, the concerned nations found it necessary to settle conflicting claims and to protect the rights of their nationals, in the absence of territorial sovereignty.

Consequently, Robert Lansing, in his article A Unique International Problem, proposed that the islands be jointly governed by the various nations, on the basis of functional sovereignty, within the framework of an international agreement. Central to Lansing's proposal was the concept of limited property rights predicated on functional sovereignty. In 1912 the parties prepared a draft convention to implement the idea.[27] World War I intervened, however, and in the changed circumstances following the war, the parties signed a treaty granting sovereignty to Norway.

A Proposal

Under a regime of functional property rights, title would arise on the basis of a principle entirely different from traditional property rights. Conferral of title would not depend upon a government's control over a specific area, but rather upon its control over the space objects and personnel at that location. Once conferred, these rights would, nevertheless, be almost identical to terrestrial property rights.

On Earth the exclusion of others from the use and enjoyment of a given area is the principal right associated with real property ownership. In space first-come, first-served occupation, and the prohibition against harmful interference with other states' activities provides states with a similar, albeit less clearly defined, right of exclusion. Property rights legislation would extend this right to a state's citizens.

Functional property rights would be subject to the limitations of Article VIII jurisdiction. These rights would terminate if activity were halted, as for example, if a space object was abandoned or returned to Earth. Finally, rights would be limited to the area occupied by the space object, and to a reasonable safety area around the facility. Hence, orbital property rights would extend only to the moving "envelope" occupied by a facility, and not to its entire orbital path.

In other respects a real property regime could be structured at a state's discretion. States would determine the conditions necessary to establish and maintain property rights. They could follow the example of the United States' Homesteading Acts, and require owners to maintain a facility (and/or conduct certain activities) in a fixed location, for a specified period of time (e.g. one to five years), to establish a property right. The regime would have to specify the period of inactivity or abandonment necessary to extinguish a property right, and the permissible deviation of an orbital facility from its proper location.

In outer space, requiring facility owners to maintain a fixed orbit offers several advantages. First, it will reduce the probability of collision. It seems likely that some sort of "space traffic control" will evolve to track and direct space objects; plotting titled orbital locations as constants would permit controllers to concentrate on space vehicles and satellites in less stable orbits. Facility owners would benefit from this arrangement if non-titled space objects (or space objects exceeding their parameters) were held presumptively liable in a collision. Secondly, fixed orbits discourage indiscriminate dumping of debris, because debris can be more easily tracked to plotted, fixed points of origin. Hence, courts would sometimes be able to assess liability for debris-caused damage.

Functional property rights permit free access to all areas of outer space and celestial bodies because they do not necessitate territorial sovereignty and its consequent appropriation of large areas of space. Safety zones may extend to a *reasonable* distance around a facility, and exist only for the security of the facility and to promote safe navigation in its vicinity.

The regime is attractive because it is so easy to implement. Nations can unilaterally enact legislation, and they can tailor that legislation to conform to their existing property laws. The regime will cost states virtually nothing to implement, yet it will encourage citizens to enter what promises to be a very lucrative field.

Participating states should additionally provide for reciprocity and/or negotiate some form of limited "mini-treaty" to coordinate national property legislation. Such a treaty would elaborate on the elements in Article VIII-- it would define the property rights conferred under Article VIII, and provide for their recordation; it would define the term "space object," with particular emphasis on the distinction between space vehicles and permanently situated space facilities; it would define the term "personnel"; and it would delineate the extent of jurisdiction and control, with particular emphasis on the physical extent of safety zones, and upon the temporal duration of jurisdiction, i.e. upon the period of abandonment necessary to extinguish jurisdiction.

Conclusion

Existing international space law permits the institution of limited, functional property rights in outer space. Article II of the 1967 Outer Space Treaty prohibits *national* and not private appropriation of spatial resources, and in particular prohibits national claims of territorial sovereignty.

This article proposes a regime of functional property

rights which would be legal under both the common law and civil law theories of property, and under Articles II and VIII of the Outer Space Treaty.

Endnotes

1. Treaty on Principles Governing the Activities of States in the Exploration and Use of Outer Space Including the Moon and Other Celectial Bodies, *done* Jan. 27, 1967, 18 U.S.T. 2410, T.I.A.S. No. 6347, 610 U.N.T.S. 205 (*entered into force* Oct. 10, 1967).

2. Agreement Governing the Activities of States on the Moon and Other Celestial Bodies, U.N. Doc. A/AC.105-L.113/Add.4 (1979).

3. *E.g.* Christol, *The Common Heritage of Mankind Provision in the 1979 Agreement Governing the Activities of States on the Moon and Other Celestial Bodies*, 14 INT'L LAW 429, 437, 448 (1980).

4. *E.g.* Prevost, *Law of Outer Space Summarized*, 19 CLEV. ST. L. REV. 595, 606 (1970); Tennen, *Outer Space: A Preserve for All Humankind*, 2 HOUS. J. INT'L L. 145, 149 (1979).

5. International Institute of Space Law, *Draft Resolution of the International Institute of Space Law Concerning the Legal Status of Celestial Bodies*, in PROC., SEVENTH COLLOQUIUM ON THE LAW OF OUTER SPACE 351 (1965); Institute de Droit International, *Resolution on the Legal Regime of Outer Space* (unanimously adopted Sept. 11, 1963), *reprinted in* C.W. JENKS, SPACE LAW 416 (1965); David Davies Memorial Institute of International Studies, *Draft Code on the Exploration and Uses of Outer Space* (1963), *reprinted in* C.W. JENKS, *supra*, at 419; International Law Association, *Resolution on Air Sovereignty and the Legal Status of Outer Space* (1960), *quoted in* C.W. JENKS, *supra*, at 167.

6. International Institute of Space Law, *supra*.

7. Gorove, *Interpreting Article II of the Outer Space Treaty*, 37 FORDHAM L. REV. 349, 351 (1969).

8. Huss, *Lets Claim the Moon--Now!*, MECHANIX ILLUSTRATED, Feb. - Mar., 1957, at 72.

9. *U.S. Rejects any Flag Planting as Legal Claim to Rule Moon*, N.Y. Times, Sept. 14, 1959, § I, at 1, col. 8, 16, col. 3; *Territory Claim Won't be Made*, N.Y. Times, Sept. 15, 1959, § I, at 1, col. 5, 20, col. 1.

10. Comment, *Superordinate Goals in Space Law*, 22 BAYLOR L. REV. 242, 247-48 (1970).

11. E.g. S. BHATT, LEGAL CONTROLS FOR OUTER SPACE 159 (1973).

12. Judgment in the Case Concerning the Legal Status of Eastern Greenland (Den. v. Nor.), 1933 P.C.I.J., Ser. A/B, No. 53, at 45-46 [hereinafter cited as Eastern Greenland Case].

13. Eastern Greenland Case, *supra* note 12; Clipperton Island Arbitration (Mex. v. Fr.), 2 R. INT'L ARB. AWARDS 1105 (1931), *translated in* 26 AM. J. INT'L L. 390, 394 (1932).

14. I G. HACKWORTH, DIGEST OF INTERNATIONAL LAW 404 (1940); A. ROSS, INTERNATIONAL LAW 147 (1947); Brooks, *National Control of Natural Planetary Bodies, Preliminary Considerations*, 32 J. AIR L. & COM. 315, 322 (1966).

15. Text of Khrushchev's Reply to Messages in Connection with his U.S. Trip, N.Y. Times, Sept. 15, 1959, § I, at 20, col. 2; M. MCDOUGAL, H. LASSWELL & I. VLASIC, LAW AND PUBLIC ORDER IN SPACE 773 (1963); Finch, *Territorial Claims to Celestial Bodies*, in SENATE COMM. ON AERONAUTICAL AND SPACE SCIENCES, 87TH CONG., 1ST SESS., LEGAL PROBLEMS OF SPACE EXPLORATION: A SYMPOSIUM 627 (Legislative Reference Service, Library of Congress 1961).

16. Case of the S.S. "Lotus," 1927 P.C.I.J., Series A, No. 10, at 18.

17. *The Bogota Declaration*, EL ESPECTADOR (Columbia), Dec. 7, 1976, at 13A, *reprinted in* 6 J. SPACE L. 193 (1978)(English translation).

18. *E.g.* Finch, *supra* note 15, at 221.

19. *E.g.* S. BHATT, *supra* note 11, at 154-55, 167-68; Brooks, *supra* note 14, at 323-24.

20. Cepelka & Gilmore, *Application of General International Law in Outer Space*, 36 J. AIR L. & COM. 30, 38-39 (1970); Christol, *supra* note 3, at 471.

21. Katzenbach, *The Law in Outer Space*, *in* SPACE: ITS IMPACT ON MAN AND SOCIETY 78 (L. Levy ed. 1965).

22. Pub. L. No. 96-283, 94 Stat. 553 (1980), *reprinted in* 19 INT'L LEGAL MATERIALS 1003 (1980); 20 INT'L LEGAL MATERIALS 1228 (1981); 21 INT'L LEGAL MATERIALS 867 (1982).

23. Act of Interim Regulation of Deep Seabed Mining, 1980 Bundesgesetzblatt I 9080, No. 50, at 1457 (W. Ger.), *reprinted in* 20 INT'L LEGAL MATERIALS 393 (1981); Deep Sea Mining (Temporary Provisions) Act 1981, ch. 53, (U. K.) *reprinted in* 20 INT'L LEGAL MATERIALS 1217 (1981); Deep Seabed Mineral Resources Exploration and Mining Act, Act No. 81-1135, Journal Officiel de la Republique Francaise du 24 decembre 1981, at 3499, (France) *reprinted in* 21 INT'L LEGAL MATERIALS 808 (1982); Law on Interim Measures

for Deep Seabed Mining, July 20, 1982, (Japan) *reprinted in* 22 INT'L LEGAL MATERIALS 102 (1983) (text issued by Ministry of International Trade and Industry of Japan).

24. I. CSABAFI, THE CONCEPT OF STATE JURISDICTION IN INTERNATIONAL SPACE LAW 136 (1971).

25. C. W. JENKS, *supra* note 5, at 297.

26. U.N. Doc. A/AC.105/C.2/SR.70 at 14 (1966); U.N. Doc. A/AC.105/PV.44 at 41 (1966).

27. Lansing, *A Unique International Problem*, 11 AM. J. INT'L L. 763, 764, 766-69 (1917).

Telecommunications and the Outer Space Treaty

Francis Lyall
Professor of Public Law
University of Aberdeen
Aberdeen AB24 3UB
Scotland, UK

Abstract

Although telecommunications were identified by early UN Space Resolutions as important areas of the emergent space technologies, the 1967 Outer Space Treaty is silent about them as such. Given that the ITU has jurisdiction in these matters, the actual development of space telecommunications in the space age, and the undesirability of splitting terrestrial and space communications as a matter of legal regime, it is not desirable greatly to modify the 1967 Treaty to cover telecommunications in any detail. A useful addition would, however, be to emphasise the importance of space telecommunications as part of the 'benefit' of space in the general world interest so as to strengthen bulwarks against abuse of the ITU system. Further, given the inability of some governments properly to discharge their obligations, a move towards some wider regulatory/executive role for some international body would be desirable.

A. Introduction

When the space age opened with the launch of Sputnik 1 on 4 October 1957, scientific research was the purpose in immediate view. However, telecommunications services soon became a major aim, justifying to governments and the public much expenditure on experiment and research. Projects SCORE and Courier were delayed repeater systems launched in 1958 and 1959, which had limited success, as did Project LOFTI, a low orbit, low frequency system. More progress was made with Telstar, Relay and Syncom, the first two being low orbit systems the last of these being, as its name indicates, using the geosynchronous orbit first envisaged by Arthur C. Clarke as far back as 1945.[1] Thereafter we saw the development of the geostationary systems which dominated the provision of world satellite telecommunications, and will continue to do so for many purposes, particularly television, albeit that the Low Earth Orbit (LEO) systems recently proposed will have their effect on telephony. The US Communications Satellite Corporation (COMSAT) was created, and INTELSAT, INMARSAT and the regional systems established first in interim and then in their definitive forms.[2]

B. The International Legal Context

The international legal context for these developments was a mixture. When the Communications Satellite Act of 1962, the authority for COMSAT, was enacted all that formally existed at the international level was a couple of UN Resolutions. The UN Committee on the Peaceful Uses of Outer Space had been busy since its inception, but was still a year away from presenting what became the Declaration of Legal Principles Governing the Activities of States in the Exploration and Use of Outer Space (1963 GA Res. 1962 (XVIII)). The Outer Space Treaty itself was five years in the future.

Nonetheless, parts of the two relevant UN Resolutions had expressed the view of the members of the United Nations on space telecommunications matters. In 1961 it was the UN view that 'communication by means of satellite should be available to the nations of the world as soon as practicable on a global and nondiscriminatory basis'.[3] Part E of the space resolution of the next

----

year, 1962, states *inter alia* the belief of the UN that 'communication by satellites offers great benefits to mankind, as it will permit the expansion of radio, telephone and television transmissions, including the broadcast of united nations activities, thus facilitating contact along the peoples of the world.'[4]

We should also note that the 1961 Resolution expressed in its Part A its commendation of the applicability of international law to outer space matters, and the principle that outer space and celestial bodies are free for exploration and use by all States in conformity with international law and are not subject to national appropriation.

The 1963 Declaration of Legal Principles Governing the Activities of States Pertaining to the Exploration and use of Outer Space (1963 GA Res. 1962 (XVIII)) took matters further. Of course it contains expression of many principles later more formally to be included in the Outer Space Treaty of 1967, which we are celebrating at this Colloquium. But I would stress the importance of the UN Resolutions of 1961, 1962 and 1963 as showing that the Principles expressed in treaty form in 1967 have thicker roots than the average international treaty.

But to go back a bit in time. The Declaration of Policy and Purpose that is sec. 102 of the US Communications Satellite Act of 1962 reflects some but not all of the aspirations of the UN for the new modality. 'Non-discriminatory access' is provided for, but for 'authorised users' (sec. 102 (c)). The provision of services to 'economically less developed countries and areas' is also intended as is the extension of the system 'as promptly as possible' to provide global coverage at the earliest practicable date' (sec. 102 (b)). As I have shown elsewhere, what was in the contemplation of many was a US owned system.[5] However, following negotiations Interim INTELSAT was established by the two linked agreements of 1964, and INTELSAT was eventually constituted in its definitive form in 1973. The 1964 documents show traces of the early UN Resolutions, while the 1973 arrangements were entered into in the light of the 1967 Treaty and other developments within international telecommunications law. INMARSAT followed a similar pattern.

But by 1963, however, telecommunications had disappeared as a separate matter for mention in the UN Resolution of that year, nor does the matter receive mention in the 1967 Treaty. In part this is because by 1963 the International Telecommunication Union had clearly taken space matters under its jurisdiction, and (perhaps) the UN had realised that it really was not technically competent to pronounce anything other than platitudes on such matters.

Even before UN Resolution 1721 the 1959 Administrative Radio Conference of the ITU (held along with the Plenipotentiary Conference which revised the ITU Convention) had defined a 'space service' and 'earth-space service'. However, its allocations of radio spectrum for space purposes made in paras 70 and 71 of the then Radio Regulations were directed to space research, not to telecommunications purposes. That development starts with the 1963 World Extraordinary Administrative Radio Conference. There frequencies were allocated, and much of the present set of space services were envisaged and started on their way with spectrum allocation and special procedures tailored more to the requirements of the new developments than were the previous terrestrial service procedures. The ITU was therefore doing what the UN had suggested it should do in Part D of Resolution 1721 of 1961. The ITU has, of course, continued to make progress. The Constitution and Convention of the ITU deal explicitly with space matters, and, of course the Table of Allocations and the procedures for notification and registration laid out in the Radio Regulations are crucial elements in the operation of space communications.

The other element in the disappearance of telecommunications as a matter for separate mention in the Principles Resolution of 1963 and the resultant 1967 Outer Space Treaty may be that it was thought that the language of 'freedom of use' and of 'benefit' was sufficient to cope with the matter of space communications. The generality of the words is capable of the specific reference. Perhaps someone who knows will inform me.

Whatever the case, the fact is that the Principles Resolution of 1963 and the Outer Space Treaty itself four years later make no explicit mention of telecommunications. The question should therefore be asked whether the time has come to change that position. Should there be an amendment to the Treaty, as envisaged by art. 15, to meet some of the worries which I am about to express, and the difficulties which the law of international space telecommunications is encountering?

We must first list these matters before turning to whether an amendment to the Outer Space Treaty would help. Some of these I have tagged in earlier papers for the IISL, Colloquia and would refer to those for a fuller statement of the matters.

C. Problems

1. First come, first served or engineered spectrum?

Traditionally the position within the ITU structure is that an assignment of a frequency which is in conformity with the Table of Allocations is entered in the Master International Frequency Register provided that there is no prior registered frequency to which it may cause interference. Later notified assignments have to be coordinated with earlier, thus giving a degree of protection to the earlier. With modifications in detail but not principle, the procedure applies to both terrestrial and to space services.

This system is known colloquially as the 'first come, first served', priority being given in order of notification. It applies not only to radio frequencies, but, in relation to space services also has application in the matter of geostationary orbital positions (slots).

The protection thus afforded to registered assignments and slots has caused concern among the newer countries who are anxious lest the most useful frequencies and slots be entirely taken up before they have the finance or opportunity to enter the fray.

Some have suggested that instead of the 'first come, first served' system it would be more sensible, and arguably more just, formally to agree a division of spectrum use in advance of actual use. The counter argument is that it would be inefficient not to make use of frequencies simply because a state having a right to a slots and appropriate frequencies lacks the capacity to make use of it.

In fact, as far as space broadcasting is concerned, the ITU has 'engineered' both frequencies and slots for satellite broadcasting, the World Administrative Radio Conference of 1985-88 allocating every country in the world a position within an orbital arc and up-link and down-link frequencies to go with it for these purposes. Telecommunications systems are not covered.

That said, the argument is not yet finished. My own preference is for the 'first come, first served' approach as making for a more efficient use of the radio spectrum and orbital positions. However, the problem of the late-comer must not be overlooked, or unduly prejudiced. Art. 1 of the Outer Space Treaty requires that all countries shall share in the benefits of space. It does not provide that these shares shall be equal, or proportionate. But surely a share must mean some minimum value in the matter. This must be defended.[6]

2. ITU and COPUOS Voting : Power without Responsibility

The problem of the different competences of the space-faring and the less-developed countries shows up also in the making of decisions. In the ITU the absolute minimum class of financial contribution is 1/640th of the maximum, yet each member state has one vote. This led *inter alia* to the constitution of the Development Sector as one of the three new Sectors into which the work of the ITU was divided in the revisions to the ITU structures agreed at Geneva in 1992.[7] The creation of the Development Sector is a departure from previous ITU concentration on technical matters. While one can have some sympathy with the desire of the less-developed nations to see the ITU channel expertise and technology to them, it is not what the ITU is basically about, and, in my view should have been left to other UN development organisations. In COPUOS, and later in the UN, we have seen the problem caused by a departure from the principle of consensus and the forcing through of the Principles regarding Direct Broadcasting in 1982.

The reason for the use of voting power unaccompanied by financial burden is the natural desire of the less-developed world to progress. The problem is that it threatens the whole institutional structures that are involved, especially as they were originally designed for quite different purposes, and makes other progress on technical matters difficult. The developed states are, naturally, not willing freely to pass on their assets and the fruit of their endeavour and investment.

3. Abuse of procedures

The procedures of the ITU were intended to maximise the efficient use of the radio spectrum and geostationary orbital resource. They have been used and arguably abused for other purposes. Thus, in 1992 the Kingdom of Tonga filed assignments for 31 geostationary slots, with the then International Frequency Registration Board (IFRB)[8] - far more than it needed for its own domestic or international telecommunications requirements - and claimed them on the 'first come, first served' basis.[9] It was not going to operate those slots to meet its own telecommunications needs.

The then International Frequency Registration Board (IFRB) asked it to justify its claims, and did persuade Tonga to reduce their number from thirty-one to six. These were duly entered on the ITU Register. Since then Tonga rented one position to a US company based in Colorado. It also bought and moved two former

Soviet satellites into two of the slots. Finally it auctioned the remaining two to other enterprises, leaving one slot unused.

Apparently Tonga claimed to have done nothing illegal. But the claim 'nothing illegal' is a defensive position. It might be argued that this is space being used for the benefit of all, and for the benefit indeed of a less developed country. If so, it is certainly not a use of space in contemplation of those who hammered out the ITU arrangements.

4. Non-compliance with Procedures

In recent years there have been instances of non-compliance with ITU procedures. Two in particular call for comment: the action of Indonesia in relation to the Tonga affair, and the instance of the Apstar launch.

In 1993 when the Tonga matter referred to above was starting, Indonesia moved one of the PALAPA satellites into one of the slots claimed by Tonga on the ground that the assignment of that slot to Tonga was wrong in law. This matter was negotiated to settlement between Tonga and Indonesia in November 1993. But that departure from procedures is unwelcome. The fear of many countries is precisely that those who can launch satellites will launch them into slots suitable for themselves, to the detriment of those who might come later. Indonesia's actions, going outwith normal procedures, foster such fears.

In the China instance, on 21 July 1994 Apstar-1 was launched by the Chinese Long March system and was scheduled to start operation from 1 September 1994 from a position at 131°E. That position is 1° away from satellites belonging to Japan and to Tonga, by then registered with the ITU. That Tongan slot is occupied by a Russian Gorizont satellite, Rimsat-1, operated by Rimsat Ltd, a company of Fort Wayne, Indiana, USA. The Japanese slot is used by the Telecommunications Advancement Organisation of Japan, which operates a CS-3A satellite for various Japanese companies and government bodies. The Apstar satellite is owned by APT Satellite of Hong Kong, and is sponsored by the Chinese government, although offering service to various organisations such as Turner Broadcasting, Time Warner and Viacom International.

Neither China nor Hong Kong had taken the matter through ITU procedures, and when I first wrote of this negotiations were at present under way on the question of radio interference between the three satellites.[10] I am not aware of what has happened.

China's action in disregarding the international procedures is a threat to the stability of the system which has been developed by the ITU. Unless the system is adhered to by all parties, it could very soon fragment.

5. Phantom Satellites

A further instance of abuse of ITU procedures is the case of the phantom satellites, that is notifications made by states of planned space systems which are at a very embryonic stage (and may well not be proceeded with at all) so as to secure a priority position within the 'first-come, first served' system. This was the subject of my paper to the Beijing Colloquium in 1996.[11] The matter has not significantly improved. The Director of the Radiocommunication Bureau recently indicated the extent of the problem in a fascinating document. By mid-1996 nearly 1800 space networks were in the ITU processes for advance publication and coordination. Of these some 1500 notified by 54 Administrations (including on behalf of 6 international satellite organisations) were planned to use the most congested space frequency bands - the C, Ku and Ka bands. Twenty four Administrations had notified less than 10 systems. Ten had notified between 10 and 20 each, eight between 20 and 30 each, five between 30 and 50 each, four between 50 and 100 each, and two had notified between 100 and 200 each. One Administration had notified over 300 systems.[12]

There are various effects of this problem. First, of course, the whole ITU system has been slowed down. A huge amount of resource, both financial and human, is required to process such a volume of submissions. Second, as indicated above, there is a duty of coordination between the several systems, the onus basically lying on the later to approach the earlier. Telecommunications administrations are therefore also overworked by the need to coordinate with earlier notified systems. And in both the ITU and in states, there is a huge waste, because, as said, many of the notified systems seem not to be likely to eventuate in working systems. Notifications are made to 'book' spectrum and orbital space, and in effect to claim such before there is an actual system. The fact is that some orbits and some spectrum space are better than others for particular purposes. Orbital positions which can 'see' continental America, or trans-Pacific, or trans-Atlantic, or Europe, are at a premium. If a state can deter others from making proposals to use such positions because it has got its notice in first, and the later state has not the personnel or assets to engage in long negotiations, it appears it will do so. The processing of these 'phantom' systems within the ITU and state procedures is therefore a complete waste of resource.

6. Lack of Supervision of Activities

Another conundrum which emerges from the various developments mentioned above is whether all states are equally able to discharge their obligations properly to supervise the space activities which are engaged in either by themselves, or by those whom they license to act. Questions have been asked in relation to the 'phantom satellite' problem as to whether all notifications to the ITU have been properly screened by the state authorities making the submission.[13] It would be interesting to have the opinion of the staff of the ITU Radiocommunication Sector as to the competence in radio and space matters of the state officials in many of the countries with which they have to deal. One wonders whether there is any correlation between the class of contribution of an ITU Member state, and the capabilities of that state in conducting or supervising space activities.

7. Lack of Supervision of Nationals: the 'appropriate state' question

Under Art. VI of the Outer Space Treaty states bear international activities for national activities in outer space, whether the activity is carried out by governmental agency or non-governmental entities. These last require authorization and continuing supervision by the 'appropriate state'. Under the Liability Convention, of course, the liability of a launcher state is involved, but how might another state come to be involved other than is a damages question? Is the formula of Art. VI not indicative of a state responsibility and duty for companies whose nationality they possess? Or can companies of one state be excused from supervision by their home state on the ground that what they are doing is authorised by another?

The UK would seem to consider that nationality is a link within the terms of the requirement of Art. VI. Thus, as a catch-all, and subject to exception, s. 3 of the UK Outer Space Act 1986 which implements the UK obligations under various of the UN space treaties, prohibits anyone through whose activities the international liability of the UK might be engaged (other than the UK itself) from procuring or carrying on any space activities without both obtaining and complying with the terms of the licence (s.3(1)). By s.2 of the Act, those required to obtain a license are UK nationals, Scottish firms and bodies incorporated under UK law, the term "UK national" including British citizens, the citizens of British Dependent Territories, British Nationals (Overseas), British Overseas citizens, British subjects under the 1981 British Nationality Act and persons who are British protected persons in terms of that Act (s.2(1)(2)), and that list may be extended (s.2(3)).

However, under s. 3(2)(b) of the 1986 Act the Secretary of State may exempt a person from the licensing requirement if satisfied that it is not needed to comply with the UK international obligations (s.3(3)). And most importantly, as noted in relation to the activities to be licensed, a UK licence is not required if other arrangements have been made between the UK and another country to secure compliance with UK obligations. In other words the UK legislation does require that the supervision of space activities by UK nationals is real. How many other states make such provision in their legislation?

What is practice in the case of companies incorporated in one state, but which are licensed by other states to conduct space activities on their behalf? Mention has been made above of US companies involved in activities licensed by the Kingdom of Tonga. This is the problem of the definition of the 'appropriate state' which has been discussed before in these Colloquia.[14] It is time that the matter was authoritatively clarified. There should be an express duty imposed on both a state licensing an activity and the state whose corporate national or non-governmental entity is involved in such activities by licence, franchise, hire or other commercial arrangement.[15] This could go beyond the question of telecommunications, but within that sphere would include a clear statement of an obligation on the home state of a telecommunications corporation to exercise proper supervision of the conduct of its nationals even when they are active outwith the formal territorial jurisdiction of the state and licensed to act by others. This might also help reduce qualms about the supervisory capacities of some states which may be tempted to enter the space business by proxy, as it were.

8. Privatisation and Competition

I would here do no more than point to the privatisation of national telecommunications services in many countries, to the establishment of many other service providers and the opening up of competition within countries. Pressures towards privatisation and competition have had their effect also on the international organisations which were created to provide the global non-discriminatory service referred to in both the UN Resolutions and their constituent treaties.[16] Within the European Union also there has been significant progress towards the fostering of competition and freedom for telecommunications service providers to operate within the Union, with little regard to traditional territorial jurisdictions.

9. World Trade in Telecommunication Services

Last, and inevitably related to the above, I must, however, mention as something separate the discussions that are going on within the World Trade Organisation *fora* as to Telecommunications Services. These matters were left unresolved by the Uruguay Round of the General Agreement on Tariffs and Trade (GATT) in 1994. The New Fourth Protocol to the General Agreement on Trade and Services done at Geneva on 15 February 1997,[17] significantly commits its parties towards permitting freedom. I do not want to go into detail on this Agreement: suffice it to say it is a step towards the dismantling of protectionism in the provision of telecommunications services.[18]

Taken with the matters indicated under 8.. above, one can see that the world is increasingly being considered as a single marketplace within which deregulations and freedom of competition is to be allowed to mould developments.

D. Resultant Concern?

In the light of the foregoing, the concern which I would draw attention to is that amid the technical and economics developments indicated, the goal of global non-discriminatory telecommunications services may be lost, or at least considerably obscured. If unopposed, the object of profit inexorably means one of two things. Either all telecommunication services will be required to be cost-effective and profit-making, with low density traffic routes becoming very expensive, or low density traffic routes will be neglected and service either not provided, or given insufficient investment to deliver an efficient, reliable and satisfactory facility. What we might call 'public utility' services may cease to be provided either completely, or for some areas of the globe, and the telecommunications needs of small-market countries may well be under-serviced.

Now, I would not lightly accede to the more clamorous demands for aid, technology transfer and the like. The history of the last few decades has produced shocking examples of waste, profligacy and the depredations of the kleptocracy. Nonetheless, there is a balance to be kept. The aspirations of the two early UN Resolutions referred to above are right. The Outer Space Treaty has the right emphasis. There should be a global telecommunications network providing service to all countries irrespective of their degree of economic or technological development.

That there are these and other emergent problems shows that there is something wrong. It raises questions as to whether treaty provisions are being dealt with in good faith, as is required by art. 26 of the Vienna Convention on the Law of Treaties,[19] a provision which most take to be declarative of customary international law, not only constitutive of a treaty-law principle.[20]

The other point to make here is that not all the indicated problems (or the others that can be thought of) are such as can be tackled by amendment to the Outer Space Treaty. Some of these problems are mirrored almost exactly in the concerns that have arisen internally within states as they have engaged in privatisation. Often the solution has been the creation of a regulatory body with appropriate powers to ensure competition. no abuse of market dominance, and the continuation of public utility services. Elsewhere I ave contemplated the development of the International Telecommunication Union into a World Communications Commission.[21] Perhaps that is a way forward to deal with some such problems.

E. An Amendment to the Treaty?

But what is to be done? Amend the Outer Space Treaty, or draft something new?

I have to say that as of September 1997 I am not sure. Some of the difficulties with space telecommunications are solvable by a return to a good faith observance of international obligations, coupled with a willingness by the ITU to flex its muscles a little.[22] Part of the solution is simply to be found in a genuine application of the principles that have already been agreed.

Need we go further? One problem with an amendment to the Outer Space Treaty is that it may well not suit all existing members of the Treaty. That could produce a 'limping' treaty, with not all members being bound by the same obligations. More importantly it might provoke some to make more extreme proposals for amendments, with the possibility that other states might consider that their interests are better served by using the provisions of art. 16 to withdrawal from the Treaty and rely on the more inchoate principles as to the use of outer space which most of us would say were now part of customary international law. Treaties are not legislation. They are agreement. They can go beyond the principles of customary international law. The Outer Space Treaty is a magnificent tree. I doubt whether its terms would be arrived at were its negotiation to be started in the present day, and I am nervous of shaking its trunk lest too many useful but not entirely well-secured fruits of the Treaty should drop to the ground. Notwithstanding recent developments,[23] the debate between the space-faring nations and others as to the meaning and application of the phrases of Art. I of the Treaty as to the exploration and use of outer space being carried out for the 'benefit and in the interests of all countries irrespective of their degree of economic or scientific development', and the horrible example of the failure of consensus and the resultant Principles on Direct Television Broadcasting do not reassure me. I am wary of reopening the Treaty by proposals for its amendment.

On the other hand it would be no bad thing to make states responsible for the activities of their corporate nationals, irrespective of whether these nationals are being licensed by another state to act on its behalf. And it would be no bad thing to incorporate some of the tenor of the recent Resolution on the 'common benefit' concept into the Treaty, although what is a 'mutually acceptable basis' is obscure to me.

But, as I end I find myself coming back to this. The problems I have identified above are not fundamental problems of the international agreements on the matter. They are problems of their implementation by states. Businessmen must be made to realise that international agreements are not merely inconvenient regulations if possible to be by-passed, disregarded and ignored. Governments must be governments, and upholders of principle, not agents of profit-seekers. The last sentence of Para 1 of the 1996 Declaration says that in the exploration and use of outer space 'particular account should be taken of the needs of developing countries'.[24] I would suggest that the interests and needs of the world as a global community also need to be stressed.

Perhaps I am calling for a religious conversion rather than a legal modulation.

NOTES

1 Arthur C. Clarke, 'Extra-terrestrial Relays. Can Rocket Stations Give World-Wide Radio Coverage?', *Wireless World*, October 1945, 303-8.

2 See F. Lyall, *Law and Space Telecommunications* (Aldershot: Dartmouth Publishing; Gower Press: Brookfield VT, 1989), chs. 2-7.

3 'International Cooperation in the Peaceful Uses of Outer Space', 1961 GA Res. 1721 (XVI) Part D.

4 'International Cooperation in the Peaceful Uses of Outer Space' 1962 GA Res. 1802 (XVII).

5 F. Lyall, above n.2, 74-9.

6 Cf. The Declaration on International Cooperation in the Exploration and Peaceful Use of Outer Space for the Benefit and in the Interest of All States, Taking into Particular Account the Needs of Developing Countries' GA Res. 51/122, 4 February 1997; cf. also M. Benko and K. Schrogl, 'The 1996 UN Declaration on "Space Benefits" Ending the North-South Debate on Space Cooperation' (1996) 39 *Proc. IISL,* 183-6.

7 F. Lyall, 'The International Telecommunication Union Reconstructed' (1994) 36 *Proc. IISL*, 78-88.

8 The duties of the former International Frequency Registration Board are now performed by the Radiocommunication Sector of the Union.

9 See M.L. Smith, 'Legal and Policy Developments in International Satellite Communication' 1991 34 *Proc. IISL,* 342-7 at 345-6; D. Riddick, 'Why does Tonga own Outer Space?', 19 *Air and Space Law*, 15-29. Cf. *Space News* vol. 4, no. 11, March 15-21, 1993 22; vol. 5, no. 4, January 24-30, 1994, 3 and 29; no. 25, June 21-27, 8.

10 See *Financial Times* (London), 2 August 1994, 1 and 10; *Space News*, Vol. 5, no. 29, July 25-31, 1994, 3 and 20; no. 31, August 8-14, 1994, 18.

11 F. Lyall, 'Paralysis by Phantom: Problems of the ITU Filing Procedures' (1996) 39 *Proc. IISL,* 187-93.

12 Some Aspects of Resolution 18 Issues, ITU Docs SC-RG4/037 and SC-RG5/032.

13 Paper cited above, n.11.

14 On the 'appropriate state' question see recently: K-H Bocksteigel, 'The Term 'Appropriate State' in International Space Law' (1994) 37 *Proc. IISL,* 77-9; W.B. Wirin, 'Practical Implications of Launching State - Appropriate State Definitions' (1994) 37 *Proc. IISL,* 109-15; G. Silvertrov, 'On the Notion of "Appropriate State" in Article VI of the Outer Space Treaty' (1991) 34 *Proc. IISL,* 326-30; and, H.L. van Traa-Engleman, 'Problems of State Responsibility in International Space Law' (1983) 26 *Proc. IISL,* 139-42.

15 Steps would also need to be taken so that the veil of incorporation and the device of multiple contracts and jurisdiction flitting can not be used to insulate the state of nationality of the ultimate owners of the corporation from this duty.

16 D. Wear, 'INTELSAT: Evolving to Meet the Challenges of a New International Telecommunications Marketplace' (1995) 38 *Proc. IISL,* 123-33.; A. Auckenthaler, 'Recent Developments at INMARSAT' (1995) 38 *Proc. IISL,* 149-59; C. Roisse, 'Recent Developments at EUTELSAT' (1995) 38 *Proc. IISL,* 160-7; F. Lyall, 'Privatisation and International Telecommunications Organisations' (1995) 38 *Proc. IISL,* 168-74.

17 World Trade Organisation Agreement on Telecommunications Services (Fourth Protocol to the General Agreement on Trade in Services) (1997) 36 *International Legal Materials*, 354-95.

18 I note that the Commission of the European Union has objected to the way in which the US proposes to implement the Agreement on Telecommunications Services under its national law; *Financial Times* (London) 6/7 September 1997, 4.

19 Convention on the Law of Treaties, Vienna, 23 May 1969, 1155 UNTS 331; (1980) UKTS 7964; (1969) 8 ILM 679; (1969) 63 AJIL 875.

20 The US has not ratified the Vienna Convention on the Law of Treaties.

21 F. Lyall, 'The International Telecommunication Union: A World Communications Commission?' (1994) 37 *Proc. IISL,* 42-47.

22 Cf. F. Lyall, ''Paralysis by Phantom: Problems of the ITU Filing Procedures', cited above n.11.

23 Cf. n.6 above.

24 See n.6 above.

IISL-97-4.17

SPACE TELECOMMUNICATIONS AND THE INTERNET: IMPLICATIONS FOR THE OUTER SPACE TREATY

Larry F. Martinez, Ph.D.*/
Department of Political Science
California State University
Long Beach, CA 90840-4605 USA
http://www.csulb.edu/~martinez

Paper Submitted to the
40th IISL Colloquium
Torino, Italy October 6-10, 1997

Abstract

The speedily expanding Internet is in the process of transforming the technological, economic, and policy bases for nation-state regulation of telecommunications, including space-based satellite networks. Deployment of the packet-switched Internet has accelerated the liberalization of telecommunications markets and has led to far-reaching regulatory restructuring and policy shifts regarding state ownership and control of networks and information flows. As space-based GMPCS networks become integral parts of the globalizing Internet infrastructure, the state-centric legal paradigm requiring state "authorization and continuing supervision" of space activities by "non-governmental entities" stipulated under Article VI of the OST and associated treaties forming the outer space legal regime will be called increasingly into question.[1] This paper examines the technological, economic/trade, and security issues that question whether the existing state-centric paradigm for regulating Internet-based GMPCS satellite systems will remain in legal phase with emerging liberalized regulatory regimes for terrestrial Internet-based infrastructures.

Introduction

Two milestones in the liberalization of world telecommunications markets have been passed during the last twelve months with far-reaching implications for international space law. In October 1996, the ITU's World Telecommunications Policy Forum adopted the principles for a Memorandum of Understanding (MOU) facilitating the use and transport of Global Mobile Personal Communications by Satellite (GMPCS) satellite terminal handsets over national borders. Among the GMPCS systems being deployed are those which will extend the Internet into low-earth orbit constellations of satellites which will interconnect seamlessly with the terrestrial Internet. In February 1997, the World Trade Organization concluded agreements liberalizing significant sectors of world telecommunications markets, providing an additional impetus for an even speedier expansion of the Internet. These

*/

set the stage for worldwide deployment of the Internet while eroding the technological, economic, and political foundations underlying the legal framework for nation-state regulation of low-earth orbit satellite communications. The goal of this paper is to stimulate discussion of the challenges Internet-based GMPCS systems pose to the Outer Space Treaty's (OST) state-centric legal regime for satellite communications, especially as it pertains to the Article VI assignment of state "authorization and continuing supervision."[2]

- **Technology Factors: Internet-based GMPCS satellite network architectures make compliance with OST Article VI stipulations for state "authorization and continuing supervision" increasingly difficult.**

System Architecture of the Internet (Network of Networks)[3]

Technology is the flipside of infrastructure economics and regulation. International law, originating during eras of agricultural and industrial primacy, acknowledged the state's preeminent legal status to control its technological infrastructure (postal services, telegraph, telephone and spectrum-based broadcasting and wireless telecommunications networks) as essential elements of state sovereignty and later, economic necessity. The parameters of state-centric international law congruently matched the technological and jurisdictional boundaries of governmental monopolists and their licensed natural monopoly infrastructures as evidenced in the OST and conventions governing the International Telecommunication Union (ITU) and other intergovernmental organizations focusing on state monopolist infrastructures (i.e., postal services, broadcasting, airlines, etc.).

This "natural monopoly" jurisdictional congruence is most clearly seen in the system architectures of geostationary satellite systems (GEOs) during the early decades of their development. GEO systems required massive front-end investments in multi-million dollar antennas, satellites, and associated equipment with the technical constraints inherent with *analog* signal processing technology (propagation, modulation, and channel access) which resulted in satellite networks operating essentially as "cables in the sky," network configurations that closely resembled existing terrestrial and submarine cable infrastructures and their concomitant natural monopoly regulatory regimes.

On a deeper technological level of analysis, analog signal processing requires the network operator to provide a discrete signal pathway (i.e., a circuit) for each pair or set of communicators. Control rested with the network provider who performed the circuit switching function within a circuit-based regulatory regime. Compliance with the OST Article VI stipulation requiring state "authorization and continuing supervision" of space-based GEO satellite circuits was straightforward as the state and the government monopolist network operator were in most cases the same entity and closely matched the network architecture. This was to radically change with the advent of digital signal processing techniques, most notable of which was the innovative packet-switched network architecture which has eroded the natural monopoly characteristics of telecommunications networks in general, and satellite communications in particular.

The Internet is a Packet-Switched Network

The origins and functioning of the Internet are shrouded in the near-myth-like veils of its Cold War inspiration, leading to the widespread belief the Internet works because it was designed to survive nuclear war. In 1964, California-based RAND Corporation researcher Paul Baran and British researcher Donald Davies independently conceived of a rather surprising solution to the military's problem of network survivability: assume that each signal's pathway would fail and build accordingly. The Baran-Davies' brainstorms split each message into sequentially numbered blocks (which Davies called *packets*), each of which would have an electronic address label slapped on by the sending computer and sent on their way. The sending computer would transmit each packet down the next available empty circuit, whereupon the packet would bounce from computer to computer using otherwise empty pathways until arriving at the destination computer. The receiving computer would then use the sequential address labels to reassemble the packets back into the original message. If message packets were missing or damaged because of pathway breakdowns, it would be a relatively simple matter to have the sending computer retransmit those packets using an alternative roadmap to the destination. [4]

The *digital* packet-switching proposal was blasphemous to the entrenched circuit-based theology rooted in the orthodoxy required by slow *analog* mechanical telephone switching centers that were geared to human conversations, not computer data bursts. The packet-switching thesis was Copernican in its eventual effect on the development of computers, far beyond its nuclear war inspiration of system survivability. Because a packet-switching system by design used whichever circuit was empty at the time a packet's transmission between computers, it achieved far higher communication efficiencies and economies to the great efficiency and cost benefit of the computer users. By sharing computer resources across a packet-switched network, they could get more work done, more quickly and efficiently. The economic and technological basis for state-operated natural monopolies has quickly eroded in the face of the Internet's explosive growth.

Analog to Digital: Convergence and Boundary Blurring

The Internet represents a major milestone in the transition from an analog-based telecommunications infrastructure to its digital successor. The vulnerability of analog transmissions to interference from unwanted electrical signals, required discrete communications pathways; hence, telephone, radio and television developed within well-defined technological and jurisdictional boundaries. *In the analog world, control rested with the network operator*. In earlier analog eras, a governmental regulatory authority ensured that a nation's telephone switches would interconnect two users with an interference-free (often "physical") synchronous electronic circuit between their telephone instruments. In contrast to analog network architectures, the Internet's digital packet-switched network architecture puts telephone, radio, and television on the same digital pipe, encoded onto data packets which are sent everywhere at the same time. Most significantly, in the asynchronous packet-switched Internet environment, the network merges with the computer, and *control shifts to the user.*[5]

In essence, in the totally-digital Internet environment, the network become just another part of the computer's interface with another component; the "Internet" looks like a disk drive. "Closed" analog systems are becoming "open" or "transparent" digital systems where network control is dispersed among users of the network(s). In contrast to the conventional analog network, where the "smarts" of the network resided inside the telephone central switch (and therefore with the monopolistic network operator), today's computer consumer operates an appliance that rivals or supercedes slower adapting military or governmental computer networks. For example, on March 20, 1997, National Public Radio reported that a computer consulting group in the San Francisco area had managed to "crack" the new digital cellular encryption codes using off-the-shelf Intel pentium-class processors.[6]

In sum, "the Internet model, rather than the older hierarchical model of the circuit-switched telephone system" is becoming the basis for the telecommunications infrastructure of the 21st Century. As the technology-mandated circuit boundaries between telephone, radio, and television disappear, so too does the politically-mandated regulatory justification for a monopolistic network provider.[7]

Observations: Internet-based GMPCS Technology and the OST

GMPCS, the next generation of communications satellites, is moving control over information flows one step further away from state purview. While Motorola's *Iridium* narrowband system is already being deployed in orbit, a great deal of attention is directed at Microsoft's *Teledesic* projects,[8] a proposal place more than 600 satellites into low earth orbits (LEO). Satellite Internet users construct their own networks with the click of a computer mouse, bypassing domestic telephone networks completely as they communicate voice, data, or video on the Internet, from anywhere to anywhere.[8] In short, a state's ability to fulfill the OST requirement for state responsibility and control is eroding fast as digital Internet packet-switched networks replace their earlier analog infrastructures. If pathways change according to the whims of a packet-switched network router, no one can define the network's boundaries at any point in time. Furthermore, users will not even be aware that they may be using a space-based communications pathway and that their use may be subject a wholly different legal regime. The Internet's packet switching technology and the GMPCS system architectures are obsoleting traditional analog-era state monopolists and the clear-cut congruence between the boundaries of their national networks and the grant of monopoly control by the state and recognized by the OST.

- ❑ **Economics/Trade Factors: Liberalization of world telecommunications markets and services is making state compliance with OST Article VI provisions increasingly problematical with respect to Internet-based GMPCS systems.**

States or their legally-licensed monopolists have operated railroads, postal systems, power grids, and telephone/broadcasting infrastructures for many decades. The market and the transactions for many of these services took place entirely within the territorial boundaries of the nation-state. Today, traditional arguments underlying these arrangement based upon economies of scope and scale are falling by the wayside as states privatize and liberalize large formerly public sectors. In addition, market transactions themselves are no longer taking place within the purview of the nation-state. The boundaries for what once was considered a domestic or foreign transactions are blurring owing to the massive flows of capital along global data networks. The Internet is already accelerating the shift of transactions for a growing range of services out of the territorial boundaries of states into an as yet unclaimed regions of cyberspace.

Privatization to Liberalization:

What is the appropriate role for the state? In this era of far-reaching alterations to both the physical and mental landscapes of what used to be familiar vistas toward the public good or political legitimacy, the issue of how or even whether governments should regulate the telecommunications/information sector is increasingly the subject of national and international debate. Two issues stand out: privatization and liberalization. *Privatization* is the process of transferring ownership of what were usually government-owned telephone and broadcast industries to private stockholders. The transfer of ownership may mean a transfer of investment, profit, and risk, but not an actual transfer of regulatory control. That may still lie with governmental ministries who may, by regulatory fiat, establish service areas, prices, and market structure. *Liberalization* refers to a reduction in government's jurisdictional competence in regulating what firms may or may not do within their allowed market or service areas. Hence, liberalization connotes a change in control and a lessened role for the state in both OST Article VI areas of authorization and continuing supervision.

Even in cases where there is some market access afforded competitors, the governmental monopolist as the initial or dominant provider benefits from the high entry costs new entrants must bear to install facilities to each home or business. Electrical power and natural gas distribution networks are generally operated as public utilities due to their natural monopoly characteristics. In these cases of natural monopoly, the public interest is promoted through authorization of a provider within a regulatory regime to monitor pricing and costing practices so that all users were able to access the network for essentially the same price regardless of the actual cost (i.e., universal service at uniform (equitable) pricing). Today, however, there is a growing acceptance that the natural monopoly argument against telecommunications privatization and/or liberalization was valid only for the economies of scope and scale present for *analog* technologies operating under conditions of network resource scarcity. As communications infrastructures shift to digital computerized networks, so-called channel scarcity and arguments for natural monopoly are falling as quickly as the prices for computing power.

Convergence

The analog world is one of channel scarcity owing to the discrete and different pathways each signal travels. As each new technology came into its new market, it used a different pathway to its users. Hence, telephones initially used a massive grid of wires and switches; radio and television an invisible set of pathways through the airwaves. In each case, to sustain the public good aspects of universal service at uniform prices, market entry was prohibited against potential "cream skimmers" who would otherwise attempt to take advantage of the natural monopolist's requirement to cross subsidize access for high cost users by charging rates significantly above actual costs to low cost users. Scarcity of channels, whether telephone lines or airwaves, reinforced the strictures of natural monopoly regulation to prevent market entry to potential cream-skimmers.[8]

Intergovernmental agreements establishing the de facto natural monopoly satellite networks owned and operated by state monopoly providers, such as INTELSAT or INMARSAT, could, in effect, ensure compliance with OST Article VI stipulations through monopoly protection mechanisms such as the INTELSAT Agreement Article XIV(d).[9] Legal scholar Carl Christol argued that the 1967 Outer Space Treaty and subsequent legal instruments "prevent juridical person, other than States - and in particular international intergovernmental organizations - from claiming

exclusive operational, as well as management rights, which are denied to States."[10]
However, in the digital era states are attempting to claim control over computer information in its digital form, ones and zeros, which may well travel down the same network pipe, be it wire, fiber optic, terrestrial cells, or satellite - constituting a massive volume of services and information flows - that defy any attempt to block market entry *once the pipe is in place*.

The "social contract" between governmental network monopolists and users which required "authorization and continuing supervision" of entrants into a natural monopoly market has fallen to the tremendous market synergy between personal computers, computer networks, and market liberalization fueling the explosion of the Internet in the early 1990s, resulting in 1996 Telecommunications Reform Act.[11] Analogous (so to speak!) developments are taking place in growing numbers countries liberalizing their financial and communications infrastructures.

- ❑ **Political Factors: National Security, Internet-based GMPCS, and Legal Boundaries**

The 1990-91 Persian Gulf War fundamentally altered perceptions about the nature of warfare and national security in the information age. As Nye and Owens observe in a 1996 *Foreign Affairs* article:

> Knowledge, more than ever before, is power. The one country that can best lead the information revolution will be more powerful than any other.[12]

Many credit American information power prowess to its liberalized information and telecommunications regulatory structure. More than it being merely an "open society," profit motivations successfully inspired American data processing, telecommunications, and broadcasting firms to expand and diversify on an unprecedented scale. However, the race to commercialize cyberspace also raises a rather uncomfortable policy and legal issue: If information technology is today a key component of military strategic/tactical power, are states capable of complying with OST Article VI to ensure that competitive activities of commercial companies do not violate international legal precepts banning aggressive uses of power?

Case in point: In June 1996, Israel asked the United States Government to restrict U.S. commercial satellite firms operating remote sensing satellites with 3-meter resolution or better from imaging Israeli and neighboring territory.[13] Israel's request is indicative of the international legal and political issues coming to a boil as information technology makes territory and the international legal principles based on territory increasingly irrelevant. Moreover, the definition of aggressive war or illegal intervention may also be obsolete in cyberspace and hence for the OST-imposed obligations on states attempting to supervise GMPCS systems. Legal cyberscholar Sean Kanuck points out that

> There is a critical distinction between "Information Age warfare," which utilizes new technologies to transform the conduct of war while still pursuing traditional military objectives, and "Information Warfare," which redefines the very nature of international conflict. Information Warfare has officially been defined as follows: Action taken in support of national security strategy to seize and maintain a decisive advantage by attacking an

> adversary's information infrastructure through exploitation, denial, and influence, while protecting friendly information systems.[14]

Or alternatively as:

> Any action to deny, exploit, corrupt, or destroy the enemy's information and its functions; protecting ourselves against those actions; and exploiting our own military information functions.[15]

Kanuck writes:

> Thus, a tripartite legal distinction exists under international law. The first class of observational, data-collection activities are simply subject to domestic regulations. The second tier of activities, pro-active efforts to influence domestic affairs short of armed engagement, are most often violations of domestic law and are also "condemned" by international law. Finally, threats or actual use of force are expressly proscribed by the United Nations Charter as well as customary international law.[16]
>
> Therein lies the critical distinction behind both the future of international conflict and the latent deficiency of public international law. As information evolves into the target itself (i.e., its destruction becomes the veritable end, rather than a means to other military objectives), the entire concept of warfare will be revolutionized.[17]

Hence, what Kanuck is arguing in analogous to the evolution of regulatory structures, i.e., from circuit-based regulation to service-based regulation. A new paradigm of international law is needed to shift from outmoded principles based on territoriality (discreet analog circuits) to one based on the digital paradigm, i.e., services.

> Even the basic military-civilian distinction under the customary laws of war (*jus in bello*) was founded on the theoretical ability to segregate physically those two types of entities in order to determine what hostilities could be perpetrated on each category. Today, it is becoming increasingly difficult to separate fully the military and civilian networks because they utilize many of the same satellites, fiber optics, computer nodes, etc.[18]

Kanuck advocates a long look at the existing prohibitions against intervention in the domestic affairs of another state as the most promising field for developing a usable legal paradigm for cyberwar.

> Interactions, and not physical territory, must become the basis of the new system; only then can aggression and intervention be redefined in terms of undesired effects and not merely the direct or indirect use of armed forces.[19]

In sum, states' ability to comply with the Article VI stipulation requiring state "authorization and continuing supervision" of non-governmental (i.e., commercial) entities utilizing the Internet-based GMPCS may be inadequate in the light of encryption techniques that will further hide who is using a GMPCS system and to what purpose.

Conclusion: Role of the State

The emergence of Internet-based GMPCS communication satellite systems utilizing privately owned terrestrial gateway earth stations, easily transportable handsets, and Internet packet-switching technology, represent a quantum leap in the seemingly inexorable march toward a privately-owned commercial Global Information Infrastructure (GII). As voice, data, and broadcasting services migrate towards an Internet-based infrastructure, traditional governmental jurisdictional boundaries will blur further, making all the more visible the apparent inability or unwillingness of states to comply with the OST Article VI obligations. The OST, promulgated during an era of governmental space programs and telecommunications monopolies, is premised upon technological, economic, and security boundaries of state jurisdiction that may no longer exist.[20] Instead, a new paradigm for managing commons resources may be required that utilizes the self-organizing parameters of the emerging Internet paradigm for international cooperation.[21]

Notes:

1. Article VI of the OST states:
States Parties to the Treaty shall bear international responsibility for national activities in outer space, including the moon and other celestial bodies, whether such activities are carried on by governmental agencies or *by non-governmental entities*, and for assuring that national activities are carried out in conformity with the provisions set forth in the present Treaty. *The activities of non-governmental entities in outer space, ... shall require authorization and continuing supervision by the appropriate State Party to the Treaty....* [emphasis added]

Treaty on the Principles Governing the Activities of States in the Exploration and Use of Outer Space, Including the Moon and other Celestial Bodies, Jan. 27, 1967, 18 U.S.T. 2410, T.I.A.S. No. 6347, 610 U.N.T.S. 205 (effective Oct. 10, 1967) [hereinafter Outer Space Treaty]. The other four treaties treat the questions of space liability, registration of objects, rescue and return of objects and astronauts, and an agreement covering activities on the moon and other solar system celestial bodies: *Convention on International Liability for Damage Caused by Space Objects* (1972) , 24 U.S.T. 2389, T.I.A.S. 7762; The Agreement on the Rescue of Astronauts, the Return of Astronauts, and Objects Launched in Outer Space (1968), 19 U.S.T. 7570, T.I.A.S. 6599; *The Convention on the Registration of Objects Launched into Outer Space* (1976), 28 U.S.T. 695, T.I.A.S. 7762; *The Agreement Governing the Activities of States on the Moon and Other Celestial Bodies* (1979), United Nations General Assembly Resolution 34/68.

2. See, "Memorandum of Understanding on GMPCS now ready for signing - GMPCS continues to make history," *ITU News*, #3 1997, pp. 5-7. Commercialization of space activities has proceeded apace in the areas of space launch vehicles, space remote sensing, space navigation. "WTO's landmark agreement on basic telecommunication services," *ITU News*, #4, 1997, pp. 34-38. While this paper focuses on liberalization of space telecommunication sectors and the GMPCS in particular, readers are urged to refer to articles by Fred Kosmo Note: The Commercialization of Space: a Regulatory Scheme That Promotes Commercial Ventures And International Responsibility. 61 *S. Cal. L. Rev.* 1055 (May, 1988). Source: Nexis-Lexis; see also, Kunihiko Tatsuzawa, "Policy and Law in Space Commercialization," in K. Tatsuzawa (ed.) *Legal Aspects of Space Commercialization*, (Tokyo: CSP Japan, Inc., 1992), pp. 10-31.

3. An earlier version of this analysis was presented in a paper to the Southwest Social Science Association

Conference in New Orleans, March 1997, while a more complete description of the Internet's underlying technology can be found in an article written in 1996 for the United States Information Administration.

4. See, Katie Hafner and Matthew Lyon, *Where Wizards Stay Up Late: The Origins of the Internet*, (New York: Simon and Schuster, 1996), pp. 53-65.

5. See, Richard Jay Solomon, "Telecommunications Technology for the Twenty-first Century," in William J. Drake (ed.), *The New Information Infrastructure: Strategies for U.S. Policy*, (Washington, DC: A Twentieth Century Fund Book), pp. 93-111; also,"GMPCS: The Regulatory Dilemma, ITU, World Telecommunications Policy Forum pamphlet, 21-23 October 1996. Teledesic Web Page: http://www.itu.int/pforum/paper2-e.htm

6. Author's notes from *National Public Radio* "All Things Considered" news broadcast on March, 20, 1997.

7. See, Revised *Report by the Secretary-General: Policy and Regulatory Issues Raised by Global Mobile Personal Communications By Satellite (GMPCS)*, International Telecommunication Union, October 23, 1996.

8. See, Gregg Daffner and Larry Martinez, "The Legal And Economic Dimensions of A Competitive Global Satellite Regime: Consequences for Developing Countries," paper presented to the International Bar Association Conference, New York, NY, in October 1990.

9. *INTELSAT Agreement*, Article XIV (d).

10. Quoted from Glenn H. Reynolds and Robert P. Merges, *Outer Space: Problems of Law and Policy*, (Boulder: Westview Press, 1989), pp. 80-81.

11. The official citation for the Act is: *Telecommunications Act of 1996*, Pub. L. No. 104-104, 110 Stat. 56 (1996).

12. See, Eliot A. Cohen, "A Revolution in Warfare," *Foreign Affairs*, March-April 1996, pp. 37-54.

13. Warren Ferster, Israel Wants Imagery Ban, *Space News*, June 17-23, 1996, p. 1.

14. Sean P. Kanuck, Information Warfare: New Challenges for Public International Law, 37 *Harv. Int'l L.J.* 272 (Winter 1996). Fn. 7, citing the U.S. Office of the Chief of Naval Operations, Dep't of the Navy, Opnavinst 3430.26 1 (Jan. 18, 1995). Source: Nexis-Lexis.

15. *Id.*, fn. 9, citing the U.S. Dep't of the Air Force, *Cornerstones of Information Warfare* 3--4 (1995). Source: Nexis-Lexis.

16. *Id.* p. 276; fn. 19: "All Members shall refrain in their international relations from the threat or use of force against the territorial integrity or political independence of any state, or in any manner inconsistent with the Purposes of the United Nations." *U.N. Charter*, Article. 2, para. 4.

17. *Id.*, p. 286.

19. *Id.*, p. 292.

20. See, Brian Kahin and Charles Nesson, *Borders in Cyberspace*, (Cambridge: MIT Press, 1997).

21. Anthony Rutkowski, "The Internet: An Abstraction in Chaos," in *The Internet as Paradigm*, monograph published by the Aspen Institute, 1997, pp. 1-22.

MUST THE SPECIAL TYPOLOGY OF AEROSPACE PLANES LEAD TO THE SUPPLEMENTATION OF THE RULES OF THE OUTER SPACE TREATY?

by
dr. Gabriella Catalano Sgrosso*
University of Rome, Italy

Abstract

The plans for the realisation of the future aerospace planes lead the jurist to face interesting questions on the applicable discipline.

After having identified the different types of aerospace planes by classifying them in three groups according to similar characteristics, from the most typical ones of the aerospace plane to the more particular ones of the aircraft, the most suitable regime for the regulation of new situations will be sought.

Considering the spatialist thesis, the functionalist one and the thesis on a specific regime, the second seems to be the most suitable, albeit adapted to the different typology of aerospace plane. The crucial points over which space and air law differ are those concerning registration, with consequences over the air- worthiness certification and on the identification of the launching State, over the statute of the crew and over liability, in the three orders of liability for damages to third parties, for transportation damages and for damages due to faults of the product.

Some solutions have been suggested in order to solve the conflicting situations the aerospace plane could face. In conclusion, the solution is not to be found so much in the modification of the Treaty on Outer Space and subsequent UN Space conventions, as it is in specific co-operation agreements among the interested parties and in the increasing privatistico regulations which are already emerging in the insurance and transport field.

*Researcher of International Law, Faculty of Economics, University of Rome, Italy; Director IISL.

Typology of aerospace planes

The planning phase of aerospace planes, carried out by various countries, is at present suffering a slow-down, and at times an abandonment, due to political and economic reasons. The jurist, however, must consider in advance the problem concerning which discipline should apply to these new objects having the hybrid characteristics of an aircraft and of a space object, and if space law, such as it arises from the Outer Space Treaty and from the following UN. Conventions, is also suitable to regulate these new situations.

The same term "aerospace plane" has not been accepted univocally, so that, in the German answer to the first question in the unfruitful questionnaire of the COPUOS Legal Sub-Committee dated March 1996[1] on the possible legal issues with regard to aerospace objects, it was preferred to adopt the more generic term "space transportation system". Until the Scientific and Technological Sub-Committee has studied the different types of space planes and given a more technical definition, it is preferable to keep to a term, even "aerospace plane", which indicates both fields of operation. On the other hand, while the Chicago Convention of 1944 gives an aerodynamic definition of the air vehicle with reference to its faculty of support in the atmosphere, neither the Outer Space Treaty nor the other conventions give a precise and univocal definition of the space object[2]. This lack of definition is probably intentional in a field where technological evolution is so fast that standard definitions might soon become obsolete.

The Space Conventions prefer to state the discipline applicable to space objects, leaving the parties to agree on the connection in case of the creation of new types. This is what happened in the IGA for the International Space Station, where it was agreed that the discipline concerning the registration and control was to be applied to the single elements of the Station[3].

The need to lower the cost of access to outer space has led to the planning alongside with non-recoverable launchers, of a series of reusable vehicles of transportation with different technological characteristics. These space planes have different functions and therefore they will necessarily undergo different legal treatments.

a) The first group includes those aerospace planes with characteristics closer to the space objects: vertical take-off, launch from earth launchers, placement in orbit- where their mission is carried out-, return on earth with horizontal landing at times in another State.

The first realisation of this kind of transportation vehicle was the American Space Shuttle used for the transportation into outer space of space labs which, in its return phase lands with a gliding flight, after 8300 Km navigation, on a previously planned American aerodrome. The United States have defined the Shuttle as a space object and therefore it is regulated by space law. The reason is functional: its mission is carried out mostly in outer space and the limited manoeuvrability during its return on Earth accustoms the Shuttle to the classical space capsule.

The other Shuttle, used only once in 1988 and now abandoned, is the Soviet Buran.

Among the future projects for transportation vehicles there is the optional ESA programme HERMES, made up of a transportation aerospace plane and of a resource module which disintegrates in the atmosphere, with a vertical take-off from Guyana and a horizontal landing. The HERMES project which should serve the Columbus module has been abandoned for the time being.

The Japanese HOPE project can also be classified within this group: a one-stage, reusable vehicle without crew, with a vertical take off and an automatic landing, after a descent to 80 Km altitude, in a pre-chosen place. It can serve the Space Station and carry payloads and also be used as an un-inhabited space lab.

The United States, after having abandoned the NASP project, are studying a one-stage Reusable Launch Vehicle (RLV). Some demonstrations were carried out within the experimental programme X33. An optional vehicle could be available before 2010.

b) The second group includes projects for supersonic planes, for "Super Concorde" by the United States and also by Europe and Japan. The "Super Concorde" will take-off like a plane, it will not use a rocket launcher, it will not have the aim to be placed in orbit and if, for a few instants, it should be in a high orbit over the extra-atmospheric space, its mission will basically be for transportation from one place to another over the planet.

c) The third group of objects foresees some multi-stage hybrid aerospace systems where some of the components have the characteristics of an aircraft carrying out their mission in the air space and some others of the space object carrying out their function in orbit. The German project SANGER foresees a hyper-sonic aircraft with a liquid hydrogen aerobic engine carrying a second stage on its back which separates at approximately 40km, at mach 7 with the mission of carrying payloads for space stations into space. The first stage lands and the second returns to Europe by gliding flight.

According to the Russian concept of a transportation aircraft, the AN-225 (or Myria) could carry a second stage detaching at approximately 9 Km. A co-operation programme between Russia and Great Britain foresees the transportation on the AN-225 of a space vehicle named HOTOL. This reusable one-stage automatic flight launcher would keep smaller payloads in the lower orbit at a lower cost[4].

Legal regime applicable to aerospace planes

The regime of air law, prevalently based on the principle of the exercise of the sovereignty of the overflown State, and that of space law, based instead on the principle of freedom of navigation are very different. The comparison between the international conventional rules regulating matters of liability, registration, statute of the crew, reveals a very different content between the rules applicable to the air vehicle and those applicable to the space vehicle.

The doctrine is divided over the criteria to be adopted for the identification of the applicable regime. The spatialist thesis would rather apply air law or space law to the aerospace plane according to its location.

The functionalist thesis is instead based on the function carried out by the vehicle, on its purpose: if the aerospace vehicle is designated to carry out a mission in orbit the rules of space law will be applied even when it is in air, whereas if its function is to link two points on earth, even by passing briefly through space, the rules of air law should be applied[5].

A third thesis supports a specific regime such as the creation of a text among the States having such vehicles, where the existing rules of air law and space law are to be adapted to the new type of aerospace plane[6].

The space approach does not seem to gather much success. First of all due to the unsolved question of the delimitation between air space and outer space lying, maybe wilfully, on the table of the legal Sub-Committee[7]. Furthermore, the acceptance of this theory would involve the application of different regimes for a same object, while the doctrine, the Space conventions and the States themselves tend towards a uniform regime, as it appears in the answers to questions 2 and 4 of the UNCOPUOS questionnaire.

The functional approach seems to be the most convincing, and at present a space vehicle with a double purpose of transporting passengers from one point to another of the Earth and of a mission in orbit does not seem feasible. However, the theory is to be adapted to the previously mentioned different typology of aerospace vehicles.

The Space Shuttle and the vehicles classified in the first group carry out their function in outer space, they must be considered as space objects and therefore they will be subject to the application of the rules of international space law for registration, liability for damage, rescue and recovery of astronauts and space objects.

The most delicate point is the passage in the air space, especially during the return phase which could take place with a horizontal landing in a different State from where it was launched. For space objects the return has until now been established in the launching State and the problem concerning the passage in the air space of another State has only been for the take-off phase. This could be defined as an "inoffensive passage right" but it has however been taken into consideration either in air or space law. This passage has never been questioned by the States leading to the possibility of the creation of a common law rule. With the advent of the aerospace plane we are moving towards an opposite tendency[8] because the crossing of the air space of another State risks being more frequent and longer lasting. The USA Space Shuttle, the Buran and other future systems will need approximately 8,000 Km from their re-entry into Earth's atmosphere until their point of landing. Crossing the atmosphere for 14-15 minutes and flying lower than 60 Km the necessary precautions will have to be taken in order to avoid collisions with other aircraft[9]. The rules for the security and traffic organisation of the overflown State must be forcibly respected and the most feasible system seems to be to reach an agreement among the States involved in the carrying out of the mission of the aerospace plane.

The second group includes those supersonic aircraft with the purpose of linking and transporting in the shortest time from one point to another over the Earth. The passage in low altitude is only due to a technological necessity. This kind of vehicle has the function of an aircraft and must undergo the national laws of the overflown States and the rules, for registration and authorisations, of the Chicago Convention on Civil Aviation of December 7th 1944 and for liability of the Carrier the rules of the Warsaw Convention on Air Transportation of October 12th 1929 and the following modifying Protocols of The Hague 1955, Guadalajara 1961, and Guatemala 1971. The Rome Convention of October 7th 1952 concerning damage caused to third parties to the surface by foreign aircraft, determining an objective responsibility of the carrier, has only obtained approximately 30 ratifications and is still missing the ones from those States which mostly carry out such air activity[10].

However, the conventions on air law have not taken into consideration the liability for possible damage caused to other space objects in the albeit short period of the crossing of the space by the supersonic aircraft. On the contrary, the opposite situation, that is to say the liability for damage caused by a space object to aircraft in flight during the phase of the crossing of the air space is regulated by the 1972 UN Convention on Liability with the determination of an absolute objective liability. This will certainly be a problem to be faced, at least by those States provided with these aircraft[11].

The third group includes multi-stage aerospace planes. A complex space system will involve greater difficulties in the determination of the applicable law and suitable solutions will have to be studied to regulate those special situations which will happen during the life of the aerospace plane.

Some authors state that even the first stage transporting and launching in air space the space object, or from whose back it takes off, must be considered as a "component part" and precisely the launching vehicle and therefore that it should have the status of a space object. However it is clear that it cannot be admitted that the first stage carrying out its function in the air space is infringing the sovereignty of the overflown States and that it must therefore respect all the rules concerning information, authorisation and security of the overflown State. The *status* of the astronauts/pilots is yet to be defined[12]. It therefore seems more appropriate to consider, in this complex system, the first stage, carrying out a function of transportation in air space and flying over the air space of other countries and landing, especially in Europe, in different States from the launch, as an aircraft and therefore subject to all the national and international rules of air law. The second stage, carrying out its function in orbit, can be considered as a space object.

The difference in the rules of air and space law on matters such as registration, certification, status of the crew and liability will involve a certain adaptation in those conflicting situations the aerospace plane could find itself in.

Difference in the present regulations and tendency to rapprochement

Art. 17 of the Chicago Convention obliges the nationality States to **register** the aircraft once and for all with an identification symbol. The 1975 Convention on Registration of Objects Launched into Outer Space establishes a national and an international registry, care of the Secretary of the United Nations, in which the space object is to be registered with a different number each time it is launched. The first question is if the aerospace plane, linking two points on earth by briefly passing through space, should be registered at each launch. Some Authors state that the incidental transit in space of a flight between two continents does not involve the application of the 1975 Convention[13]. A greater certainty of identification, with the registration even in the international registry, since the orbital parameters will be different each time, leads towards the applicability of the Convention itself[14].

In order to guarantee the respect of important security measures, stated in enclosure 8 of the Chicago Convention, each aircraft must obtain an **air- worthiness certificate.** In order to maintain the air- worthiness certificate the builder must prove that the conception of the aircraft is in accordance with the established rules, that the model of the aircraft is in accordance with the certified type and that all the requirements stated in the maintenance manual are fulfilled. International space law, instead, does not require any certification even for manned flights. The American Shuttle, as a space object, does not need to obtain an air-worthiness certificate but the NASA is obliged to respect a certain umber of FAA regulations, especially those concerning air circulation. On the other hand the FAA must ensure sufficient air space for the return of the Shuttle. There is therefore the question if the transportation aerospace plane, which will use the air space for a longer period and which will land in different States from the launching one, should be provided, essentially for security reasons, with a similar air- worthiness certificate. The doctrine seems to be positively oriented suggesting a sort of domestic certification by the single States or by the European Space Agency[15] similar to the American regime of licences issued to those carrying out commercial launches[16].

The identification of the **launching State**, could be critical, also for the purpose of registration, in the case of an aerospace plane of the third group taking off from or launched into space by a transportation aircraft. The 1975 Convention describes the launching State as the State which launches or procures the launching of a space object or a State from whose territory or facility a space object is launched. Three launching States could be identified: the one to whom the second stage going into outer space belongs, the State owning the transportation aircraft or the State of the air space if the detachment takes place in an air space under the sovereignty of a third country. The problem can be solved, once again, only with the agreement of the involved States. On the other hand article II of the 1975 Convention on registration states that "where there are two or more launching States in respect of any such space object, they shall jointly determine which one of them shall register the object .... without prejudice to appropriate agreements concluded or to be concluded among the launching States on jurisdiction and control over the space object and over any personnel thereof".

Some more differences may be found in the **crew statute**. While art. 32 of the Chicago Convention establishes that the captain and the crew on board an aircraft must obtain a license according to specific conditions, in space law the 1968 Convention on the Rescue of Astronauts, despite considering them to be "envoys of mankind" and therefore enjoying specific guarantees and assistance by all the States, does not establish any particular statute[17]. The astronauts undergo such a selective procedure that their preparation is undoubtedly equal to an aircraft pilot, and therefore in order to render the regime more uniform it would be useful to require the personnel of the aerospace plane to obtain a special licence.

It is presumed that there will be various kinds of passengers on the aerospace plane. Those who will be part of the mission, that is to say the specialists, the engineers, the scientists, such as in the Space Shuttle, will be considered envoys of humanity, but those passengers who are only being transported may not be treated the same as the astronauts and they will be envisaged in the framework of air transportation.

The Chicago Convention establishes the captain's powers and his relationship with the other people on board. The captain of a space object enjoys a similar statute[18], but it would be necessary to give a conventional definition of the position of the captain of an aerospace plane. On the other hand there are some previous rules regulating abnormal situations. The "Regulation on the security of the personnel and the powers of the captain" was created for the crew of the Space Shuttle[19], and for the International Space Station the States party established, in the IGA and in the mentioned Memorandum, a "code of conduct" also concerning the command hierarchy, the powers and responsibility of the captain[20].

As for the aerospace plane it will be necessary to establish the relationship between the captain on board and the personnel on ground in order to properly share the decisional powers. In fact, while the captain of an aircraft, albeit taking into consideration the opinion of the personnel on ground, makes his decisions independently, the captain of a space object obeys the orders of the mission Director on ground.

The regime of **liability** applicable to space objects is rather different from the one established for aircraft. In the case of an aircraft the Warsaw Convention on air transportation establishes in detail the liability conditions of the carrier towards passengers based on very specific principles: presumption of liability of the carrier, limitations for compensation, jurisdictional competence. The responsibility for damages caused to third parties on the surface by aircraft is regulated by the 1952 Rome Convention which establishes an objective liability of the carrier and defining maximum limits. This convention did not receive many ratifications and therefore national law is usually applied to solve controversies and also for the cases of collisions between aircraft.

While the regime of air law is a privatistico regime centred on the physical figure of the carrier, the regime of space law is focused on the State being the subject authorised to request compensation. The 1972 space law Convention however, only establishes the liability for damages: liability for fault, in the event of a collision between space objects, and absolute objective liability in the event of damage to aircraft in flight or caused on the earth's surface, but no liability concerning transportation.

There is however a certain rapprochement between the two regimes. There is in fact a tendency to overcome the compensation limits in air regulation which is more restrictive than the space one. The 1955 Hague Protocol and the 1961 Guadalajara Protocol have already increased the maximum of the compensation for damage to persons, the United States have established derogations from the Convention and the Japanese companies have stated that they do not wish to avail of the compensation limits.

At present the Institutions of the European Union are examining, with the procedure of co-operation between Council and Parliament, a suggestion for the regulation of the Council on the liability of the air carrier in case of accidents[21]. Some countries have in fact considered it necessary to adapt the liability limits of their carriers in order to take into consideration the risk reduction within the framework of present transportation, the higher level of compensation for the accident victims and also of the evolution of the cost of life. The suggestion for the regulation aims to abolish the monetary limits of liability in case of gross negligence of the carrier; to establish a maximum for objective liability up to the concurrence of an amount equivalent to 100.000 "Special withdrawal rights" (approximately 120.000 ECU) even if the carrier has taken all the necessary measures to avoid the accident; to establish the dispositions for the payment of the advances to the victims; and to oblige the carriers not belonging to the Community, to inform passengers about their conditions of transportation when purchasing the ticket.

Many of these aspects have already been accepted by a certain number of world airlines in the framework of the agreement among IATA carriers on the liability towards passengers approved in Kuala Lumpur on October 31st 1995. In order to avoid fragmentation of the ratifications the Community is introducing a regulation for the member States[22].

Space law ignores liability for damages to transported people or things, caused by space transportation activities. Some cases have been taken into consideration in the jurisprudence of the United States[23] but it cannot be stated that a private law regime has been established on the matter. When the cases of air transportation will be more frequent with the advent of the aerospace plane a discipline for this kind of liability will be foreseen in the transportation contracts according to the model of the Warsaw Convention, the amending protocols and the eventual community regulations. This will lead to a phenomenon already taking place for naval, sea and air transportation: the uniforming of transportation with the creation of a common discipline.

The damage due to faults in the product, which could cause a liability of the maker and of his subcontractors, may also be verified. Nowadays the scheme for the contractual waiver is commonly placed in the launch contracts stipulated by the United States, in those stipulated by Arianespace and by the Chinese launching industry. The Commercial Space Launch Act, adopted by the United States Congress in 1984 and modified in 1986, states that the essential requirement to establish a regime of mutual waiver of the appeals consists in the fact that each company having obtained the concession for the launching, its clients, the Government and the relevant contractors and sub-contractors must accept to undertake the risk of each loss, material or physical damage or death consequent to the realisation of a space programme[24]. Because the market for launching activities and space transportation is opening to private parties, of whose activities the States are responsible (according to the Outer Space Treaty of 1967), it has become compulsory for the companies operating in commercial launches to stipulate insurance coverage for liability towards third parties. In the United States the subject requesting the governmental licence must prove the necessary provisions for a mechanism of insurance protection for liability towards third parties[25] have been taken. It can therefore be remarked that, with the advent of commercialisation by private parties of the activities of launching and space transportation, the regulation of the phenomenon of liability for damages will be more and more included in the national or international contracts or in the internal legislation of the States.

Conclusions

Some final considerations must be made in order to answer the initial question if, in view of the realisation of the aerospace plane, any changes should be made to the Treaty on Outer Space and to the subsequent UN. Conventions.

Despite accepting the functional theory according to which it will be the function of the aircraft and the place where it will mainly be operating that will determine the applicable legal regime, we have seen that some measures are necessary in order to solve the conflicting situations in which the aerospace plane could find itself. However, it does not seem possible to suggest for this reason neither the revision of the present conventions of space law, nor the stipulation of an *ad hoc* multi-lateral agreement for the aerospace plane[26]. The amount of time necessary for the stipulation and incorporation of the changes or of the new legal instruments would be excessive.

While at the beginning of the space activities the States were more inclined to entrust the regulation of the matter to international law, this attitude has deeply changed in time. The UN. are unable to lead to the stipulation of agreements; at the most, some principles of resolutions have been produced by the General Assembly. The States do not wish to give up their exclusive competence on the subject and in order not to slow down the space activities further they prefer to regulate the situations with specific agreements between the parties, as for the International Space Station.

It is foreseeable that the over-flight and the landing of the aerospace planes in zones with a different sovereignty from the launching State will be the object of special agreements between the involved States. An example of an agreement of this kind may be the Agreement of Space Co-operation, signed on July 11th 1991, between the United States and Spain where it is established that in case of need the American Space Shuttle may fly over the Spanish air space and land in a Spanish base. In the agreement the authorised Spanish bases are specified, as are the procedures in the event of an emergency in another location of the Spanish territory, and the observance of the NOTAM and of the Spanish aeronautic services is required. The liability of the United States for damage caused on Spanish territory is also determined[27].

Other national legislations, among which the German, extend the application of the rules established for aircraft also to space objects[28].

In a regional framework such as the European one it could be considered to entrust the European Organisation for the Safety of Air Navigation (EUROCONTROL) with the duty to promote co-operation, for air safety, among the European countries interested in the carrying out of the mission of an aerospace plane. Recently, in the final act, dated June 27th 1997, of the Conference for the revision of the Convention on EUROCONTROL, it was stressed that the policy of the Organisation includes, among its aims, "... those of standardisation, planning, performance and safety regulations; the technical and financial selection of major framework programmes for co-operation; external relations with States and organisations and applications for accession to this Convention". Furthermore, among the duties of the Council, there is the duty to "determine the rules and procedures applicable to standards, specifications and practices for air traffic management systems and services"[29].

Finally, as previously stressed, with the advent of private operators even in the field of space transportation, there is the prospect of a privatistico kind of regulation which will be included in the transportation contracts and in the insurance contracts in order to cover the different cases of liability for damage, without neglecting the observance of internal dispositions issued *ad hoc* by the States responsible for the activities of private parties in space.

NOTES

[1] Doc. ESA/IRC (96) 15.P.J. Annexes. Document de Travail. Paris, 24 April 1996. Only ten answers came from the member States among which only two from ESA countries (Italy and Germany)

[2]GOROVE, Toward a clarification on the term "space object", an International legal policy alternative, in Journal of space law 1993, p.11

[3]ZANGHI', Aerospace Object, in Outlook on Space Law over the next 30 Years, ESA, The Hague, London, Boston 1997, p. 116

[4]For a description of the different projects of aerospace planes see "Actes du Colloque: L'avion Spatial et le droit", in Revue Française de Droit Aérien et Spatial, 1991; CHRISTOL, The Aerospace Plane: its legal and political future, in Space Policy, 1993, vol. 9, n. 1, p. 35; HASHIMOTO, The Space Plane and International Space Law, in Proc. of the 35th Colloquium on the Law of Outer Space, Washington, 1992, p. 378; Doc. ESA/IRC (96) 15., see note 1

[5] CHRISTOL's allocative theory is slightly different, The Aerospace Plane, see note 4, states: "....that hybrid vehicles are to be examined on the basis of their purpose as construed in the light of the accompanying element of intent ... for the application in specific situations of either air law or space law."

[6]KAYSER, Aux confins de l'air et de l'espace d'accursius à l'avion spatial, in Annals of Air and Space Law, 1994, vol. XIX, p. 479

[7] For the problem of the boundaries see CATALANO SGROSSO, La responsabilità degli Stati per le attività svolte nello spazio extra-atmosferico, Padova, 1990, p. 23

[8]VERESHCHETIN, "Utilisation de l'avion spatial et droit de l'espace", in Actes du Colloque "L'avion spatial et le droit", see note 4, p. 515 and p. 524

[9]See the German answer to question 3 of the UNCOPUOS questionnaire

[10] For the text of the above mentioned Conventions see BALLARINO BUSTI, Diritto aeronautico e spaziale, Milano 1988, pp. 769, 608, 833, 314

[11]ZANGHI, Aerospace object, see note 3, p. 122, asserts that in the meantime the 1972 Convention should be applied

[12]HASHIMOTO, The Space Plane and International Space Law, note 3, p. 379

[13]KRAMER, Registration of Space Objects, in Outer Space Law, edited by CATALANO SGROSSO, , New Developments and Prospects, Padova 1994, p. 237

[14]Both the Italian and the German answers to question 9 of the UNCOPUOS questionnaire seem to be oriented towards the registration of aerospace planes in accordance with the 1975 Convention

[15]MOTLIVAULT, Immatricolation et certification, in Actes du Colloque, see note 4, p. 495

[16]KAYSER, Au confins de l'air et de l'espace, see note 6, p. 475

[17]A project for the definition of a statute of the astronaut has been deposited with the United Nations Space Committee but it has yet to be discussed; in doctrine see LAFFERRANDERIE, Pour une charte de l'astronaute, in An. Air and Space Law 1987, p.263, PEYREFITTE, Droit de l'espace, Paris 1993, p. 185

[18]VENET, Statut juridique et responsabilité du commandant de bord, in Rev. Franç. droit aér. et spat. 1990, p. 169

[19]Space Transportation System, 44 Fed. Reg. 131 (July 6, 1979) and Fed. Reg. 57 (March 7, 1980)

[20]Art. 11.1, Intergovernmental System, Washington, 29 Sep. 1988 and art. 11.5 MOU, see CATALANO SGROSSO, La responsabilità ... see note 7, pp. 110-116 to p.163

[21]COM (95) 0724-C4-0209/96, 95/0359 (SYN) in G.U.C.E. n. C 320/30 28 Oct. 1996

[22]See the motivation of the Parliament in the Recommendation for the second reading: DOC\IT\RR\327297 of May 21 1997

[23]KAYSER, Aux confins de l'air et de l'espace, see note 6, p. 478, note 35

[24]STRAUBEL, The Commercial Space Act; the Regulation of Private Space Transportation, in J. of Air Law and Commerce, 1987, p. 322; MILLER, Licencing and Regulating US Commercial Space Launches, in J. of Law and Technology, 1988, p. 45

[25]CATALANO SGROSSO, Insurance Implications about Commercial and Industrial Activities in Outer Space, in proc. of the 36th Coll. on the Law of Outer Space, Graz 1993, p. 187

[26]In this sense the German answer to the UNCOPUOS Questionnaire (question 3) is for the moment against the creation of a special regime for air transportation systems

[27]BOE n. 124, p. 16104, May 25th 1994, modified in BOE n. 269. p. 34672, November 10th 1994 (R. 1994/3138); see report FARAMINAN, Analisi giuridica del concetto di oggetto spaziale, presented at the International Congress "The Legal Regulation of aero-space Transportation Vehicles", Rome, Feb. 28th - March 1st 1997

[28] Germany replied to question number 8 ("Are there any national and/or international legal norms with respect to the passage of space objects after re-entry into the Earth atmosphere?) stating: "..According to German Law, the Federal German Aviation Code is not only applicable to aircraft but also for other objects such as, for example, space craft and rockets..."

[29]Art. 1.2, III, IV, V and art. 7.2, m of the Final Act of the Diplomatic Conference on the Protocol consolidating the Eurocontrol International Convention relating to co-operating for the Safety of Air Navigation of December 13th 1960, as variously amended, Brussels June 27th 1997, amended version CNS/56

LEGAL AND POLICY ISSUES RAISED BY THE PROPOSED NOTION OF "AEROSPACE OBJECT"[+]

Stephen Gorove[*]

ABSTRACT

This presentation first takes note of the use of the term "aerospace" in the general context of air and space law and traces briefly the background that gave rise not long ago to the introduction of a novel expression "aerospace object," a phrase distinct from the notion of "space object," commonly used since the dawn of the space age.

The paper analyzes what has been suggested as a possible definition of "aerospace object" and points to the need for further clarification in light of envisaged commercial developments in very fast, long-distance, earth transportation so that appropriate policy evaluations and choices can be made with respect to the applicability or inapplicability of norms of air and space law.

If an aerospace object's basic purpose and function is earth-bound transportation and is seen in the light of an aircraft that briefly uses what may be termed cosmic speed, a multitude of legal issues surface. Basic in importance are the policy choices governing the flight of one state's space object through the airspace of another state. Such choices should take into account a whole gamut of variables, including the existing legal framework of air and space law, past state practices and precedents, national security considerations, and even effects of possible violations of cardinal norms of space law if such can be envisaged. At the same time, such aerospace objects even in free airspace should conform to international air law regulations and requirements such as safety, much as aerospace objects in earth-bound transportation should comply with space debris mitigation, rules of the road, and other requirements while operating briefly around the fringes of outer space. Careful solutions must be pondered by air traffic and space lawyers with respect to legal norms applicable to space objects ascending or descending through the airspace of a foreign state.

While the definition of aerospace object is currently based on just two criteria, *i.e.,* the capability of an object to travel through outer space and its capability to remain in the airspace for a certain period of time, as aerospace objects become more sophisticated in design, it is clear that their definition and the effect of such on legal and policy issues for air and space law may have to be reevaluated in light of additional criteria.

Introduction and Retrospect

The use of the phrase "aerospace object" has surfaced in U.N. discussions thrusting on international legal technicians and policy makers a set of challenging issues and alternative choices in the fields of air and space law. Prior to its sudden emergence, the phrase has rarely been encountered in the legal literature although the word "aerospace" has been used in joint combination with "law" to make up the phrase "aerospace law."

If an attempt were made to shed light on the background of the term "aerospace law" and trace the possible reasons for its emergence, one would undoubtedly come across the writings of such a well-known authority as John C. Cooper who suggested analogies from the law of the sea and referred to territorial airspace, contiguous zone and the space beyond.[1]

Similarly, Nicolas Mateesco Matte compared the territorial sea to the territorial air and used the expression of "Aerospace Law" as the title of his 1969 book in which he restated his earlier held views, opposing arbitrary legal boundaries between airspace and outer space and

---

[*] Professor of Law Emeritus, University of Mississippi Law Center, Director of Space Law and Policy Studies and Chairman, Ed. Bd. JOURNAL OF SPACE LAW, University, Mississippi. Member, International Academy of Astronautics. Hon. Director (Vice-Pres.), International Institute of Space Law.

championing the establishment of an aerospace law based on a new functional theory.[2]

There can be little doubt that analogies drawn from the law of the sea have contributed to the tendency to place air and space law under the unifying umbrella of "aerospace law." Of course, this was a hard thing to sell because with entry into force of the fundamental charter of space law, the 1967 Outer Space Treaty,[3] the contours of a legal system distinct from air law seemed to have emerged. With the passage of time and in light of four additional international treaties and a number of resolutions dealing with the distinct law of outer space, the contention that there could be a fusion of air and space law under the heading of "aerospace law" became even more remote. A quick glance at the vast literature reveals an overwhelming number of writers who have preferred to use the now generally accepted phrase "space law."

Aerospace Object and the U.N. Questionnaire

The coupling of the terms "aerospace" and "object" in the joint expressions of "aerospace object" surfaced first in the Legal Subcommittee of the U.N. Committee on the Peaceful Uses of Outer Space (COPUOS) in connection with the definition and delimitation of outer space.[4] This issue has been on the agenda of the Subcommittee since 1967 due to the fact that advocates of the spatial and functional theories could not agree whether a boundary should be established internationally, at a height of approximately 100-110 km above sea level, as proposed by the Soviet Union, or whether such delineation was unnecessary, serving no useful purpose, as asserted by several western spacefaring nations, including the United States. Notwith-standing this deadlock, practical developments over the last 30 years seem to have confirmed that the area where artificial satellites and other man-made objects are in orbit around the earth and beyond is outer space. However, the actual boundary line between airspace and outer space remains internationally undeter-mined to date.

In 1991, the Soviet Union made an attempt to overcome the long-standing impasse and during the general discussion of the Legal Subcommittee of the COPUOS agenda item on the definition and delimitation of outer space, with the support of some other delegations, suggested that the Subcommittee should commence an "exchange of views on the international legal aspects of future exploitation of aerospace systems."[5] As a follow-up in 1992, the Russian Federation, continuing the membership of the Soviet Union in the United Nations, submitted a working paper entitled "Questions concerning the legal regime for aerospace objects."[6] It was after this initiative that, in 1993, the Chairman of the respective Working Group circulated an informal working paper entitled "Draft questionnaire concerning aerospace objects."[7] Two years later, at the thirty-fourth session of the Subcommittee, the Working Group finalized the text of the Questionnaire on possible legal issues with regard to aerospace objects" (hereinafter "Questionnaire") in an effort to clarify issues concerning the definition and delimitation of outer space and recommended that it should be circulated to the States Members of COPUOS.[8]

Definition and Meaning of "Aerospace Object"

It was the circulated Questionnaire that raised the issue whether an "aerospace object" could be defined as "an object which is capable both of traveling through outer space and of using its aerodynamic properties to remain in airspace for a certain period of time."[9]

While the purpose of the Questionnaire was to help the Committee in finding a common ground regarding the definition and delimitation of outer space, it became apparent from several of the responses received that while the definition was acceptable for working purposes, it needed further refinement and clarification. As an example, reference was made to the phrase "for certain period of time." Some delegations expressed the view that the definition should provide only for functional, man-made objects as opposed to space debris or natural objects. It was also pointed out that while the use of the term "aerospace systems" or "space transportation systems" may have appeared more appropriate than the uncommonly used phrase "aerospace object," the term "space transportation systems" as used in the report of COPUOS and its Scientific and Technical Subcommittee had a wider meaning, covering both the transportation systems of the space-shuttle-type vehicles and the usual rocket carriers. Therefore, that term would not be appropriate for describing the hybrid systems that might be used for both air flight and missions in outer space.[10]

One other legitimate query requiring clarification was the question whether the definition was meant to apply exclusively to a type of vehicle like the space shuttle whose main function has been transportation of people and cargo into space but which has also been using its aerodynamic properties when returning to land on earth, in a way reminiscent of the landing of an aircraft. The use of the term "aerospace object" in lieu of "space object" left this interpretation somewhat doubtful. In view of this, it was more likely that, apart from objects launched into outer space, the definition had also intended to cover the proposed aerospace-plane-type vehicle the primary purpose of which was point-to-point transportation on earth (the carriage of a payload and/or passengers from one point on the earth to another) though for a brief period of time such vehicle was expected to travel through the fringes of outer space.

Several of the responses received were in line with the foregoing interpretation. For instance, the Czech Republic in its answer indicated that "aerospace object" may cover different types of aerospace vehicles, some of which are still in the design and planning stage, while others have been suspended or even abandoned.[11] Italy also observed that the definition which considers the twofold capacity of the aerospace object is closely linked with the developing technology.[12] Most importantly, the Russian Federation, which brought up the issue of "aerospace object," stated unequivocally that there are two basic programs (purposes) for using aerospace objects, namely:

1. undertaking a flight from one point on the earth to another (for this purpose the object may undertake part of its flight in outer space, not attaining cosmic speed); and

2. delivering a crew and/or payload in outer space and back to the earth (its aerodynamic properties at the time of take-off and landing enable the object to remain in airspace for a certain period of time).[13]

If, as these responses suggest, the Questionnaire's proposed definition is not limited to the space-shuttle-type vehicle but also covers the aerospace plane which in light of expected commercial developments aims at a very fast, long-distance earth transportation, a whole range of issues must be examined so that appropriate policy evaluations and choices can be made with respect to the applicability or inapplicability of norms of air and space law in factual scenarios which relate to both the aerospace plane and the space-shuttle-type vehicle.

In line with the preceding assumption that the Questionnaire's purported definition covers both the space-shuttle-type vehicle and the briefly circumscribed aerospace plane, our attention will now turn to the determination of the legal regime that should govern the two distinctly different aerospace objects.

Aerospace Object as an Aerospace Plane

The development and eventual utilization of the aerospace plane is expected to herald the introduction of an advanced space transportation system consisting of a vehicle which would be capable of taking off horizontally and proceeding directly single-stage into outer space.[14] It would have the potential of spawning a new generation of commercial aircraft with the ability to span intercontinental ranges in a matter of minutes. A flight from New York to Tokyo may take only a couple of hours compared to the currently required time of 16 hours or more.[15]

The program relating to the development of the aerospace plane reflects a combination of aeronautical and space technologies; its utility could be gauged from the vehicle's capability of global unrefueled operation and of reaching any point on the earth in two hours or less. While at this stage of scientific research and experimentation, it is not possible to determine with certainty the configuration and eventual capabilities of future aerospace planes, for purposes of our inquiry, it will be assumed that early versions of the plane under discussion will be used as terrestrial transportation devices with the capability of taking off from a point on earth, flying at will in the airspace and traversing through the fringes of outer space without completing an orbit, for the sole purpose of reaching another point on earth.

While there are many legal and policy issues which arise in the wake of the development of the aerospace plane, the central policy issue will be to determine what laws, domestic and international, should be applied to this versatile vehicle in different factual scenarios. The main issues relate to the definition and delimitation of airspace and outer space, the status of astronauts, and the issues of liability, registration and jurisdiction.[16] Will the policy choice be to apply air law to the aerospace plane while traveling through the fringes of outer space or will the choice be to apply space law necessitating the application of space law rules embodied in the relevant international conventions? Will the personnel of the craft be regarded as astronauts, *i.e.*, "envoys of mankind" to whom the special privileges extended by the Agreement on the Rescue and Return of Astronauts and Space Objects would be applicable?[17] Will the law of the underlying state be applicable to an aerospace plane in the airspace in areas currently not utilized by conventional aircraft in view of the fact that the upward extent of national sovereignty has internationally not been determined as yet? Will space law govern an object orbiting the earth at a height of 30 km if new technology enables it to remain in orbit at that height?[18]

In formulating responses to these questions, as a general guideline, it may be suggested that if the aerospace object is used as an aerospace plane for the primary purpose of operating as an aircraft engaged in earth-bound transportation and only incidentally reaches the fringes of outer space, air law should be applicable to it. However, it stands to reason that such objects may be expected to comply with space debris mitigation, rules of the road, and other requirements while operating briefly around the fringes of outer space.

More problematic would be to determine the law applicable to the aerospace plane in areas which are below outer space but which are above areas currently used by aircraft and recognized as national airspace. While the general guideline might still be useful, international agreement or another form accommodation may be necessary to resolve any dispute that might arise.

It is also doubtful, although not necessarily impossible, that new technology could lead to the acceptance of lowering the current height of the area which is regarded as outer space from approximately 100 km to 30 km.

Aerospace Object as a Space-Shuttle-Type Vehicle, i.e., a "Space Object"

The issue of whether the policy choice should be to apply rules of air law or space law in connection with a technological innovation is not entirely new. At the time when the space shuttle was born, policy makers and lawyers were already faced with a similarly vexing issue which arose because the shuttle ascends into outer space with the assistance of rockets just as does a conventional spacecraft and descends from outer space by gliding through the atmosphere and touching down on a runway in a manner reminiscent of the landing of an aircraft. If the policy choice was arrived at because the vehicle's primary function and purpose was to operate as a device in outer space, this would in fact mean that the choice was to regard the vehicle as a "space object" with all the attendant legal consequences that follow therefrom.

The notion of "space object" has been central to the international law of outer space. Since the dawn of the space age, it has been the most frequent concept encountered in international agreements, U.N. resolutions, domestic laws, executive pronouncements, and court cases. Notwithstanding its crucial position, only a partial definition of this phrase may be found in the Liability and Registration conventions, both of which state that the term "space object" includes "component" parts of a space object as well as its "launch vehicle" and "parts" thereof.[19]

After a consideration of such vital issues as the relevance and purpose of launching, the pre-launch and landing phases, the relevance of outer space, the issue whether to regard extraterrestrial materials as space objects and the meaning of an object, this writer has suggested that a space object be defined as

> an object launched or attempted to be launched in orbit around the earth or beyond. Such object (or a part of it) is a space object (or a part of it) from the time of its launch or attempted launch, through its ascent from earth to outer space or while in outer space, as well as during its orbit, deorbit, reentry and landing on earth.[20]

If so defined, the space-shuttle-type vehicle would clearly qualify as a space object.

The foregoing conclusion was reinforced by the overall purpose and functions of the shuttle and was also fully borne out by an earlier review of the Federal Aviation Act of 1958, the National Aeronautics and Space Act of 1958, the underlying Congressional intent, the relevant legislative history, as well as NASA practice. To this was added an authoritative statement of the Chief Counsel of the Federal Aeronautics Administration, to the effect that space law had to be applied to the space shuttle. This determination was in line both with international air law incorporated in the Paris Convention of 1919 and the Chicago Convention of 1944, as well as with international space law embodied in the Outer Space Treaty of 1967 and the subsequent major international space law conventions.[21]

Since the policy choice has been to regard the space-shuttle-type craft as a space object, all the rules applicable to such objects under international space law apply to such objects. Most crucial among factual scenarios in which such rules apply is the flight by a space-shuttle-type vehicle through the sovereign airspace of another state.

In its response to Question 7 of the UN Questionnaire which raised the issue of whether there are "precedents with respect to the passage of aerospace objects after re-entry into the Earth's atmosphere" and whether international customary law exists with respect to such passage, the Russian Federation stated that there have been relatively few instances of space objects flying over territories of foreign states. As one of such instances, it referred to the flight of the Space Shuttle Atlantis in March 1990 about which the United States communicated information to the USSR a few hours before the overflight as a matter of courtesy.[22] Germany referred to the flight, on November 15, 1988, of the Soviet "Buran" which, after reentry into the Earth's atmosphere, overflew foreign countries for the purpose of touchdown in Baikonur. Germany did not believe that international customary law existed with respect to the passage of space transportation systems over foreign territory, since no international practice on this respect existed and it did not regard this occasion as relevant for the formation of international customary law, especially since the former Soviet Union which was the launching state did not exist anymore.[23]

In view of the relatively few relevant flights that have been noted in the literature, exclusive of accidental situations, it is perhaps not surprising to find that the initial responses of other states, which did not, as of then, include a response from the United States, failed to reveal sufficient support for the conclusion that the right of passage for an ascending or descending space object has been generally recognized as a customary rule of international law. At the same time, there has been an indication that an explicit admission of the right of innocent passage which was not prejudicial to the peace, good order or security of the subjacent states and a more detailed regulation of the exercise of this right should be considered as a way for the legalization of the actual practice.[24] This point was also intimated by the response of the Russian Federation in its statement that provisions of international customary law with respect to the passage of aerospace objects after re-entry into the Earth's atmosphere are currently being "elaborated."[25]

If one turns from the preceding initial sample of governmental responses to the views of distinguished authorities, a recent survey appears to reveal to this writer widely divergent views.[26] At one end of the spectrum, are the positions of Cheng, Dembling and Terekhov denying the existence of international customary law with respect to the passage of aerospace objects through foreign airspace. Less explicit in their denial are the views of Kopal, Haanappel and Masson-Zwaan, with some equivocation by Vereshchetin and Danilenko and a limited recognition of the right by Lachs. At the other end of the spectrum of competent opinions are the assertions by Finch and Christol that there are such rights.[27]

One explanation for the lack of uniformity, apart from strongly entrenched beliefs, may also have been the time element. During a ten year period, or sometimes even less, as suggested by learned allusions to the notion of "instant custom," perceptions, approaches, and attitudes can change. This writer's own position has also been influenced by actions or inactions in the world arena over a period of time and will no doubt be affected again in the future. Seen in such a light, he stated in 1988:

> The principle of the freedom of exploration and use of outer space, a cardinal principle of the 1967 Outer Space Treaty, in a sense implies the freedom to go into outer space and also the freedom to return to earth from outer space. Because of the very limited number of space flights that might have traversed through the airspace of foreign states the exact nature and scope of this freedom has so far not been determined by internation-al customary law.[28]

In 1993, this author made the following observation:

> States have not objected to the
>
> flight of artificial earth satellites above their territories in outer space nor to the ascent or descent of foreign space objects though undoubtedly some of these may have passed through their territorial air spaces. It is not certain how many times such a passage may have occurred since the upper boundary of national territorial air space so far has not been determined by international agreement or international customary law. If there is an international customary law, it is based on common perceptions and shared expectations of international authoritative decision-makers regarding such passage and supported by the cardinal principle of freedom of exploration and use of outer space embedded in the Outer Space Treaty of 1967 and generally recognized to the extent and in line with existing state practice."[29]

On the same occasion, he added:

> to the extent that States have not objected to the flight of artificial satellites above their territories in outer space nor to the ascent or descent of space objects through their national air spaces in the situations where such have occurred, a limited international custom with legal implications seems to have emerged.[30]

As late as 1996 this writer's position which, in his view, other learned colleagues appeared to have shared, may be briefly restated in the following terms. If the space-shuttle-type of aerospace object was used for the primary purpose to operate as a device in outer space, space law should apply to it. Once the primary purpose of the object is determined, the corresponding legal regime applicable to it should continue to be applied for the duration of the object's flight, whether in the airspace or outer space, at a particular time. Attempting to proceed otherwise would lead to conflicting interpretations with respect to the applicable law and would greatly confuse the problem.

If the primary function of the aerospace object was to operate as a spacecraft, then air law would not be applicable to it except in situations in which the craft returns in a non-accidental situation to a non-launching state. Aerospace objects launched into outer space are subject to the rules governing the registration of objects so long as the primary purpose of the object has been to operate as a spacecraft. Such an object should be governed by the national laws of the launching state, or if it was launched from a platform in outer space, it should be governed by outer space rules. As long as the object's primary function was to operate as a spacecraft - its safe passage to and from outer space has now attained the status of international customary law.[31]

Within the confines of this presentation, it is not possible to list even in a schematic manner all the relevant factors which prompted this writer to alter his earlier positions but attention may be drawn to the flight of the Soviet "Buran,"[32] about which no advance notice appears to have been given and no permission was requested or granted. Another notable occasion was the flight of the Space Shuttle Atlantis[33] about which a few hours notice was given by the U.S. to the U.S.S.R. only as a "matter of courtesy." This was accepted as such without any charge of a violation of territorial sovereignty that was frequently made in the past in connection with overflights of another kind. Nor was there any warning about avoidance of such an overflight in the future. Additionally, an agreement was also reached establishing that the fact that this information was furnished should not be deemed to set a precedent.[34]

As to the argument of traditionalists that the emergence of international customary law is normally a relatively slow process, it may be pointed out that many publicists do not reject the notion of "instant" international customary law in relation to space activities. As observed by the International Court of Justice in the North Sea Continental Shelf case: "The passage of only a short period of time is not necessarily, or, of itself, a bar to the formation of a new rule of customary law."[35]

Conclusion and a Glance at the Future

The preceding overview of some of the legal and policy choices associated with the notion of "aerospace object" suggests that the choice whether to recognize, in whatever form or extent, the existence of international customary law or to pursue an adoption of guidelines in the form of a UN resolution or perhaps to go the route of international treaty making, should take into account a whole gamut of variables, including the existing legal framework of air and space law, past state practices and precedents, national security, and other considerations,[36] and even effects of possible violations of cardinal norms of space law, if such can be envisaged.

Could a state lawfully deny another state's spacecraft the right of innocent passage at a height of 40-90 km in the space above its territory? Would this violate the fundamental freedom of the exploration and use of outer space? Should the answer be influenced by an analogy to the law of the sea where, in the absence of mutual agreement or international convention, a land-locked state has no independent right for access to the sea and claim innocent passage through the territory of a coastal state notwithstanding the principle of the freedom of the seas?[37] Should this be our policy choice for interpreting the freedom of exploration and use of outer space enshrined as a fundamental principle in the 1967 Outer Space Treaty? A courageous negative answer to this will be a challenge for air and space lawyers in the 21st century. However, in pondering their answer, decision makers should be reminded of the words of the late Judge Manfred H. Lachs who cautioned that

> [T]he interdependence of the traffic in the air and outer space should not subject the activities of states to unnecessary limitations. To survive in the world today states need to open the frontiers of the air to other states unless they prefer to live in complete isolation, where very few, if any, could survive and develop.[38]

If the right of innocent passage to and from outer space, as a matter of international customary law is contested or challenged, a U.N. resolution, as a preferable minimum, could help in allaying any doubt.

Before closing this presentation, a final *caveat* ought to be stressed. While the definition of aerospace object is currently based on just two criteria, *i.e.,* the capability of an object to travel through outer space and its capability to remain in the airspace for a certain period of time, as aerospace objects become more sophisticated in design, their definition and the effect of such on legal and policy issues for air and space law may have to be reevaluated in light of additional criteria. Additionally, if future technological developments were to create a hybrid vehicle capable of moving freely in the air like an aircraft and also moving at will in outer space, a consideration of new laws, both domestic and international, may become necessary in order to adjust legal regulations to the latest scientific and technological innovations.

NOTES

1. *See* John C. Cooper, *Legal Problems of Upper Space,* 50 PROC. AM. SOC'Y INT'L L. 85-93 (1956); *idem, Flight-Space and the Satellites,* 7 INT'L & COMP. L. Q. 82, at 89 (1958). By his proposal, Professor Cooper, in essence, revived the old zone idea which was advocated half a century earlier and was based on maritime analogy.
2. Matte retained the same title in his subsequent book. *See* NICOLAS MATEESCO MATTE, AEROSPACE LAW - FROM SCIENTIFIC EXPLORATION TO COMMERCIAL UTILIZATION (1977).
3. Treaty on Principles Governing the Activities of States in the Exploration and Use of Outer Space, Including the Moon and Other Celestial Bodies, Jan. 27, 1967, 18 U.S.T. 2410, T.I.A.S. No. 6347, 610 U.N.T.S. 205 (entered into

force for the United States Oct. 10, 1967) [hereinafter "Outer Space Treaty"].

4. U.N. Doc. A/AC.105/484, at annex II, para. 9 (1991).

5. *Id.*

6. *See* U.N. Docs. A/AC.105/C.2/L.189 (1992)

7. U.N. Doc. A/AC.105/C.2/1993/CRP.1.

8. U.N. Doc. A/AC.105/C.2/1995/CRP.3/ Rev. 3 of 31 Mar. 1995, reproduced in U.N. Doc. A/AC.105/607, para. 38 and annex I, app. (1995). The text of the Questionnaire also appears in 23 J. SPACE L. 223 (1995).

9. *Id.*

10. U.N. Doc. A/AC.105/639, at 14-15 (1996).

11. U.N. Doc. A/AC. 105/635, at 10 (1996). Similarly, in addition to the U.S. space-shuttle-type vehicle, Germany noted references to future space transportation systems such as, for example, HERMES (ESA), HOTOL (U.K..), HOPE (Japan), S/NGER (Germany) and NASP, the United States Space Plane, which are still in the planning phase and for some of which the financing is disputed or has already been canceled. *Id.* at 10-11.

12. *Id.* at 11.

13. U.N. Doc. A/AC. 105/635/Add.1, at 4-5 (1996).

14. For a comprehensive analysis, see *Stephen Gorove, Legal and Policy Issues of the Aerospace Plane,* 16 J. SPACE L. 147 (1988).

15. *See* The National Aerospace Plane Program, Joint Hearing Before the Subcommittee on Transportation, Aviation and Materials of the Committee on Science, Space and Technology, and the Subcommittee on Research and Development of the Committee on Armed services, U.S. House of Representatives, 100th Cong., 1st Sess. (March 11, 1987), p. 22.

16. For details, see STEPHEN GOROVE, DEVELOPMENTS IN SPACE LAW - ISSUES AND POLICIES 355-56 (1991).

17. Agreement on the Rescue of Astronauts, the Return of Astronauts, and the Return of Objects Launched into Outer Space, April 22, 1968, 19 U.S.T. 7570, T.I.A.S. No. 6599, 672 U.N.T.S. 119 (entered into force for the United States Dec. 3, 1968).

18. That scientific and technological innovations can affect orbiting capabilities appears to be borne out by a recent application of Sky Station International to the Federal Communication Commission (FCC) to create a new Global Stratospheric Telecommunications Service (GSTS) by using a revolutionary technology that holds each of the proposed 250 Sky Station platforms stationary at a 30 km altitude. *See* Request to Establish New GSTS Service, Additional Comments and Petition for Rulemaking, FCC, ET Docket No. 94-124, Mar. 20, 1996. This development suggests the necessity of exercising continued caution to avoid premature determination of demarcation lines.

19. *See* Art. I(d) of the Convention on International Liability for Damage Caused by Space Objects, March 29, 1972, 24 U.S.T. 2389, T.I.A.S. No. 7762, 961 U.N.T.S. 187 (entered into force for the United States Oct. 9, 1973); Art. I(c). of the Convention on the Registration of Objects Launched into Outer Space, opened for signature Jan. 14, 1975, 28 U.S.T. 695, T.I.A.S. No. 8480, 1023 U.N.T.S. 15 (entered into force for the United States Sept. 15, 1976).

20. Stephen Gorove, *Toward a Clarification of the Term 'Space Object' -- An International Legal and Policy Imperative?,* 21 J. SPACE L. 25-26 (1993). The fact that the partial definition of "space object" refers back to itself when speaking of "component parts" of a "space object" and "its" launch vehicle leaves unanswered the fundamental issue of what is or is not a space object or under what circumstances an object becomes or ceases to be a "space object" and the question of the applicability of the relevant space treaty provisions. *Id.* at 12.

21. *See* the SPACE SHUTTLE AND THE LAW 2-3 (Stephen Gorove ed. 1980).

22. U.N. Doc. A/AC.105/635/Add.1, at 4-5 (1996).

23. U.N. Doc. A/AC.105/635/Add.1, at 10-11 (1996).

24. U.N. Doc. A/AC.105/635/Add.1, at 4-5 (1996).

25. According to Terekhov, the phrase "is being elaborated," which appears in U.N. Doc. A/AC.105/635/Add.1, at 5 (1996), is an imprecise translation from Russian; the better translation is "evolving." *See* Andrei D. Terekhov, *Passage of Space Objects Though Foreign Airspace: An International Custom?,* 25 J. SPACE L. 1, at 9 (1997).

26. *Id.* at 4-8.

27. *Id.*

28. *Stephen Gorove, Legal and Policy Issues of the Aerospace Plane,* 16 J. SPACE L. 147, at 148 (1988).
29. PROC. AMERICAN BRANCH OF THE INTERNATIONAL LAW ASSOCIATION, REPORT OF THE SPACE LAW COMMITTEE 105, at 110-111 (1993-1994).
30. *Id.* at 114.
31. *See* Stephen Gorove, *Legal and Policy Issues Raised by the U.N. Questionnaire on Aerospace Objects*, 24 J. SPACE L. 52-53 (1996).
32. *See* note 23 and text preceding it.
33. *See* note 22 and text preceding it.
34. U.N. Doc. A/AC. 105/635/Add.1, at 5 (1996).
35. 1969 I.C.J. 43. *Cf.* discussion of "instant custom" by He Qizhi, *The Outer Space Treaty in Perspective,* 25 J. SPACE L. 93, at 97 (1997), *supra.*
36. The factors that underlie, in general, decision making in the world community are well-known. They operate within the context of the overall conditions and trends prevailing therein. For an overview of the processes of interaction, claim and decision, see MYRES S. MCDOUGAL, HAROLD D. LASSWELL AND IVAN A. VLASIC, LAW AND PUBLIC ORDER IN SPACE 3-137 (1963).
37. While the Geneva Convention appears in favor of ensuring free transit for States having no sea-coast through the territory of a coastal state, this right is predicated on mutual agreement. Under Art. 3 of the Convention:

(1) In order to enjoy the freedom of the seas on equal terms with coastal States, States having no sea-coast should have free access to the sea. To this end States situated between the sea and a State having no sea-coast shall by common agreement with the latter and in conformity with existing international convention accord:

(a) To the State having no sea-coast, on a basis of reciprocity, free transit through their territory; and

(b) To ships flying the flag of that State treatment equal to that accorded to their own ships, or to the ships of any other States, as regards access to seaports and the use of such ports.

(2) States situated between the sea and a State having no sea-coast shall settle, by mutual agreement with the latter, and taking into account the rights of the coastal State or State of transit and the special conditions of the State having no sea-coast, all matters relating to freedom of transit and equal treatment in ports, in case such States are not already parties to existing international conventions.

Convention on the High Seas, Geneva, Apr. 29, 1958, *Entered into force*, Sept. 30, 1962; 13 U.S.T. 2312, 450 U.N.T.S. 82.
38. Manfred Lachs, *Freedom of the Air - the Way to Outer Space, in* AIR AND SPACE LAW: *DE LEGE FERENDA*. 244 (T. L. Masson-Zwaan and P.M.J. Mendes de Leon, eds. 1992).

ADDENDA

REPORT OF THE DISCUSSIONS HELD AFTER THE SESSIONS OF THE 40th COLLOQUIUM ON THE LAW OF OUTER SPACE

SESSION 1
BACKGROUND AND HISTORY OF THE OUTER SPACE TREATY

In the course of the discussions which followed the presentation of the papers, the following issues were addressed, most of which were related to the future of the Moon Agreement:

- To the question why developing countries have not taken more interest in the implementation of the Moon Agreement, *Mr Jasentuliyana* replied that not much more could be done as long as the Agreement wasn't ratified and implemented by the most interested states, which is unlikely unless some of the treaty's most «sensitive» and ambiguous provisions were modified. Mr. Jasentuliyana also drew a parallel between the Moon Agreement and the Convention on the Law of the Sea which was belatedly ratified by industrialised states - after the text of the Convention was amended.

- As to what amendments would be necessary for the Moon Agreement to be revived, *Mr von der Dunk* argued that if the provision regarding the common heritage of mankind were purely and simply deleted, perhaps the industrialised states would accept to ratify the Moon Agreement .

- *Mr. J. Monserrat* (Brazil) wondered whether it is judicious or even possible to modify the Agreement, or whether one should just let space activities take their course without the benefit of a Moon treaty, and Amb. Jankowitsch (OECD/Austria) argued that rules are necessary to ensure the orderly development of space activities. Of course, we should proceed by steering a middle course between elaborating and implementing legal rules and letting the intervening parties compete with each other in conformity with the rules, as competition is necessary to encourage the further developent of space activities.

SESSION 2
CONCEPTS OF SPACE LAW AND THE OUTER SPACE TREATY

- *Prof.Christol* (USA) commented on the paper by *Dr Terekhov* (Russia), and recalled the numerous discussions about the consensus decision making process in the UN, and one of the first papers on that topic, written by Dr. E. Galloway. He held that the discussions about the choice to use "shall" or "should" are interesting, but agreed with the author's conclusion that they are not useful to define the legal status of a document.

- *Mr.White* (USA) enquired about the necessity for more precise international regulation for the exploitation of space resources. *Mr. von der Dunk* (Netherlands) argued that, in the light of growing private activity in this field, and considering the delays in international law-making, it might be better if States would formulate rules at national level to control these activities.

- With regard to *Prof.Kerrest*'s paper and the terminological problems of defining "responsibility" and "liability", *Mr Wirin* (USA) claimed a sense of "majesty" for the concept of responsibility in the Outer Space Treaty. He underlined a possible distinction between responsibility connected to the future and liability connected, instead, to the past, but, in his opinion, the most important concept is the "sense of responsibility" of each State for activities, official or private, in outer space.

- *Ms K. Gorove* (USA) commented on *Mr Poulantzas'* proposal to entrust a Chamber of the ICJ with the settlement of space law disputes, and recalled that in 1993 a Environmental Law Chamber had been created, and that this had possibly precluded the establishment of a special Environmental Court.

- Finally, *Judge A.Koroma* of the International Court of Justice underlined the interest of the themes dealt with by the various papers. Regarding the creation of an "Outer Space Chamber", he argued that the ICJ would certainely examine the matter if the need for such a chamber arose. He reminded that the Court would consider the entire spectrum of international law, and not limit its considerations to space law.

SESSION 3
APPLICATIONS & IMPLEMENTATION OF THE OUTER SPACE TREATY

- *Prof. C.Q. Christol* (USA) found some useful suggestions in *Mr. B. Smith's* presentation on "Problems and Realities in Applying the Provisions of the Outer Space Treaty to Intellectual Property Issues", and wondered whether they might be worthy of consideration by UNISPACE III. He then asked whether Mr. Smith considered that Article VI of the Outer Space Treaty, providing for authorization and control, and violation of the substantive provisions of the Outer Space Treaty by TRW's patent had ever impeded science, to which the answer was "no". *Prof. K.H. Böckstiegel* (Germany) wondered whether the Outer Space Treaty, by presuming and even establishing the free use of outer space, had not already been violated as such by the United States' legislative actions, to which Mr. Smith answered with an emphatic "yes". *Dr. Doyle* finally pointed at the analogy - to some extent - of the patent to ITU's allocation of certain slots and orbits to states, which was however an allocation occurring at the international level by an intergovernmental body with almost global membership.

- Regarding the paper "Mars 96 Planetary Protection Program and Implementations for Mars Environment Preservation", *Dr. L.I. Tennen* (USA) asked the author whether the mission involved life protection experiments on board, to which *Mr. Debus* answered "no", inter alia because the decontamination required resulted (hopefully!) in an environment impossible for sustaining any life.

- With regard to the paper by *Mr. Y. Hashimoto* on "The Legality of Military Observation from Outer Space", *Prof. G. Gal* (Hungary) generally agreed with the observation that military reconnaissance has been allowed, and pointed inter alia to the ABM Treaty in this respect. He then, however, asked to what extent such a bilateral treaty could legalize as such the military activities under consideration. Also, he wondered to what extent the provision by a third state of important data to one if the parties in an armed conflict could be considered legal or illegal. *Mr. Hashimoto* replied firstly that bilateral agreements, while as such of course not binding upon third states, could considerably contribute to the establishment of relevant customary law, particularly if it involved the two most important states from the perspective of global military power and any global treaty on the subject was absent. Secondly, he pointed out that the non-discrimination-requirement made one-sided provision of reconnaissance data in an armed conflict illegal.

- The paper by *Prof. A. Kerrest* (France) on "Launching Spacecraft from the Sea and the Outer Space Treaty: the Sea Launch Project" raised an interesting and heated discussion. *Mr. W. Wirin* (USA) proposed to have the slide showing the list of states involved in Sea Launched again on the overhead projector, and then to ask the audience to 'vote' off-hand, at each particular state, whether the involvement of that state in Sea Launch would suffice for qualifying it as a launching state for cases of damage arising as a consequence of Sea Launch operations. This was done, and if the ensuing 'vote' did one thing, it was confirming that amongst space lawyers little agreement exists so far on the precise scope of the term 'launching state' for liability purposes. *Prof. Christol* asked what the legal relevance of Long Beach being the 'home-port' of the Sea Launch venture would be, to which *Prof. Kerrest* answered that it would be the flag of ship and launch platform which would count under international law. Yet, the 'vote' just taken confirmed that nevertheless even this form of involvement was interpreted by some to make (in this case) the United States a 'launching state'.

- Commenting on the paper by *Prof. P.B. Larsen* (USA) on "Legal Issues in Augmentation of Global Navigation Satellite Systems (GNSS)", *Dr. E. Galloway* (USA) wondered whether the ITU (or another global institution similar to it) would not present the best option for arriving at a coherent international legal regime for these operations. *Prof. Larsen* agreed that ITU had some role to play, but considered the analogy with remote sensing as dealt with at the international level more adequate. Dr. Galloway reiterated, that one overarching international authority with the necessary expertise would be required to realize an internationally workable environment for future GNSS. In addition, *Mr. Kinnell* of INMARSAT pointed out that legal issues regarding either the use of EGNOS, or WAAS, or both, were already being discussed within INMARSAT amongst other fora. Finally, *Mr. F.G. von der Dunk* answered the question of Dr. Galloway in some more detail, by pointing out that within the multiple discussions being presently undertaken on the operational GNSS systems and the augmentation systems, as well as on future systems and a coherent global regime therefore, a prominent topic was that of establishing a separate global GNSS Agency which should guarantee a just and workable balance between the various interests involved.

SESSION 4
THE FUTURE APPLICATIONS OF THE OUTER SPACE TREATY

- *Prof. S. Gorove* (USA) asked whether the definition of (aero-) space-object (in particular with regard to the Aerospace plane) will remain an issue for discussion within the Legal Sub-Committee of UNCOPUOS. *Dr. Schrogl* answered that this topic, which had been put forward by the SU, mainly in view of specific questions regarding the Buran-project and the necessity to approach the landing site through the airspace of third states, has been discussed within the framework of the delimitation-item. Several delegations were of the opinion that no special passage right should be created for such aerospace-objects. Although the Buran project has been terminated, the topic is likely to remain under discussion, if only because in UNCOPUOS-practice the removal of topics is much more difficult than the addition of new topics. *Judge V. Vereshchetin* (ICJ/Russia) later stipulated that the original proposal did not only relate to the Buran-project, but to various other planned systems as well. Also, he added, the Buran project is not dead, it just does not exist anymore, and other projects are under way.

- *Prof. C. Christol* asked Dr. Perek's opinion on the issue of space objects which had become debris and the procedure which had to be followed to determine as such, and, what to do with these objects that were no longer space objects. *Dr. L. Perek* answered that fragments should be separated from objects. This however, poses no major problem. Much can be done through the use of tracking systems, as there are technical ways to determine whether one deals with a fragment or an entire object. The bigger problem is to determine which satellites have ceased to be active. In fact, this is a question to which only the owner knows the answer. This will have to be solved through means of the Registration Convention. Dr. Perek had noted the difference between his approach and the approach suggested by dr. Hoskova, which he attributed to different points of departure. He suggested they work together in order to try to reach a common position.

- *Mr White Jr.* asked Ms. Cramer whether materials to be found on the surface of the Moon were evenly distributed or concentrated in certain specific areas only, since this could be of relevance to the issue of property rights. *Ms. Cramer* answered that the most relevant material, Helium 3, is evenly distributed over the surface of the Moon.

- The last question was from *Mr F. Smith* (UK) who had noticed that the discussion on property rights mostly concerned the Moon, and he wondered what the situation was with respect to asteroids. *Ms. Cramer* replied that there has been mention of plans to claim passing asteroids with the prospect of their exploitation.

<u>*Session Rapporteurs:*</u>
Jean-François Renaud
(University of Paris XI, for session 1)
Marialetizia Longo
(University of Rome, for session 2)
Frans von der Dunk
(International Institute of Air and Space Law, Leiden University, for session 3)
Olivier Ribbelink
(T.M.C. Asser Institute, The Hague/University of Amsterdam, for session 4)

Tanja Masson-Zwaan
IISL Secretary/Colloquium Coordinator

INTERNATIONAL INSTITUTE OF SPACE LAW
of the
INTERNATIONAL ASTRONAUTICAL FEDERATION

Standing Committee on the Status of International Agreements Relating to Activities in Outer Space

Chairman:
Andrei D. Terekhov
Senior Legal Officer
Office of Legal Affairs
United Nations

ANNUAL REPORT - 1997

Background

At the meeting of the IISL Board of Directors held at Brighton, United Kingdom, in October 1987, the Board decided to create a standing committee of the Institute to prepare and submit at each annual Colloquium a report on the status of agreements relating to activities in outer space.

Institute members who can contribute substantively to the work of the committee are invited to serve on it. Membership is open to any interested IISL member. The committee members help collect, review and verify data concerning selected space agreements. Agreements of general global applicability or regional applicability are selected for reporting. Bilateral agreements are not included in the report. In general, the agreements included in the present report, and their status, are believed to be valid as of 1 January 1997.

Coded entries are used in the table as follows:

R	=	ratification, accession or succession
S	=	signature only
D	=	declaration of acceptance of rights and obligations

When no entry is made in a column opposite a state's name, that state has not signed that agreement, or is not a party to it, or that state has withdrawn from the agreement.

COMMITTEE MEMBERSHIP

Prof. Dr. K.-H. Böckstiegel
Institut für Luft- und Weltraumrecht
University of Cologne
Albertus-Magnus-Platz 5000
Cologne, 41 GERMANY

Prof. Carl Q. Christol
1041 Anoka Place
Pacific Palisades, CA 90272
USA

Prof. Dr. Aldo Armando Cocca
Lavalle 1527, piso 10
1048 Buenos Aires
ARGENTINA

Prof. Dr. Isabella
Diederiks-Verschoor
President Emeritus, IISL
Leestraat 43, 3743 EH Baarn
NETHERLANDS

Dr. Stephen E. Doyle
3431 Bridget Brae
Shingle Springs, CA 95682
USA

Dr. Eilene Galloway
4612 29th Place, NW
Washington, D.C. 20008
USA

Prof. Dr. Andrzej Gorbiel
ul. Marysinska 90 m.69
91-850 Lodz
POLAND

Prof. Dr. Vladimir Kopal
IAF General Counsel
Vidlicová 2200
16000 Prague 6 - Dejvice
CZECH REPUBLIC

Prof. Francis Lyall
Department of Public Law
University of Aberdeen
Aberdeen, AB92UB SCOTLAND

Mr. G. Lafferranderie
Legal Adviser, ESA
8-10, rue Mario Nikis
Paris 75738 FRANCE

Dr. HE Qizhi
Legal Advisor
Department of Treaty & Law
Ministry of Foreign Affairs
Beijing CHINA

Dr. W. von Noorden
Legal Adviser, INMARSAT
40 Melton St., Euston Square
London NW1 2EQ ENGLAND

Prof. Dr. Neri Sybesma-Knol
Vrije Universiteit Brussel
Pleinlaan 2, 1050 Brussels
BELGIUM

Dr. A. D. Terekhov (Chairman)
Office of Legal Affairs
Room 3430G, UN Secretariat
New York, NY 10017 USA

AGREEMENTS INCLUDED

1. **1963 NTBT** - Treaty Banning Nuclear Weapon Tests in the Atmosphere, in Outer Space and Under Water; done at Moscow on 5 August 1963; entered into force on 10 October 1963. (14 UST 1313; TIAS 5433; 480 UNTS 43)

2. **1967 OST** - Treaty on Principles Governing the Activities of States in the Exploration and Use of Outer Space, including the Moon and Other Celestial Bodies; done at Washington, London and Moscow on 27 January 1967; entered into force on 10 October 1967. (18 UST 2410; TIAS 6347; 610 UNTS 205)

3. **1968 ARRA** - Agreement on the Rescue of Astronauts, the Return of Astronauts and the Return of Objects Launched into Outer Space; done at Washington, London and Moscow on 22 April 1968; entered into force on 3 December 1968. (19 UST 7570; TIAS 6599; 672 UNTS 119)

4. **1972 LIAB** - Convention on International Liability for Damage Caused by Space Objects; done at Washington, London and Moscow on 29 March 1972; entered into force on 1 September 1972. (24 UST 2389; TIAS 7762; 961 UNTS 187)

5. **1975 REG** - Convention on Registration of Objects Launched into Outer Space; done at New York on 14 January 1975; entered into force on 15 September 1976. (28 UST 695; TIAS 8480; 1023 UNTS 15)

6. **1979 MOON** - Agreement Governing the Activities of States on the Moon and Other Celestial Bodies, done at New York on 18 December 1979; entered into force on 11 July 1984. (18 ILM 1434; UN doc. A/RES/34/68 of 14 December 1979, Annex)

7. **1971 INTL** - Agreement Relating to the International Telecommunications Satellite Organization (INTELSAT) with annexes, and Operating Agreement Relating to the International Telecommunications Satellite Organization with annex; done at Washington on 20 August 1971; entered into force on 12 February 1973. (23 UST 3813, 4091; TIAS 7532)

8. **1976 INMR** - Convention on the International Maritime Satellite Organization (INMARSAT) with annex, and the Operating Agreement on the International Maritime Satellite Organization with annex; done at London on 3 September 1976; entered into force on 16 July 1976. (31 UST 1, 135; TIAS 9605)

9. 1971 INTR - Agreement on the Establishment of the "Intersputnik" International System and Organization of Space Communications; done at Moscow on 15 November 1971; entered into force on 12 July 1972. (862 UNTS 3)

10. 1976 INTC - Agreement on Cooperation in the Exploration and Use of Outer Space for Peaceful Purposes (INTERCOSMOS); done at Moscow on 13 July 1976; entered into force on 25 March 1977. (16 ILM 1)

11. 1976 ARBS - The Agreement of the Arab Corporation for Space Communications (ARABSAT); done at Cairo on Wednesday, 14 Rabi Al Akhar 1396 H., corresponding to April 14, 1976; entered into force on July 16, 1976. (Space Law and Related Documents, U.S. Senate, 101st Congress, 2nd Session, 395 (1990))

12. 1975 ESA - Convention for the establishment of a European Space Agency (ESA) with annexes; done at Paris on 30 May 1975; entered into force on 30 October 1980. (14 ILM 864)

13. 1982 EUTL - Convention establishing the European telecommunications satellite organization (EUTELSAT); done at Paris on 15 July 1982; entered into force on 1 September 1985. (XI Annals of Air and Space Law 416 (1986); UK Misc. No.4, Cmnd. 9154 (1984))

14. 1983 EUMT - Convention for the establishment of a European organization for the exploitation of meteorological satellites (EUMETSAT); done at Geneva on 24 May 1983; entered into force on 19 June 1986. (Federal Republic of Germany Bundesgesetzblatt, Jahrang 1987, Teil 11, p. 256 (1987). This Convention has been published in the national bulletins of the ratifying States)

15. 1974 BRUS - Convention relating to the distribution of programme-carrying signals transmitted by satellite; done at Brussels on 21 May 1974; entered into force on 25 August 1979. (1144 UNTS 3)

The texts of all the above agreements, except 1974 BRUS, 1982 EUTL and 1983 EUMT, were published in Space Law and Related Documents, U.S. Senate, 101st Congress, 2nd Session, June 1990.

Status of International Agreements Relating to Activities in Outer Space

STATE	TREATY (1) 1963 NTBT	(2) 1967 OST	(3) 1968 ARRA	(4) 1972 LIAB	(5) 1975 REG	(6) 1979 MOON	(7) 1971 INTL	(8) 1976 INMR	(9) 1971 INTR	(10) 1976 INTC	(11) 1976 ARBS	(12) 1975 ESA	(13) 1982 EUTL	(14) 1983 EUMT	(15) 1974 BRUS
Afghanistan	R	R					R		R						
Albania													R		
Algeria	S	R		S			R	R			R				
Andorra													R		
Angola							R								
Antigua and Barbuda	R	R	R	R	R										
Argentina	R	R	R	R	R		R	R							S
Armenia	R						R						R		R
Australia	R	R	R	R	R	R	R	R							R
Austria	R	R	R	R	R	R	R					R	R	R	R
Azerbaijan							R						R		
Bahamas	R	R	R				R	R							
Bahrain							R	R			R				
Bangladesh	R	R					R	R							
Barbados		R	R				R								
Belarus	R	R	R	R	R				R				R		
Belgium	R	R	R	R	R		R	R				R	R	R	S
Benin	R	R		R			R								
Bhutan	R						R								
Bolivia	R	S	S				R								
Bosnia and Herzegovina							R						R		R
Botswana	R	S	R	R			R								
Brazil	R	R	R	R			R	R							S
Brunei Darussalam							R	R							
Bulgaria	R	R	R	R	R		R	R	R	R			R		
Burkina Faso	S	R					R								
Burundi	S	S		S	S		S								
Cambodia				S											
Cameroon	S	S	R				R	R							
Canada	R	R	R	R	R		R	R				(1)			
Cape Verde	R						R								

STATE	TREATY (1) 1963 NTBT	(2) 1967 OST	(3) 1968 ARRA	(4) 1972 LIAB	(5) 1975 REG	(6) 1979 MOON	(7) 1971 INTL	(8) 1976 INMR	(9) 1971 INTR	(10) 1976 INTC	(11) 1976 ARBS	(12) 1975 ESA	(13) 1982 EUTL	(14) 1983 EUMT	(15) 1974 BRUS
Central African Republic	R	S		S			R								
Chad	R						R								
Chile	R	R	R	R	R	R	R	R							
China		R	R	R	R		R	R							
Colombia	R	S	S	S			R	R							
Congo			S				R								
Costa Rica	R		S	S			R	R							
Cote d'Ivoire	R						R								S
Croatia	R						R	R					R		R
Cuba		R	R	R	R			R	R	R					
Cyprus	R	R	R	R	R		R	R					R		S
Czech Republic	R	R	R	R	R		R	R	R	R			R		
Dem. People's Rep. of Korea									R						
Denmark	R	R	R	R	R		R	R				R	R	R	
Djibouti											R				
Dominican Republic	R	R	S	R			R								
Ecuador	R	R	R	R			R								
Egypt	R	R	R	S			R	R			R				
El Salvador	R	R	R	S			R								
Equatorial Guinea	R	R					R								
Ethiopia	S	S					R								
Fiji	R	R	R	R			R								
Finland	R	R	R	R			R	R				R	R	R	
France		R	R	R	R	S	R	R				R	R	R	S
Gabon	R		R	R			R	R							
Gambia	R	S	R	S											
Georgia									R				R		
Germany	R	R	R	R	R		R	R	R			R	R	R	R
Ghana	R	S	S	S			R	R							
Greece	R	R	R	R			R	R					R	R	R
Guatemala	R			S		S	R								

STATE	TREATY (1) 1963 NTBT	(2) 1967 OST	(3) 1968 ARRA	(4) 1972 LIAB	(5) 1975 REG	(6) 1979 MOON	(7) 1971 INTL	(8) 1976 INMR	(9) 1971 INTR	(10) 1976 INTC	(11) 1976 ARBS	(12) 1975 ESA	(13) 1982 EUTL	(14) 1983 EUMT	(15) 1974 BRUS
Guinea							R								
Guinea-Bissau	R	R	R												
Guyana		S	R												
Haiti	S	S	S	S			R								
Holy See		S					R						R		
Honduras	R	S		S			R								
Hungary	R	R	R	R	R		R	R	R	R			R		
Iceland	R	R	R	S			R	R					R		
India	R	R	R	R	R	S	R	R							
Indonesia	R	S					R	R							
Iran (Islamic Republic of)	R	S	R	R	S		R	R							
Iraq	R	R	R	R			R	R			R				
Ireland	R	R	R	R			R					R	R	R	
Israel	R	R	R	R			R	R							S
Italy	R	R	R	R			R	R				R	R	R	R
Jamaica	R	R	S				R								
Japan	R	R	R	R	R		R	R							
Jordan	R	S	S	S			R				R				
Kazakstan							R		R						
Kenya	R	R		R			R								R
Kuwait	R	R	R	R			R	R			R				
Kyrgyzstan							R		R						
Lao People's Dem. Republic	R	R	R	R					R						
Latvia								R					R		
Lebanon	R	R	R	S			R	R			R				S
Lesotho		S	S												
Liberia	R							R							
Libyan Arab Jamahiriya	R	R					R				R				
Liechtenstein				R			R						R		
Lithuania													R		
Luxembourg	R	S	S	R			R						R		

TREATY

STATE	(1) 1963 NTBT	(2) 1967 OST	(3) 1968 ARRA	(4) 1972 LIAB	(5) 1975 REG	(6) 1979 MOON	(7) 1971 INTL	(8) 1976 INMR	(9) 1971 INTR	(10) 1976 INTC	(11) 1976 ARBS	(12) 1975 ESA	(13) 1982 EUTL	(14) 1983 EUMT	(15) 1974 BRUS
Madagascar	R	R	R				R								
Malawi	R						R								
Malaysia	R	S	S				R	R							
Maldives			R												
Mali	S	R		R			R								
Malta	R		S	R			R	R					R		
Marshall Islands								R							
Mauritania	R						R				R				
Mauritius	R	R	R				R	R							
Mexico	R	R	R	R	R	R	R	R							R
Micronesia (Fed. States of)							R								
Monaco			S				R	R					R		
Mongolia	R	R	R	R	R				R	R					
Morocco	R	R	R	R		R	R				R				R
Mozambique							R	R							
Myanmar	R	R	S												
Nepal	R	R	R	S			R								
Netherlands	R	R	R	R	R	R	R	R				R	R	R	
New Zealand	R	R	R	R			R	R							
Nicaragua	R	S	S	S	S		R		R						R
Niger	R	R	R	R	R		R								
Nigeria	R	R	R				R	R							
Norway	R	R	R	R	R		R	R				R	R	R	
Oman				S			R	R			R				
Pakistan	R	R	R	R	R	R	R	R							
Panama	R	S		R			R	R							R
Papua New Guinea	R	R	R	R			R								
Paraguay	S						R								
Peru	R	R	R	S	R	S	R	R							R
Philippines	R	S	S	S		R	R	R							
Poland	R	R	R	R	R		R	R	R	R			R		

STATE	TREATY (1) 1963 NTBT	(2) 1967 OST	(3) 1968 ARRA	(4) 1972 LIAB	(5) 1975 REG	(6) 1979 MOON	(7) 1971 INTL	(8) 1976 INMR	(9) 1971 INTR	(10) 1976 INTC	(11) 1976 ARBS	(12) 1975 ESA	(13) 1982 EUTL	(14) 1983 EUMT	(15) 1974 BRUS
Portugal	S		R				R	R					R	R	R
Qatar				R			R	R			R				
Republic of Korea	R	R	R	R	R		R	R							
Republic of Moldova													R		
Romania	R	R	R	R		S	R	R	R	R			R		
Russian Federation	R	R	R	R	R		R	R	R	R			R		R
Rwanda	R	S	S	S			R								
Samoa	R														
San Marino	R	R	R										R		
Saudi Arabia		R		R			R	R			R				
Senegal	R		S	R			R	R							S
Seychelles	R	R	R	R	R										
Sierra Leone	R	R	S	S											
Singapore	R	R	R	R	S		R	R							
Slovakia	R	R	R	R	R			R	R	R			R		
Slovenia	R		R	R											R
Somalia	S	S	S				R				R				
South Africa	R	R	R	S			R	R							
Spain	R	R		R	R		R	R				R	R	R	S
Sri Lanka	R	R		R			R	R							
Sudan	R						R				R				
Suriname	R														
Swaziland	R		R				R								
Sweden	R	R	R	R	R		R	R				R	R	R	
Switzerland	R	R	R	R	R		R	R				R	R	R	R
Syrian Arab Republic	R	R	R	R			R		R		R				
Tajikistan							R		R						
Thailand	R	R	R				R								
Togo	R	R		R			R								
Tonga	R	R	R												
Trinidad and Tobago	R	S		R			R								R

STATE	TREATY (1) 1963 NTBT	(2) 1967 OST	(3) 1968 ARRA	(4) 1972 LIAB	(5) 1975 REG	(6) 1979 MOON	(7) 1971 INTL	(8) 1976 INMR	(9) 1971 INTR	(10) 1976 INTC	(11) 1976 ARBS	(12) 1975 ESA	(13) 1982 EUTL	(14) 1983 EUMT	(15) 1974 BRUS
Tunisia	R	R	R	R			R	R			R				
Turkey	R	R	S				R	R					R	R	
Turkmenistan									R						
Uganda	R	R					R								
Ukraine	R	R	R	R	R			R					R		
United Arab Emirates							R	R			R				
United Kingdom	R	R	R	R	R		R	R				R	R	R	
United Rep. of Tanzania	R			S			R								
United States of America	R	R	R	R	R		R	R							R
Uruguay	R	R	R	R	R	R	R								
Venezuela	R	R	S	R			R								
Viet Nam		R					R		R	R					
Yemen	R	R	R				R		R		R				
Yugoslavia	R	S	R	R	R		R	R					R		R
Zaire	R	S	S	S			R								
Zambia	R	R	R	R			R								
Zimbabwe							R								
ORGANIZATION															
European Space Agency (ESA)			D	D	D										
European Telecommunications Satellite Org. (EUTELSAT)				D											
TOTALS:	R-123	R-93	R-83	R-76	R-39	R-9	R-139	R-80	R-23	R-10	R-20	R-14	R-45	R-17	R-21
	S-10	S-27	S-24	S-26	S-4	S-5									S-10
			D-1	D-2	D-1										

Note: (1) Canada has a Cooperation Agreement with ESA, but is not a Member.

ADDENDUM

IISL-ECSL Symposium Celebrating the 30th Anniversary of the Outer Space Treaty

Held on occasion of the thirty-sixth session of the Legal Subcommittee of COPUOS
Vienna International Centre, 1 April 1997

INTRODUCTION

by

E. Fasan
Coordinator

On April 1, 1997 a Symposium Celebrating the 30th Anniversary of the Outer Space Treaty was held shortly after the opening of the 36th Session of the Legal Subcommittee of the UN Committee on the Peaceful Uses of Outer Space. The Symposium was organized by the International Institute of Space Law (IISL) and the European Center for Space Law (ECSL).

On request of the President of the IISL, *Dr. Nandasiri Jasentuliyana*, the session was coordinated by *Dr. Ernst Fasan*, Honorary Director of IISL, and *Mr. Jitendra Thaker*, Legal Affairs Officer to the UN Office for Outer Space Affairs served as rapporteur.

Fasan welcomed the President of the Legal Subcommittee, *Dr. Vaclav Mikulka*, and many of its distinguished delegates and guests. He introduced the four speakers in the order of the agenda, namely: *Prof. Yuri Kolosov*, of the Moscow State Institute of International Law; *Dr. Ram Jakhu*, of the International Space University; and *Dr. Frans von der Dunk*, of the Institute of Air and Space Law of Leiden University.

After presentation of the papers by the three speakers, there followed a commentary on the topic by *Prof. Vladimir Kopal*, of the Charles University in Prague, and a "questions and answers" period.

Thereafter, *Dr. Gabriel Lafferranderie*, Chairman of the ECSL, and Legal Adviser of ESA, presented the United Nations and the COPUOS delegations copies of the book, "Outlook on Space Law Over the Next 30 Years". The book had been published by the ECSL to celebrate the 30th Anniversary of the Outer Space Treaty. Jasentuliyana received the book on behalf of the UN, and congratulated *Lafferranderie* and the authors. He closed the Symposium with thanks to the speakers and participants.

In conclusion, the Interim Chairman of the Legal Subcommittee, *Kopal*, expressed his appreciation for the information given and thanked the IISL and the ECSL for organizing the event.

A little reception closed the program.

In its Report of April 14, 1997 (A/AC.105/674), the Legal Subcommittee noted with satisfaction that this Symposium had been held, and agreed that IISL and ECSL should again be invited to hold a Symposium on Space Law at the Subcommittee's 37th session.

IISL-ECSL Symposium Celebrating the 30th Anniversary of the Outer Space Treaty
Held on occasion of the thirty-sixth session of the Legal Subcommittee of COPUOS
Vienna International Centre, 1 April 1997

BACKGROUND AND HISTORY OF THE OUTER SPACE TREATIES

by

Y. Kolosov
Professor, Moscow State Institute of International Law

The international law of outer space started emerging in the 1950's. A whole range of international instruments have been elaborated since then to regulate the exploration and use of outer space.

The crucial role in this field of law belongs to the UN Outer Space Committee with its Legal and Scientific -and-Technical Subcommittees[1].

Tribute should also be paid to a great number of lawyers from many countries who paved the way towards the successful activities of politicians and diplomats within the United Nations. I would like to mention just several names: A. Araujo, K. Böckstiegl, J. Busak, M. Bourely, A. Cocca, B. Cheng, J. Cooper, P. Chaumont, C. Christol, I. Diederiks-Verschoor, E. Fasan, F. Fiorio, D. Goedhuis, A. Gorbiel, S. Gorove, E. Galloway, G. Gal, I. Herczeg, A. Haley, N. Jasentuliyana, C. Jenks, V. Kopal, E. Korovin, M. Lachs, N. Matt, M. Marcoff, E. Pépin, R. Quadri, G. Valladao, E. Vasilevskaya, V. Verestchetin, G. Zhukov and many others.

The essential role in the progressive development of the outer space law play international scientific institutions among which the International Institute of Space Law and the International Academy of Astronautics must be mentioned.

Joint efforts of scientists and diplomats brought to miraculous results. Within the years of 1959 and 1996, i.e. merely three and a half decades the COPUOS drafted and the UN General Assembly adopted five major outer space treaties and five sets of principles.

Thus, one can make several conclusions. The COPUOS has been the principal body in the process of codification and progressive development of the international outer space law which is one of the goals of the United Nations in accordance with Article 13 (a) of its Charter.

The international outer space law has been mainly developing as a treaty law, leaving little place for customary rules of law.

The tempo of the development of the international outer space law is second to none.

On the occasion of the thirtieth anniversary of the Outer Space Treaty of 1967 and the fortieth anniversary of the launching of the first man-made object into outer space in the USSR on 4 October, 1957, it is quite opportune to recollect some facts from the background and history of the outer space treaties.

The evolution of the outer space treaties was not an easy one. To understand the issues of controvercies and the emergence of the consensus one must study the UN records. The systematic analysis of the preparatory work may be found in the manual on space law compiled and edited by N. Jasentuliyana and R.S.K. Lee[2].

The adoption of the Outer Space Treaty in December 19, 1966, and its entry into force in October 10, 1967, was preceded by the adoption of the Declaration on Legal Principles Governing the Activities of States in the Exploration and Use of Outer Space, adopted by the UN General Assembly in December 13, 1963[3]. The adoption of the Declaration was in its turn preceded by UNGA resolutions 1472 (XIV) of 12 December, 1959, and 1721 (XVI) of 20 December, 1961. They defined the competence of the COPUOS and formulated the initial basic principles of the exploration and uses of outer space.

Those initial principles were the following:

"(a) **International law, including the Charter of the United Nations** applies to outer space and celestial bodies;
(b) Outer space and celestial bodies are free for exploration and use by all States in conformity with international law and are not subject to national appropriation".

In 1962 (UNGA Res. 1802 (XVII) of December 14) the next step was made -the applicability of international law, including the Charter of the United Nations, was extended to the exploration and use of outer space (preambular paragraph 2) (italics added -Y.K.).

The Declaration comprised the principles of the 1961 resolution whereas the Outer Space Treaty reflected the provisions of the Declaration.

Thus, a conclusion can be made that the history of the Outer Space Treaty began in 1961.

Among the "predecessors" of the Outer Space Treaty mention should be also made of the USA-USSR agreement on the ban of deployment of nuclear weapons or other weapons of mass destruction in the outer space which was endorsed by the UNGA resolution 1884 (XVIII) of 17 Octobeer 1963. It was reflected in Article IV of the Treaty.

The first draft treaty was submitted by the USSR in June 16, 1966. Next day the USA submitted its own version of the draft treaty, which referred only to the Moon and other celestial bodies.

The negotiations in the Legal Sub-Committee and at the XXI-st session of the UN General Assembly brought to the agreement on the text of the Outer Space Treaty in December 19, 1966[4].

This could have been hardly possible if there was no Declaration of 1963.

The Treaty was opened for signature in 27 January 1967 simultaneously in the capitals of the three depositaries -in London, Moscow and Washington.

The day of entry into force of the Outer Space Treaty has become the birthday of the outer space law. As from this date the new stage of the development of the outer space law started. The fundamental principles of the exploration and use of outer space, including the Moon and other celestial bodies, have acquired the form of legally binding rules. From that point further development of these principles was begun.

The next four treaties are the Agreement on the Rescue of Astronauts, the Return of Astronauts, and the Return of Objects Launched into Outer Space; the Convention on International Liability for Damage Caused by Space Objects; the Convention on Registration of Objects Launched into Outer Space; and the Agreement Governing the Activities of States on the Moon and Other Celestial Bodies. They develop the following articles of the Outer Space Treaty: Articles V and VIII (the Agreement on the Rescue of Astronauts), Article VII (the Liability Convention), several articles mentioning registration of objects, including articles V and VIII (the Registration Convention), and Articles IV, XII and others (the Moon Agreement).

The Agreement on the Rescue of Astronauts has its own background. In March 20, 1962 the Chairman of the Council of Ministers of the USSR in his message to the President of the USA suggested the elaboration of an international agreement on the assistance in search and rescue of space objects in case of emergency landing, including the rescue of astronauts[5].

In March 9, 1964 the USSR delegation in the COPUOS submitted a draft agreement on the rescue of astronauts and space objects in case of an accident or an emergency landing.

In 1964 draft agreements on this subject were also submitted by the USA and jointly by Australia and Canada.

The UN General Assembly urged the COPUOS to continue the elaboration, inter alia, of an international agreement on the rescue of astronauts[6].

The draft was agreed upon by the Legal Sub-Committee of the COPUOS on 15 December 1967 during the XXII-nd session of the UN General Assembly. The agreement was opened for signature in April 22, 1968, and entered into force in December 3, 1968.

The work on the clraft was finalized in a relatively short period of time due to a couple of factors. There was no controversy on the issue between the two space powers. The subject matter was of a highly humanitarian value and the goal of the agreement was a noble one. Another explanation was that the majority in the COPUOS were interested in adopting as soon as possible the convention on the international liability for damage caused by space objects. They rightly believed that finalization of the Rescue Agreement would enable expediting the adoption of the liability convention.

The Convention on International Liability for Damage Caused by Space Objects (the Liability Convention) is a unique treaty which in fact is the first and the only special international instrument dealing with the State liability.

In the process of drafting the Outer Space Treaty there was a disagreement on the right to conduct activities in outer space. The USA delegation suggested to recognize that States as well as their non-governmental agencies have the right to carry out activities in outer space. The USSR delegation suggested that States only may be the actors of space activities. The compromise was reached on the basis of the formula according to which there was no special provision on the issue of actors of space activities. Instead, in Article VI of the Outer Space Treaty in the indirect form there was a recognition of the right of non-governmental agencies to conduct outer space activities, but States as such shall bear international responsibility for any national activities in outer space. More than that, the activities of non-governmental entities in outer space require authorization and continuing supervision by appropriate States Parties to the Treaty.

It was also agreed in Article VII of the Outer Space Treaty that States will be internationally liable for damage caused by space objects or their component parts on the earth, in air or in outer space.

Therefore in the Liability Convention it is stated that "a launching State shall be absolutely liable to pay compensation for damage caused by its space object" (Article II).

Another controversial issue was the upper limit of liability. Some delegations from the developing countries insisted on establishing of a very high limit (several hundred billion dollars) which was unacceptable for launching States. In the end of lengthy deliberations no limit at all was fixed in the Liability Convention.

The legal force of the Claims Commission established under Article XIV of the Liability Convention was still another controversial issue. In Article XIX it was agreed that the decisions of the Commission shall be final and binding it the parties to the dispute so agreed. Otherwise the Commission shall adopt a recommendation which the parties should consider in good faith.

The elaboration of the Liability Convention was conducted on the basis of several drafts submitted in 1964 by Hungary, the USA and Belgium. In 1967, India and Italy also submitted their drafts on this issue. The work started in the Legal Sub-Committee of the COPUOS in 1964. In 1971 the work was finalized and the 26th session of the UN General Assembly approved the text of the Convention[7]. It was opened for signature in March 29, 1972, in London, Moscow and Washington. It entered into force on 30 August, 1972.

It may be noted that the General Assembly adopted the Liability Convention by vote (93 in favour and 4 abstentions, no delegation voted against it).

The Convention on Registration of Objects Launched into Outer Space was adopted in 1974[8]. It entered into force in September 15, 1976.

Initially the system of the international registration of-space objects was established by the UN General Assembly resolution 1721 (B) of December 20, 1961. The UN called upon launching States to furniish promptly to the COPUOS, through the Secretary-General, the information for the registration of launchings. The Secretary-General was invited to maintain a register of this information.

The actual informnation about launchings started coming to the UN since 1962 on a voluntary basis.

In 1968 France submitted to the Legal Sub-Committee the draft convention on registration of space objects. In 1972 it was substituted by a joint Franco-Canadian draft.

After the adoption of the Registration Convention the submission of such information became mandatory.

As at 31 December 1995 there were only 39 States Parties to this convention, but such launching States as the USA, Russia, China and France are among them. The European Space Agency declared acceptance of the rights and obligations of the Convention in accordance with its Article VII.

In 1966 the USSR submitted a draft treaty governing the exploration of the Moon and other celestial bodies.

1969 American astronauts landed on the Moon.

In 1969 Poland and Argentina proposed in the Legal Sub-Committee their drafts of rules relating to man's activities on the surface of the Moon and other celestial bodies. Later, Argentina, France and Poland prepared a joint text on this issue.

In 1970 Argentina proposed a draft agreement on the principles governing activities in the use of natural resources of the Moon and other celestial bodies.

In 1972 the Soviet Union submitted to the Committee on the Peaceful Uses of Outer Space a draft treaty relating to the Moon. There were some doubts expressed as to the need of such a treaty since the Outer Space Treaty of 1967 contained specific provisions relating to the legal regime of the Moon. In the process of drafting the agreement some innovations were added. The most important one dealt with the natural resources of the Moon.

One of the issues under discussion was whether a moratorium should be placed on the exploitation of the natural resources of the Moon. The agreement was achieved through the adoption of the formula according to which "the Moon and its natural resources are the common heritage of mankind" (Article 11 (para. 1) of the Agreement Governing the Activities of States on the Moon and Other Celestial Bodies).

This formula means the non-appropriation of the Moon, any part of it or its natural resources, the equality of all States Parties for the purposes of exploration and use of the Moon, the establishment of a special international regime governing the exploitation of the natural resources of the Moon when such exploitation is about to become feasible.

Sometimes the discussion acquired doctrinal dimensions. For example, there was an interesting exchange of opinions between the delegations of the USSR and Argentina on the meaning of the notion "common heritage of all mankind"[9]. The difficult compromise became possible due to ardorous work of the delegation of Austria.

Another problem related to the right to use certain orbits and trajectories.

In December 5, 1979, the UN General Assembly adopted the final Agreement[10]. In December 18, 1979, it was opened for signature.

The delegations of the USA and the USSR in the COPUOS were among the very last to support the consensus on the final text of the Agreement. Neither the USA nor the USSR signed this Agreement[11]. It took four and a half years for the Agreement to enter into force on July 11, 1984, after it received five ratifications. As at 31 December 1995 there were only nine States Parties to the Agreement and five more Signatories.

In 1980 Dr. Ernst Fasan mentioned that the Moon Agreement left some unsolved problems and concluded that "legal development may not stop here but ought to go ahead in a pace reasonably in accordance with further scientific development[12].

At this moment I would like to wish all the success to the Legal Sub-Committee and to the COPUOS on the way of further progressive development of the outer space law.

Today one can definitely speak about the existence of a new branch of international law.

The five outer space treaties are the foundation of this branch. They serve as a basis of international cooperation in the exploration and use of outer space, one of the global provinces of mankind.

ENDNOTES

1. The Committee on the Peaceful Uses of Outer Space was established on 12 December, 1959, by the UN General Assembly resolution 1472 (XIV). The two Sub-Committees were instituted in 1962.

2. Manual on Space Law. Volumes I-IV.

Compiled and Edited by N. Jasentuliyana and R.S.K. Lee. 1979-1981. Oceana Publications, Inc./Dobbs Ferry, New York.

3. United Nations General Assembly Resolution 1962 (XVIII), 13 Dec., 1963.
4. United Nations General Assembly Resolution 2222 (XXI), 19 Dec., 1966.

5. Newspaper "Pravda", 22 March, 1962.

6. United Nations General Assembly Resolution. 2130 (XX), 21 Dec., 1965.

7. United Nations General Assembly Resolution 2777 (XXVI), 29 Nov., 1971.

8. United Nations General Assembly Resolution 3235 (XXIX), 12 Nov., 1974.

9. See United Nations Doc. A/AC.105/101, 28 March, 1973 and 17 April 1973.

10. United Nations General Assembly Resolution 34/68, 5 December, 1979.

11. See the analysis of the Moon Treaty in "Agreement governing the Activities of States on the Moon and Other Celestial Bodies. 96th Congress, 2nd Session, Committee Print, US Senate, May 1980.

12. E. Fasan. Some legal problems regarding the Moon. "Proceedings of the 23rd Colloquium on the Law of Outer Space". International Institute of Space Law of the International Astronautical Federation. September 21-28, 1980, Tokyo, p. 10.

IISL-ECSL Symposium Celebrating the 30th Anniversary of the Outer Space Treaty
Held on occasion of the thirty-sixth session of the Legal Subcommittee of COPUOS
Vienna International Centre, 1 April 1997

APPLICATION AND IMPLEMENTATION OF THE 1967 OUTER SPACE TREATY [1*]

by

Ram S. Jakhu
Professor, International Space University

Introduction

It is a matter of great pleasure and honor for me to be present at this internationally organized Legal Symposium Celebrating the 30th Anniversary of the 1967 Outer Space Treaty[2] - the constitution of outer space. I congratulate the convenors of the Symposium and express my deep appreciation to Dr. Ernst Fasan and Dr. Nandasiri Jasentuliyana for extending me an invitation to participate in this Symposium.

Anniversaries of important events and institutions are always good occasions to critically appreciate as well as to take stock of their achievements. In this context, I would like to join those scholars from all over the world who have almost unanimously been declaring the 1967 Treaty a big success in creating an appropriate order in outer space. I will say it has been perhaps the finest achievement of diplomacy and statesmanship of those involved in its negotiation and drafting, especially during the height of cold war when the world was deeply polarized in two diagonally opposed political and economic systems. In those very tense days, when the human civilization was almost at the brink of extinction, the 1963 Resolution of the UN General Assembly on outer space which eventually was transformed into the 1967 Treaty, showed a ray of hope that the human beings are capable of finding peaceful solu-tions to global problems. The Legal Sub-committee of the UN Comm-ittee on Peaceful Uses of Outer Space (COPOUS) should be congra-tulated for the conclusion of this very significant piece of interna-tional legislation on outer space.

Probably, the most important achievement of the 1967 Treaty, in my opinion, has been that it laid down *a priori* a fundamental legal principle of freedom of exploration and use of outer space by all States. At the same time it categorically and unambiguously denied the application of sovereignty, especially the traditional territorial sovereignty, to outer space and celestial bodies. I am sure that the learned audience does not need to be reminded that territorial soverei-gnty has been, and to a large extent even today, the main cause of almost all catastrophic wars that were fought by the so-called "social animals". Undoub-tedly, this principle has been well respected over the last thirty years. In my opinion, there is no apparent serious threat to this principle at least in the near future, though a very small minority of authors attempt to interpret some provisions of the 1967 Treaty allowing indirect appropriation of outer space or a part of it. I will discuss this issue shortly.

The list of the achievements of the 1967 Treaty is long and I do not intend to describe them. I believe that the best way to celebrate the 30th anniversary of the 1967 Treaty is to point out the areas where improvements must be made so that it can serve its purpose for the next thirty years and beyond. As pointed out earlier, the Treaty has been successful during the past thirty years. The question now arises whether the Treaty in its entirety will remain useful and valid in the future. The answer to this question is obviously in the negative and certainly not in its present form. One apparent reason for this is that the magnitude and nature of the space activities at the time of conclusion of this Treaty, and those envisioned at that time, have changed and will alter dramatically and significantly in the future. In the beginning space activities were carried out by States and that too by a very limited elite group of States. The number of space-faring nations is quickly increasing. But what is more important is that in the space arena States are rapidly being replaced by private entities in pursuant to new politico-economic trends of privati-zation, commercialization and globalisation of almost all human activities and space activities are not an exception. Therefore, the international treaty (i.e. the 1967 Treaty), which was concluded essentially to govern the activities of the States and that too essentially of scientific and exploratory nature, is under question whe-ther it still remains relevant in the new era of space activities. In my view, if the basis of a law has changed, the law itself must change in order to remain valid and useful.

Since the areas for improvements of the 1967 Treaty as well as the other space treaties[3] are extensive and can not be adequately covered in the limited time at my disposal, I will concentrate only on some of those issues which directly relate to the subject of my presentation; i.e. the application and implemen-tation of the 1967 Treaty. These issues essentially are : where, to whom and to what does the 1967 Treaty apply ? How is the 1967 Treaty applied and implemented ?

Where does the 1967 Treaty apply ?

It is obvious that the Treaty applies to activities in outer space, the Moon and other Celestial bodies. However, what has not been settled in the Treaty is where does outer space begin. The boundary between air space and outer space, both having different and inconsistent legal regimes (exclusive soverei-gnty over air space and freedom of outer space), has been the longest unresolved legal problem. This problem has been discussed in-side and out-side the UN, it is still on the agenda of the Legal Sub-committee and there is no sign of any clear resolution either. The issue has been discussed exten-sively in the Subcommittee and in legal literature for about four decades. Some authors are of the opinion that the height of the lowest perigee of a satellite (or 100 or 110 Kms above the sea level) is being recognized under customary international law as the demar-cation line between air space and outer space. However, this is still controversial. There are mainly two schools of thought on this issue. The advocates of the first (spatialists) stress the need of a clear internationally agreed upon demarcation between air space and outer space, thereby activities would be regulated according to the legal regime application to the "place" where they occur. On the other hand, supporters of the second approach (functionalists) see no need for such demarcation because all activities should be regulated according to their nature and purpose rather than the "place" of their occurrence. In my view, both sides have some convincing arguments to support their positions. However, as space activities increase and new techno-logies develop to place various new devices in outer space at the height lower than the lowest peri-gee of a satellite, it would become inevitable to have a clear under-standing of where does outer space begin.

To whom does the 1967 Treaty apply ?

The 1967 Treaty is the most adhered to treaty that governs outer space and outer space activities. There are over ninety States Parties to the Treaty. It is needless to say that the Treaty applies to this group of States which comprise of space-faring and non-space-faring countries. How-ever, a question arises whether or not the legal principles of the Treaty have become a part of customary international law and thereby apply to all States ? This remains a controversial issue.

Normally, when a treaty applies to a State it is considered also applicable to both public and private persons belonging to that State. A large majority of legal scholars believe that the 1967 Treaty applies to, and regulate the activities of, private entities in accordance with Article VI of the Treaty. The Article in part provides that :

> States Parties to the Treaty shall bear international responsibility for national activities in outer space, including the moon and other celestial bodies, whether such

activities are carried on by governmental agencies or by non-governmental entities, and for assuring that national activities are carried out in conformity with the pro-visions set forth in the present Treaty. The activities of non-governmental entities in outer space, including the moon and other celestial bodies, shall require authori-zation and continuing super-vision by the appropriate State Party to the Treaty.

However, a small minority of authors[4] argue that private entities can appropriate outer space or a part of it though their States are prohibited to do so essentially because Article VI, in their view, "mainly concerns national activities in outer space" and Article II of the Treaty prohibits only "national appropriation".

Without going into too detailed analysis of these two schools of thought, one can say that the views of the minority are not legally tenable. Firstly, the negotiating history and the wording of Article VI make abundantly clear that private entities can not do what their States are prohibited to do. The Soviet Union while negotiating this Treaty accepted the involvement of private entities in the exploration and use of outer space only once it was assured that these entities will participate only once authorized by appropriate States which will continuously supervise their activities.[5] Without such an assurance agreement on this issue would have not been possible. Secondly, the States Parties to the Treaty are under clear obligation to ensure that space activities of the private entities are in conformity with the provisions of the Treaty.[6] Thirdly, by allowing private entities to appropriate outer space or its part would defeat the very purpose of Article II which contains compre-hensive provisions prohibiting appropriation. Moreover, any act of public or private entity which is contrary to Article II will defeat the purpose of Article I par. 2 which lays down the most fundamental principle of space law; i.e. the freedom of outer space. Finally, State practice as expressed in the appropriate national laws of some of the important space-faring nations, like the Russian Federa-tion, South Africa, Sweden, the United Kingdom and the United States[7], contain clear provisions assuming State responsibility in ensuring that the activities of their private entities are carried out in accordance with the provisions of the applicable international treaties, including the 1967 Treaty.

In my opinion, though the minority view holds no legal basis, the existence of this controversy implies that the provisions of the Treaty with respect to its appli-cation to private entities are not beyond doubt. Secondly, what could become more controversial is the level and legal mechanism of "authorization and continuos supervision" by appropriate State since these terms are not defined in the Treaty and are open to different interpretations. Thirdly, there is a confusion about the terms "launching State" and "appropriate State".

Under Article VII of the 1967 Treaty, the "launching State" is internationally liable for damage caused by its space objects. The Article does not actually define the term "launching State", but contains the same criteria for being a launching State as in Article I (a) of the Registration Convention and Article I (c) of the Liability Convention; i.e. the term "launching State" means (1) a State which <u>launches</u> or <u>procures</u> the launching of a space object and (2) a State from whose <u>territory</u> or <u>facility</u> a space object is launched. In analyzing this term, one author correctly states that "even more than four 'launching' states may be involved with regard to one space object if one state launches from the facility of another state which is on the territory of yet another state and if several states are considered to 'procure' the launching".[8] Similarly, the term "appropriate State" is not defined in the 1967 Treaty and is open to various interpretations. The earlier quoted Article VI of the 1967 Treaty holds the "appropriate State" responsible for "authorization and continuos supervision" of non-governmental entities. There is possibility of multiple "appropriate States" with respect to a particular private entity and "appropriate State" would not necessarily be the "launching State" of a space object of that entity.[9] This brief discussion shows that since the subjects of the 1967 Treaty are not well defined, its application would not be effective.

To what does the 1967 Treaty apply?

Obviously the 1967 Treaty applies to space objects and their component parts. However, the term "space object" is no where defined in the Treaty and thus is open to various interpretations.[10] There remain numerous and important questions unanswered. What is a space object ? What is a component part of a space object ? Is space debris a space object ? Is it only when an object is actually launched into outer space becomes a space object ? What about the objects or things manufactured in outer space with or without the composition of space objects taken from the Earth ? These issues need to be resolved clearly before one can expect an effective application of the 1967 Treaty.

How is the 1967 Treaty imple-mented and applied?

The implementation of the 1967 Treaty is being effected both at the international and national levels. At the international level, the Legal Sub-committee has elabo-rated several general legal principles contained in the Treaty in four international agreements and several resolutions. It is not the purpose of my presentation to analyze the provisions of these agreements and resolutions. In fact, they have already been extensively discussed in legal literature and deliberated in numerous fora. However, I would like to make only the following observations:

(1) Firstly, the most important and relevant treaties for the conduct of commercial space activities by private entities are the 1972 Liability Convention and the 1979 Moon Agreement. Both these treaties contain provisions which are confusing, inconsistent and inadequate; thus are not parti-cularly encouraging for any significant private investment. The ineffectiveness of the Liability Convention was shown in the case of COSMOS 954 accident. The confusion of terms like "launching State" and "appropriate State" is another factor. The Moon Agree-ment, though drafted nearly two decades ago, has been ratified only by nine States and none of them is a major space-faring nation. This is so because of a number of reasons, one of which is that the Agreement contains not fully defined concepts like "common heritage of man-kind".

(2) Secondly, the UNGA resolutions dealing specifically with legal regimes of direct TV broadcasting[11] and remote sensing[12] can hardly be said to create any solid, comprehensive and effec-tively implementable legal regime for the regulation of relevant space activities. This is so not only because they are simply non-binding resolutions but also on crucial issues their provisions are such that they create more confusion than resolving them. The latest resolution on the sharing of outer space benefits[13], in my opinion, from a legal point of view is in fact a step backward. It could be interpreted to have transformed the provisions of Article I par. 1 of the 1967 Treaty[14] into non-binding resolution without any legal mechanism and guidance in the sharing of the benefits of the exploration and use of outer space.

(3) Thirdly, the international process of elaboration and imple-mentation of the 1967 Treaty is too slow and cumbersome.

At the national level, the 1967 Treaty has been implemented and is being applied through national legal systems. Some of the States have adopted specific laws that give effect to the 1967 Treaty and also provide detailed procedures for the application of its provisions to particular space activities. The United States is undoubtedly the leader in the adoption of appropriate national legislation. As noted earlier, the Russian Federation, South Africa, Sweden, and the United Kingdom have promulgated specific laws imple-menting their international obligations under the 1967 Treaty. Depending upon their legal systems, some other States rely on their existing laws to give effect to the 1967 Treaty. However, one general observation can be made that States Parties to the 1967 Treaty in general are slow in enacting their national laws, even when such legislations are necessary under their constitutional law. For example, in Canada an appropriate domestic law is necessary to implement an international treaty to which Canada is a party. Without such a law the provisions of that treaty can not be applied or enforced domestically. At present, no such law exits in Canada. Similar situation prevails in a large majority of States, including some of the space-faring countries.

Conclusions and recommendations

In my presentation, I tried to enlist briefly some of the issues that need to be resolved. It is in fact a small inventory of issues which must be resolved to maintain law and order in outer space and consequently certainty of legal regime. This certainty is a pre-requisite for commercial activities, especially those by private entities.

The main purpose of the 1967 Treaty was to establish funda-mental legal "principles" that would govern space activities of the States. It has, in my view, succeeded in that mission. One must not try to find in the Treaty what was not intended to be included; i.e. legal "rules" to regulate all aspects of space activities. However, in order to effectively regulate space activities, specific, clear and sufficient legal rules are needed to be agreed upon. The 1967 Treaty is inadequate to address the relevant and fundamental issues involved in the new wave of space activities. The Treaty needs to be updated either by amending it or by supplementing it through additional protocols. Such task should be undertaken by the Legal Subcommittee. However, the activities of the Legal Subcom-mittee in the drafting of legally binding instruments for the last almost two decade have been less encouraging. Unless the Legal Subcommittee or its parent committee (i.e. the COPUOS) significantly transforms itself to become more active and efficient body, one can not hope quick revision of the 1967 Treaty through this body. Alternatively, one may look at the International Law Commission for such a task. But that body too is overburdened with drafting some of the important legal instruments. There prevails, in general, a negative feeling about convening special conferences, like the Law of the Sea Conference which negotiated and drafted the 1982 Law of the Sea Convention. Perhaps, a small group of scholars representing major space-faring nations and various regions of the world could be entrusted with the task of examining all the space treaties and eventually to create a comprehensive draft treaty which should include (1) the fundamental legal principles which are already adopted and well respected, (2) clear rules of law that would govern all space activities, (3) unambiguous definitions of the terms used, (4) an efficient dispute settlement mecha-nism and (5) sufficient provisions for its amend-ment. Such a draft treaty would then be presented to the Legal Subcommittee or a specially convened Conference on the Law of Outer Space for finetun-ing and adoption. That approach, perhaps, could produce some desirable results in a relatively short period of time and hopefully salvage ths last frontier from becoming a bone of contention, and a source of devastating pollution. It is my wish that outer space should remain and be fully respected and used as an enhancer of the quality of life for everyone on our planet called Earth. May be it is a wishful thinking!

ENDNOTES

2/ Treaty on Principles Governing the Activities of States in the Exploration and Use of Outer Space, including the Moon and other Celestial Bodies; entered into force on 10 October 1967. At present, there are 93 States Parties to this Treaty.

3/ Agreement on the Rescue of Astronauts, the Return of Astronauts and the Return of Objects Launched into Outer Space (1968), entered into force on 3 December 1968; Convention on Registration of Objects Launched into Outer Space (1975); Convention on International Liability for Damage Caused by Space Objects (1972), entered into force on 1 September 1972, Agreement on the Activities of States on the Moon and other Celestial Bodies (1979), entered into force on 11 July 1984.

4/ See Wassenbergh, H., "Responsibility

and Liability for Non-Governmental Activities in Outer Space", in ECSL Summer Course on Space Law and Policy : Basic Materials, Martinus Nijhoff Publishers, 1994, pp. 197 et seq. Also, Gorove, S., "Interpreting Article II of the Outer Space Treaty", in 37, Fordham Law Review, 1969, p. 349 at p. 351.

5/ For details see, Matte, N.M., Aerospace Law, Sweet & Maxwell Limited, London, 1969, at p. 309.

6/ Under Article VI of the 1967 Treaty, "a nation which becomes a party to the treaty agrees to be responsible for space activities carried on by one of its governmental agencies as well as by any nongovernmental entity. For the United States, this means that the government would accept responsibility for the activities of NASA as well as those of the Communications Satellite Corporation (COMSAT), etc. Furthermore, the government would see that such activities conform to the treaty's provisions, and also authorise and continuously supervise the space activities of nongovernmental entities. The relationship between the US Government and COMSAT is already defined in the U.S. Communications Satellite Act of 1962 (Public Law 87-624 (76 Stat. 419)) and in the President's Executive Order of January 4, 1965 on carrying out provisions of the COMSAT Act of 1962 concerning government supervision, including international aspects and the role of the Secretary of State. .....This article is designed to ensure responsibility for space activities, inherently international in nature, at the governmental level" : the Staff Report on the Treaty on Principles Governing the Activities of States in the Exploration and Use of Outer Space, including the Moon and other Celestial Bodies: Analysis and Background Data (1967, pp. 27-28) that was prepared to provide information on the legislative evaluation of the provisions of the 1967 Treaty for the Committee on Aeronautical and Space Sciences of the US Senate and to be used by the Senate during its consideration of the Treaty for the purpose of advising the US President whether or not to ratify the Treaty. See also Dembling, Paul G., "Treaty on Principles Governing the Activities of States in the Exploration and Use of Outer Space, including the Moon and other Celestial Bodies" in Jasentuliyana, N. and Lee, R. (eds.), Manual on Space Law, vol. 1, (1979), p.1, at p. 17.

7/ The Law of Russian Federation on Space Activity (20 August 1993) Article 4 (1) of which states that "Space activity shall be carried out in conformity with the following principles: .... international responsibility of the state for space activity under its jurisdiction". Under section 11 of the 1993 Space Act of South Africa (Act no 14917 of 23 June 1993), a licence is required for "the participation by any juristic person incorporated or registered in the Republic [of South Africa], in space activities : (i) entailing obligations to the State in terms of international conventions, treaties or agreements entered into or ratified by the Government of the Republic". Furthermore, "A licence shall be issued subject to such conditions as the Council may determine for that particular licence, taking into account : .....(c) the international obligations and responsibilities of the Republic". The Swedish Act on Space Activities (1982:963) in its Section 6 specifies that "If the Swedish State on account of undertakings in international agreements has been liable for damage which has come about as a result of space activities carried on by persons who have carried on the space activity shall reimburse the State what has been disbursed on account of the above-mentioned undertakings, unless special reasons tell against this". The Decree on Space Activities (1982:1069), which was issued under this Act, in its Section 4 states that "The National Board for Space Activities shall keep a register of the space objects for which Sweden is to be considered the launching State in accordance with Article 1 of the Convention on registration of objects launched into outer space of 14 January, 1975". The 1986 United Kingdom Act on Space Activities (1986 Ch. 38) was enacted "to confer licensing and other powers on the Secretary of State to secure compliance with the international obligations of the United Kingdom with respect to the launching and operation of space objects and the carrying on of other activities in outer space by persons connected with this country". Under section 3.(1) of the Act, "A person to whom this Act applies shall not, subject to the following provisions, carry on an activity to which this Act applies except under the authority of a licence granted by the Secretary of State". Section 5 of the Act specifies that, a licence may be granted subject to such conditions, as the Secretary of State thinks fit, and in particular, may contain conditions (e) requiring the licensee to conduct his operations in such a way as to ..(iii) avoid any breach of the United Kingdom's international obligations". The US Act to Facilitate Commercial Space Launches, and for Other Purposes of 1984, as Amended 1988 (Public Law 98.575, 98th Congress, H.R. 3942, October 30, 1984. 98 Stat. 3055), in its Section 6 (a) (1) states that "No person shall launch a launch vehicle or operate a launch site within the United States, unless authorised by a license issued or transferred under this Act". Section 16 of the Act requires "Each person who launches a launch vehicle or operates a launch site under a license issued or transferred under this Act shall have in effect liability insurance at least in such amount as is considered by the Secretary to be necessary for such launch or operation, considering the international obligations of the United States". Further more section 21 (d) states that "The Secretary shall carry out this Act consistent with any obligation assumed by the United States in any treaty, convention, or agreement that may be in force between the United States and any foreign nation. In carrying out this Act, the Secretary shall consider applicable laws and requirements of any foreign nation".

8/ Bockstiegel, K.-H., "The Term 'Launching State' in International Space Law", Proceedings of the thirty-seventh Colloquium on the Law of Outer Space, 1994, p. 80 at p. 81.

9/ For details, see Bockstiegel, K.-H., "The Term 'Appropriate State' in International Space Law", Proceedings of the thirty-seventh Colloquium on the Law of Outer Space, 1994, pp. 77 et seq. and Wirin, W.B., " Practical Implications of Launching State - Appropriate State Definitions", Proceedings of the thirty-seventh Colloquium on the Law of Outer Space, 1994, pp. 109 et seq.

10/ For a detailed discussion, see Gorove, S., "Definitional Issues Pertaining to 'Space Object'", Proceedings of the thirty-seventh Colloquium on the Law of Outer Space, 1994, pp. 87 et seq., Cheng, B., "Nationality for Spacecraft ?", in ECSL Summer Course on Space Law and Policy : Basic Materials, Martinus Nijhoff Publishers, 1994, pp. 67 et seq.

11/ Principles Governing the Use by States of Artificial Earth Satellites for International Direct Television Broadcasting (1982), UNGA Resolution No. A/RES/37/92.

12/ Principles Relating to Remote Sensing of the Earth from Outer Space (1986), UNGA Resolution No. 41/65.

13/ Declaration on International Cooperation in the Exploration and Use of Outer Space for the Benefit and in the Interests of All States, Taking into Particular Account the Needs of Developing Countries, UN Doc. A/AC.105/L.211 (11 June 1996).

14/ For a detailed analysis of Article I par. 1, see Jakhu, Ram S, "Developing Countries and the Fundamental Principles of International Space Law", in Girardot, R.G., et al (eds.), New Directions in International Law, (Frankfurt, 1982), pp.351-373.

IISL-ECSL Symposium Celebrating the 30th Anniversary of the Outer Space Treaty
Held on occasion of the thirty-sixth session of the Legal Subcommittee of COPUOS
Vienna International Centre, 1 April 1997

FUTURE DEVELOPMENTS RELATING TO OUTER SPACE TREATIES

by

F. von der Dunk
Co-Director, Institute of Air and Space Law of Leiden University

1. Introduction

When it comes to join in celebrating the thirtieth birthday of the Outer Space Treaty, and to take a look at the future of this 'Magna Charta' for outer space and the legal framework built upon it, this is not only a pleasure but a challenge as well. It obviously involves taking a look at present and future developments in the practice of space and space-related activities.

In my view, the most comprehensive development is the increasing involvement of private enterprise. Consequently, we would need to ask ourselves where and how this development relates to, and impacts upon, the present legal framework for outer space activities as provided by the Outer Space Treaty and the remainder of the *corpus juris spatialis internationalis*. Let me try to be modest however. I will not purport to deal with this issue comprehensively here and now, only discuss one element of it which I think is of crucial importance and at the same time very illustrative from the legal point of view.

2. The Outer Space Treaty, Private Enterprise and National Space Legislations

The Outer Space Treaty basically addresses states. It does itself establish only the framework for dealing with non-governmental activities, by means of the closely interrelated Articles VI, VII and VIII. In essence, this framework lays the burden of establishing legal rights and duties for private enterprise squarely upon the shoulders of states, which should establish national space legislations for that purpose.

The problem however is that a number of crucial definitions and interpretations of the structure thus realized have not been defined to any precise extent at the international level. Yet, they are of crucial importance for the ways in which national space legislations *vis-à-vis* private enterprise will actually take shape. In addition, *vice versa* national space legislations can have their own impact on the relevance, authority and clarity of international space law.

Already as of now, five states - the United States, Sweden, the United Kingdom, Russia and South Africa - have established a national body of space legislation in the true sense of the word, i.e. dealing with private involvement in space activities on the national level by providing for rights and duties for the entities concerned. In one case, actually there are currently three separate national regimes in force for the respective areas of space activities where private enterprise represents a major factor: satellite communications, launching and remote sensing. This makes seven regimes altogether.

All five national legal frameworks concerned, all seven regimes involved have their own things to say with regard to the various terms and concepts provided by the aforementioned three Articles of the Outer Space Treaty. For reasons of time, I will by way of example limit myself to only one crucial aspect of the scope of those national regulations, namely that of the categories of actors which are covered by them. To whom are the license obligations which form the core of the system of binding private entities to (international) space law made to apply?

Apart from the requirement under Article VI of the Outer Space Treaty to authorize and continuously supervise "national activities in outer space" of non-governmental entities, also interesting in this regard is the role of Article VII of the Outer Space Treaty, as well as the Liability Convention. The international liability for damage of a state as established by these texts obviously provides a strong incentive to legislate on the national level *vis-à-vis* private entities (potentially) involved.

3. The United States: Communications Act of 1934

In respect of telecommunications in general, in the United States the Communications Act was enacted in 1934. The Federal Communications Commission (FCC) declared in 1970 that the Act was to be applied to space telecommunications as well. Section 301 provides for a licensing obligation for any person undertaking telecommunications activities "from [a] place in (...) the United States". Thus, the territorial jurisdiction of the United States is exercised.

Depending on the interpretation of "national activities" of Article VI of the Outer Space Treaty, however, potentially international responsibility of the United States for United States companies operating outside United States territory could also arise. Such responsibility would then not be covered by national measures. The United States, in other words, interprets the term "national activities" as solely meaning 'activities undertaking from the territory of' for the purpose of this law.

On the other hand, liability as far as regulated by the Liability Convention depends upon involvement in the launch of the communications satellite and not on its operations *per se*. Thus, any damage caused by such operations can incur United States liability only to the extent that the United States is a launching state. As a result, domestic consequences thereof also depend upon the national legislation relevant to private launch activities.

4. United States: Commercial Space Launch Act of 1984/1988

As to this second category of space activities, the Commercial Space Launch Act of 1984 is applicable, including the Amendments of 1988. The Act deals with the issue of application of the license obligations in Section 6(a). Firstly, the Act applies to all persons undertaking these activities within the United States. Secondly, it applies to United States citizens, here meaning individual citizens as well as juridical persons incorporated in the United States, undertaking these activities outside the United States. Thirdly, it applies in principle to non-United States-incorporated juridical persons under a controlling interest of any United States national or United States-incorporated juridical person. This applies, provided the entity undertakes the activities in question outside the United States as well as outside any other state's territory, unless by agreement jurisdiction and control over the activities has been handed over to another state.

Here, the United States has used its jurisdiction in a comprehensive manner. Both with respect to its nationals and with respect to its territory, legal control over the launching activities of private entities is exercised. Consequently, when it comes to launching activities to the extent they do fall under the scope of Article VI of the Outer Space Treaty, the United States considers "national activities" to encompass both 'activities by its nationals' and 'activities undertaken from its territory'.

5. United States: Remote Sensing Acts of 1984 and 1992

Regarding remote sensing finally, for the United States the 1984 Land Remote Sensing Commercialization Act and the 1992 Land Remote Sensing Policy Act which replaced it merit analysis. The 1984 Act under its Section 402 applied to any private person "who is subject to the jurisdiction and control of the United States" operating a remote sensing satellite system. This "jurisdiction and control" encompassed United States citizens, corporations and firms organized under United States law, and private entities "having substantial connections with the United States or deriving substantial benefit from United States law". This seems to indicate that the United States, for purposes of remote sensing, considered "national activities in outer space" to comprise at least both 'activities of its nationals' and 'activities undertaken from its territory'.

The 1992 Act replacing the 1984 one applies to private persons "subject to the jurisdiction *or* control of the United States", according to Section 202(a). Private persons merely controlled by the United States, without falling under its jurisdiction as such, are now also falling under the applicable legal regime. As a consequence, a foreign company undertaking private remote sensing activities from outside United States territory, but nevertheless controlled by the United States, also requires a United States license. By such a license, the United States actually transforms the activities concerned into "national activities" as relevant under Article VI of the Outer Space Treaty - not on the basis of territory or nationality, but on the basis of the license as such. Hence, it would also assume international responsibility for them.

6. Sweden: Act on Space Activities of 1982

In accordance with Section 2, the Swedish Act on Space Activities of 1982 firstly applies to all space activities to the extent they are undertaken from Swedish soil. Secondly, it applies to all space activities undertaken by Swedish natural or juridical persons "anywhere else". This forms a concise formulation of all activities undertaken within the territorial jurisdiction respectively jurisdiction over nationals of Sweden. Sweden really in quite simple terms interprets "national activities" as meaning activities either undertaken from its territory or by its nationals.

Thereby, as a matter of fact the possibility of Sweden being held liable for damage caused by privately launched space objects from its territory is also taken care of. Amongst the four options for being labelled a launching state under Article VII of the Outer Space Treaty and the Liability Convention, the territorial criterion provides by far the most comprehensive and risky one, begging for national legislation. If on the other hand Sweden launches itself a space object with private entities involved, or procures a launch of such an object, or allows a governmental facility to be used for such a launch, it is actively and consciously engaged in the launching. Hence, it has other means such as contract clauses at its disposal in order to deal with its potential liability on the international level.

7. United Kingdom: Outer Space Act of 1986

The United Kingdom's 1986 Outer Space Act by virtue of Section 2 applies to "United Kingdom nationals, Scottish firms, and bodies incorporated under the law of any part of the United Kingdom". Under Section 1, this is applicable to the activities of those persons "whether carried on in the United Kingdom or elsewhere". As a consequence, the United Kingdom appears to interpret "national activities" as solely referring to 'activities of nationals'.

In view of this exclusive reliance on jurisdiction over nationals, activities undertaken by non-nationals from British soil do not fall within the scope of the Act. For launching activities, this might be a satisfactory solution, since no spaceport yet exists on British territory from which to undertake launching activities. Private space communication or remote sensing activities, however, might very well be conducted from British tracking-and-control centres; nevertheless, the United Kingdom did not choose to bring those under the ambit of the Outer Space Act. Will it be held responsible under Article VI of the Outer Space Treaty in applicable cases nevertheless?

8. Russian Federation: Law on Space Activities of 1993

With regard to the Russian Law on Space Activities of 1993, Article 1(1) provides that it applies to "space activities under the jurisdiction of the Russian Federation". This jurisdiction includes both territorial and nationality-based jurisdiction, as becomes clear when reference is had to a number of other provisions - most clearly with respect to the licensing regime itself. The use of the term jurisdiction moreover is explicitly linked to the international responsibility of Russia.

Hence, although a number of additional provisions leave room for uncertainties or discretion of relevant authorities, in general the Russian interpretation of "national activities" seems to follow the Swedish one, with all consequences flowing from this with respect to international responsibility and international liability potentially incurred by the Russian Federation.

9. South Africa: Space Affairs Act of 1993

South Africa's 1993 Space Affairs Act asserts both the territorial jurisdiction and the jurisdiction over nationals of South Africa by means of its licensing system. The former however, under Section 11(1)(a) and (c), relates only to the activities of launching itself and - presumably - operating a launch facility. An absence in principle of, for instance, satellite communications or remote sensing activities undertaken by non-nationals from South African territory results. It can only be repaired by consequent use of a stop-gap clause provided by Section 11(1)(e), providing for discretion on the side of the government to include such activities.

The assertion of jurisdiction over nationals on the other hand is comprehensive under Section 11(1)(b) and (d)(i). Any juridical or natural person with South African nationality undertaking launch activities outside of South Africa automatically triggers the applicability of the Act. The same applies to any other space activities entailing obligations for South Africa under applicable international treaties - i.e. for instance satellite communications and satellite remote sensing. In view of the special application of these provisions to launching activities, it is clear that South Africa primarily undertook to cover its potential international liability. Article VI of the Outer Space Treaty, when solely related to satellite communications and remote sensing activities, is therefore interpreted by South Africa as applying in principle to those activities of its nationals only.

10. Conclusions

By way of conclusion, the ever increasing involvement of private enterprise in mankind's space endeavour has a direct legal component related to the future of the Outer Space Treaty and the whole *corpus juris spatialis internationalis*. States are held internationally responsible and/or liable, even in cases where private entities are the real and exclusive actors undertaking the activities at issue. States should, consequently, authorize and continuously supervise such space activities; and should, where applicable, exercise the jurisdiction and control they retain over objects launched into outer space for that purpose.

The previous presented a summary analysis of one of the most fundamental issues in this respect, reflecting one crucial term of Article VI of the Outer Space Treaty. Which activities constitute, for a particular state, the "national activities" for which it may be held internationally responsible? A closely related question, left unanswered here, would be which activities that state should exercise authorization and continuing supervision over as the "appropriate state"? Also, the discussion on how liability interferes with these notions has been left aside.

Nevertheless, already the present analysis should provide sufficient indication for the prevailing measure of divergence in the actual application of jurisdiction. This in turn results in divergent interpretations of "national activities", as it is used by Article VI of the Outer Space Treaty, at the national level.

Four of the seven regimes implicitly define "national activities" of a state as comprising both 'activities undertaken from its territory' and 'activities undertaken by its nationals'. This concerns the United States' Commercial Space Launch Act respectively the remote sensing regime, as well as the Russian Law and the Swedish Act on Space Activities. Only the last one mentioned, however, does so in an unequivocal manner.

Furthermore, the South African Space Affairs Act makes a distinction between launching activities, where the aforementioned interpretation is applied, and satellite communications and remote sensing, where "national activities" are basically considered to be equivalent to 'activities undertaken by nationals'. This latter interpretation is more clearly and fundamentally upheld by the United Kingdom's Outer Space Act; whereas the United States Communications Act finally, equally clearly and fundamentally yet diametrically opposed, defines "national (satellite communications) activities" as 'activities undertaken from its territory'.

Finally, in view of ongoing discussions at the theoretical level, little doubt will exist that other fundamental terms and concepts of Articles VI, VII and VIII of the Outer Space Treaty relevant for the present topic have not been defined with sufficient precision on the international level either. Which geographical area is considered "outer space", as the responsibility of states under Article VI attaches to activities in that area? When does one become a state "procuring" the launch of a space object? Does a state become a launching state under Article VII of the Outer Space Treaty, and under the Liability Convention, if a private entity with its nationality, alternatively operating from its territory, launches or procures the launch of a space object? What is the relationship between the responsibility of Article VI and the liability of Article VII? What role plays the jurisdiction as operative under Article VIII with respect to these issues?

I feel the time has come to move beyond the discussions on the international level and to establish authoritative interpretations achieving worldwide acceptance. If not, states who find themselves confronted with the need - whether objective or subjective - to deal with private enterprise on the national level will take their own measures. By doing so, they provide their own interpretations, explicit or implied, with the obvious consequence that, on the international level, inconsistencies, gaps and overlaps will appear, with all due confusions and complications. There is an unequivocal interest in preventing national discretion to interpret and define the relevant terms and concepts from becoming the norm, and a matter of right almost by customary force.

In other words: it is my contention that the future of the international space law treaties for a substantial part lies in legislation related to space activities on the national level, as a consequence of the current place of private enterprise under international space law. Such a result of the juxtaposition of international law and national law should not so much be seen as a paradox. Rather, it should lead to a well-balanced and structured synthesis, of the thesis of the international interest in application of an essentially uniform legal regime to space and space activities, and the anti-thesis of the national need to deal in legal terms with private involvement therein, worthy of stimulation but requiring control of its potentially detrimental effects. The task which should therefore figure prominently on the international agenda is that of clarification of international space law concepts to facilitate more efficient regulation of private space activities on the national level.

IISL-ECSL Symposium Celebrating the 30th Anniversary of the Outer Space Treaty
Held on occasion of the thirty-sixth session of the Legal Subcommittee of COPUOS
Vienna International Centre, 1 April 1997

SOME COMMENTS ON THE SIGNIFICANCE OF THE 1967 OUTER SPACE TREATY FOR THE ESTABLISHMENT AND DEVELOPMENT OF SPACE LAW OF OUR TIMES

by

Vladimir Kopal
Professor of International Law, Charles University, Prague

I have got a difficult task to comment on three substantive contributions to the IISL/ECSL Symposium on Celebrating the 30th Anniversary of the 1967 Outer Space Treaty, which differ both in their approaches to the subject and in their content. Moreover, Dr. G. Lafferranderie, Chairman of the European Centre for Space Law, is about to make a presentation of a book on "Outlook on Space Law over the Next 30 Years", which represents a major collection of essays devoted to the anniversary of the Outer Space Treaty. All these papers have brought many valuable observations and suggestions dealing with the past and future of the Treaty. My intention is to comment only on some of these thoughts and to leave the other for the discussion. But prior to offering some substantive comments, let me make a historical remark.

In the first paper, Prof. Y.M. Kolosov paid a tribute to a great number of lawyers from many countries who paved the way towards the successful outcome of the United Nations efforts in establishing the foundations of a legal order for space activities. He mentioned that his listing of a number of names was not complete and I would like to proposed still some other names to be added to his list. First, let me recall the name of Prof. Vladimir Mandl, who was author of the first monographical study dealing with the subject of space law published as early as 1932. Dr. Mandl as an attorney-at-law in the city of Plzen (Pilsen) and later on became Professor of Industrial Law at the Czech Technical University in Prague. His study on space law was published in Germany under the heading *"Das Weltraum-Recht: Ein Problem der Raumfahrt"* (The Law of Outer Space: A Problem of Space Flights), because its author wished to draw the attention of the international specialized community to these issues. Mandl commanded the German language because following his studies at the Law Faculty of Charles University in Prague, he made his post-graduate studies at the University of Erlangen in Germany. Mandl's study of space law issues was based on the assumption that the prospects of reaching outer space by means of rockets would raise new questions not settled by air law, which regulated only the legal regime of airspace. As he wrote, "beyond these Earth coastal spaces is space which does not have any relations whatsoever with our globe, nor with individual parts of the Earth surface either, which no longer has any pertinence to the Earth and is therefore free on any state power, coelum liberum. In this area, spaceship traffic is completely free." (Cf. p. 33 of Mandl's study)

Recently, an English translation of Dr. Mandl's monographical study was provided by the care of Dr. Stephen Doyle, a former representative of the United States to COPUOS Legal Subcommittee, who effectively participated in its work for several years. It will be included in extenso in Dr. Doyle's book on early learning on space law to be published soon.

Similarly, the name of Wolf Heinrich, Prince of Hanover, should be recalled in this context. Prince of Hanover wrote the first doctoral dissertation on Air Law and Space in 1953, still before the launching of the first satellites, and defended it successfully at the Georg-August University in Göttingen,

Germany.

The list and role of international and national non-governmental institutions in the development of space law and drafting of appropriate legal agreements for this purpose should be also enlarged. In the Institute of International Law (l'Institute de Droit international), at its 1963 session at Brussels, the legal regime of outer space was considered and a resolution was adopted, including the issues which were later on covered in the United Nations Declaration of Legal Principles Governing the Activities of States in the Exploration and Use of Outer Space as approved by the United Nations General Assembly on 13 December 1963. The International Law Association (ILA), another important international non-governmental organization dealing with different subjects of the development and codification of international law, worked out, through its space Law Committee, a number of drafts, particularly on the topics of Settlement of Space Law Disputes and of the Protection of the Environment from Damage Caused by Space Debris. Also the Hague Academy of International Law included from time to time special courses on space law in its programmes of lectures, e.g. the course presented by Prof. Manfred Lachs of Poland, the first Chairman of COPUOS Legal Subcommittee and later on Judge and President of the International Court of Justice.

The work of a number of national institutions, such as the Institute of Air and Space Law, McGill University in Montreal, Canada, which was initiated by Prof. John Cobb Cooper of the United States, who was later on succeeded by Prof. Eugène Pépin of France, should not be omitted in our assessment of efforts that paved the way to the United Nations legislative efforts in the field of space activities. Similarly, the work of the Institute of Air and Space Law of the University of Cologne, Germany, originally founded by one of the pioneers in these disciplines, Prof. Alex Mayer, and now led by Prof. Karl-Heinz Böckstiegel, should be recalled when discussing the beginnings of space law.

Certainly, the 30th anniversary of the conclusion of the Outer Space Treaty gives an opportunity for celebrating this important international legal instrument and also for an assessment of the law of outer space as a whole. In this evaluation, all positive elements should be taken into account, but also a number of weak spots and gaps in the existing legal order for space activities should be brought to the foreground.

Without doubt, the 1967 Outer Space Treaty has been one of the significant law-making treaties which contributed in an outstanding manner to the progressive development of international law of our times. For it enshrined a set of fundamental principles to govern space activities, which have been recognized by the present international community and established a firm basis for the development of space law for the subsequent decades. Freedom of exploration and use by all States of outer space, including the Moon and other celestial bodies, for the benefit and in the interests of all countries, non-appropriation of any part of outer space, assistance to astronauts in the event of accidents, international responsibility for all space activities and liability for damage caused by such activities to other States Parties to the Treaty or to their natural or juridical persons, registration of space objects and jurisdiction and control over such objects based on such registration, cooperation and mutual assistance in the exploration and use of outer space, as well as important steps towards ensuring the maintenance of outer space, the Moon and other celestial bodies for peaceful purposes - these are the essential elements of the Outer Space Treaty. Prof. Kolosov showed us today how these principles were agreed upon and Prof. Ram Jakhu analyzed them in greater detail in his thoughtful paper.

On the other hand, the foregoing speakers made it clear that not all issues were resolved and some of them were only bridged by rather vague compromise formulas. Moreover, new problems emerged since then. For example, an important category of questions had arisen in connection with the growing volume of private activities in outer space. Both Prof. Jakhu and Dr. von der Dunk demonstrated what kind of issues are at stake. The latter offered a valuable comparison of national laws dealing with this subject in which the principles of nationality and territoriality for establishing the national jurisdiction over such activities have been applied. He also indicated some discrepancies existing between individual national laws in this respect.

At present, "space law" consists of two layers.

The first one is represented by the international law of outer space that governs activities of international persons (States and international intergovernmental organizations). International space law established by the United Nations has created a basis that should be respected by all nations and international organizations, and also by individuals. A growing number of bilateral and plurilateral treaties mostly devoted to international cooperation between two or more States and for international organizations complete the international layer of space law.

The second layer has been filled by national laws regulating space activities of individual nations and their nationals, both physical and juridical persons. The national laws should remain in harmony with international obligations arising from the international law of outer space that has been recognized by the States concerned.

This growth and character of space law is more or less similar to the development of air law, the law of the sea/maritime and the law of the environment.

The 1967 Outer Space Treaty laid down the cornerstones of the whole building of space law. As such, it does not need any revision, its main principles are valid and useful. We probably could not agree on any other principles to replace those enshrined in the Treaty. At the same time, some provisions of the Outer Space Treaty, and also those of the other United Nations Space Treaties, need clarification and their application should be adapted in the light of new phenomena and issues. Basic principles of the Outer Space Treaty should be supplemented and developed by further instruments. During the years that followed the conclusion of the 1967 Outer Space Treaty, a number of examples have been given on how to proceed further in the progressive development of international space law. Four other United Nations treaties were concluded, by which some individual principles of the main Treaty were successfully elaborated in more specific instruments. Why not to continue in this process by using the same method?

In my opinion, the Legal Subcommittee should now start studying the newly emerging legal issues and report on them regularly to the Committee on the Peaceful Uses of Outer Space and through it to the General Assembly. In these endeavours, the Legal Subcommittee should closely cooperate with the Scientific and Technical Subcommittee and vice versa. After all, a close cooperation and interaction of both Subcommittees was one of the aims of COPUOS when it was establishing them as its subsidiary bodies. One of the early subjects to be studied by both Subcommittees might be the protection of space environment against the generation of space debris. This issue has already been included in the agenda of the Scientific and Technical Subcommittee which adopted its work plan for studies of this subject that is being implemented. The work of the Scientific and Technical Subcommittee should be followed soon by the studies of legal aspects involved in the Legal Subcommittee. A thorough analysis and evaluation of the relevant provisions of the 1967 Outer Space Treaty and other United Nations space treaties, whether and to what extent they were applicable to the issue of space debris, would be an appropriate beginning of these studies. The 1967 Outer Space Treaty and its subsequent instruments aimed at developing international cooperation in an protecting of space activities carried out by means of space objects as active and functional tools of space programmes. But the up-to-date legal instruments did not care about the proliferation of remnants of such objects which have ended their missions. Is it not logical to think about appropriate remedies, including some adequate legal measures to ensure safety of life and property in space navigation?

1997 has been the year of several anniversaries related to the development of space activities and space law. As already mentioned, the space era started 40 years ago by launching the first artificial satellites of the Earth. This symposium has been devoted to the 30th anniversary of the conclusion of the

Outer Space Treaty. Perhaps another anniversary - a modest one but not without any significance - should be also recorded, namely that 35 years ago both Subcommittees of the United Nations Committee on the Peaceful Uses of Outer Space were established and started their regular work by holding their first sessions in Geneva, in the spring of 1962. In the Legal Subcommittee, this was the beginning of the legislative process in the field of international law of outer space which continues until these days. We should now think in a constructive way on how to improve the further develop its work in order to reach a rewarding success in our future endeavours.

IISL-ECSL Symposium Celebrating the 30th Anniversary of the Outer Space Treaty
Held on occasion of the thirty-sixth session of the Legal Subcommittee of COPUOS
Vienna International Centre, 1 April 1997

OUTLOOK ON SPACE LAW OVER THE NEXT 30 YEARS

by

G. Lafferranderie
Chairman of the ECSL and Legal Adviser of the ESA

Dear Mr. Deputy to the Director General of the United Nations Office at Vienna, and Director of the Office for Outer Space Affairs,
Dear Mr. Chairman,
Dear delegates,
Dear friend,

1. It is for me a signal privilege to be with you today at this symposium intended to mark the thirtieth anniversary of the entry in force of the Outer Space Treaty (OST) and to make a modest contribution to this celebration in the form of this book. To present to you the results of ECSL's research is for me a great honour and something of a challenge. You may well find us somewhat presumptuous in presenting to you this essay in which we try to imagine what the Treaty would look like if we had today to formulate the basic concepts, draft a text and negotiate it for the thirty years to come. It is not my intention to sing the Treaty's praises or highlight its weaknesses; it is easy to criticize, as we all know.

2. Space activity has come a long way since the Outer Space Treaty was adopted. If we compare the first questions that were asked in its early stage by the UN Committee on Outer Space, with those that arise today - or can reasonably expect to arise - there is certainly quite a gulf. But to my mind that is not what matters. What is important above all is to consider the purpose of the underlying legal rule that space activities should be pursued for the benefit of all, of all mankind. The ability to ask questions is a distinctive human trait. Why should we have to wait until circumstances force us to use that ability ?

Since it was adopted the Treaty has been much analysed and criticized. Was there any need for yet another review of its content? I have to confess that our secret desire, and I say this with all due modesty, was to try to bring a little fresh thinking to the debate, from a different angle. It seemed to us preferable to work from the assumption - one I am sure you will accept - that the OST forms the basis of space law and to give thought to the recent and continuing changes in the legal area.

3. How is space law evolving in an environment itself undergoing change? First of all the players are of course no longer the same. At the time the Treaty was written the two superpowers, the USA and the USSR, were the unchallenged leaders in the space domain and the prime movers of the legal texts. And although many other States were taking a close interest in the utilization and exploration of this new frontier, they were as yet no more than passive onlookers.

The two original space powers have now been joined by many other players. Space activity is no longer the exclusive privilege of states. For a number of years now private firms have also been active in the space field. COPUOS of course continues to be active and declarations and statements of principles are still adopted. But these are now far from being the sole regulatory source in this area of activity. Account now has to be taken of many other sources of space law.

A number of states, such as the United States, the United Kingdom Sweden, South Africa, Japan, Russia and Argentina have developed or will develop national space legislation. An ever-expanding network of multilateral and bilateral relations is taking shape. Other international organisations within the UN family such as the ICAO, UNESCO, WIPO, WMO and organisations more specifically concerned with space, examples being Intelsat, Eutelsat, Inmarsat, ESA, Eumetsat, Arabsat, Intercosmos, and finally scientific organisations like COSPAR, Astronomical Union IAF (and IISL), intervene in the codification of space law. These various types of relation all give rise to contracts, agreements and understandings. Many legal disciplines now come into play. We are no longer talking about public international law but also about private international law commercial law criminal law, civil law, patent law...... The regulatory framework for space activity has become exceedingly complex.

This proliferation of bodies and law flows from the rapid increase in the range of activities and their newly acquired accessibility. Another inevitable development is the emergence of new forms of law, such as soft law. What sort of direction should we be taking ?

4. On this book

It consists of more than thirty contributions by European authors, members of ECSL, some of them academics, some business professionals, all having some kind of involvement in space law and is devoted to a substantial number of issues (you will recognize the main chapters of the Treaty):

- the actors
- the status of Outer Space, including celestial bodies, and its use
- space objects
- man in space
- liability
- registration, jurisdiction and control, ownership
- settlement of disputes
- forms of cooperation
- environment of Earth and space
- remote sensing
- commercial activities

The last part of the book is devoted to the role of the Legal Subcommittee.

5. On that basis we built a collective effort to reflect on current and emerging activities. What today is the relationship between space activities and the Charter of space law? What is the status, the place of these rules of law when confronted with the new situations and players to which we have referred? While the authors raise many questions they suggest few answers, that was never really our intention, simply a stimulus for further reflection. Nobody thinks the recognised international space law that has been applied for many years now has become obsolete. That would be to call our entire legal basis into question. But it may well be that this international corpus of law calls for some adaptation, enlargement, additions, in the light of new realities. We have left our crystal ball at home for we know there are no limits to human imagination and inventiveness. We can be sure that new forms of space transport will be devised, that humans will go back to the Moon and will set foot on Mars, new domains will be opened (microgravity), new benefits for the conduct of terrestrial activities: worldwide communications (information highways etc.), navigation satellites, etc. Facilities for living and working in space will become commonplace; with all the attendant technical consequences, some of which will be undesirable. We will of course have to join forces in addressing such problems as space debris, intellectual property rights, patents, code of conduce and new forms of international cooperation. Coming back to the question of the purpose of space law and hence to Article I of the OST, we will have to consider the ethnics of space activities and how best to resist growing pressure to depart from accepted principles. How can we hope to ensure the continuation, for instance, of scientific work and in particular astronomy and astrophysics, if we fail to take timely actions to regulate certain space activities inconsistent with that work ? How to protect privacy of our life ? (I limit myself to civil activities).

6. We must pull together on this task and we must see COPUOS as not only the father of the Space Law but also as the focal point for coordinating, at legal level, the acts and responsibilities of the various players, organising exchanges of views in order for all to work with a common objective and full knowledge. If this

focal point, this force for coordination and progress ceased to exist, the benefits hitherto derived from the Treaty could be placed in jeopardy.

It is hope that this book, intended to highlight new space-related issues, certainly not exhaustive, will arouse interest among interest among the "space audience" which you are representing today. It is now an honour for me, as Chairman of the European Centre for Space Law, to present the book to you, Mr. Jasentuliyana, as a token of gratitude for the work already done by the Legal Subcommittee and as an encouragement for the years to come.

I would also like to express special thanks to Mr. Mikulka, our Chairman, and yourselves, ladies and gentlemen, for your kind attention.

IISL-ECSL Symposium Celebrating the 30th Anniversary of the Outer Space Treaty
Held on occasion of the thirty-sixth session of the Legal Subcommittee of COPUOS
Vienna International Centre, 1 April 1997

VOTE OF THANKS

by

N. Jasentuliyana
Deputy to the Director-General,
United Nations Office at Vienna; and
Director, Office for Outer Space Affairs

It gives me great pleasure to accept the book, "Outlook on Space Law over the Next 30 Years", on behalf of the United Nations. The book, as you know, was prepared to celebrate the 30th Anniversary of the 1967 Outer Space Treaty. The "Magna Carta" of space, as the Treaty is sometimes so called, set out a basic legal framework for the international regulation of space activities and its provisions have served as the building blocks of the other space treaties and sets of legal principles drafted by the Legal Subcommittee. Thus, over the past thirty years, the Outer Space Treaty has become the guiding light by which space activities are conducted by States. Its importance can be gauged by the fact that some of its provisions can be said to have become part of the fabric of international customary law.

This valuable publication comes at the right time. We stand at the threshold of the 21st Century. The Cold War, during which the space law treaties and most of the legal principles were drafted, is over. Space activities are being conducted by a greater number of countries. Private space enterprise is flourishing and is continuing to grow at a tremendous rate. All this calls for a new look at space law. That is what, in essence, Dr. Lafferrenderie's book has done. It has set the stage for space law in the new millennium.

On behalf of the United Nations, I would like to congratulate Dr. Lafferrenderie and the contributors to the book for their valuable, and timely, input into the subject of space law and also to the work of this Subcommittee.

MANFRED LACHS SPACE LAW MOOT COURT COMPETITION 1997

OPENSKEY vs. ANTIPAPADIA

Case Concerning Commercial Very High Resolution Remote Sensing Systems

1. INTRODUCTION

The finals of the 6th Manfred Lachs Space Law Moot Court Competition were held on 9 October 1997. The competition was realized with the help of the University of Turin, the Local Organizing Committee of the IAF, KLM Royal Dutch Airlines, the European Centre for Space Law (ECSL), and the Association of US Members of the IISL (AUSMIISL). Preliminary competitions were held in Europe and the USA, and the winners of those preliminaries met in the final round in Turin. The University of Paris XI (France) and the University of North Carolina (USA) competed in the case "Openskey vs. Antipapadia", dealing with Very High Resolution (VHR) remote sensing systems (written by Harry Tuinder, Marco Ferrazzani and Frans von der Dunk.). The honourable court was composed of Judge Koroma (President) Judge Rezek and Judge Vereshchetin of the International Court of Justice. The team of the University of Paris won the competition. Its members were Jean-François Renaud, Ranjani Srinivasan. and Amine Laachani. The members of the University of North Carolina team were Christina Benson and Scott Syfert. The Law Offices of Sterns and Tennen provided two awards to Outstanding Oralists Ms Benson and Ms Srinivasan. The Journal of Space Law offered a one-year subscription for the team with the Best Memorial, the French team.

2. THE PROBLEM*

General Issues Presented:

- Whether the downgrading of spatial resolution of a remote sensing satellite image in the case of a specific country is in conformity with international law;
- Whether remote sensing satellite data may be intercepted by a state without approval of the operator state of the satellite;
- Whether remote sensing data which proves to be incorrect can result in liability under international law of the state which is responsible for the satellite operation.

Statement of Facts

The Republic of Starstripe licensed the private company Goldstar to construct and sell the commercial remote sensing satellite Golden Eye to the Kingdom of Openskey. Golden Eye is capable of delivering Very High Resolution (VHR) data with a ground resolution between 1 and 5 meters.

The instruments on board Golden Eye have an optical capacity and an active synthetic aperture radar (SAR) capability and there is limited on board processing and recording of data. Also the satellite is equipped with an optical telecommunication transponder for intersatellite links which makes high bit data relay possible with the command station.

Golden Eye was successfully launched by the State Secondandia from the territory of the Republic Starstripe and subsequently was sold to the private company Superview Inc., which has its headquarters in Openskey, but is incorporated elsewhere.

Superview Inc. is a typical large multinational company and has shareholders all over the world. Superview Inc. is licensed by Openskey to carry out commercial VHR activities in conformity with international law including space law. Superview, Inc. pays Openskey 10% of the revenue it receives from the sale of sensed data. Openskey has a national space law which contains the provisions in the Annex. Agreements have been concluded between Superview Inc. and the governments of Papadia and Antipapadia to receive data directly from Golden Eye. In these agreements, which also were approved by the Foreign Ministry of Openskey, technical assistance is provided by Superview Inc. to Papadia and Antipapadia, and ground stations are delivered by Openskey to the respective countries. Both agreements refer to the conditions of the license of Superview Inc. as an integral part of the agreement.

Papadia and Antipapadia, which were separated from each other after a lengthy civil war, are located in the same geographic region and are both comprised of a group of small islands. The sea area in which they are situated contains many minerals, and large oil fields are supposedly located in the disputed territorial sea between Papadia and Antipapadia. Two oil companies, Drillwell Inc. and Sinkbetter, Inc., have long-standing close contacts with the governments of Papadia and Antipapadia respectively, due to the existence of these strategic oil fields. Drillwell Inc. is incorporated under the law of Papadia, and Sinkbetter Inc. is incorporated under the law of Antipapadia.

The first large client of Papadia for VHR data is Drillwell, Inc. which, based on the data delivered through the ground station of Papadia, started to search for oil in a small coastal area in the middle of the disputed territorial sea between Papadia and Antipapadia.

Antipapadia at the same time contracted with Sinkbetter Inc. and Sinkbetter, based on the data gathered through the ground station of Antipapadia, starts an oil search some 12 kilometers (8 miles) east of the area where Drillwell Inc. is searching. This area, as well, is situated in the middle of the disputed territorial sea between Papadia and Antipapadia. Both Drillwell and Sinkbetter are licensed through the respective Papadian and Antipapadian ministries to carry out exploratory drilling and subsequently to exploit any oil fields they find in the disputed areas.

The ground stations of Papadia and Antipapadia for receiving the VHR data from Golden Eye are technically different. Papadia, which still maintains close political and economical ties with its former colonial power Starstripe, purchased from Openskey the latest technology and software for processing all the data. Antipapadia, which after its separation from Papadia became part of the movement of progressive former colonial countries, has only limited access to space technology, due to limited economic resources. Consequently it was only able to purchase much less advanced hardware and software for processing the VHR data. Moreover, in the license of Goldstar to export Golden Eye to Openskey, a provision was included which gave Starstripe the right to demand a downgrading of the data in furtherance of the national security interests of Starstripe.This provision was also included in the licence from Openskey to Superview, Inc. and was made known to Papadia and Antipapadia when they signed the agreements with Superview, Inc.

Tensions have arisen between Papadia and Antipapadia due to the plans of both oil companies to start exploratory research in the disputed sea areas by sending special ships for experimental drilling. Both countries send warships to the disputed zones. A situation of international tension is reported by the press.

The data received from Golden Eye, according to press reports, also has military significance and enables both countries to monitor the other country's military activities. Papadia requests Starstripe to intervene in the conflict and also requests the Secretary General of the United Nations to discuss the situation in the Security Council. The Security Council subsequently adopts a Resolution calling upon the two countries to refrain from any further controversial actions in the disputed area and to enter directly into negotiations concerning the disputed sea area. The Resolution does not mention space activities nor the use of satellite derived information.

Openskey, after discussions with Starstripe, orders Superview Inc. to switch-off the satellite signals intended for the ground station of Antipapadia. Superview complied with this order.

At the same time, the oil companies Drillwell Inc. and Sinkbetter Inc. published the first results of their test drilling. It appears that Drillwell found an undersea oilfield which will justify commercial exploitation. It is located entirely in the disputed area. Sinkbetter, on the contrary, did not locate any oilfield. After having made an investigation with experts from Antipapadia, the conclusion was drawn that due to the technical inferiority of the ground station and processing facilities, it could not carry out its research with the data obtained through the Antipapadian ground station. It was also discovered that the data received by Antipapadia had been deliberately and constantly downgraded by Superview Inc. on request of the Government of Openskey. In the process of downgrading, some of the data received in Antipapadia was carelessly but unintentionally transformed in such a way that with the processing software Antipapadia possessed the data was being represented incorrectly.

Antipapadia, after having received the experts' report, decided to buy a state-of-the-art mobile ground station. In order to be sure to receive the VHR data in good order, Antipapadia started secretly to operate the mobile ground station from the territorial waters of Papadia. Antipapadia delivered VHR data to Sinkbetter, which, according to press releases, will take one more month to acquire enough data to start new research. At the same time, the Government of Antipapadia published pictures of secret Papadian military bases to the international press, which it claims clearly show the military build-up by Papadia. These pictures were derived from the satellite data obtained from Golden Eye through Antipapadia's recently acquired mobile ground station.

After these publications, Starstripe requests Openskey to switch off Golden Eye when it is in the coverage area of the ground stations in Papadia and Antipapadia until the disputed reception of data by Antipapadia ends, and until Papadia and Antipapadia resolve their territorial dispute. Based on this request, Openskey decides to suspend the license of Superview Inc. to operate Golden Eye, and its control is taken over by the army of Openskey. Neither Papadia nor Antipapadia have received any data from Golden Eye since this decision.

Openskey and Antipapadia have decided to bring their dispute before the International Court of Justice for resolution of the issues stated below. There are no issues as to the Court's jurisdiction.

Issues before the ICJ

1) Whether Antipapadia is violating international law by intercepting and publicly distributing the signals of Golden Eye, and, if so,

a) whether Antipapadia is liable to Openskey for the loss of revenue suffered by Opensky due to the switching-off of the satellite while within the coverage areas of the Papadian and Antipapadian ground stations; and,

b) whether Antipapadia should stop making this information public and should destroy or return to Openskey all data received.

2) Whether Openskey violated international law by switching-off Golden Eye while it was within the coverage of the Antipapadian ground station;

3) Whether Openskey violated international law by delivering to Antipapadia a ground station with hardware and software technically inferior to that purchased by and used in the Papadian ground station.

4) Whether Openskey violated international law by the intentional downgrading and unintentional transformation of Golden Eye data transmitted to Antipapadia, and, if so,

a) whether Openskey is liable for the costs incurred by Antipapadia in the unsuccessful exploratory research; and,

b) whether Openskey should compensate Antipapadia for the loss of expected oil revenues suffered by Antipapadia.

Instructions to the students:

You should prepare one memorial for the Applicant (Openskey) and one memorial for the respondent (Antipapadia). You should assume that all of the states referred to in this case are parties to all of the relevant international treaties and conventions and have adopted the United Nations General Assembly Resolution concerning "Principles Relating to Remote Sensing of the Earth from Outer Space" (G.A. Res. 47/68). None of the

States referred to in this case are parties to the UNCLOS III Treaty.

Annex
Openskey Law on Space Activities (Excerpts)

Preamble:

Having Regard the Increasing Commercial Uses of Outer Space and the Obligations for the States Party to the United Nations Space Treaties and Resolutions;

Taking Into Account the Articles of the Outer Space Treaty and especially Articles VI, VII and VIII;

Recalling the United Nations General Assembly Principles Relating to Remote Sensing of the Earth from Outer Space;

Believing that this Law will help strengthen the Leadership of Openskey;

Article 1
This Law applies to activities in outer space (space activities). In addition to activities carried out entirely in outer space, also included in space activities are the launching of objects into outer space and all measures to manoeuvre or in any other way affect objects launched into outer space.

Article 2
Space activities may not be carried out from Openskey's territory by any party other than the Openskey state, without a license. Nor may an Openskey natural or juridical person carry on space activities anywhere else without a license.

Article 3
A license to carry on space activities is granted by the Openskey government.
A license may be restricted in the way deemed appropriate with regard to the circumstances. It may also be subject to required conditions with regard to control of the activity or for other reasons. Inspection of the space activities of license holders is exercised by the authority decided by the Government.

---

3. WINNING BRIEFS

A. MEMORIAL FOR OPENSKEY

AGENTS

Ranjani Srinivasan,
Jean-François Renaud & Amine Laachani

ARGUMENT

The legal instruments governing the use outer space for peaceful purposes are drawn from various fields of law. Because the possibilities of outer space have yet to be fully explored and the implications of outer space research have yet to be understood, it is necessary to consider international law from a variety of angles to determine how to the apply existing law. In the case of Openskey versus Antipapadia, a case concerning very high resolution commercial remote sensing systems, we encounter issues that have yet to be dealt with from a legal perspective. One this is sure, however. Remote sensing data serve a very important purpose, increasing the extent of our management of the natural environment surrounding us. The state of Openskey is committed to directing remote sensing data towards the good they can serve the human community, and to increasing the number of users of this data.

Request for provisional measures to be taken by the Court

Before this Court makes any decisions regarding the merits of the case concerning very high resolution commercial remote sensing, the state of Openskey respectfully requests the honorable International Court of Justice to award preliminary measures of security.[1] Because Golden Eye data intercepted by the new mobile ground station of Antipapadia are currently being publicly distributed, and these data pose a potential threat to international peace and security, the public distribution of these data must be immediately halted before the situation of regional and international security is further exacerbated. The threat to international peace and security is evident and urgent; both countries have already sent warships to the disputed zone. Continuing the public distribution of stolen data will only heighten the tensions and could disturb international or regional public order.

In an effort to protect international peace and stability during this procedure before the respected International Court of Justice, the state of Openskey respectfully requests the adoption of this provisional measure of security.

Question 1: *Did Antipapadia violate international law by intercepting and publicly distributing the signals of Golden Eye?*

The state of Antipapadia clearly violated international law by the interception and public distribution of Golden Eye signals according to several areas of binding international law. These include property law (I), legal instruments concerning communications (II) and sovereignty principles (III). Furthermore, Antipapadia posed a serious threat to international peace and security (IV).

I. Antipapadia's actions were in breach of Openskey's property rights

Article VIII of the Treaty on Principles Governing the Activities of States in the Exploration and Use of Outer Space Including the Moon and Other Celestial Bodies of 1967 (hereinafter referred to as OST) establishes a state's jurisdiction and control over a satellite launched into space: "(A) State Party to the Treaty on whose registry an object launched into outer space is carried shall retain jurisdiction and control over such object, ..." The facts make clear that Openskey has adopted the OST, and even highlights Article VIII of the OST in its national law on space activities.[2] Because Golden Eye belongs to an Openskey judicial person[3], Openskey is the State party to the OST on whose registry appears Golden Eye. Therefore Openskey retains jurisdiction and control over Golden Eye. Since jurisdiction is an expression used to designate the double capacity of the state of determining the rules of law applicable and applying them,[4] not only does Openskey determine the applicable law, but exercises jurisdiction and control over the satellite.

Maintaining jurisdiction and control over the satellite is a form of ownership over the same. Because it exercises jurisdiction and control over the satellite, any product of the satellite belongs to the State Party to the OST. For example, if a satellite or component is found, it is returned to the State Party on whose registry appears this satellite.[5]

Concerning Golden Eye, data are gathered by optical sensors and SAR capability, both of which are physically in the satellite. The data are then transmitted from the satellite to the ground station. Openskey has jurisdiction over the satellite, its instruments, and in addition has jurisdiction and control over the products of the satellite, which are data. In virtue of Article VIII of the OST, Openskey owns the data gathered and processed by the satellite. In intercepting data not intended for it, the state of Antipapadia is in complete violation of Openskey's property rights and its jurisdiction over Golden Eye data. The state of Antipapadia acted in complete violation of Article VIII of the OST, and therefore of international law.

II. Antipapadia's violation of international law is further characterized as a violation of communication law

The state of Antipapadia practiced a violation of communication law, a general term here employed to include copyright law (A) and information law (B).

A. In violation of copyright law

Two main conceptions of international copyright law can be distinguished. The first is an Anglo-Saxon notion of copyright, codified by the Universal Copyright Convention (UCC), signed in 1952 and amended in 1971. The second is a "continental" concept of "droit d'auteur", which finds its roots in the Bern Convention for the Protection of Literary and Artistic Works, signed in 1886 and amended several times thereafter (hereafter referred to as the Bern Convention). Over the past few decades and through the process of internationalization of both concepts, they have essentially merged. As Openskey exercises jurisdiction and control over Golden Eye, it determines the law applicable to the same. However, it is considered that the Bern Convention provides a higher standard of copyright protection that the UCC.[6] The UCC provides that if a nation is a member of both conventions, the terms of the Bern Convention will govern, and if a nation is a member of a Bern Convention, it cannot withdraw and be only a member of the UCC, loosing both memberships at the same time.[7] It is then judicious to consider the terms of the Bern Convention, which Openskey has determined to be the applicable law, to the case in point. Furthermore, the case indicates that all relevant treaties have been adopted. The Bern Convention one such highly relevant treaty.

According to copyright law, Golden Eye data are protected both intrinsically and as databases. The state of Antipapadia violated both aspects of copyright law.

1. Golden Eye primary data are protected

Primary satellite data are protected under copyright law as original works in virtue of the Bern Convention, which provides for the protection of authors' rights for their artistic and literary works, defined in Article 2.[8]

Primary satellite data respond to the conditions for copyright protection imposed by the Bern Convention. These conditions require the existence of a human author and the originality of the work. First, the collection of remote sensing data implies a direct human intervention through the programming of the satellite from the Earth and the selection of images.

The second condition, the originality of the work, supposes that work wears the mark of its author. Modern satellites such as Golden Eye inherently have numerous defined parameters such as the viewing angles or the spectral bands, all of which permit technicians and others to mark the data with human originality while determining what data to be collected. "Beyond any doubt, remote sensing raw data is unique and original...."[9] The conditions of originality and authorship of Golden Eye data are therefore fulfilled, meaning that the data are protected.

The Bern Convention requires in addition that the work must be fixed in a material support. In practice, the transmission of raw data from the satellite to the ground station is accepted as a viable support, therefore protected.[10]

The author, s/he who exercises jurisdiction and control over the satellite, in this case Openskey, is entitled to certain protections of its work. Most importantly, the unauthorized reproduction[11] and public communication are duly prohibited.[12] Golden Eye data are therefore protected under the Bern Convention, and Openskey is the legitimate author of data benefitting from copyright protection. The state of Antipapadia violated Openskey's rights as an author, which are the copyright protections extended to Golden Eye data. In intercepting copyrighted data not belonging to it and proceeding with an unauthorized public communication of these data, the state of Antipapadia violated the provisions of the Bern Convention and therefore of international copyright law. The state of Openskey is the legitimate author of the data. Only Openskey may authorize the reception and public distribution of data. The state of Antipapadia completely transgressed these copyright protections afforded to Golden Eye data in virtue of the Bern Convention.

2. Golden Eye data are protected as databases

Not only are Golden Eye data protected intrinsically, but also can benefit from copyright protection as constituents of databases according to Article 5 of the WIPO Copyright Treaty of 1996.[13] Golden Eye data benefit from Article 5's definition from two perspectives.[14]

First, satellite data represent a potential image in a file format. This file is represented in the form of a 1, 0 grouping. The combination of digital binary data in the file constitutes a database by itself. Furthermore, to be protected, the database must demonstrate originality, meaning that it is organized, or that its elements are humanly selected. In the case of Golden Eye data, this condition is fulfilled. The order of digital binary data is chosen with due regard for the future image to be created from the data. The organization of the file is not random, rather, the voluntary action of its author.

Second, the satellite transmits a multitude of satellite photos, and therefore a multitude of satellite images expressed as files. The grouping of these files is constitutive of a database, stocked in the satellite and therefore organized with due regard for the site chosen to be observed. The condition of originality is duly fulfilled.

Golden Eye data are therefore protected as databases.

Because Golden Eye data are protected works, both intrinsically or as databases, the interception and distribution of data with the express authorization of the author, which is Openskey, are internationally illegal acts. The state of Antipapadia violated all copyright protections rendered to Golden Eye data and to Openskey as its author, thereby transgressing international law.

B. In violation of "information law"

The denomination of "information law" encompasses the liberty of information and regulations imposed by the International Telecommunications Union.

Golden Eye data are not information in the sense of human rights principles. Human rights address the freedom of expression and the liberty of information. In this case, information is a social liberty comprising political, economic, philosophical and cultural messages. Golden Eye data are not messages in this sense. Rather, they are signals which are not exploitable and non comprehensible until further treated. Because the intercepted Golden Eye are not information, they are not protected by the liberty of information.

Furthermore, the transmission of data from a satellite to a ground station is technically a telecommunications application, governed by telecommunications law. The interception of data violated the rule governing privacy of telecommunications. Principle III of the RS Principles indicates that "(R)emote sensing activities shall be conducted in accordance with (...) the relevant instruments of the ITU." The Golden Eye data intercepted by Antipapadia were intended for the Papadian ground station and not for Antipapadia's mobile ground station. The transmission of these data constitutes a private communication between Golden Eye and the Papadian ground station, in which Antipapadia interfered. Private communication is defined by the International Telecommunications Convention.[15] Antipapadia consciously eavesdropped on and intercepted this private communication, which is a violation of the International Telecommunications Convention. This action underlines Antipapadia's bad faith and could be characterized as piracy of a private communication.

The interception and public distribution of Golden Eye data cannot benefit from a human rights label of liberty of information. As it intercepted a telecommunications application, the state of Antipapadia stands is violation of the International Telecommunications Convention governing private communications.

III. In intercepting and distributing Golden Eye data, Antipapadia transgressed rules related to sovereignty

As initially defined by Max Huber[16], the principle of sovereignty is at the base of all international relations. The UN Charter recognizes this principle which governs all relations within the organization.[17] Rendering this principle *erga omnes* in international public law, sovereign equality is a general principle of international law, meaning that it is an explicit and direct source of positive law[18], applicable independently of any conventional confirmation.

In intercepting and publicly distributing Golden Eye signals, the state of Antipapadia violated two corollary principles to sovereignty: territorial integrity (A) and non intervention in the affairs of another state (B).

A. In violation of territorial integrity

The principle of territorial integrity is enunciated in the UN Charter[19], and repeated in several binding international instruments.[20] Territorial integrity prohibits a state's penetration of another state's territory without the authorization of the second state.

Though it was established in the statement of facts that none of the states mentioned have adopted UNCLOS III, the treaty itself is a codification of conventional practices. Practice demonstrates that territory includes, among others, territorial waters.[21] Furthermore, faced with the issue maritime delimitation, the ICJ has always opted to follow the model codified in the 1982 UNCLOS III.[22] Being a non signatory of UNCLOS III does not erase borders, clearly customary as well as conventional, of territorial seas.

The state of Antipapadia entered the acknowledged territorial waters of its neighboring state Papadia and remained in these waters to operate its mobile ground station with a goal to intercept a private communication. The Antipapadian mobile ground station entered Papadia's territorial waters without authorization and without flying its flag to indicate its presence. This violation of territorial integrity con moreover be interpreted as an act of piracy according to customary law.

Moreover, the facts of the case illustrate that satellite signals, while "switched off" to the Antipapadian ground station, were available on the total coverage area of both Papadia and Antipapadia, but downgraded.[23] The state of Antipapadia was in no way forced to invade the territorial waters of its neighbor, as the same downgraded signals could have been received from Antipapadian territory, in particular with its new mobile ground station. Any defence claiming that Antipapadia was forced to invade Papadia is false, as the signals were available over the total coverage zone and could be captured from any point in this zone through Antipapadia's new mobile ground station. This means that Antipapadia is unable to claim that *force majeure* conditions forced it to commit this internationally illegal act.

B. In violation of non intervention

The UN Charter illustrates clearly that a state is prohibited from intervening, directly or indirectly, in the internal affairs of another state.[24] Internal affairs are domains tied to a state's exercise of sovereignty, including public policy, choice of government and military affairs.

The state of Antipapadia intercepted and diffused information regarding the military affairs of another state. Military affairs, in particular military data, are customarily considered to be part of a state's *reserved domain*, intimately tied to a state's internal affairs. Clearly practicing intervention through the distribution of Golden Eye data, the state of Antipapadia violated an essential corollary to sovereignty and a fundamental base of international law.

Extraneous as it may seem to discuss the preceding points in the case of Openskey versus Antipapadia, there exists nonetheless a rule of international positive law which was violated. This violation illustrates the level of illegality and bad faith to which the state of Antipapadia was willing to proceed in order to procure data not intended for nor belonging to it.

IV. Antipapadia posed a threat to international peace and security

A situation of tension between the neighboring states of Papadia and Antipapadia was evaluated by the UN Security Council and reported by the international press. The Secretary General of the UN, following a request from Papadia, used his diplomatic functions according to Article 99 of UN Charter to avert the situation of crisis to the UN Security Council.[25]

The UN Security Council acts on matters relating to Chapter VII of the UN Charter: "Action With Respect to Threats to the Peace, Breaches of the Peace, and Acts of Aggression." Article 24 of the UN Charter confers the responsibility for the safeguard of international peace and security to the Security Council.[26] The Papadia - Antipapadia Resolution is clearly taken under the auspices of safeguarding international peace and security. Chapter VII is not an auspice lightly used; only grave situations threatening peace and security are considered. The fact the

Secretary General and the Security Council acted on the situation involving the states of Papadia and Antipapadia demonstrates that a clear threat to international peace and security did exist.

A. In transgression of the Security Council Resolution

The Papadia - Antipapadia Security Council Resolution requested both states to "refrain from any further controversial actions in the disputed area and enter directly into negotiations concerning the disputed sea area."[27] The state of Antipapadia did not refrain from two very controversial actions, one of which was carried out in the disputed territorial sea. The interception of data from Papadia's territorial waters is clearly a controversial action addressed by the resolution, as it involved an invasion into Papadian territory. Antipapadia clearly undertook a "controversial action(s) in the disputed area." The second, the public distribution of data, is clearly susceptible to heighten the tensions between the two states.
Because the Resolution was taken in a situation addressed by Article 24 of the UN Charter, it can be established that the legal force of the resolution is drawn from Article 25 of the UN Charter. Based on Article 25 of the UN Charter, the Security Council is empowered to pass binding resolutions when acting within its domain of competence, which is the safeguard of international peace and security.[28] The state of Antipapadia, in intercepting and distributing Golden Eye data, violated a binding resolution, which is a direct source of international law.

B. In transgression of maintaining peace

The conditions of Antipapadia's interception and diffusion of militarily sensitive information regarding a state with which it has tense relations constitute a provocation, susceptible to foster a military escalation which could result in an armed international conflict. A clear threat to peace, prohibited by Article 2 of the UN Charter[29], can be ascertained.
The preamble of the UN Charter confirms moreover that all nations should "practice tolerance and live tougher in peace with one another as good neighbors...." While the preamble of an international treaty normally has a political rather than legal value, the ICJ confirmed in the 1962 South West African Case that a preamble can have a legal value in the interpretation of a text.[30] In this case, the UN Charter is a text which proposes peace rather than conflict, good neighborliness rather than strife. The state of Antipapadia breached this essential condition for peace through its interception and distribution of Golden Eye signals.
The state of Antipapadia carried out its actions of interception and distribution of Golden Eye data in flagrant disrespect of international peace and security and in bad faith. It in addition violated the sovereignty of another state; such an invasion of cannot and should not be tolerated. International peace and security were clearly threatened by the state of Antipapadia.

In conclusion, the clear violations of international law committed by the state of Antipapadia have previously been characterized on various levels, including property law, communication law, rules relative to sovereignty and, most serious, in flagrant disrespect of the stability that the international community collectively endeavors to safeguard. An important issue remains to be addressed, however. The state of Openskey suffered harm as a result of these irresponsible and illegal acts undertaken by the state of Antipapadia.

Question 1a: *Is Antipapadia liable to Openskey for the loss of revenue suffered by Openskey due to the switching off of the satellite signal while within the coverage areas of the Papadia and Antipapadian ground stations?*

The harm caused to the state of Openskey by the actions undertaken by the state of Antipapadia can be characterized as financial harm. The state of Antipapadia knowingly engaged in internationally illicit acts in the interception and distribution of remote sensing data not intended for it.
Openskey was entirely justified in switching-off Golden Eye while within the coverage zones of both the Papadian and Antipapadian ground stations (I). It is further entitled to claim lost revenues (II). Finally, due to its internationally illicit acts, Antipapadia is responsible for all financial harm suffered (III).

I. The switch-off of Golden Eye while within the coverage areas of the Papadian and Antipapadian ground stations is justified

To the state of Openskey, the threat to international peace and security was clear. The first "switch-off" of satellite signals, a "switch-off" intended solely for the Antipapadian ground station and a recourse to which Openskey was fully justified [31], proved insufficient to ensure that Golden Eye data would not be used in a manner detrimental to the international community. The state of Antipapadia engaged nonetheless in dangerous and risky activities, activities which could have led to military escalation, activities which threatened international peace and security. In requesting the switch-off Golden Eye's signals to both ground stations, the state of Openskey ensured that the satellite data could not be used by either state to harm the other.
Furthermore, Openskey is entitled to request a switch-off of Golden Eye. Having jurisdiction and control over the satellite in virtue of Article VIII of the OST, Openskey can determine not only applicable law but what actions to undertake. Openskey National Law on Space Activities permits it to control the activity of the satellite; the switch-off is one such action.[32] Finally, other commercial remote sensing operators do not accept any liability for the interruption of service.[33] EOSAT, NASDA, SPOT Image, ESA and Radarsat are prime examples.[34]
An analogy to the switch off of Golden Eye to both ground stations can be drawn to the actions of SPOT Image during the Persian Gulf Crisis of 1991. SPOT Image normally practices a non discriminatory "openskies" policy, which has also been adopted by other remote sensing systems. However, from August 8, 1990,

the situation changed completely as a result of the Persian Gulf Crisis. SPOT Image decided to no longer sell data to any country of the Persian Gulf area and to renounce the policy of "open skies" such that a risk of furnishing data that might compromise the situation would be avoided. In essence, service was suspended to the entire region of the Middle East.[35]

Just as SPOT Image reacted on a crisis by suspending service, Openskey felt that a complete switch-off of Golden Eye was the best way to ensure the safety of the region and protect the peaceful status quo.

II. Openskey is entitled to claim all lost revenues

It is evident that the majority of lost revenue belongs to Superview Inc. Openskey receives 10% of that revenue. Two corollary principles enable the state of Openskey to claim the entirety of lost revenues, Article VI of the OST and diplomatic protection.

Article VI of the OST provides that "(s)tate parties to the Treaty shall bear international responsibility for national activities in outer space (...) whether such activities are carried on by governmental agencies or non-governmental entities (...) The activities of non-governmental entities in outer space (...) shall require authorization and supervision by the appropriate State Party to the Treaty." The state of Openskey bears international responsibility for the activities of Golden Eye. As the bearer of international responsibility, the state of Openskey must be prepared to represent Superview Inc. in international fora such as this one, where only state parties may present their disputes.[36] The case of Openskey versus Antipapadia implies the actions undertaken by both the state of Openskey and a judicial person of Openskey. Because a judicial person of Openskey can be represented only by the state of Openskey before the ICJ, the state of Openskey, bearing international responsibility for its space activities, is entitled to represent Superview Inc.'s interests, which are ultimately those of Openskey, before the Court.

Furthermore, the state of Openskey is exercising its diplomatic protection over Superview Inc. The criteria for establishing diplomatic protection are flexible, as determined by the ICJ in 1970.[37] The Court insisted that a clear connection must be drawn between the corporate entity and the state exercising diplomatic protection over it. In the domain of space activities, actors are varied and their activities are carried out predominantly in territory not subject to appropriation.[38] Therefore, diplomatic protection can be more difficult to establish. Superview Inc. is licensed by the state of Openskey to carry out space activities, and is a confirmed judicial person of this state according to Openskey national law.[39] While it is true that Superview Inc. is not incorporated in Openskey, it does have its headquarters in Openskey. Superview Inc.'s shareholders are all over the world, as is the case with many multinational companies. This said, the satellite Golden Eye was exported to the state of Openskey through the intermediary of Superview Inc. All space activities of Golden Eye, carried out by Superview Inc., must be approved by the state of Openskey.[40] Article VIII of the OST sheds further light on the connection between the private entity and the State, as it established Openskey's jurisdiction and control over Golden Eye. These lost revenues incurred due to the switch off of the satellite affect both an Openskey judicial person, Superview Inc., and the state of Openskey. Based on these strong connections, in particular the fact that Openskey exercises control and jurisdiction over the satellite itself indicates that the state of Openskey is entitled to exercise diplomatic protection over Superview Inc., and claim all revenues lost by both the state and its judicial person.

III. Antipapadia is responsible for all financial harm suffered

Both the ICJ and the PCIJ have had experience with this type of question. A state's international responsibility is implied if an internationally illegal action has been committed and is attributable to a state. A harm must then have been suffered. Finally, there must be a link between the illicit act and the harm suffered.[41]

The loss of revenue for which Antipapadia is responsible is the revenue gained by Superview Inc. from the sale of sensed data to both Papadia and Antipapadia. Antipapadia is responsible for the total switch-off of Golden Eye while over coverage zone of both states, in light of its gross violations of international law in the interception and distribution of Golden Eye signals not intended for it. The illegal actions of the state of Antipapadia engendered this switch-off, which was undertaken in the interest of international peace and security, but created a loss of important revenues.

The international illicit acts committed by Antipapadia were the interception and public distribution of Golden Eye data not intended for it. This act included invasion of another state's territorial waters. Illegality, of course, proposes two qualifying factors which would exonerate the state from its responsibility: *force majeure* and legitimate counter measure. As neither qualification is applicable to Antipapadia's action, it can therefore be established that an international illegal act was committed by Antipapadia.

The interception and distribution of data is directly attributed to the Government of Antipapadia. It purchased the mobile ground station and undertook to intercept signals from Papadia's territorial waters. The Government of Antipapadia also distributed the data to the international press for publication.[42]

As a result of the international illegal acts committed by the state of Antipapadia, Openskey was obliged to switch off the Golden Eye while over the coverage area of both states, such that Antipapadia would not reproduce such an action. The clear link between the illegal act and the harm suffered can be determined.

From the moment Superview Inc. complied with the order to switch off Golden Eye, Superview Inc. and the state of Openskey, lost revenue. It is undeniable that the link between Antipapadia's illicit act and the loss of revenue exists. Financial harm was and continues to be suffered by Superview Inc. and Openskey as long as Golden Eye remains switched off, meaning until the time the two states resolve their territorial dispute.

The state of Antipapadia is responsible for the all loss of revenue suffered by Openskey. The obligation of reparation of violations is automatic in character and confirmed in international law by the PCIJ in its decision of September 13, 1928 regarding the Chorzow Factory. "Reparation must, as far as possible, wipe out all consequences of the illegal act and re-establish the situation which would in all probability have existed if that act had not been committed."[43] The principle of *restituo in integrum* must be applied to this situation, and the state of Antipapadia is responsible for all loss of revenue suffered by Openskey as a result of the switch-off.

Question 1b: *Should Antipapadia stop making this information public and destroy or return to Openskey all data received?*

The data should no longer be exploited

Continuing with the idea of *restituo in integrum*, the situation will not be entirely restored by the sole reimbursement of lost revenue due to the switch-off of the satellite signals. Antipapadia's illicit acts imply further harm caused and therefore reparations to be made. What is important in this case is that the data are no longer exploited, and can therefore cause no more harm. The best guarantee of this is to destroy these data. As this can be an expensive undertaking, Openskey proposes two options. In the first, Openskey offers its services to help Antipapadia destroy these data. If this is not acceptable, Openskey will accept the return of the data and the good faith of Antipapadia that no copies have been made. Openskey will then undertake to destroy the data with its own means.

Question 2: *Did Openskey violate international law by "switching-off" Golden Eye while it was within the coverage of the Antipapadian ground station?*

The state of Openskey acted within its legal rights and responsibilities in requesting Superview Inc. to "switch-off" satellite signals intended for the state of Antipapadia. The action is entirely justified by the scope of the Openskey National Law on Space Activities and international law. It was taken in the interests of national and international security.

While a satellite such as Golden Eye can be "switched-off" to one country and continue to deliver signals to another even when the two countries are in as close proximity as Antipapadia and Papadia, this is not the case in the present situation. The Golden Eye coverage area or satellite footprint covers both countries, which is why satellite signals were available over the entire coverage area, even after the "switch-off".[44] The "switch-off" of satellite signals intended for the Antiapapadian ground station was actually a severe downgrading of data rather than a complete "black out" as the term may suggest. The data were downgraded enough such that Antipapadia, with its technically infer ground station, could no longer receive them. However, the more advanced the ground station, the higher the likelihood of reception. Therefore, the term "switch-off" is here employed to mean a severe downgrading such that Antipapadia did not receive signals with its ground station. Once Antipapadia did buy a state of the art mobile ground station, it should have able to receive these severely downgraded data from any point in the coverage area, and did from Papadia's territory. Whether in Papadian or Antipapadian territory, the signals were equally downgraded.

Openskey did not violate international law in the "switch-off" of satellite signals intended for the Antipapadian ground station. First, agreements passed authorized the present this severe downgrading resulting in what seemed to be a "switch-off". (I). Second, Openskey declines all liability regarding the quality and continuity of data (II). Finally, the principles of access to data must be balanced by security imperatives (III). Openskey did, however, practice equal treatment to both countries by downgrading the entirety of data available on the whole coverage area to the same extent.

I. Openskey had the right to "switch-off" the satellite signals

Openskey was entitled to downgrade data in virtue of agreements passed. Downgrading in virtue of national security is a common practice. Further, the application of such clauses were necessitated by the situation.

Concerning the practice of downgrading, it is common one among commercial remote sensing systems, and downgrading is not a practice contrary to international law. Rather it protects all of the actors involved in commercial remote sensing, in particular the observed state. For example, the Draft Agreement Between the United States National Oceanic and Atmospheric Administration and the European Organization for the Exploitation of Meteorological Satellites on An Initial Joint Polar-Orbiting Operational Satellite Systems, updated on September 26, 1995 includes a provision permitting the downgrading of data in the interest of international security.[45] In addition, in the GPS system, downgrading is a common and accepted practice.

The provision regarding the downgrading of data on the request of Openskey, in full conformity with common and accepted industry practices, is a part of the agreement between Superview Inc. and Antipapadia.[46] Furthermore, this action was undertaken after discussions with Starstripe. Starstripe also had the right to request a downgrading of data in furtherance of its national security interests. This provision was included the the licence to Goldstar to export Golden Eye to Openskey. It in addition was included in the licence between Openskey and Superview Inc., and was in turn integrated into the agreement between Antipapadia and Superview Inc. Therefore, Antipapadia was fully aware of the possibility of downgrading and who could request it.[47]

Furthermore, Article VIII of the OST gives the state of Openskey jurisdiction and control over the satellite. Openskey determines the law and actions applicable to the satellite. Downgrading was an action undertaken by Openskey falling within its entire jurisdiction.

Concerning the application of these clauses, implications related to national security cannot be avoided in a transaction involving the sale of remote sensing data of between 1 and 5 meter resolution. Compared to existing remote sensing systems, this VHR data is very precise and could reveal much more than is intended. All countries have the right to define and protect their national security according to the current geopolitical situation. National security imperatives are defined by a state and do not necessitate justification, confirmed by the spirit of the UN Charter, Article 2, paragraph 7. The elaboration of a state's national security imperatives falls within the competence of that state and only that state.

Furthermore, Openskey acted in the interest of international security. An international situation of crisis was established by the UN Security Council Resolution, and the international community obviously had an interest in the peaceful resolution of the conflict. As a recognized state of international crisis, the state of Openskey had an interest in "switching-off" the satellite signals intended for the Antipapadian ground station in response to the situation and for the benefit of peace.

The state of Openskey, after discussions with Starstripe, decided to downgrade data severely enough such that the technically inferior ground station would not receive data, while the advanced station would continue to do so. The data were equally downgraded for both Antipapadia and Papadia. In doing this, Openskey was able to play an active role in maintaining international peace and security while also providing non discriminatory service to both countries. Therefore, Openskey practiced equal treatment of its clients.

If Starstripe requested the downgrading, which it had the right to do, Openskey acted in good faith towards its obligations to Starstripe. Because Openskey respected international law, acted in good faith and within its clear rights regarding all of its agreements, it is not in any violation of international law.

II. Openskey declines all liability regarding continuity and quality of the data

Most commercial and experimental remote sensing systems accept no responsibility for a temporary interruption of services, nor for the quality of data upon reaching the ground station. EOSAT, ESA, SPOT and NASDA contracts do not provide warranties of continuity of the satellite service or quality of data.[48] The practice of these four prominent operators, originating in various parts of the world, emphasizes a regular practice in development in the field of remote sensing. The practice underlines an evolving custom in the field. It emanates from the practices of states or actors most interested.[49] Remote sensing is a rapidly evolving field, and legal practices develop quickly as temporal aspect is sped up. The four remote sensing systems cited above use the same provisions regarding the exclusion of their liability concerning the quality and continuity of data.

Based on the developing contractual practice in the field of remote sensing, the state of Openskey is not responsible for ensuring the continuity of service to the Antipapadian ground station.

III. Access to data must be balanced by security imperatives

The 1986 RS Principles enunciate in Principle XII the concept of access to data. "As soon as the primary data and the processed data concerning the territory under its jurisdiction are produced, the sensed State shall have access to them on a non-discriminatory basis and on reasonable cost terms." The state of Openskey fully believes in the goals of the RS Principles.

The RS Principles are a code of standard behavior in the remote sensing industry. The General Assembly resolution 47/68, after years of work within the UN Subcommittee on the Peaceful Uses of Outer Space, was taken in the spirit of promoting a global remote sensing community between observers and the observed. From a legal perspective, the resolution was taken in virtue of article 10 of the UN Charter, which authorizes the General Assembly to pass simple recommendations without any binding legal value. The fact that the resolution was adopted by consensus does not render to them the required *opino juris* necessary to create customary law. Moreover, the Resolution was subject to unilateral declarations of states particularly interested[50]. These declarations certainly subtract from any possible *opinio juris.* While many of the Principles annexed to the resolution are repeated in binding international legal instruments such as the OST, making them obligatory in virtue of their conventional confirmation, Principle XII regarding access to data has not yet been subject to conventional or other confirmation, which would bring it into the scope of applicable law. In addition, no customary practice has confirmed the principle of access to data in a commercial situation such as the one in the case of Openskey v. Antipapadia. Therefore, Openskey is not bound by Principle XII of the RS Principles, providing for immediate access to data regarding territory under Antipapadia's jurisdiction. The Principle in question does not have a conventional or customary confirmation which would transform it to an source of binding international law. While it is true that in the domain of meteorology, a strong emphasis is placed on the immediate access to data in virtue of the involvement of the WMO. However, the present case deals with a purely commercial situation excluding the involvement of the WMO.

Furthermore, until the situation of peace and security was aggravated, the state of Openskey through Superview Inc. practiced complete access to data. While the RS Principles undertake to protect the rights of the sensed state by providing immediate access to data, the spirit of the resolution is to promote peace, management of natural resources, the protection of the Earth's natural environment and the protection of mankind from natural disasters. This recalls in addition the spirit guiding the 1967 OST.

Based on the foregoing, the state of Openskey did not violate international law in requesting the "switch-off" of satellite signals intended for the Antipapadian ground station. Rather, Openskey acted in conformity with its own national legislation, developing contractual practices, and attempted to safeguard international peace and security. Most importantly, Openskey acted in good faith, offering non discriminatory treatment to both clients and respecting its international obligations. Rather than violating international law, Openskey opted to take an active role in safeguarding international peace and security.

Question 3: *Did Openskey violate international law by delivering to Antipapadia a ground station with hardware and software technically inferior to that purchased by and used in the Papadian ground station?*

In the facts of the case it is very clear that Antipapadia, "...was only able to *purchase* much less advanced hardware and software for processing the VHR data." This is as a result of Antipapadia's limited economic resources and therefore limited access to space technology. Papadia "*purchased* from Openskey the latest technology and software for processing the data." (Emphasis added) Both countries clearly purchased the equipment they received. Furthermore, Antipapadia was fully aware that the ground station purchased by it was technically inferior to that of Papadia.[51]

Incidentally, when Antipapadia purchased a ground station from the state of Openskey, it did not have the financial resources to afford a technologically advanced ground station. However, when it seemed to Antipapadia that a new state of the art mobile ground station was necessary, it procured the resources, which could be construed as an action undertaken in bad faith.

Openskey clearly did not violate international law in delivering to Antipapadia a technologically inferior ground station. It respected the inherent laws of commercial freedoms (I) and acted in conformity with the law governing the international sale of goods (II). Finally, any claimed right to development does not create any legal obligations for the state of Openskey (III).

I. Openskey respected the inherent laws of commercial freedom

"Thus the basic law of the open market may prevail: those who cannot pay cannot buy."[52] Though it is true that developing countries are at a disadvantage regarding international commerce, the basic rules of commercial freedoms cannot be ignored.

Considering the thesis of commercial freedom, all states have the right to engage in international commerce on a non-discriminatory basis. This means that all states engage in international commerce based on the same rules, without favoritism or bias. For example, the recent WTO Agreement contains a provision addressing the most favored nation status.[53] All states must accord a most favored nation status to any state with which it engages commerce if the same status is given to another state for the same product, meaning that non-discriminatory treatment must be practiced universally. The state of Openskey delivered to Antipapadia a ground station within its economic means and corresponding to its needs. The goods were according to the needs it was meant to serve, were delivered as agreed and followed the law of commercial freedom.

Openskey offered both Papadia and Antipapadia equal treatment. Both obtained that which they could afford, without favoritism or bias shown to either. Moreover, the ground station seemed to function properly. Sinkbetter Inc., based on data gathered through the Antipapadian ground station, was able to begin an oil search in the same general area as that of Drillwell. Inc.[54]

It is further important to understand that data directly from Golden Eye is raw data. They must be processed to produce usable information. As Sinkbetter Inc. based their search on raw data gathered through the Antipapadian ground station, it had to further process the data to locate oil resources. The data could have been further processed through Sinkbetter Inc.'s materials to render maps indicating where oil resources may be located. Therefore, Sinkbetter Inc.'s data processing materials play a role in the outcome of the final information used to locate oil resources.

The state of Openskey was and continues to be very interested in including a growing number of countries, regardless of their economic development, into the remote sensing community, of its own volition and in virtue of Article 1 of the OST. It is for this reason that Openskey initially contracted with Antipapadia through its intermediary of Superview Inc. However, the state of Openskey follows both the rules governing international contracts and those covering commercial freedom. The fact that Papadia was able to purchase more advanced materials does not engender a breach of law on Openskey's part. The fact that it had the financial means, regardless of its source, does not oblige Openskey to make financial or pricing concessions for Antipapadia because it is unable to purchase the same materials as its neighbor.

II. Openskey acted in conformity with law governing the international sale of goods

Almost all nations of the world have signed the UN Convention on Contracts for the International Sale of Goods of April 11, 1980. Because the parties to this case have adopted all relevant international treaties except UNCLOS III, it can be assumed that both are party to this treaty. Chapter II of the Treaty details the obligations of the seller, with which the state of Openskey acted in complete conformity, including fitness for a particular purpose. Article 35 indicates that the seller "must deliver the goods which are of the quantity, quality and description required by the contract..." Antipapadia was aware that, due to its limited financial resources, it was purchasing a ground station of lower technological standards than that purchased by Papadia. The quality of the ground station was expressly made known to the state of Antipapadia when purchasing it.[55] The goods, the hardware and software to process the data, were conform to that which was agreed.

Furthermore, regarding fitness for a particular purpose, the ground station delivered to Antipapadia is capable of receiving the majority of Golden Eye data. The goods were then conform to the contract because they were fit for the purpose of collection and processing of remote sensing data, in accordance with Article 35, paragraph 2 (b) of the UN Convention on Contracts for the International Sale of Goods.[56] Antipapadia, fully aware of its economic situation, should have recognized that the ground station it purchased with limited economic means would not have extended capabilities.

III. The right to development does not create obligations for Openskey

Principles II, V, VI, VII and VII of the RS principles mention the idea of cooperation and consideration of the limitations, both economic and technical, of developing countries. These principles, important to the development of a global remote sensing community, are goals rather than binding legal principles. The RS Principles, by themselves, dictate no more than a code of good practice. The state of Openskey is dedicated to realizing the goals enunciated in the RS Principles. However, these ends do not justify the creation of obligations on the part of Openskey regarding the ground station.

The New International Economic Order tried to integrate the right to development into positive international law. However, the principle of the right to development has been resigned to simple UN General Assembly recommendations, which posses no binding legal value in virtue of Article 10 of the UN Charter.[57]

Most importantly, the right to development, to which all states are entitled, exists in itself, and Openskey concluded agreements with Antipapadia and Papadia to foster their development in the remote sensing industry. However, this right does not create any legal ramifications such as obligations to provide discriminatory service or favorable pricing structures. The right to development does not oblige Openskey to sell materials at an advantageous price. Both Papadia and Antipapadia purchased ground stations which they could afford, neither benefitting from special treatment from the state of Openskey. Openskey once again acted in good faith by providing non discriminatory access to space technology.

Therefore, any claim of a right to development which creates obligations towards Openskey is not founded. Moreover, Antipapadia was well able to afford a state of the art mobile ground station during the course of events. Any plea that Openskey did not respect supposed obligations derived from the right to development would only underline Antipapadia's bad faith, as it did purchase an advanced ground station soon after.

In conclusion, Openskey supports all efforts of all countries to enter the remote sensing community and reap the benefits thereof. However, the state of Openskey practices non discriminatory treatment, whether in the sale of goods or data. It respected the inherent laws of commercial freedom and obeyed the law governing the international sale of goods. Finally, it is clear that the right to development does not create any obligations towards Openskey. A technically inferior ground station is the result of Antipapadia's lack of economic means, and in no way a sign of Openskey's violation of international law.

Question 4: *Did Openskey violate international law by the intentional downgrading and unintentional transformation of Golden Eye data transmitted to Antipapadia?*

The state of Openskey in no way violated international law in the downgrading and unintentional transformation of Golden Eye data transmitted to Antipapadia. Openskey had the right to request the downgrading of data, as clauses relative to national security are applicable (I). Furthermore, current remote sensing practices demonstrate that a satellite operator is not responsible for the quality of data (II). Openskey's good faith, reflected in its actions throughout, is clear in its claim that the data were unintentionally transformed (III).

I. Clauses relative to national security are applicable

According to the facts of the case, "(A)greements were concluded between Superview and the governments of Papadia and Antipapadia. (...) Both agreements refer to the *conditions of the license of Superview Inc.* as an integral part of the agreement. (...) (I)n the license of Goldstar to export Golden Eye to Openskey, a provision was included which gave Starstripe the right to demand a downgrading of the data in furtherance of the national security interests of Starstripe. *This provision was also included in the license from Openskey to Superview, Inc.*"[58] (Emphasis added) Both agreements, between Superview Inc. and Papadia and that between Superview Inc. and Antipapadia integrated Superview Inc.'s conditions of license. These conditions of license include a provision permitting Starstripe to request the downgrading of data in furtherance of its national security. Antipapadia's acceptation of such a provision is reflected in its approval of the agreements to receive remote sensing data directly from Golden Eye. Both Papadia and Antipapadia were aware and agreed that Starstripe could request a downgrading of data in furtherance of its own national security interests.

Furthermore, Openskey also had the right to demand a downgrading of data in furtherance of its national security concerns.[59] This provision was included in the license between Openskey and Superview Inc., and therefore integrated in the agreement between Superview Inc. and Antipapadia.

Whether the downgrading was requested by Openskey or by Starstripe, both had the right to request it. Furthermore, Antipapadia was aware that both states could request the downgrading of data.

The Government of Openskey, on its own or on the prompting of Starstripe, obviously felt that data intended for the Antipapadian ground station should be downgraded. One reason is the security issues which may have arisen. All states have the right to determine their security imperatives, and the state of Openskey determined that remote sensing data intended for the Antipapadian ground station should be downgraded. National security interests do not need to be justified by disclosing them. Rather, national security imperatives are the competence of a state's *reserved domain.* Suffice it to say that Openskey and Starstripe felt their national security threatened by the situation brewing in the region. The issues in contention are whether Openskey first had the right to request the downgrading of data, and if the request was justified. The answer to both is yes. The agreement passed between Superview Inc. and Antipapadia permits the states of Openskey and Starstripe to downgrade data in furtherance of their national security interests. Antipapadia, in signing the agreement, agreed to and was aware of this provision.

II. Remote sensing data are not guaranteed to be accurate

Downgrading is a common practice in the remote sensing industry, just as well in the case of a specific country as to all ground stations served by the remote sensing system. Major remote sensing operators internationally engage in the practice, as was detailed above in question 2.[60] Moreover, the practice has been confirmed by legal practioners in the field.[61]

Due to its evolutionary state, remote sensing data are generally not guaranteed to be accurate. Moreover, EOSAT, SPOT Image, NASDA, ESA and Radarsat do not provide warranties of data accuracy or of suitability of imagery or data for any use. Further, they do not guarantee the quality of the data.[62] Because the contract between Superview Inc. and Antipapadia is one of commercial exploitation of remote sensing data, an analogous situation can be discerned in the current practice of SPOT Image, a private French commercial remote sensing system. SPOT Image sells remote sensing data obtained by SPOT satellites and processed through their ground stations. In the delivery of remote sensing data to a ground station, certain risks are present. SPOT Image denies any obligation of result. SPOT Image adopts a policy of "best efforts," meaning that it engages to make the best effort to ensure the intended product. However, it assumes no responsibility for the end product or accuracy of data arriving in the hands of a client.[63]

In the case of Openskey versus Antipapadia, Superview Inc. sells remote sensing data to Antipapadia who processes the data through a ground station purchased from Openskey for the technical assistance is carried out by Superview Inc. Adopting this current practice, the state of Openskey made its "best efforts" to deliver accurate data to Antipapadia. The transformation of data, unintentional as it was, is a natural risk of the industry, and a risk for which Openskey declines all responsibility. Therefore, the state of Openskey in no way violated international law in the unintentional transformation of data, rather it followed the general practice of commercial remote sensing organizations.

III. Remote sensing activities were carried out in good faith

Good faith is a general principle of international public law. According to Article 26 of the 1969 Vienna Convention, "(E)very treaty in force is binding upon the parties to it and must be performed by them in good faith." In including this article in the Vienna Convention, the Commission for International Law underlined that this article emphasizes the fundamental principle of the law of treaties. In addition to the Vienna Convention, the UN Charter further establishes the principle of good faith[64], and Article 7 of the UN Convention on Contracts for the International Sale of Goods of April 11 1980 addresses good faith.[65] The execution in good faith and the respect of *pacta sunt servanda* are linked and constitute two complementary aspects of the same principle.

In commercial relations, good faith means that engagements are undertaken and carried out without ruse, fraud, infringement on law and loyalty to the engagements taken. Regarding the transformation of some remote sensing data, the state of Openskey was unaware that the downgrading of data would cause such a transformation. Had it been aware, this state would have remedied the situation immediately. The downgrading of data, as with all actions carried out by Openskey, was undertaken in good faith.

Because the state of Openskey did carry out its remote sensing activities in good faith, it cannot be held responsible for the unintentional transformation of data. Remote sensing, in particular with a satellite with Golden Eye's capabilities, is a very new field, and it is often difficult to predict the nature of problems which may occur in the collection and transmission of data.

Because the remote sensing activities, including the downgrading of data, were carried out in good faith and with no intention to deliver misrepresented data, the state of Openskey is not responsible for the unintentional transformation of some data.

Based on the foregoing arguments, the state of Openskey did not violate international law in the intentional downgrading unintentional transformation of Golden Eye data transmitted to Antipapadia. Openskey had the full right to downgrade data and was justified in doing so. Openskey further follows accepted commercial practices in the industry in not guaranteeing the quality, continuity or fitness of the data for any particular purpose. Finally, the downgrading, just as all of Openskey's actions, was undertaken in good faith.

***Question 4a:** Is Openskey liable for costs incurred by Antipapadia in the unsuccessful oil research?*

No claim implicating Openskey's responsibility is valid

When determining an issue of liability, it is necessary to juxtapose the conditions for reparations of damage with the facts. The PCIJ, in its decision of September 13, 1928 regarding the Chorzow Factory Case, and the ICJ, in its decision of February 5, 1970 regarding Barcelona Traction, confirmed the three conditions necessary for the reparation of damages. First an international illegal act must be ascertained. Second, the illegal act must be attributable to a state. Third, a clear link between the illicit acts and harm suffered must be discerned.

In this situation, the unification of these three conditions cannot be established. The state of Openskey was fully within its rights to request the downgrading of the data intended for the Antipapadian ground station. As for the unintentional transformation of some data, remote sensing data is not guaranteed accuracy according to current commercial remote sensing practices; it is not responsible for an unintentional transformation of data.. In addition, Openskey acted in good faith. It is clear that no internationally illegal act was committed. The most important condition for the reparation of damages is not fulfilled. Without the unification of the three conditions, revenues lost in virtue of Openskey's actions cannot be claimed.

Moreover, returning to the analogy previously drawn between SPOT Image and Superview Inc., SPOT contracts, as well as those of EOSAT, NASDA, ESA and Radarsat, explicitly deny the data's fitness for any particular purpose or type of research. Furthermore, SPOT contracts absolve it of any liability as a result of losses or damages based the utilization of data. Finally, the receiver of the data waives its right to any recourse against the data provider in the case of losses or damages.[66] Similarly, Superview Inc. does not guaranty the data's fitness for any purpose, in this case, the identification of Antipapadian oil resources. It therefore cannot be held liable for any losses or damages based on the exploitation of remote sensing data. Moreover, the state of Antipapadia is not justified in claiming any losses or damages, as it waived its right to do so in signing an agreement with Superview Inc.

Moreover, it is important to keep in mind that remote sensing raw data are not exploitable until processed. In order to locate oil resources, Sinkbetter Inc. may have processed some of the data with their cartographic materials. Some processing may have been carried out by it in order to cater the Golden Eye data to their specific goal. Sinkbetter Inc.'s materials may have played a role in the transformation of data, and should be looked into further.

The state of Openskey is not liable for any costs incurred by the state of Antipapadia in the unsuccessful exploratory research. Furthermore, the state of Antipapadia is not entitled to claim these damages suffered due to Golden Eye data.

Question 4b: *Should Openskey compensate Antipapadia for the loss of expected oil revenues suffered by Antipapadia?*

Hypothetical revenues are not the responsibility of Openskey

Antipapadia's future oil revenues are purely hypothetical revenues. The only oil field found thus far is located entirely in the disputed territorial sea between Papadia and Antipapadia. This oil field may not belong to the state of Antipapadia; this question can be answered only after negotiations are undertaken by both states to resolve their territorial disputes.

Given that the state of Openskey is absolved of any responsibility regarding the downgrading and unintentional transformation of some data, it does not accept any responsibility for any hypothetical future revenues which may or may not be attributed to the state of Antipapadia. Furthermore, Superview Inc. cannot be held responsible for these hypothetical future losses suffered due to remote sensing data, according to current commercial remote sensing practice. Finally, before considering future revenues and damages that may be claimed thereof, the state of Antipapadia must resolve its territorial dispute with Papadia, as was requested in the UN Security Council Resolution.

The state of Openskey is in no way responsible or liable for any loss or hypothetical loss, either from unsuccessful exploratory drilling nor from expected oil revenues.

SUBMISSIONS TO THE COURT

For the foregoing reasons, the Government of Openskey, Applicant, respectfully requests the Court to adjudge and declare that:

1. The state of Antipapadia violated international law in intercepting and publicly distributing Golden Eye data:

a. the state of Openskey exercises jurisdiction and control over Golden Eye;

b. copyright provisions are applicable to Golden Eye data;

c. the state of Antipapadia transgressed principles of sovereignty;

d. the state of Antipapadia posed a threat to international peace and security.

In consequence, Antipapadia is responsible to Openskey for all lost revenues due to the switch-off of Golden Eye while over the coverage area of the Papadian and Antipapadian ground station:

a. the switch-off of Golden Eye is justified;

b. the state of Openskey is entitled to claim lost revenues;

c. the state of Antipapadia is responsible for all financial harm suffered.

The intercepted Golden Eye data should no longer be exploitable.

2. The state of Openskey did not violate international law by "switching-off" Golden Eye signals intended for the Antipapadian ground station:

a. the state of Openskey had the right to "switch-off" Golden Eye signals intended for the Antipapadian ground station;

b. the state of Openskey declines all liability for the continuity and quality of data;

c. access to data must be balanced by international peace and security imperatives.

3. The state of Openskey did not violate international law by the delivering a ground station with hardware and software technically inferior to that purchased by and used in the Papadian ground station:

a. the state of Openskey respected the inherent laws of commercial freedom;

b. the state of Openskey acted in conformity with the law governing the international sale of goods;

c. the right to development does not create obligations for the state of Openskey.

4. The state of Openskey did not violate international law by the intentional downgrading and unintentional transformation of Golden Eye data transmitted to Antipapadia:

a. clauses relative to national security are applicable;

b. remote sensing data are not guaranteed to be accurate;

c. remote sensing activities were carried out in good faith.

Therefore, no claims implicating the state of Openskey's

responsibility are valid, and any hypothetical future oil revenues are not the responsibility of the state of Openskey.

---

1 ICJ Statute art. 41, para. 1: "1. The Court shall have the power to indicate, if it considers that circumstances so require, any provisional measures which ought to be taken to preserve the respective rights of their party. "

2 Manfred Lachs Moot Court Space Law Contest 1997 Case; Preamble, Openskey National Law on Space Activities

3 "Superview Inc. is licensed by Openskey to carry out commercial VHR activities in conformity with international law including space law." It is clear that, regardless of the location of Superview Inc.'s headquarters and its shareholders, Superview Inc. is a judicial person and licensee of the state of Openskey. See also page 35, Question 1a for further discussion of Superview Inc.'s connection to Openskey

4 Couston, M., Droit Spatial Economique (1994) at 11.

5 Treaty on Principles Governing the Activities of States in the Exploration and Use of Outer Space Including the Moon and Other Celestial Bodies, January 27, 1967, U.K.T.S. 10 (1968) Cmnd. 3519 (hereafter referred to as OST), art. 8 "... Such objects or component parts found beyond the limits of the State Party to the Treaty on whose registry they are carried shall be returned to that State Party, ..."

6 Salin, P., Selected Legal Aspects of Commercial Remote Sensing (1991) at 86

7 UCC art. XII, para. 2

8 Bern Convention, art. 2, "Literary and artistic works shall include every production in the literary, scientific and artistic domain, whatever may be the mode or form of this expression such as (...) photographic works to which ar assimilated works expressed by a process analogous to photography (...), maps and three dimensional works relative to geography ...."

9 Salin, P., Selected Legal Aspects of Commercial Remote Sensing (1991) at 92.

10 Salin, P., Selected Legal Aspects of Commercial Remote Sensing (1991) at 92.

11 Bern Convention art 9, , "gives the author the exclusive right to authorize the reproduction of these works in any manner or form."

12 Bern Convention art 11bis, para. 1.iii, , "Authors of literary and artistic works shall enjoy the exclusive right of authorizing (...) the public communication by (...) images (...) of the work."

13 WIPO Copyright Treaty art. 5 "(the protection of) compilations of data (databases)or other materials in any form by reason of the selection or arrangement of their contents constitute intellectual creations."

14 For further information, see Gaudrat, P., *Commercialisation des données publiques- Rapport pour l'Observatoire Juridique des technologies de l'Information*, (1994).

15 International Telecommunications Convention, 1982 Nairobi, art. 37 .

16 Island of Palmas (Netherlands v. United States), 1928 Permanent Court of Arbitration (April 4) "The sovereignty in relations between states signifies independence." (award of sole arbitrator Max Huber)

17 UN Charter art 2, para. 2, " The organization is based on the sovereign equality of all its Members."

18 UN Charter (Statute of the ICJ) art 38, "The Court, whose function is to decide in accordance with international law such disputes as are submitted to it shall apply : a. international conventions ...; b. international custom ...; c. the general principles of law ...; d. (...) judicial decisions and the teachings of the most qualified publicists...."

19 UN Charter art. 2, para. 4, "All Members shall refrain in their international relations from the threat or use of force against the territorial integrity or political independence of any state, or in any other manner inconsistent with the Purposes of the United Nations."

20 Military and Paramilitary Activities in and against Nicaragua (Nicaragua v. U.S.), 1986 ICJ (June 27).

21 UN Convention on the Law of the Sea, Montego Bay, December 10, 1982, 21 I.L.M. 1261 (1982). Military and Paramilitary Activities in and against Nicaragua (Nicaragua v. U.S.), 1986 ICJ (June 27). The Court affirmed that territorial waters are a part of a State's territory.

22 See Continental Shelf (Tunisia v. Libya), 1982 (February 24); Continental Shelf (Libya v. Malta), 1985 (June 3).

23 Manfred Lachs Moot Court Space Law Competition 1997, Questions to the Court, question # 8

24 UN Charter art.2 para. 7, "Nothing contained in the present Charter shall authorize the United Nations to intervene in matters which are essentially within the domestic jurisdiction of any state or shall require the Members to submit such matters to settlement under the present charter; ...."

25 UN Charter art.99 "The Secretary General may bring to the attention of the Security Council any matter which in his opinion may threaten the maintenance of international peace and security."

26 UN Charter, art. 24 para. 1, "In order to ensure prompt and effective action by the United Nations, its Members confer on the Security Council primary responsibility for the maintenance of international peace and security, and agree that in carrying out its duty under this responsibility, the Security Council acts on their behalf."

27 Manfred Lachs Moot Court Space Law Competition 1997 Case, page 3

28 UN Charter art.Art. 25: "The Members of the United Nations agree to accept and carry out the decisions of the Security Council in accordance with the present Charter."

29 UN Charter art. 2 para 4, "All Members shall refrain in their international relations from the threat or use of force against the territorial integrity or political independence of any state, or in any other manner inconsistent with the Purposes of the United Nations."

30 South West African Case (Ethiopia and Liberia v. South Africa), 1962 I.C.J. (December 21).

31 For further discussion see below page 40, Question

2

32 Openskey National Law on Space Activities, art. 3 "A License (...) (m)ay be subject to required conditions with regard to control of the activity or for other reasons."

33 Salin, P., Legal Aspects of Commercial Remote Sensing (1991) at 55.

34 Salin, P., Legal Aspects of Commercial Remote Sensing (1991) at 55.

35 Martin, P.M., Droit des Activités Spatiales (1992) at 186-187.

36 UN Charter (ICJ Statute) art. 34 para.1 "Only states may be parties in cases before the Court."

37 Barcelona Traction Power and Light Company (Belgium v. Spain), 1970 ICJ (February 5) "However, in the particular field of diplomatic protection of corporate entities, no absolute test of the "genuine connection" (between the corporate entity and the State) has found general acceptance." paragraph 70 of the decision.

38 OST art. 2 "Outer space, including the moon and other celestial bodies, is not subject to national appropriation by claim of sovereignty, by means of use or occupation, or by any other means."

39 Manfred Lachs Moot Court Space Law Competition 1997 Case, Openskey National Law on Space Activities art. 2 , "Nor may any Openskey natural or judicial person carry on space activities anywhere else (than Openskey's territory) without a license."

40 For example, the sale of sensed data to Papadia and Antipapadia was approved by the Openskey Foreign Ministry.

41 Cape Horn Pigeon Arbitration (U.S.v. Russia), 1902 (November 29); Chorzow Factory (Germany v. Poland), 1928 P.C.I.J. (ser. A) No. 17 (September 13); Barcelona Traction Light and Power Company (Belgium v. Spain) 1970, I.C.J. (February 5).

42 Manfred Lachs Moot Court Space Law Competition 1997, Case, page 3.

43 Chorzow Factory (Germany v. Poland), 1928 P.C.I.J. (ser. A) No. 17 (September 13).

44 Manfred Lachs Moot Court Space Law Competition 1997, questions to the Court, question 4

45 Annex 1 to EUMETSAT / NOAA Cooperation Agreement

46 Manfred Lachs Moot Court Space Law Competition 1997, questions to the Court, question 17

47 For further discussion of the right to downgrade, see below page 48-50, Question 4

48 EOSAT- Earth Observation Satellite System (U.S.), ESA- European Space Agency (Europe), SPOT Image (France), NASDA National (Japan).

49 North Sea Continental Shelf (Federal Republic of Germany v. Denmark; Federal Republic of Germany v. the Netherlands) 1969 I.C.J., (February 20). The Court insisted that a custom develops from the practice of states most interested.

50 For example, the United Soviet Socialist Republic and the United States of America.

51 Manfred Lachs Space Law Moot Court Competition, 1997, Questions to the Court, question 9

52 Krepon, Michael et al., Commercial Observation Satellites and International Security (1990) at 37.

53 General Agreement on Trade and Tariffs, art. II (most favored nation status) April 21, 1994.

54 Manfred Lachs Moot Court Space Law Competition, 1997 Case page 2.

55 Manfred Lachs Moot Court Space Law Competition, 1997, Questions to the Court, question 11.

56 UN Convention on Contracts for the International Sale of Goods, art. 35, para. 2(b), "Except where the parties have agreed otherwise, the goods do not conform with the contract unless they: (...) (b) are fit for any particular purpose expressly or impliedly made known to the seller ...."

57 For example, the Charter of Economic Rights and Duties of States 1974, the Declaration on the Establishment of a New International Economic Order 1974, the Declaration on the Right to Development 1986.

58 Manfred Lachs Moot Court Space Law Competition, 1997 Case page 2.

59 Manfred Lachs Moot Court Space Law Competition, 1997 Questions to the Court, question 15.

60 For further discussion, see above p. 39, Question 2.

61 Interview with Professor Ram Jakhu, Director, MSS Program, *International Space University*, Strasbourg, February 7, 1997; interview with Professor Nicolai Tolyarenko, Senior Faculty, MSS Program, *International Space University*, Strasbourg, February 7, 1997; interview with M.Y.S. Prasad, Counsellor (Space), Indian Space Research Organization, *Embassy of India*, Paris, January 28, 1997.

62 Salin, P., Legal Aspects of Commercial Remote Sensing (1991) at 58 and 83.

63 Couston, Droit Spatial Economique at 233-235 (1994)

64 UN Charter art. 2, para. 2, "All Members, in order to ensure to all of them the rights and benefits resulting from membership, shall fulfill in good faith the obligations as assumed by them in accordance with the present Charter."

65 UN Convention on the International Sale of Goods, April 11, 1980, art. 7 "In the interpretation of this Convention, regard is to be had to its international character and to the need to promote uniformity in its application and the observance of general good faith in international trade."

66 Couston, Droit Spatial Economique (1994) at 233-235.

B. MEMORIAL FOR ANTIPAPADIA

AGENTS

Christina Benson & Scott Syfert

ARGUMENT

I. Antipapadia did not violate international law either by intercepting or publicly distributing the signals of Golden Eye.

In the case before the Court, Antipapadia is accused of two separate wrongful acts. First, Openskey asks this Court to find that Antipapadia violated international law by intercepting the satellite transmissions of Golden Eye.[1] Secondly, Openskey asks this Court to hold that "publicly distributing the signals of Golden Eye" violated international law.[2] We will discuss each claim in wturn.

A. Antipapadia's Violation of Papadia's Territorial Sovereignty Was Justified.

1. Openskey lacks standing to assert the breach of Papadia's territorial sovereignty.

As a threshold issue, Openskey has no standing to independently assert the violation of Papadia's sovereignty. Generally, states are liable for breaches of an international obligation that directly injure another state.[3] But further, as this Court discussed in *Barcelona Traction*[4] and the *East Timor Case,*[5] there is a subclass of obligations towards the international commu nity as a whole. These *erga omnes* obligations include acts of aggression, genocide, slavery, and certain other basic human rights, and confer standing upon any nation asserting they have been violated. Territorial jurisdiction has never been held by this Court to be such an obligation, nor is such standing conferred by any of the treaties Antipapadia has signed. *Ergo,* Openskey has no standing to raise the issue of breach of Papadia's territorial sovereignty.

2.Antipapadia's incursion into Papadia's territory is excusable as a proportional action taken in self-defense.

Antipapadia concedes that she did enter the territorial waters of Papadia in order to obtain the satellite photographs that provide proof of Papadia's military buildup. However, these actions were justified under international law. Antipapadia's actions were a proportional response for the purpose of self-defense. The right of self-defense is one of the oldest principles of customary international law. The right existed prior to the U.N. Charter,[6] but was reaffirmed under Article 51, which states that "nothing in the present Charter shall impair the inherent right of individual or collective self-defense."[7] Most publicists agree that the Charter did not eliminate the right of anticipatory self-defense, in spite of the reference to "armed attack" in Article 51.[8] Ian Brownlie, for example, points out that Article 51 does not "impair" the existing customary right of self-defense.[9] Indeed, maintaining peace and security in an age of weapons of mass destruction requires nations to anticipate threats and to act accordingly.[10]

Any action taken in self-defense must meet the following requirements: (1) it may be used only in vindication of a denied right against States that have breached international law;[11] (2) there must be a serious and imminent danger to the essential rights of the state taking action in self-defense;[12] and (3) the action taken must be proportional to the danger at hand.[13] These requirements, elucidated in the *Caroline Case,*[14] are generally referred to as the principles of *necessity* and *proportionality.*[15] Thus, the concept of self-defense rests on the notion of protection, rather than punishment or revenge.[16]

3. Antipapadia's actions were both necessary & proportional.

Antipapadia's actions were justified under international law. First, Antipapadia reasonably believed her actions were necessary to protect herself from aggression. Recall that Papadia and Antipapadia had fought a "lengthy civil war" against one another.[17] Tempers were again high, and warships had been called into the disputed area.[18] Indeed, the scale of the crisis was such that the U.N. Security Council felt compelled to intervene. It was at this moment of heightened international tension that the satellite signal from Golden Eye, which allowed Antipapadia to monitor Papadia's troop movements, were cut off, leaving Antipapadia blind.[19] It was reasonable of Antipapadia to believe such an action was the prelude to aggressive action against her; and indeed, pictures subsequently acquired show there *was* such a buildup.[20] In such circumstances, Antipapadia believed regaining access to the satellite signals from Golden Eye was necessary to protect herself from armed attack.

Her interception of the satellite signal was also proportional. First, Antipapadia's incursion into Papadian territorial waters only occurred after the collective breaches of the Outer Space Treaty by Papadia, Openskey and Starstripe. Second, at no time did Antipapadia resort to the threat or use of force. No one was harmed by her actions; no property was damaged; no shots were fired. Throughout the period at issue, no violent or aggressive action of any kind was taken to obtain the data. Antipapadia merely intercepted the data in an effort to verify the level of threat posed by Papadia, and to preempt an armed conflict by bringing the crisis to the attention of the international community for a peaceful resolution. Seen thus, her actions were both necessary and proportional, and as such, justified under international law.

4. Territorial incursions to gain verification and obtain evidence are permitted under international law as promoting peaceful conflict resolution.

Customary international law results from a general and consistent practice of states followed by them from a sense of legal obligation.[21] State practice supports the view that surveillance actions taken in anticipatory self-defense are legal and even commendable. The international community condoned the use of communications satellites and military overflights to monitor Iraq's military readiness, both before the onset of the Gulf War and after the war ended.[22] Such forms of reconnaissance reduced the danger of armed aggression, and thus legitimized the U.S. claim of anticipatory self-defense.[23] The United States used similar reconnaissance techniques to identify the construction of Soviet ballistic missile sites in Cuba in 1962, and used this information to justify a naval quarantine of the island as an action in self-defense.[24]

State practice also reflects a high degree of tolerance for actions taken in self-defense when the purpose is to verify or gain evidence of a breach of international law by the party against whom such action is taken. In the *Corfu Channel* case, the British Royal Navy entered Albanian territorial waters to obtain evidence that Albania was responsible for the explosion of mines that had caused damage to its ships several weeks earlier.[25] Although this Court acknowledged that Albania's territorial sovereignty had been violated, it admitted the evidence of mines and used it to hold Albania liable for the damage to British ships.[26] Similarly, in the case of *United States v. Iran*, the U.S. violated Iranian borders in a failed attempt to rescue the hostages being held in the U.S. Embassy in Teheran. This Court declined to find the rescue attempt impermissible, and awarded reparations to the United States without any deductions to compensate Iran for the incursion into its territory.[27]

B. Antipapadia Did Not Violate International Law by Intercepting the Signals of Golden Eye.

1. Collection of remote sensing data is permitted under the principle of free and equal access to outer space.

The principle of free and equal use of outer space is established under the 1967 Outer Space Treaty and in customary international law.[28] Principle IV of the Principles of Remote Sensing ("PRS") incorporates and further clarifies this principle with regard to remote sensing activities.[29] State acceptance of this principle is evidenced by more than 800 agreements reached within the U.S. Landsat program for the collection and distribution of remote sensing data among various countries.[30] The French Spot-Image program had signed more than 40 protocol programs with various countries as of 1980, which embraced the principle of equal access.[31] Foreign reconnaissance satellites have been widely used by numerous international scientific agencies for more than 20 years, and not one has ever been destroyed by any State.[32]

2. Espionage and foreign intelligence gathering is legal in international law.

Antipapadia's incursion into Papadian territory could also be characterized as "espionage" or "foreign intelligence gathering." As such, it would be protected as a "general practice accepted as law" under Article 38(1)(b).[33] International law permits sending spies across borders to conduct surveillance activities,[34] and leaves it to the domestic law of each nation to punish spies who are caught.[35] Every nation in the world engages in espionage and has organizations devoted to that end. Every codification of the laws of war, including the 1874 Declaration of Brussels,[36] the 1899 Hague Rules,[37] the 1907 Hague Rules,[38] and the Geneva Conventions of 1949[39] and 1977,[40] has recognized the legitimacy of espionage.

Under Article 38(1)(d) of the ICJ Statute, this Court is permitted to look to the works of "the teachings of the most highly qualified publicists"[41] as a subsidiary means of determining the rule of law. The legality of espionage was documented as early as the seventeenth century, when Grotius wrote, "[S]pies, whose sending is beyond doubt permitted by the law of nations—such as the spies Moses sent out . . . if caught are usually treated most severely.[42] Another more recent publicist noted, "the law of nations permits the sending of spies . . . intelligence activities are now accepted as a common, even inherent, attribute of the modern state."[43]

Furthermore, under Article 38(1)(d) of its Statute, this Court may consider the judicial decisions of various nations as a means of determining the rule of law.[44] In the cases of *U.S. v. Brown*[45] and *U.S. v. Abbie Hoffman,*[46] U.S. federal courts held that foreign intelligence activities are in accord with international law. In the *Hoffman* case, the court recognized a "thirty year history of authorizations by Presidents recognizing the necessity for such surveillance where national defense is involved."[47]

In the absence of a duty not to engage in some conduct, it is presumed that sovereign states have the right to act in accordance with their perceived self-interest.[48] The Lotus case provides that restrictions on the actions of states are not to be lightly presumed, and acts not expressly forbidden, are presumed legal.[49] Thus, although clandestine intelligence gathering may be considered an unfriendly act, it does not violate international law.[50]

C. Antipapadia Did Not Violate International Law by Publicly Distributing the Data.

Antipapadia did not violate her obligations under the Outer Space Treaty or the PRS by publicly distributing the data. Distribution of remote sensing data is allowed under the principle of equal and nondiscriminatory access. Antipapadia did not interfere with the ability of other States participating in the remote sensing program to receive and process data from Golden Eye. The data relating to Papadian territory was released publicly and thereby made available to all states, including the sensed state. Thus, Antipapadia in no way violated Papadia's rights as a sensed State under the PRS.

1. The "Open Skies" principle permits free data distribution.

There is no duty in international law to return remote sensing data, as Openskey requests. Customary international law, as codified in the PRS, allows no restrictions on data distribution. The 1986 PRS were arrived at through a lengthy process of negotiation and consensus building to erect a legal framework for remote sensing.[51] The main issue of contention concerned the legality of free data distribution. Eventually, the "open skies" principle was agreed upon. The PRS ended the "prior consent" regime of remote sensing, by deliberately omitting any reference to the right of a sensed state to either veto remote sensing activities by sensing states, "or to restrict the dissemination of such data to third parties."[52]

The "open skies" principle can also be found in the general and consistent practice of states, which requires unrestricted data distribution. For example, the U.S. Land Remote Sensing Policy Act of 1992 requires equal access for all customers with no restrictions to third parties.[53] The GLOBE Program, a series of bilateral treaties between the U.S. and Costa Rica,[54] Moldova,[55] Kazakhstan,[56] Korea,[57] Sweden,[58] and Ecuador,[59] respectively, allows data to be distributed, "worldwide without restriction as to [its] use or redistribution." A similar treaty between the U.S. and India explicitly provides for "unrestricted public availability of all Landsat data," in conformance with general Landsat dissemination practices.[60] Similar agreements exists with China,[61] Brazil,[62] and Zaire.[63] Therefore, it was entirely legal for Antipapadia to distribute the data under customary international law. In effect, Openskey asks this Court to ignore the PRS and reinstate the "prior consent" regime that existed prior to 1986, in defiance of both international law and the COPUOS negotiations.

2. Openskey has no intellectual property rights in the remote sensing data.

Openskey has no right to have the data returned for a second reason. As a corollary of the "open skies" principles, no intellectual property rights attach to remote sensing data. In 1984, for example, NASA refused to allow the Messerschmitt-Beelkow-Blohm/ENRO consortium to fly its remote sensing pallet, because they intended to retain proprietary rights in the data in violation of the "open skies" policy.[64] Similarly, a standard feature of NASA contracts is the absence of copyright clauses,[65] illustrating that state practice does not recognize proprietary rights in remote sensing data.

Similarly, none of the relevant treaties or conventions governing copyright or intellectual property protects remote sensing data. Both the Brussels Convention of 1974[66] and the Rome Convention of 1960[67] apply to literary or artistic works, and imply a public reception and distribution of sounds or images fixed on some medium. Remote sensing data, either processed or unprocessed, does not fit within this definition. In 1986 the Austrian Supreme Court ruled as a matter of law in Sky Channel, that the fact that data transmissions are encrypted proves they are not intended for public reception,[68] and therefore neither Convention would apply to remote sensing data. By their own terms, therefore, neither Convention is applicable.

For similar reasons the Berne Convention[69] does not apply. The unenhanced data from Golden Eye is raw material containing no originality or uniqueness of expression, and does not fall within the purview of Berne's Article 2.[70] For this reason, one publicists notes, "remote sensing imagery cannot be protected on the basis of the provisions of [the Berne] Convention."[71] Nor is raw data copyrightable, making the Universal Copyright Convention equally inapplicable.[72] Only the Draft Directive of the European Commission on the Legal Protection of Databases[73] is arguably relevant in this case, but it is neither in force nor binding. (Moreover, any protection in the data under any of these treaties would accrue not to Openskey, but to Antipapadia as the party that enhanced and processed the data. Antipapadia does not need to rely on this protection, however, as the data was not illegally taken.)[74] States recognize that data obtained through espionage or surveillance is not protectable as intellectual property. For this reason they invest millions of dollars in encryption technology to protect the integrity of their systems and their data. They know that information lost, fairly or unfairly, is simply not recoverable as intellectual property.

D. Antipapadia Owes No Damages to Openskey for Lost Revenues.

Antipapadia's actions in intercepting and obtaining the data were in accordance with international law and thus Antipapadia cannot incur any responsibility for damages.

1. Openskey's damages are remote and indirect.

Even if Antipapadia's acts had been wrongful, damages could not be awarded because her acts were not the proximate cause of Openskey's losses.[75] Indirect damages are not recoverable under international legal principles.[76] For example, in the Alabama Arbitration,[77] American claims regarding the cost of pursuing a Confederate ship built in Britain were non-recoverable for two reasons. First, they were indistinguishable from general war expenses.[78] Second, prospective earnings were speculative.[79] Similar results were reached in The Newshwang case,[80] as well as the Naulilaa Claims,[81] where losses which were not a direct result of fault, were non-recoverable.

In the *Naulilaa Arbitration*, the tribunal rejected any causal link between German actions and a native uprising in a Portuguese colonial territory as too remote.[82] Similarly, in this case there is no showing that Antipapadia's acts of intercepting and distributing the data resulted in switching off the satellite transmissions to Antipapadia and Papadia. Indeed, the evidence contradicts this conclusion. Openskey had already switched off the signals to Antipapadia *before* the signals from Golden Eye were intercepted and distributed. If Openskey had fulfilled her treaty obligation by continuing to transmit data to both countries equally, Antipapadia would never have had to resort to intercepting the data. Furthermore, it was Starstripe who insisted that the data transmissions be switched off. Any damages arising from this action would be a contractual issue between Openskey and Starstripe and not attributable to Antipapadia. When ruling on the issue of remoteness of damages, international courts must consider such parallel or subsequent causes (or *novus actus interveniens*) as negating responsibility for damages.[83] Such is the case here.

2. Openskey suffered no compensable harm.

In order for liability to obtain a state must suffer some sort of direct injury to the state interest. Generally, this involves property damage, injury to nationals, or, as in the Cosmos 954, harm to a state interest (*e.g.*, violation of airspace) coupled with a straightforward threat to the population's health and safety.[84] The list of potential injury is by no means exclusive, but in this case it is difficult to see what compensable injury Openskey has suffered at all. Openskey's territory was not violated; Openskey set the terms of the contract; Openskey made the decision to shut off Golden Eye. Further, the data Antipapadia released to the press concerned *Papadia*, not Openskey. In sum, the only harm Openskey has suffered has been the loss of information which she was contractually obligated to deliver to Antipapadia in any event. Neither embarrassment or aggravation alone is compensable under international liability principles.

II. Openskey violated international law by switching off Golden Eye in the coverage area of the Antipapadian ground station.

A. <u>Openskey Breached Her Duty to Ensure that Outer Space is Used Exclusively for Peaceful Purposes by Using Golden Eye to Aid Aggression.</u>

1. Outer Space must be used solely for peaceful purposes.

The peaceful use of space has been the cornerstone of international space law since its inception, and is a fundamental tenet of customary law.[85] Indeed, the U.N. committee to oversee space issues is aptly named the Committee on *Peaceful* Uses of Outer Space ("COPUOS"). The term "peaceful" can be found in virtually every U.N. document devoted to outer space.[86] The Outer Space Treaty recognizes "the common interest of all mankind in the progress of the exploration and use of outer space for *peaceful purposes*."[87]

Nor is this language merely hortatory. Article 31 of the Vienna Convention on the Law of Treaties ("Vienna Convention") provides that ambiguous treaty language should be interpreted in the context of, and consistently with, other treaty provisions, including the preamble, annexes and other documents referenced in the treaty.[88] Thus, this Court may consider these items in determining what uses of space are consistent with this "peaceful purposes" requirement. Article IV of the Outer Space Treaty requires the demilitarization of space and the use of space to maintain "international peace and security."[89] Article III incorporates the Charter of the United Nations, which prohibits the "threat or use of force" under Article 2(4).[90] The Charter also requires States to settle international disputes "by peaceful means," without endangering "international peace and security."[91]

Prominent scholars of international space law have interpreted Articles III and IV of the Outer Space Treaty as not only prohibiting the militarization of space, but also as prohibiting *any* use of space for aggressive purposes.[92] Professor Dembling, former General Counsel of NASA states, "[a]ny military use of outer space must be *restricted to non-aggressive purposes* in view of Article III, which makes applicable . . . the [U.N.] Charter."[93] One publicist notes that the "peaceful purposes" language of Article IV, "brings in the notions of both *intent* and of *consequences*; the activity must not be designed to terminate in some use of force contrary to international law."[94] The use of remote sensing systems to aid aggression runs contrary to these principles.

2. Openskey allowed Golden Eye to be used for aggressive purposes.

Of course, not all military uses of outer space are prohibited. Remote sensing satellites have been extensively used for military reconnaissance, communications, and command-and-control purposes. Nevertheless, it is equally clear that any use of such satellites for the purpose of inciting, facilitating or encouraging armed conflict, or to intervene in the affairs or interests of another state, is expressly prohibited. In other words, the use of remote sensing systems to usurp natural resources from another nation, to launch a secret attack, or to bully a weaker nation would all violate these proscriptions. That is what has occurred here.

Openskey manipulated Golden Eye for the wrongful purpose of providing a military and strategic advantage to Papadia. Whether Papadia's ultimate purpose was to seize the disputed oil fields or to attack Antipapadia is irrelevant. In either event, Openskey connived with Papadia to use outer space aggressively to the detriment of a developing nation.

B. <u>Switching Off Golden Eye Constituted an Impermissible Threat of Force</u>.

The requirement that space must be used exclusively for "peaceful purposes" incorporates the concept of nonintervention and the prohibition on the "threat or use of force" as established in the U.N. Charter and under customary international law.[95] In the *Nicaragua* case, this Court held that the principle of nonintervention has risen to the level of *jus cogens*, and thus no derogation from this duty is permitted.[96] The Court concluded that there is no right "for States to intervene, *directly or indirectly, with or without armed force* . . . in another State.[97] The Court focused on the underlying intent of the United States, finding that the U.S. would have illegally intervened merely by supplying funds to the Nicaraguan Contras, because its intent in doing so was to destabilize the Nicaraguan government.[98] This ruling reflects that the use of armed military force is not a necessary prerequisite for finding an intervention illegal under international law. Antipapdia concedes that switching off Golden Eye likely does not rise to the level of impermissible intervention. However, such an action, particularly in the circumstances in this case, in which war appeared imminent, would constitute a "threat of force," actionable under international law. Such a threat of force is not only illegal in and of itself, but justified Antipapadia's incursion into Papadia's territory as an act of self-defense, discussed *supra*.

C. <u>Customary International Law Regards Switching off Reconnaissance Satellites As Aggressive</u>.

1. Switching off Golden Eye was a veiled act of aggression.

Under Article 38 of its Statute this Court may apply international custom, or "a general practice accepted as law."[99] The use of remote sensing satellites to monitor potential opponents is a general practice and is regarded in the international community as necessary to prevent war, by making surprise attacks impracticable. The London Session of the David Davies Memorial Institute Draft Code of the Rules of Outer Space noted that military surveillance or reconnaissance satellites, "contribute to an open world and so increase rather than diminish security."[100] The General Assembly has adopted numerous resolutions condoning the use of "confidence and security building" measures of verification.[101] To this end, the U.S.-Soviet Anti-Ballistic Missile Treaty of 1972 ("ABM Treaty") recognized the mutual existence of remote sensing satellites and considered them a safeguard in the balance of power and arms control.[102] For this reasons, Article 12(2) of the ABM Treaty prohibited both parties from interfering with the others' remote sensing satellites. Such interference, which would include switching off a satellite signal, would have been regarded as an aggressive act.

Terminating a military reconnaissance signal is a "threat of force" impermissible under customary international law. It this case it was particularly provocative, as the U.N. Security Council resolution calling upon both nations to refrain from "controversial actions".[103] Because of this illegal act, Antipapadia was forced to acquire verification of the military buildup in Papadia in order to comply with U.N .Charter requirements regarding the pacific settlement of disputes under Chapter VI, and responses to breaches of the peace under Chapter VII.[104]

2. Openskey behaved provocatively & in bad faith.

The only motive served by switching off signals to Antipapadia, while continuing full service to Papadia, was to give Papadia a strategic and military advantage in the conflict with Antipapadia. Starstripe is a close economic and political ally of Papadia, and it is stipulated that Papadia asked Starstripe to "intervene" in the conflict against Antipapadia on her behalf.[105] Thus, the purpose underlying the decision to switch off Antipapadia's access to Golden Eye was to illegally intervene in the conflict between Papadia and Antipapadia.

Further, Openskey exploited Antipapadia by preventing her access to natural resources in which has a legitimate right. Openskey acted in bad faith by leading Antipapadia to believe that she was receiving valuable data, when the data had actually been downgraded and transformed by Openskey without Antipapadia's knowledge. While Starstripe reserved the right under its export license to demand a "downgrading" of data in furtherance of its own "national security interests," this cannot explain Openskey's actions. The license makes no reference to any right to completely switch off satellite transmissions. The right of "downgrading" does not include switching off, but rather refers to reducing the technical level or quality of the data being transmitted. Furthermore, neither Openskey nor Starstripe can claim any legitimate "national security interest" in this matter, as the conflict was between Papadia and Antipapadia.

III. Openskey breached her duty to provide equal and nondiscriminatory access to data from Golden Eye by delivering to Antipapadia a ground station with hardware and software technically inferior to that delivered to Papadia.

A. Openskey Breached Her Duty To Provide Equal and Nondiscriminatory Access to Data From Golden Eye.

Antipapadia and Openskey are both parties "to all the relevant international treaties,"[106] which would include the Outer Space Treaty.[107] Under the principle of *pacta sunt servanda* set forth in Article 26 of the Vienna Convention on the Law of Treaties, treaty provisions are legally binding on Openskey and must be performed in good faith.[108]

1. The Outer Space Treaty establishes a duty of equal and nondiscriminatory access that specifically applies to remote sensing activities.

Article I of the Outer Space Treaty provides that outer space shall be used by States on a free and equal basis "without discrimination of any kind."[109] This Article gives special consideration to ensuring equality for developing countries.[110] Remote sensing of the earth by satellite constitutes a "use of outer space" within the meaning of the Outer Space Treaty. The specific nature of the treaty duty to provide equal and nondiscriminatory access is further codified in Principles II and IV of the Principles of Remote Sensing.[111] Principle IV reflects that the PRS were intended to further define and codify existing obligations contained in Article I of the Outer Space Treaty.[112] Principle II strengthens the applicability of the nondiscrimination obligation when the interests of developing countries are at stake, expressly adding the clause, "taking into particular consideration the needs of developing countries."[113] Principle XII sets forth the specific actions States must take to comply with this duty. States conducting remote sensing activities must meet five explicit requirements under Principle XII to fulfill their duty: they must (1) provide the satellite data to the sensed state without delay;[114] (2) provide the sensed state with both primary and processed data;[115] (3) provide such data on a nondiscriminatory basis; (4) provide access to the data on "reasonable cost terms"; and (5) provide any additional analyzed information in the possession of another State to the sensed state on the same basis and terms.[116] The State conducting remote sensing activities must fulfill these five requirements with "particular regard being given to the needs and interests of the developing countries."[117]

2. The duty to provide equal and nondiscriminatory access is also recognized under customary international law.

The duties set forth under the Outer Space Treaty and defined under the PRS are merely a codification of customary international law. Under Article 38 of the Statute of the International Court of Justice, an international custom is defined as, "evidence of a general practice accepted as law"[118] A rule becomes international custom when states conform to it out of a sense of legal obligation, or *opinio juris*.[119] In the *North Sea Continental Shelf Cases*, this Court found that the acceptance of a practice by a representative majority of states over a period of time was sufficient to create a rule of customary law.[120] In that case, a custom recognizing exclusive economic zones had emerged in a period of only ten years.[121]

States with active space programs consistently recognize and enforce the rule of nondiscriminatory access to data gained from remote sensing satellites. For example, U.S. domestic law requires private entities operating remote sensing systems to make data available to other countries, "as soon as such data are available and under reasonable cost terms."[122] This Act expressly prohibits any private contractual arrangements, "regarding delivery, format, pricing or technical considerations which would favor one customer or class of customers over another."[123] The U.S. policy also requires equal access for "all users," and not merely for a State whose territory is being sensed.[124]

3. Openskey breached this duty of nondiscrimination.

Superview Inc. is a private commercial satellite operator licensed by the government of Openskey to carry out commercial VHR activities using the Golden Eye satellite. Openskey reviewed and approved the agreements between Superview, Inc. and the governments of Papadia and Antipapadia, and delivered the ground station hardware and software to each country. Therefore, Openskey knowingly and intentionally delivered a ground station to Antipapadia that was technically inferior to the "latest technology" being delivered to Papadia.[125] There is no indication in the Compromis that Openskey informed Antipapadia that her ground station was inferior to the one delivered to Papadia, or that Openskey offered Antipapadia a more advanced ground station. Openskey may have assumed Antipapadia could not afford more advanced technology because she is a developing country with "limited economic resources."[126] However, the fact that Antipapadia was able to "buy a state-of-the-art mobile ground station" from another source on the open market soon after reflects that Openskey charged Antipapadia an unreasonable price for this equipment and software.[127] Thus, Openskey failed to make the system available "on a nondiscriminatory basis and on reasonable cost terms" as required under PRS Principle XII.[128]

The duty to provide nondiscriminatory access to remote sensing data applies to primary data, processed data, and any other additional analyzed information gained via satellite.[129] By delivering an inferior system and then downgrading the data transmitted to Antipapadia, Openskey prevented Antipapdia from receiving the same primary, processed and analyzed data as Papadia to identify natural resources in which both states have rights and interests.

B. <u>Openskey Engaged In An "Abuse of Rights" Which Denied Antipapadia the Opportunity to Locate Natural Resources.</u>

1. International law recognizes Antipapadia's rights in the maritime area in which Openskey is conducting remote sensing activities.

Papadia and Antipapadia are island nations located in the same geographic region and separated by a disputed territorial sea.[130] Although neither state is a party to the UNCLOS III Treaty, it is generally recognized under customary norms of the international law of the sea [131] that States composed of inhabited islands are entitled to claim up to a 12-nautical-mile territorial sea, a continental shelf, and a 200-nautical-mile exclusive economic zone.[132] International tribunals have recognized such maritime boundaries under customary international law.[133] For example, in the case of *Libya v. Malta*, concerning the delimitation of disputed seas between a small island state and a large continental state, this Court determined that the exclusive economic zone had emerged as part of customary international law.[134] These maritime boundaries are subject to bilateral delimitation, however, where the claims overlap with claims made by other islands or states, as in the present case.

Under customary law, coastal states have sovereignty over internal and territorial waters, including natural resources contained therein.[135] The principle of full and permanent sovereignty over natural resources is firmly established in customary law, as exemplified by state practice and by landmark U.N. resolutions such as the Charter of Economic Rights and Duties of States.[136] In addition to sovereignty over its internal and territorial waters, a State has exclusive rights to the natural resources contained within its exclusive economic zone.[137] As this Court ruled in the *North Sea Cases*, where boundaries are disputed, natural resources should be equitably
divided.[138] Judge Philip Jessup stated in his separate opinion that the maritime boundary in that case should be delimited by focusing on the real basis of the dispute - access to hydrocarbons in the seabed of the North Sea.[139]

Similarly, the principle of "autonomous equity" has been universally applied in international adjudications to decide maritime boundary disputes. Under this principle, the Court must take into consideration all factors relevant to fairness and equity in deciding each particular case. [140] In *Tunisia v. Libya*, Judge Jiminez de Arechaga explained the importance of equity in such disputes, stating, "the judicial application of equitable principles means that a court should render justice in the concrete case, by means of a decision shaped by and adjusted to the relevant factual matrix of the case."[141] Similarly, in the *North Sea Cases*, this Court held that the interests of all nations must be taken into consideration.[142] In the *Jan-Mayen* case,[143] this Court considered the division of natural resources to be an important equitable factor in delimiting the maritime boundary between a small Norwegian island

and Greenland.[144] In that case, the Court identified the prime commercial fishing area and divided it in half, giving both states equal access to the fishery.[145]

Therefore, based on the holdings in *Jan-Mayen* and the *North Sea Cases*, it is probable that both Papadia and Antipapadia should receive an equitable share of the natural resources contained in the disputed territory. While the facts contained in the compromis are limited, what facts we do have indicate that Antipapadia has as great a right to these natural resources as any other claimant, and at least as much right as Papadia. We know, for example, that Drillwell and Sinkbetter searched for oil near a "small coastal area" only eight miles from one another in the disputed territorial sea.[146] These facts suggest that the disputed territory is small enough for each State to have a legitimate, yet overlapping, claim to the oil fields. *Ergo*, both states likely have legitimate and legally enforceable rights to an equitable share of the oil contained within the disputed territory. Even if this turns out not to be the case, the equitable principles elucidated in *Jan Mayen* and *North Sea* would prevent a richer, more technologically advanced nation from misusing these advantages to hobble a developing nation in its search for its rightful share of natural resources. Such an action must constitute an abuse of rights, as discussed *infra*.

2. The disputed territorial sea is part of Antipapadia's "territory" within the meaning of the PRS and the Outer Space Treaty.

In setting forth obligations of the State conducting remote sensing activities, Principles XII and XIII of the PRS make reference to the "territory" and "jurisdiction" of the sensed state.[147] While the terms "territory" and "jurisdiction" are not explained within the text of either the PRS or the Outer Space Treaty, it is clear that these terms were used to protect the natural resources of developing nations. Article 32 of the Vienna Convention permits the use of supplementary means of interpreting a treaty, where the language is ambiguous or would lead to an unreasonable result.[148] The travaux préparatoires are useful in this regard.

During the negotiation of the PRS, many countries sought to add language to expressly include the sensing of marine areas within the remote sensing regime. The joint articles submitted by Brazil and Argentina made specific reference to sensing of "resources located in maritime areas under [the] jurisdiction of the sensed state,"[149] while draft principles submitted by Mexico[150] and Chile[151] contained almost identical language. Drafts by Argentina and Brazil[152] further provided for all States to receive data regarding natural resources of "maritime areas *outside* national sovereignty or jurisdiction," while drafts submitted by the Soviet Union[153] and France[154] included data access for "areas outside the national jurisdiction of any state."[155] Similarly, the domestic policy of the United States has always been to ensure equal access to data for all States, and not merely for States whose territory is sensed.[156] Therefore, in the final draft of Principle XII, "territory" was changed to read, "territory under . . . the jurisdiction" of the sensed state.[157] The drafters intended this change to reflect a broader, more flexible view of the notion of "territory" with regard to which sensed States are owed the specific duties enumerated under Principle XII. The word "territory" should be interpreted as being consistent with two key points emphasized throughout both the Outer Space Treaty and the PRS: (1) the principle of equal access and use of space by all States; and (2) the special needs of developing countries to have equal access and control over their natural resources.[158]

Antipapadia is a developing country with legitimate rights in the natural resources located in the disputed territorial sea. The intent of the Outer Space Treaty and PRS is to protect developing nations and their natural resources from being exploited by more economically developed nations such as Openskey and Papadia. That is exactly what has occurred in the present case. Antipapadia, a lesser developed nation, was denied a fair chance to find and exploit these natural resources because of the connivance of Openskey and Papadia. For this, Openskey should be held responsible. As space law publicist and ICJ Justice V. S. Vereshchetin wisely put it, "[f]reedom of outer space should not be used as a pretext for violating sovereign rights of states on the earth."[159]

3. Openskey engaged in an "abuse of rights" by interfering with Antipapadia's opportunity to locate and exploit these natural resources

Article IX of the Outer Space Treaty provides that states must pay "due regard to the corresponding interests" of other parties.[160] This otherwise vague provision is given specific legal content in the context of remote sensing through Principle IV of the PRS,[161] which Openskey signed. This principle requires that remote sensing activities, "*shall* be conducted on the basis of respect for the principle of full and permanent sovereignty of all States and peoples over their own wealth and natural resources."[162] While arguably some of the provisions of the PRS lack the required legal specificity to be regarded as customary international law, Principle IV (the longest and most detailed provision) was meant to embody a binding principle of international law by the drafters - the "abuse of rights" doctrine. Thus, read in conjunction with the specific treaty requirements of Article IX of the Outer Space Treaty, Principle IV of the PRS establishes liability for those nations which abuse their remote sensing rights.

Principle IV does not create a new right in international law, but rather makes the customary "abuse of rights" doctrine applicable to remote sensing activities. This principle was explicitly recognized in Article 3 of the Montevideo Convention,[163] Article 28 of the Geneva Convention on the High Seas,[164] and has been applied in the Trail Smelter Arbitration,[165] the German Interests in Polish Upper Silesia Case (Merits),[166] and the Anglo-Norwegian Fisheries Case.[167] Likewise, in the Free Zones Case, the Permanent Court found that while the French could legitimately enact fiscal legislation in the free zones (which were French territory), a reservation protected abuse of this right to the harm of others.[168] Publicists Ian Brownlie, Sir Hersch Lauterpacht, and Dr. F.A. Mann have noted that abuse of rights is a general principle of international law.[169]

The abuse of rights doctrine is particularly relevant in this case. First, as Brownlie notes, the abuse of rights doctrine is the alter ego of the principle of nondiscrimination,[170] and Openskey's discrimination in favor of Papadia is at the heart of this case. Second, the totality of Openskey's conduct was inequitable, discriminatory, and unjust. Openskey knew that Antipapadia's purpose for purchasing the equipment was to locate the oil fields, yet she delivered equipment and software that was unfit for this purpose. This action interfered with Antipapadia's legitimate rights and interests in the natural resources contained in the disputed territorial sea. The low quality of the data received and the inferior technology made it impossible for Antipapadia to locate the oil fields. Through this abuse of rights, Openskey violated Antipapadia's opportunity to find and use these nonrenewable natural resources.

IV. Openskey owes damages to Antipapadia for the intentional downgrading and unintentional transformation of data transmitted to Antipapadia

A. <u>Openskey is Liable for Damages under the Liability Convention</u>

1. The Liability Convention imposes absolute liability on Openskey for damages caused by Golden Eye.

As the State party who "launched or procured the launching of an object into outer space," Openskey fits the definition of a "launching party" contained both in Article VII of the Outer Space Treaty[171] and in Article I(c)(i) of the Convention on International Liability for Damage Caused by Space Objects ("Liability Convention").[172] In the present case, Openskey "procured the launching" of Golden Eye by purchasing it from Goldstar. Golden Eye, including the component parts, is a "space object" within the meaning of Article I(d).[173] Thus, under Article II of this Treaty, Openskey is absolutely liable to Antipapadia for the damages caused by Golden Eye.[174]

Article I includes within the definition of damages any "loss of or damage to property of States."[175] Such damage includes not only the outright taking of property, but also any "unreasonable interference . . . [such] that the owner thereof will not be able to use, enjoy or dispose of the property within a reasonable time after the inception of such interference."[176] The malfunctioning of Golden Eye caused such damage to the property of Antipapadia, and to the property of Sinkbetter as a juridical entity in Antipapadia.

2. The damages suffered by Antipapadia and Sinkbetter are compensable under the Liability Convention.

On first reading, the Liability Convention does not seem to be the proper vehicle for Antipapadia's claims. However, a look at the preparatory work of the Convention reveals otherwise. The General Assembly resolution calling for the drafting of such a convention expressed concern that, "the remedies for damage caused by space objects are inadequate for the needs of the nations and peoples of the world."[177] The *travaux préparatoires* of the Liability Convention reflect that it was intended to include damages caused not only by the space object itself, but also "by the functioning of scientific instruments on board."[178] At the suggestion of the United States, the measure of damages adopted under Article XII was based on *Chorzow Factory*, which provided that reparation should restore the claimant to the condition *ex ante*.[179] This formula was approved by negotiators.[180] Article XII specifically provides that damages shall be determined, "in accordance with international law and the principles of law and equity."[181] The term "equity" is used in its popular sense to signify the right of States to receive compensation for moral injury,[182] as well as actual physical harm. Moral injury includes injuries to the dignity or sovereignty of a State, such as that caused by the breach of a treaty obligation, regardless of the presence of material damages.[183] Thus, as Carl Christol points out, the Convention not only allows recovery for direct damages, "caused" by space objects, but "moral injury" as well.[184]

B. <u>Openskey is Liable for Both the Costs of Exploratory Research and for Lost Oil Revenues.</u>

The proper measure of damages under *Chorzow Factory* is to restore Antipapadia *ex ante*. Similarly, under Article 35(2) of the U.N. Convention on Contracts for the International Sale of Goods, the seller must provide goods fit for the purpose made known to the seller at the time of purchase.[185] Here, had Golden Eye functioned correctly, Sinkbetter would have located and extracted the oil. The distortion of data from Golden Eye interfered with Antipapadia's use of the ground station, software, and data processing equipment. This distortion rendered the equipment utterly useless. It required Antipapadia to purchase a new mobile ground station in order to receive clear and usable satellite signals for the exploratory research. Because of false reliance on distorted data from Golden Eye, Sinkbetter undertook unnecessary and fruitless exploration.

Moreover, it is almost certain that Sinkbetter's efforts to locate the oil would have been successful had she received clear and undistorted satellite data. When transmitting correctly, Golden Eye has a VHR capability with resolutions as high as 1.5 meters.[186] This exceptional accuracy easily enables a user to locate petroleum deposits; and indeed, using clear and undistorted data permitted the Drillwell to find such deposits. These facts establish a direct causal link between the malfunctioning of Golden Eye and the damages incurred by Antipapadia and Sinkbetter. Thus, these damages are neither remote nor indirect under international standards, and should be awarded.

---

1 Compromis at 3.

2 *Id.*

3 East Timor *(Australia v. Portugal), 1995 I.C.J. 90, 214 (June 30, 1995).*

4 Barcelona Traction, Light and Power Company (Belgium v. Spain) 1970 I.C.J. 3 (1962).

5 East Timor, *supra* note 3, at 214.

6 Robert L. Bridge, *International Law and Military Activities in Outer Space*, 13 AKRON L. REV. 649 (1979); *See also* IAN BROWNLIE, INT'L L. & THE USE OF FORCE BY STATES 362 n. 16 (1963); Fitzmaurice, *The General Principles of Int'l L. Considered from the Standpoint of the Rule of Law*, 92 RECUEIL DES COURS 1, 171 (1957).

7 Charter of the United Nations, art. 51, 1. U.N.T.S. (1945).

8 *Id.*

9 BROWNLIE, *supra* NOTE 6, at 362 n.16.

10 Bridge, *supra* note 6, at 649.

11 D.W. BOWETT, SELF-DEFENSE IN INT'L L. 9 (1958).

12 *Letter from Secretary of State Daniel Webster to British Ambassador Fox* (April 24, 1841), *reprinted in* INTERNATIONAL LAW 1219-22 (Carter & Trimble eds. 1991).

13 BOWETT, *supra* note 11, at 269.

14 Caroline Incident, 29 B.F.S.P. 1129, 1138.

15 JAMES MORENOFF, WORLD PEACE THROUGH SPACE LAW 220 (1973).

16 *Id.*

17 Compromis at 2.

18 *Id.*

19 *Id.* at 3.

20 *Id.*

21 RESTATEMENT (THIRD) OF THE FOR. RELATIONS LAW OF THE UNITED STATES § 102(2) (1987).

22 Morenoff, *supra* note 15, at 233.

23 *Id.*

24 M. DONELAN & M. GRIEVE, INTERNATIONAL DISPUTES: CASE HISTORIES 1945-1970, at 234 (1973).

25 Corfu Channel (U.K. v. Albania), 1949 I.C.J. 27 (1949).

26 *Id.* See also Reisman & Freedman, *The Plaintiff's Dilemma: Illegally Obtained Evidence and Admissibility in International Adjudication* 76 A.J.I.L. 737, 746-48 (1982).

27 United States v. Iran, 1980 I.C.J. Rep. 3, 43-44 (1980).

28 Treaty on Principles Governing the Activities of States in the Exploration and Use of Outer Space, Including the Moon and Other Celestial Bodies, art. 1, Jan. 27, 1967, 18 U.S.T. 2410, 610 U.N.T.S. 205 [hereinafter "Outer Space Treaty"].

29 Principles Relating to Remote Sensing of the Earth From Outer Space, G.A. Res. 41/65, U.N. GAOR, 95th Plenary Meeting, prin. IV (Dec. 3, 1986) [hereinafter "PRS"].

30 CARL Q. CHRISTOL, REMOTE SENSING AND INTERNATIONAL LAW 380 (A.A.S.L. 1980).

31 *Id.*

32 Leopold Peyrefitte, *The Legal Regime of Remote Sensing of the Earth from Space*, *in* PROC. FROM THE THIRTY-FOURTH COLLOQUIUM ON THE L. OF OUTER SPACE, AMERICAN INSTITUTE OF AERONAUTICS AND ASTRONAUTICS 286, 290 (Oct. 5-11, 1991).

33 Statute of the International Court of Justice, art. 38 (1)(b), 59 Stat. 1031 (1945) [hereinafter "ICJ Statute"].

34 Nathaniel Ward, *Espionage and the Forfeiture of Diplomatic Immunity*, 11 INT'L 657 (1977).

35 U.S. DEPT. OF THE ARMY, THE LAW OF LAND AND WARFARE 33 (1956).

36 Dec. of Brussels Concerning the Laws and Customs of War, *adopted by* Conference of Brussels, art. 36, Aug. 27, 1874.

37 Final Act of the Int'l Peace Conference, *signed at* the Hague, July 29, 1899.

38 Regulations Respecting the Laws and Customs of War on Land, art. 24, *annexed to* the Convention of 1899, *supra* note 37.

39 Convention Relative to the Protection of Civilian Persons in Time of War, Part I (General Provisions), art. 5, Aug. 12, 1949 (4th Geneva Convention).

40 Protocols Additional of 8 June 1977 to the Geneva Conventions of 12 Aug. 1949, Protocol I Relating to the Protection of Victims of Non-International Armed Conflicts, June 8, 1977, *reprinted in* 16 I.L.M. 1391 (1977).

41 Statute of the ICJ, *supra note 33, art. 38(1)(d).*

42 HUGO GROTIUS, LAW OF WAR & PEACE, Book III, ch. IV xviii 655 (F. Kelsey translation, 1925).

43 Geoffrey B. Demarest, *Espionage in Int'l L.*, 24 DENV. J. INT'L L. & POL'Y 321, 331 (1996).

44 Statute of the ICJ, *supra note 33, art. 38(1)(d).*

45 U.S. v. Brown, 484 F.2d 418 (5th Cir., 1973).

46 U.S. v. Abbie Hoffman, 334 F. Supp 504 (D.C. 1971).

47 *Id.* at 506.

48 IAN BROWNLIE, 1 STATE RESPONSIBILITY 30 (1983).

49 S.S. Lotus (France v. Turk), 1927 P.C.I.J., Ser. A., no. 10 (1927).

50 Demarest, *supra* note 43, at 347.

51 CARL Q. CHRISTOL, THE MODERN INTERNATIONAL LAW OF OUTER SPACE (1982).

52 JEFFERSON HANE WEAVER, LESSONS IN MULTILATERAL NEGOTIATIONS: CREATING A REMOTE SENSING REGIME, 7 TEMP. INT'L & COMP. L. J. 29, 53 (1993).

53 15 U.S.C. § 5651-5672 *(West Supp. 1995) (Land Remote Sensing Policy Act of 1992).*

54 Agreement Between the U.S. and Costa Rica for the Cooperation of the GLOBE Program, U.S.-Costa Rica, Apr. 22, 1996, art. 5, KAV No. 4587, St. Dept. No. 96-80, 1996 WL 294841.

55 Agreement Between the U.S. and Moldova for the Cooperation of the GLOBE Program, U.S.-Moldova,

Jan. 30, 1995, art. 5, KAV No. 4178, St. Dept. No. 95-99, 1996 WL 149241.

56 Agreement Between the U.S. and Kazakhstan for the Cooperation of the GLOBE Program, U.S.-Kazakhstan, Mar. 27, 1995, art. 5, KAV No. 4217, St. Dept. No. 95-92 1996 WL 323714.

57 Agreement Between the Nat. Oceanic & Atmospheric Admin. of the U.S. and the Min. of Education of the Rep. of Korea for the Cooperation of the GLOBE Program, U.S.-Korea, Apr. 21, 1995, art. 4, KAV No. 4223, St. Dept. No. 95-108 1996 WL 370388.

58 Agreement Between the Nat. Oceanic & Atmospheric Admin. of the U.S. and the Nat. Agency for the Education of Sweden for the Cooperation of the GLOBE Program, U.S.-Sweden, Aug. 23, 1995, art. 4, KAV No. 4420, St. Dept. No. 95-210, 1996 WL 596169.

59 Agreement Between the Nat. Oceanic & Atmospheric Admin. of the U.S. and the Education Min. of Ecuador for the Cooperation of the GLOBE Program, U.S.-Ecuador, Apr. 21, 1995, art. 4, KAV No. 4618 St. Dept. No. 96-108 1996 WL 424938.

60 Remote Sensing: Acquisition of Data, U.S.-India, Jan. 3, 1978, T.I.A.S. No. 9074, at 4690.

61 Memorandum of Understanding Between the U.S. NASA & Chinese Academy of Sciences, U.S.-China, Jan. 12, 1981, T.I.A.S. No. 11054, at 2.

62 Memorandum of Understanding Between the U.S. Govt. & Govt. Fed. Rep. of Brazil Concerning the Landsat System, U.S.-Brazil, May 8, 1984, T.I.A.S. N. 11252, at 2.

63 ERTS-Zaire-U.S. NASA on Ground Station & Satellite Data, U.S.-Zaire, Jan. 12, 1981, art. 6, *in* 14 I.L.M. 1474, 1476 (1975).

64 NATHAN C. GOLDMAN, AMERICAN SPACE LAW 104 (1st ed. 1988).

65 *Id.*

66 Convention Relating to the Distribution of Program Carrying Signals Transmitted by Satellite, *entered into force*, Aug. 25, 1979, 1144 U.N.T.S. 3.

67 Int'l Convention for the Protection of Performers, Producers of Phonograms and Broadcasting Organizations, Oct. 26, 1961, 496 U.N.T.S. 43.

68 Re Sky Channel (No. 2), Oberster Gerichtshof, [1987] EEC 449, Feb. 4, 1986.

69 Berne Convention for the Protection of Literary and Artistic Works, *done* Sept. 9, 1886, *last revised* July 24, 1971, 828 U.N.T.S. 221.

70 *Id.* art. 2.

71 Patrick Salin, *Proprietary Aspects of Commercial Remote-Sensing Imagery,* 13 J. INT'L L. BUS. 349, 358 (1992).

72 NATHAN GOLDMAN, AMERICAN SPACE LAW 102 (2d ed. 1996) (noting "[t]he Universal Copyright Convention does not apply" to remote sensing imagery).

73 Comm. of the European Communities, Initial Proposal, COM (92), Brussels, at 4, May 1992.

74 See notes 1-50, *supra* and accompanying text.

75 10 ENCYCLOPEDIA OF PUBLIC INTERNATIONAL LAW 931 (1987).

76 WHITEMAN, 3 DAMAGES IN INT'L LAW 1801 (1943).

77 Alabama Arbitration (U.S. v. U.K.) (1872), *in 1 MOORE, INT'L ARBITRATIONS 653.*

78 CLIVE PARRY & JOHN P. GRANT, ENCYCLOPAEDIC DICTIONARY OF INT'L LAW 15 (1986).

79 *Id.*

80 The Newshwang (U.S. v. U.K.), 6 R.I.A.A 64 (1921).

81 Naulilaa Claims (Portugal v. Germany), 2 R.I.A.A. 1013, 1037 (1928).

82 *Id.*

83 ENCYCLOPEDIA OF PUBLIC INTERNATIONAL LAW, *supra* note 75, at 932.

84 BROWNLIE, *supra* note 48, at 237, 239.

85 Ivan A. Vlasic, *The Legal Aspects of Peaceful and Non-Peaceful Uses of Outer Space, in* PEACEFUL AND NON-PEACEFUL USES OF SPACE, PROBLEMS OF DEFINITION FOR THE PREVENTION OF AN ARMS RACE 45 (Bhupendra Jasani ed. 1991).

86 *See e.g.,* Declaration of Legal Principles Governing the Activities in the Exploration and Use of Outer Space, U.N. Res. 1962, (Dec. 13, 1963); U.N. GAOR, 33rd Sess., U.N. Doc. A/33/162 (1978); *Legal Implications of the Remote Sensing of the Earth by Satellites,* U.N. Doc. A/AC.105/118 (1973); Principles Governing the use of States of Artificial Earth Satellites for Direct Television broadcasting, U.N. Doc. A/RES/37/92 (Feb. 4, 1983); PRS, *supra* note 29.

87 Outer Space Treaty, *supra* note 28, pmbl, art. III [emphasis added].

88 Vienna Convention on the Law of Treaties, U.N. Doc. A/CONF.129/15, art. 31 (March 20, 1986) [hereinafter "Vienna Convention"].

89 Outer Space Treaty, *supra* note 28, art. 3.

90 *Id.*; U.N. Charter, *supra* note 7, art. 2.

91 U.N. Charter, *supra* note 7, art. 2.

92 S. HOUSTON LAY & HOWARD J. TUBENFELD, THE LAW RELATING TO THE ACTIVITIES OF MAN IN OUTER SPACE 97 n.3 (1970).

93 Dembling, *The Evolution of the Outer Space Treaty,* 33 J. OF AIR L. & COM. 419, 433-434 (1967) [emphasis added].

94 J.E.S. FAWCETT, OUTER SPACE NEW CHALLENGES TO LAW AND POLICY 109 (1984).

95 Dec. on Enhancement of the Effectiveness of the Prin. of Refraining from the Threat or Use of Force in Int'l Rel., U.N. Doc. A/RES/42/22, March 17, 1988, *reprinted in* 27 I.L.M. 1674 (1988).

96 Nicaragua v. United States, 1986 I.C.J. 14 (1986).

97 *Id.*

98 *Id.*

99 ICJ Statute, *supra* note 33, art. 38(1)(b).

100 *Commentary*, London Session of the David Davies Memorial Institute of Int'l Studies Draft Code of the Rules of Outer Space, art. 2(5).

101 *Resolution on the Prevention of an Arms Race in Outer Space*, U.N. Doc. A/RES/51/44 (Jan. 7, 1997); *See also*, U.N. Doc. A/RES/50/69 (Jan. 9, 1996); U.N. Doc. A/RES/45/55 B (Dec. 4, 1990); U.N. Doc. A/RES/47/51 (Dec. 9, 1992); U.N. Doc. A/RES/448/74 A (Dec. 16, 1993).

102 Hubert Thierry, *IDS et Droit de L' Espace*, in DROIT DE L'ESPACE 361 (ed. Pedone 1988).

103 Compromis at 3.

104 U.N. Charter, *supra* note 7, art 33-51.

105 Compromis at 3.

106 *Id.* at 4.

107 Outer Space Treaty, *supra* note 28, art. 1.

108 Vienna Convention, *supra* note 88,art. 26.

109 Outer Space Treaty, *supra* note 28, art. 1.

110 *Id.*

111 PRS, *supra* note 29, prin. IV.

112 *Id.* at prin. IV. ("Remote sensing activities shall be conducted in accordance with the principles contained in Article I of the [Outer Space] Treaty . . . which . . . stipulates the principle of freedom of exploration and use of outer space on the basis of equality").

113 *Id.* at prin. II.

114 *Id.* at prin. XII.

115 *Id.* (also referred to as "raw" and "enhanced" data).

116 *Id.*

117 *Id.*

118 ICJ Statute, *supra* note 33, art. 28.

119 RESTATEMENT, *supra* note 21, § 102 cmt. c.

120 North Sea Continental Shelf (Denmark v. FRG Holland v. FRG), 1969 I.C.J. 63 (1969).

121 *Id.*

122 15 U.S.C.S. § 5622.

123 *Id.* § 5651.

124 *Id.*

125 Compromis at 2.

126 *Id.*

127 *Id.* at 3.

128 PRS, *supra* note 29, prin. XII.

129 *Id.*

130 Compromis at 2.

131 C. SYMMONS, THE MARITIME ZONES OF ISLANDS IN INTERNATIONAL LAW 1-151 (1979).

132 United Nations Convention on the Law of the Sea, Dec. 10, 1982, arts. 2-4, U.N. Doc. A/CONF.62/122 (1982), *reprinted in* 21 I.L.M. 1261, 1272 (1982) [hereinafter "UNCLOS III"].

133 Conciliation Commission on the Continental Shelf Area between Iceland and Jan Mayen, Report and Recommendations to the Governments of Iceland and Norway, *reprinted in* 20 I.L.M. 797 (1981); Agreement on the Continental Shelf between Iceland and Jan Mayen, done Oct. 22, 1982, Iceland-Norway, *reprinted in* 21 I.L.M. 1222 (1982); Richardson, *Jan Mayen in Perspective,* 82 A.J.I.L. 443 (1988).

134 Libya v. Malta, 1985 ICJ 38, para. 44 (1985).

135 UNCLOS III, *supra* note 132, art. 2.

136 Charter of Economic Rights and Duties of States, GA Res. 3281 (1974), U.N. Doc. A/9631.

137 *See*, UNCLOS III, *supra* note 132, art. 121; Libya v. Malta, 1985 ICJ 38 (1985).

138 North Sea Continental Shelf Cases, *supra* note 120.

139 *Id.* at 66-84 (Jessup, J., Separate Op.).

140 INT'L MARITIME BOUNDARIES xix (Jonathan I. Charney & Lewis M. Alexander eds. 1993).

141 Tunisia v. Libya, 1982 I.C.J. Rep. At 106, para. 24 (Jiminez de Arechaga, J., Separate Op.).

142 North Sea Cases, *supra* note 120.

143 Jan Mayen (Denmark v. Norway), 1993 ICJ Rep. 48-52 (1993).

144 *Id.* at 72, 79-81, paras. 76, 92.

145 *Id.*

146 *Id.*

147 PRS, *supra* note 29, at prin. XII-XIII.

148 Vienna Convention, *supra* note 88, art. 32.

149 Gennady M. Danilenko, *Principles Relating to Remote Sensing of the Earth from Space: Territorial Sphere of Application, in* PROCEEDINGS FROM THE THIRTIETH COLLOQUIUM ON THE L. OF OUTER SPACE, INT'L INSTITUTE OF SPACE LAW 289 (Oct. 10-17, 1987).

150 *Draft Proposal of Mexico on the Prin. of Remote Sensing*, U.N. Doc. WC/RS (1981)/WP 2 - U.N. Doc. A/AC. 105/288 Annex I (1981).

151 *Draft Proposal of Chile on the Prin. of Remote Sensing*, U.N. Doc. WG/RS/WP I. U.N. Doc. A/AC.105/133, Annex I, at 19-20 (1984).

152 *Joint Draft Proposal of Argentina and Brazil on the Prin. of Remote Sensing*, U.N. Doc. A/C.I/1047 (1974).

153 *Draft Proposal of USSR on the Prin. of Remote Sensing*, U.N. Doc. A/C.I/1047 (1974).

154 *Draft Proposal of France on the Prin. of Remote Sensing*, U.N. Doc. A/AC. 105/C.2/L.99 - U.N. Doc. A/AC.105/133, Annex IV, (1974).

155 *Id.*

156 15 U.S.C.S. § 5622, 5651-5672.

157 PRS, *supra* note 29, prin. XII.

158 *Id.* prin II, IX, XII, and XIII.

159 V. S. Vereshchetin, *On the Prin. of State Sovereignty in Int'l Space L., 11 ANNALS AIR & SPACE L. 429, 436 (1977).*

160 Outer Space Treaty, *supra note 28, art. IX.*

161 PRS, *supra note 29, prin. IV.*

162 *Id.*

163 Convention on the Rights and Duties of States (Montevideo Convention), *adopted by Seventh Int'l Conf. of American States, Dec. 26, 1933, 165 L.N.T.S. 19, art. 3.*
164 Geneva Convention on the High Seas, 450 U.N.T.S. 82, art. 28 (1958).
165 Trail Smelter Arbitration (Canada v. U.S.A.), 3 R.I.A.A. 1905 (1941).
166 German Interests in Polish Upper Silesia (Merits) (Poland v. Germany), 1926 P.C.I.J. Ser. A, No. 7 (May 25, 1926).
167 Fisheries Case (U.K. v. Norway), 1951 I.C.J. 116 (1951).
168 Free Zones Case (France v. Switzerland), P.C.I.J. Ser. A., no. 7 (1926).
169 BROWNLIE, *supra* note 48, at 51.
170 *Id.* at 70.
171 Outer Space Treaty, *supra* note 29, art. VII.
172 Convention on Int'l Liability for Damages Caused by Space Objects, *adopted by* COUPOS June 29 1971 [hereinafter "Liability Convention"].
173 *Id.* art. I(d).
174 *Id.* at art. II.
175 *Id.* at art. I(a).
176 Harvard Draft Conv. on Responsibilities of States, Art. 10, para. 3, Draft No. 12, April 1961.
177 G.A. Res. 2733B (XXV), Dec. 16, 1970.
178 N. MATTE, AEROSPACE LAW 157 (1977); *citing* U.N. Docs. A/AC.105/C.2/SR.94-97 (1968).
179 Chorzow Factory (Germany v. Poland), 1928 P.C.I.J. Ser. A, No. 17 at 47 (Sept. 13, 1928).
180 Matte, *supra* note 178.
181 Liability Convention, *supra* note 172, art. XII.
182 Alexander, *Measuring Damages Under the Convention on Int'l Liability for Damage Caused by Space Objects*, 6 J. SPACE L. 151, 153 (1978).
183 L. OPPENHEIM, 1 INTERNATIONAL LAW 352 (8th ed. Lauterpacht 1955).
184 Carl Q. Christol, *Int'l Liab. for Dmg. Caused by Space Objects*, 74 A.J.I.L. 346, 363 (1980).
185 U.N. Convention on Contracts for the Int'l Sale of Goods, Apr. 10, 1980, U.N. Doc. A/CONF.97/18 (1980), *reprinted in* 19 I.L.M. 668 (1980).
186 Compromis at 1.

INTERNATIONAL INSTITUTE OF SPACE LAW
OF THE INTERNATIONAL ASTRONAUTICAL FEDERATION

INSTITUT INTERNATIONAL DE DROIT SPATIAL
DE LA FEDERATION ASTRONAUTIQUE INTERNATIONALE

3-5, RUE MARIO NIKIS
75015 PARIS - FRANCE

Tél. 33-1-45.67.42.60
Telefax 33-1-42.73.21.20

STATUTES

ARTICLE I

The name of the Institute shall be the International Institute of Space Law of the International Astronautical Federation.

ARTICLE II

The purposes and objectives of the Institute shall include:

(a) The providing of advice to the President of the International Astronautical Federation when requested;
(b) Cooperation with appropriate international organizations and national institutions in the field of space law;
(c) The holding of meetings, colloquia and competitions on juridical and social science aspects of space activities, thereby providing forums for individuals from different legal systems and regions of the world to engage in the free expression and exchange of ideas relevant to space law and policy;
(d) The preparation or commissioning of studies and reports;
(e) The publication of books, proceedings, reports and other papers;
(f) The granting and presentation of awards and certificates;
(g) The conduct of such other activities as may be considered desirable in fostering the development of space law and studies of legal and social science aspects of the exploration and use of outer space.

ARTICLE III

Section 1

Individuals distinguished by their contributions to or of proven interest in the field of space law or other social science aspects related to space activities, may be elected as members of the Institute by the Board of Directors upon an application recommended in each case by a Director, or by three members of the Institute.

Section 2

Corporate bodies and other institutions actively engaged or otherwise interested in the field of space law or other social science aspects related to space activities, may be elected by the Board of Directors as institutional members of the Institute upon an application recommended in each case by a Director, or by three members of the Institute. Corporate bodies and other institutions that are members of the International Astronautical Federation (IAF) may be elected as institutional members of the Institute so long as no prejudice to the IAF is thereby incurred

Section 3

Election to membership of the Institute shall be by a majority vote of the Board of Directors in attendance at a regular or special meeting of the Board, or by a majority of affirmative answers of the members of the Board if the vote is held by mail.

ARTICLE IV

Section 1

All members of the Institute shall pay annual membership dues in accordance with the Bye-laws. These dues are expressed in terms of units.

Section 2

The Board of Directors shall establish the value of the unit in which membership dues are expressed. Alterations in the value of the unit shall take effect for each member or institutional member at the end of the period for which the relevant membership dues have been paid.

Section 3

Members of the Institute who are in default of their membership fee shall lose the right to vote in an annual or other meeting. Members of the Institute may be removed from the membership of the Institute by decision of the Board of Directors, if they are in arrears of membership dues for more than two years.

Section 4

On cause shown, the Board of Directors may reduce or remit the membership fee for a particular member.

Section 5

The Board of Directors may remove from membership of the Institute any member whom they judge to have violated the principles of the Statutes. Such decisions shall be taken by a two thirds vote of members of the Board present at the meeting at which the decision is taken.

ARTICLE V

Section 1

An annual meeting of the Institute shall be held each year, and special meetings may also be held, at such times and places as the Board of Directors may determine.

Section 2

The Board of Directors shall submit to each annual meeting a report on the activities of the Institute. The report shall outline the events and accomplishments of the preceding year and put forward the policy and projects of the Institute for the future.

Section 3

The annual meeting shall elect from among the members of the Institute, by a majority vote of the members present, officers and members of the Board of Directors, with due regard to the provisions of Article VI, Sections 2 and 3.

ARTICLE VI

Section 1

The governing body of the Institute is the Board of Directors. The officers of the Institute are the President, the two Vice-Presidents, the Secretary and the Treasurer. The President, Presidents Emeriti, two Vice-Presidents, and twelve other members shall be the voting members of the Board. Any Honorary Directors of the Institute and the Presidents of the International Astronautical Federation and of the International Academy of Astronautics ex officio shall be non-voting members of the Board of Directors.

Section 2

The Institute will seek that the membership of the Board of Directors shall reflect and represent the different regions of the world and different legal systems, due regard being paid especially to those countries in which significant space programmes or considerable efforts to develop space law have been made.

Section 3

The officers and members of the Board of Directors shall be elected for a period of three years, and may be re-elected. Members of the Nomination Committee acting in accordance with the Bye-laws, and members of the Institute when electing officers and other members of the Board, shall take account of the desirability of ensuring both a continuity and an adequate rotation of membership within the Board.

Section 4

In the case of a vacancy occurring in the Board between two annual meetings, the Board of Directors may appoint an acting Director to fill the vacancy until the expiry of the relevant term, subject to the approval of the appointment by the next annual meeting.

Section 5

An outgoing President, who has served in that capacity over several years and has contributed substantially to the development and the activities of the Institute, may be elected for life as a President Emeritus.

Section 6

Persons who have made outstanding contributions to international cooperation in the peaceful uses of outer space, or to the development of space law, as well as officers of the Institute and members of the Board of Directors who have served for several succeeding terms and have contributed substantially to the work of the Institute, may be elected as Honorary Directors. They may attend meetings of the Board of Directors in a consultative capacity, but without a right to vote.

ARTICLE VII

Section 1

The President of the Institute shall guide and supervise the activities of the Institute pursuant to its Statutes and in accordance with its Bye-laws. The President shall represent the Institute in its external relations, including those with the International Astronautical Federation.

Section 2

The Board of Directors shall:

(a) Carry out the purposes and objectives of the Institute as set forth in Article II of these Statutes;
(b) Implement the resolutions and directives adopted at the annual meetings of the Institute;
(c) Create standing committees with appropriate terms of reference and powers for their functions;
(d) Create ad hoc committees and working groups for specific purposes and functions;
(e) Appoint members of the Institute to fill vacancies occurring in the membership of the Board of Directors in accordance with Art. VI (4) of these Statutes;
(f) Supervise the correspondence of the Institute and provide for the safekeeping of the archives thereof;
(g) Establish a secretariat as appropriate, designate its duties and supervise all its activities;
(h) Arrange for meetings and colloquia;
(i) Arrange for the publication of books, proceedings, reports and other papers;
(j) Make recommendations concerning the awarding of medals, certificates and prizes;
(k) Approve budgets and supervise accounts;
(l) Accept donations and legacies, and funds from private sources, and contributions from national and international organizations, governments and agencies;
(m) Approve the annual reports of the President and the Treasurer to be presented at the annual meetings of the Institute, and the annual report of the President to be presented at the plenary meeting of the General Assembly of the Federation.

Section 3

At least eight of the voting members of the Board of Directors present at its meeting, including the President or Acting President shall constitute a quorum. Where in exceptional cases a matter has to be decided by mail, the same number of answers shall be required for a valid decision of the Board if a vote is held by mail.

Section 4

Decisions of the Board of Directors shall be taken, so far as possible, by consensus. Where a decision is taken by vote, Directors may cast their votes by mail or other written means as to any action proposed to be taken by the Board. Except where these Statutes and relative Bye-laws provide otherwise, decisions by vote of the Board shall be valid when taken by a majority of members present, or indicating their view in writing. In the case of a tie, the President shall have a casting as well as a deliberative vote.

Section 5

The ordinary affairs of the Board of Directors and of the Institute shall be conducted in accordance with the Bye-laws of the Institute. These Bye-laws shall be adopted by the Institute at an annual or special meeting by a majority vote of its members present upon a recommendation of the Board of Directors.

Section 6

The membership of standing committees, of ad hoc committees and of working groups, shall be determined by the President in consultation with the Board of Directors.

Section 7

The President, in consultation with the Board shall, by the end of April of each year, appoint the Chairman and members of the Nomination Committee responsible for nominating candidates for election as officers or other Board members whose terms of office are to begin after each annual meeting. The Committee shall be composed of not less than three, but not more than five members. The Report of the Nomination Committee shall be submitted to the President not later than one month before the annual meeting.

ARTICLE VIII

Section 1

These Statutes and Bye-laws shall be adopted at a meeting of the Board of Directors by a two-thirds majority vote of all members of the Board, subject to confirmation of this adoption at the next annual meeting of the Institute by a simple majority of members of the Institute present at that meeting.

Section 2

These Statutes and relative Bye-laws may be amended at any meeting of the Board of Directors by a two-thirds majority vote of all members of the Board, subject to confirmation of the amendment at the next annual meeting of the Institute by a simple majority of members of the Institute present at that meeting. All members of the Institute shall be informed of the intention to present such an amendment for confirmation at least six weeks prior to the next annual meeting.

Section 3

The Institute shall not enact Statutes or Bye-laws or amendments to either which are inconsistent with the provisions of the Constitution of the International Astronautical Federation, or its resolutions pertaining to the Institute.

BYE-LAWS

1. The President of the Institute shall preside at meetings of the Institute and of the Board of Directors. If unable to do so, the President may appoint one of the Vice-Presidents, and, if either of them is unable to do so, another member of the Board, to preside at a particular meeting. If the President is unable so to appoint or has not done so in time, a Vice-President, or if necessary another member of the Board, appointed by a majority of the members of the Board present, shall so preside and have the powers of the President for the purpose.

2. The Board of Directors shall endeavour to take decisions by consensus, but, if necessary, voting in the Board of Directors shall be by a show of hands, the President or Secretary informing the meeting of the views of those not present who have notified their view in writing on a matter. Such written views shall, wherever possible, be treated as votes on the matter under discussion.

3. The Board of Directors shall cause minutes to be kept of its meetings and of all actions taken by the Board. Such minutes shall be deposited with the Secretariat of the International Astronautical Federation and any member of the Institute in good standing may inspect them at any reasonable time.

4. A Nomination Committee, itself appointed by the President in consultation with the Board, shall make recommendations to the annual meeting as to elections to the offices of the Institute and membership of the Board of Directors. A person may be elected to the Board of Directors even though his or her name is not included in the list of recommendations made by the Nomination Committee. The Board and members attending shall be informed of the recommendations of the Nomination Committee which shall be made available at the latest by forty-eight hours prior to the annual meeting. In making its recommendations the Nomination Committee shall have regard to the distinction of the contribution made to space law by

potential nominees, as well as to the provisions of Article VI Sections 2 and 3 of the Statutes of the Institute.

5. The responsibilities of the Secretary of the Institute include:

(a) Maintaining the list of members of the Institute;
(b) Maintaining all necessary records, including financial records;
(c) Making arrangements for meetings of the Board;
(d) Taking the minutes of the meetings of the Board, and the annual and other meetings of the Institute;
(e) Depositing copies of the minutes of the Board with the International Astronautical Federation, and with the approval of the Board, making suitable arrangements for, the custody of the records of the Institute;
(f) In consultation with the Board and with the Secretariat of the International Astronautical Federation, making arrangements for the annual and other meetings of the Institute;
(g) Coordinating the Institute's involvement in the preparation of and making arrangements for any moot court or other competition involving or run by the Institute;
(h) Conducting the routine correspondence of the Institute;
(i) Communicating with members of the Institute, including preparing and seeing to the circulation of the Newsletter.

6. The responsibilities of the Treasurer of the Institute include:

(a) Ingathering membership dues and receiving donations and other monies received for or arising from the work of the Institute;
(b) Operating appropriate bank accounts on behalf of the Institute;
(c) Preparing appropriate budgets for approval by the Board;
(d) Paying bills incurred by the Institute or its officers in the carrying out of the purposes of the Institute from monies held by the Institute.

6. The President of the Institute shall refer an application for membership of the Institute to a membership committee, which he shall appoint, and the Board shall be informed of the report of that committee before it decides on the election of the applicant.

7. The Secretary of the Institute shall be informed in writing of any proposals for the amendment of the Statutes or Bye-laws of the Institute, other than those made by the Board, at least twelve weeks prior to the annual meeting, and any such proposal must be supported in writing by at least twenty individual fully paid-up members of the Institute.

8. The annual membership dues payable by an individual member of the Institute shall be one unit of contribution dues. Honorary Directors, if not otherwise members of the Institute, are exempt from the requirement to pay annual membership dues. Corporate or institutional members of the Institute shall pay three or more units of contribution dues of membership dues a year, the level of contribution being selected by the Institutional member in agreement with the Board of Directors.

9. A corporate or institutional member paying three contribution units per year has the right to nominate up to three representatives to vote in the annual meeting. Each additional contribution unit allows the institutional member to nominate one additional such representative.

10. A member of the Institute remains in good standing in the Institute provided that membership dues are not in arrear, and that a current mailing address has been received by the Secretary.

Author Index